Protective and Decorative Coatings for Metals

Protective and Decorative Coatings for Metals

A wide ranging survey of inorganic and mechanical processes, properties and applications

by

H. Silman BSc CEng FRIC FIChemE FIM FIMF

G. Isserlis BSc FIM FICorrT FIMF

A. F. Averill DipTech BSc MSc PhD AUMIST

FINISHING PUBLICATIONS LTD.
28 High Street, Teddington, Middlesex, England

1978

ISBN 0 904477-03-7

Printed in Great Britain
by Chas. Luff & Co. Ltd., Slough

FOREWORD

In this book the authors have aimed at providing a full account of the various processes and plant employed in the production of decorative/protective and functional coatings on metals (apart from organic finishes) on a basis of the essential scientific background. In this sense it provides the answers to both the 'how' and 'why' of metal finishing technology as it stands at the present time.

Some knowledge of 'why' is required of all those concerned with any aspect of metal finishing, as the growing range and sophistication of the operations involved make their smooth working difficult, if not impossible, without adequate scientific control, whilst the selection and specification of an appropriate finishing system cannot be properly made without a knowledge of the principles upon which it is based, its advantages, and its limitations. It goes without saying that for research and development scientific knowledge and reasoning are indispensible.

Although the book is entirely new, it is based on the guide-lines established by one of the present authors in his book on 'Chemical and Electroplated Finishes' (by H. Silman) in which the subject was covered from the viewpoint of practical experience of the large scale application of electroplating and related processes in the automobile industry—a major user of all types of finishes presenting exceptional problems of quality and durability in service. The years which have passed since the publication of the second edition of this book (and its German and Spanish editions) have seen such rapid and extensive developments with respect to both established and new processes, as well as in the design and construction of automatic plant, that it was felt desirable to allow the technology to reach a degree of stability before launching this book, which not only aims to bring the subject up to date, but covers a considerably wider field than its precursor, and includes a much fuller treatment of the theoretical principles involved.

By a combination of their respective industrial and academic backgrounds, the authors hope that this book will be of value to engineers and designers, as well as to scientists and technologists working in the industry, and in research in this interesting and important field. For there are very few artifacts of our society which do not in some way depend on surface treatment. A positive attempt has also been made to ensure that the book will cover the essential requirements of those studying for the examinations leading to the professional qualifications of the Institute of Metal Finishing.

H.S.
G.I.
A.F.A.

DEDICATION

To the craftsmen and women
whose work, though individually unrecorded,
has contributed greatly to the science and art
of metal finishing.

CONTENTS

CHAPTER 1

Theoretical aspects of electrodeposition

ELECTRODEPOSITION, commonly known as electroplating, is the deposition of metal coatings on to conducting surfaces by making the articles to be plated the cathode in an appropriate electrolyte containing heavy metal ions. A large variety of items, such as motor and engineering components, domestic appliances and other articles, are plated with metal films with the object of

(a) protecting them against corrosion and imparting a pleasing appearance; or
(b) endowing them with specific properties, e.g. hardness, wear resistance, anti-frictional, electrical, magnetic properties, etc.

In order to ensure good adhesion of the deposit to the substrate, there must be an intimate linkage between the atoms of both materials, and for this reason the surface of the substrate must be free from scale, grease, oxide films, etc. Therefore appropriate pre-treatment of items prior to plating is an essential operation.

The main electrodeposited metals and their applications are shown in Table 1-1. Electrodeposited metals and alloys are invariably crystalline, and their microstructure, and hence their properties, depend on the composition of the electrolyte and the operating parameters (bath temperature, cathode and anode current density, pH of electrolyte, agitation, etc.). However, crystal growth can rarely proceed under conditions even approaching equilibrium; hence there is insufficient time for all the metal ions to diffuse to equilibrium positions before being reduced and accommodated in the lattice of the growing crystals; the result is that electrodeposits usually exhibit lattice distortions. These can often be very severe and influence the properties of the deposit.

After removal of any scale and after subsequent suitable mechanical preparation, the work is degreased, usually in a solvent (e.g. trichloroethylene vapour). This is followed by electrolytic cleaning (cathodic, anodic or both) in an alkaline solution, after which it is thoroughly rinsed and then plated. The anodes used in the plating

TABLE 1.1. The principal electrodeposited metals and their applications.

Coating	*Basis Metal*	*Applications*
Brass and Bronze	Steel, Zinc alloy	Decorative finish. To facilitate metal-to-rubber bonding (brass).
Cadmium	Steel, Cast iron, Brass	Corrosion protection for nuts, bolts, screws, hydraulic fittings, engine parts, electrical components. To improve solderability.
Chromium	Iron, Steel, Zinc and Copper alloys	Decoration and corrosion protection for automobile and domestic hardware (over nickel). Wear resistant surfaces for piston rings, cylinder liners (direct).
	Thermoplastic materials	Decoration and protection (over copper and nickel).
Copper	Steel, Zinc alloy	Undercoat for nickel and chromium. Decorative ware. Printed circuits. Printing cylinders and plates. Masking against carburising. Making electro-formed articles.
	Thermoplastic materials	Decoration and protection. To give electrical conductivity.
Gold	Copper and Zinc alloys. Silver.	Jewellery. Watch cases. Electrical and electronic components.
Lead	Steel, Copper alloys	Bearings. Accumulator fittings. Fire extinguishers.
Nickel	Steel, Copper and Zinc alloys	Undercoat for chromium. Recovery of worn parts. Loose leaf book rings and files. Typewriter parts. Door furniture. Camera parts. Automobile and sanitary fittings. Making electro-formed articles.
	Thermoplastic materials	Decoration and protection (over copper).
Palladium, Platinum, Rhodium	Copper alloys. Silver	Jewellery; electrical and electronic components.
Silver	Copper alloys; Nickel-silver	Cutlery. Jewellery. Electrical contacts. Bearings.
Tin, Tin-nickel, Tin-zinc alloys	Steel	Tinplate. Bearings. Copper wire. Electrical equipment (for solderability).
Zinc	Iron and Steel	Corrosion protection for nuts, screws, bolts; tubes and electrical conduits; hydraulic fittings; engine and motor components; car window control gear; sheet steel and wire.

solution can be either soluble or insoluble. Soluble anodes, made of the same metal as that being deposited, partly or entirely maintain the metal ion concentration in the electrolyte by dissolution, whereas in those cases where insoluble anodes have to be used (as e.g. in chromium plating), the metal ion content of the bath must be maintained by frequent additions of appropriate metal compounds.

Transformer-rectifiers are normally used as sources of direct current, but sometimes, as for instance in the small-scale plating of noble metals, accumulators may occasionally be employed.

1.1. Discharge of metal ions in electrodeposition processes

Although the techniques of electroplating are well established and processes can run very smoothly provided that adequate control is exercised, the mechanism of ion reduction and accommodation in the lattice is extremely complex, and only a simple outline of the factors involved can be given here.

Metal cations do not exist in solution in the simple state (M^{z+}) but have other ions or molecules associated with them. The metal ions are often sheathed by a group of water molecules and such species are referred to as simple hydrated metal ions. The water molecules do not affect the electrical charge on the ion so that a metal ion when hydrated is represented by

$$M(H_2O)_x^{z+}$$

where x is the number of water molecules involved in the primary hydration sheath. If we consider a case where metal ions are present in this simple hydrated form, the reaction steps involved in the discharge or deposition process are:

(a) transport of hydrated metal ions to the cathode surface.
(b) partial dehydration of the hydrate
(c) adsorption of the partially dehydrated ions at the cathode surface, and diffusion along the surface to low energy positions
(d) complete dehydration followed by discharge:

$$M^{z+} + ze \rightarrow M$$

(e) accommodation of the deposited atoms at active sites on the substrate, from which growth of the deposit continues.

Although there are many examples of metal deposition from solutions containing simple hydrated metal ions, the use of electrolytes in which the metal is present in the form of a complex is much more common in electroplating practice. Metals such as copper, silver, gold, zinc and cadmium are commonly plated from cyanide electrolytes, whilst pyrophosphate and fluoborate complexes are also employed. These complex ions are virtually always negatively charged, so that it is difficult to visualise the fundamental process of deposition. Because the complexes are nearly always highly stable, it is extremely unlikely in most cases that deposition occurs from simple ions following dissociation of the complex. An exception, however, is the deposition of silver from cyanide solution. The silver cyanide complex is unusual in having a low stability, and deposition takes place from Ag^+ ions[1] according to the scheme

$$\text{(a)} \quad [Ag(CN)_2]^- \rightarrow Ag^+ + 2CN^-$$
$$\text{(b)} \quad Ag^+ + e \rightarrow Ag$$

In other cases deposition probably occurs by discharge of a complex with a low co-ordination number[2, 3]. This low co-ordination number complex forms by dissociation of the predominant complex,

$$[M(X^-)_x]^{(x-z)-} \rightarrow [M(X^-)_{x-y}]^{(x-y-z)-} + yX^-$$

where X^- is the complexant and z is the number of positive charges on the simple metal cation. Since, according to this view, the low co-ordination number complex will be present in relatively low concentration, this explains the high concentration overpotentials associated with deposition from complex ions.

1.2. Mode of growth of electrodeposits

From an overall crystallographic point of view, an electrodeposit will tend to grow on a substrate in such a manner that one of its lattice parameters will try to match that of the underlying metal. This is possible provided that the two lattices are similar in dimensions (the lattice parameter of the deposited metal must be within the limits —2.4 and +12.5% of the substrate metal) and that the substrate surface has not suffered too great a distortion by preliminary machining or polishing.

Such a mode of growth is referred to as " epitaxy ". Under favourable conditions, epitaxial growth can continue up to some 3 microns[4]. However, the greater the difference between the lattice parameters of substrate metal and deposit, the greater the degree of distortion of the latter, and the higher the cathode current density used in deposition, the sooner will epitaxial growth cease and give way to a type of growth which is determined by the operating parameters and electrolyte composition.

Epitaxy is not necessarily a prerequisite for good adhesion of a deposit. Mechanical deformation of the substrate often prevents such a mode of growth. For instance, in the case of zinc deposits on heavily polished steel, epitaxial growth is impossible. Instead, alloy formation occurs at the interface between deposit and substrate by diffusion of the two types of atom into each other's lattices to a depth of a few atomic diameters[5]. The cohesion between the two metals will be very strong, and in tensile testing, parting will occur by rupture of the zinc rather than by detachment at the interface.

1.3. Throwing power of an electrolyte

One of the most important properties of an electroplating solution is its macro-throwing power, i.e. its ability to deposit a metal coating of even thickness on a shaped cathode surface. The significance of this property is evident since it is closely associated with the degree of protection the coating is capable of giving to the underlying metal. A convenient way of determining the throwing power (i.e. the metal distribution) is by means of the Haring cell, briefly described below. Here, the metal distribution over the cathode surface is measured by direct weighing. Graphical methods, based on polarisation studies, and those involving the plotting of the electric fields over the cathode surface, are also available, but they are less reliable than the direct methods.

The chief factors affecting throwing power are the degree to which the cathode polarises with increase in current density, the electrical conductivity of the electrolyte and the relationship between cathode current efficiency and cathode current density. The steeper the slope of the cathode polarisation curve and the greater the conductivity of the electrolyte, the more uniform will be the current distribution over the cathode surface. The more the cathode current efficiency falls with increase in current density, the higher will be the uniformity of metal distribution. Any factor influencing the above conditions will therefore have a bearing on throwing power.

The temperature of the electrolyte has a pronounced effect on throwing power, since a rise in temperature increases the bath conductivity but reduces the cathode polarisation. Depending on which of these two factors is predominant, the throwing power will either increase or decrease with rise in temperature; in most cases it decreases.

Agitation of the electrolyte, by supplying metal ions to the cathode, reduces cathode polarisation and hence diminishes the throwing power.

In electrolytes containing simple metal ions, the cathode polarisation will usually be low. In consequence, such electrolytes can be used only for plating products of relatively simple shape and they produce deposits of relatively coarse crystal structure, which is associated with a low hardness and low strength. Addition of small quantities of surface-active substances (surfactants) will often increase cathode polarisation and refine the crystal size. Geometrical factors, such as shape and size of cell and the presence of magnetic fields[6], also have a bearing on current distribution.

In order to ensure satisfactory throwing power, not only must appropriate bath compositions and operating parameters be chosen, but resort is often had to purely mechanical devices, e.g., the use of internal anodes for the plating of internal surfaces, the shape of such anodes being adapted to those of the surfaces being plated. Additional (dummy) cathodes may be placed on either side of a flat article in order to "rob" excess current from the edges. Screens made of a non-conducting material are also sometimes used to prevent excess current from reaching specific portions of the articles being plated.

Finally, so-called "chance" factors play a part in determining throwing power. these include the nature of the basis metal or substrate, the condition of the surface (whether passive or active), its compositional and structural homogeneity, the type of pre-treatment and the electrodes used.

1.4. Metal distribution Formulae

The Haring cell, shown in Fig. 1.1, is used for carrying out throwing power measurements. It consists of a rectangular cell containing two sheet metal cathodes, which fill the entire cross-section at both ends, and one wire mesh or perforated anode. The latter is placed between the cathodes so that its distance from one of the cathodes is a fraction (normally one fifth) of its distance from the other. If polarisation were negligible as compared with the potential drop in the electrolyte and if the cathode current efficiency were 100%, the metal distribution would be determined by the inter-electrode distances, viz., the weight of metal deposited on the nearer cathode (C_n) would be five times greater than that deposited on the more distant cathode (C_f). Under such conditions, the electrolyte would behave in accordance with Ohm's Law and the metal distribution would be proportional to the current distribution. In this hypothetical case, the current distribution is referred to as *primary distribution*. However, when the 'contact resistance' at the interface between electrolyte and cathode is high as compared with the resistance of the electrolyte, due to polarization, the resulting current distribution is known as *secondary distribution*. The latter, which is encountered in practice, ensures a more uniform metal distribution on a shaped cathode than if the electrolyte were to behave in accordance with Ohm's Law.

In the Haring and Blum formula, the throwing power is evaluated from the following equation:

$$\text{T.P.} = \frac{K - C}{K} \times 100\% \tag{1}$$

where

$$C = \frac{C_n}{C_f} = \text{metal distribution ratio,}$$

and K is the ratio of the distances from the more distant and the nearer cathodes to the anode. Thus, K is the current distribution ratio. In the above particular arrangement, its value is 5. The optimum metal distribution would be that in which equal weights of metal would be deposited on the two cathodes, i.e. when

$$C = \frac{C_n}{C_f} = 1$$

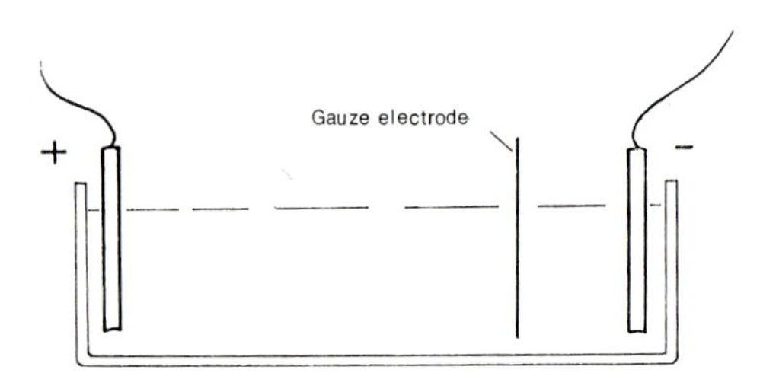

Fig. 1.1. Haring cell.

Using Haring and Blum's formula, the throwing power will be 80% under these conditions, which represents maximum throwing power, whereas if the weight of metal deposited on the nearer cathode were to be five times that deposited on the more distant one, the throwing power would be zero.

In order to equate the optimum throwing power to 100%, Heathley[6] proposed the following modification:

$$\text{T.P.} = \frac{K - C}{K - 1} \times 100\% \tag{2}$$

Using this formula, the throwing power will vary between 100% and $-\infty$.

Field's formula[7, 8] is the one generally used in the U.K., since the values it gives range from +100% to −100% irrespective of the value of K. According to this formula

$$\text{T.P.} = \frac{K - C}{K + C - 2} \times 100\% \tag{3}$$

1.5. Microthrowing power and levelling

By microthrowing power is meant the ability of an electrolyte to plate into microrecesses at the same rate at which plating occurs at peaks. A simple method of assessing microthrowing power is to take a cross section of a plated microgrooved specimen and measure the deposit thickness in the groove and at the surface by means of a microscope (Fig. 1.2). If the microthrowing power is poor, then it must be

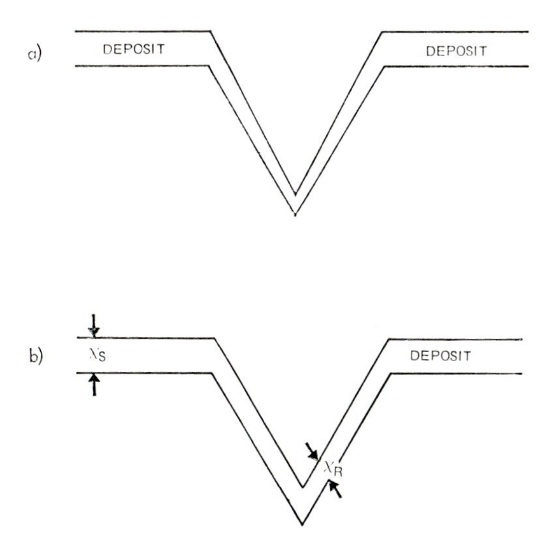

Fig. 1.2. Profiles of microgrooves plated in electrolytes having (a) poor microthrowing power and (b) good microthrowing power. For good microthrowing power $x_R \simeq x_S$.

inferred that the diffusion of charged metal complexes to the cathode surface is a controlling factor. This occurs when the concentration of dischargeable metal ion species in solution is low, as, for example, in cyanide or other complex solutions.

In this case, the cathode layer is soon depleted of metal ions so that a diffusion gradient is established. Metal ions may have to travel over considerably longer distances to reach microrecesses than peaks or recess shoulders, so that fewer ions are available for plating in the recesses. Thus, electrolytes in which diffusion or concentration polarisation is high have poor microthrowing power properties associated with them. This is in contrast to the macrothrowing power which increases with increase in concentration polarisation.

In the case of simple ion solutions, where the metal ion concentration is high and cathode polarisation low, the microthrowing power will be high, whereas the macrothrowing power will be poor.

An electroplating solution is said to exhibit *levelling* if it yields deposits which are smoother that the surface to which they are applied. This often implies that more metal is deposited in the recesses than on the peaks of a rough surface (*cf* Fig. 1.3), or, more specifically, that the deposition rate within a recess is greater than at its edges. This levelling action is due to the presence of one or more specific minor constituents (levelling agents) which are adsorbed onto the cathode surface and are codeposited with it. The removal of levelling agent from solution by codeposition on the cathode leads to a diffusion layer. The diffusion path to the groove shoulder is shorter than that to the sides, so that more levelling agent is codeposited at the edges or shoulders. The higher concentration of levelling agent results in increased cathode polarisation and this encourages the diversion of current to recesses, so that levelling occurs[10, 11].

1.6. Codeposition of foreign substances

Coatings deposited from commercial plating baths nearly always contain impurities, e.g. organic and inorganic colloids, anions and uncharged particles. The degree of contamination of the deposit depends on the rate of reaction of the deposited metal with the surrounding medium. On this basis, metals can be divided into three groups, namely:

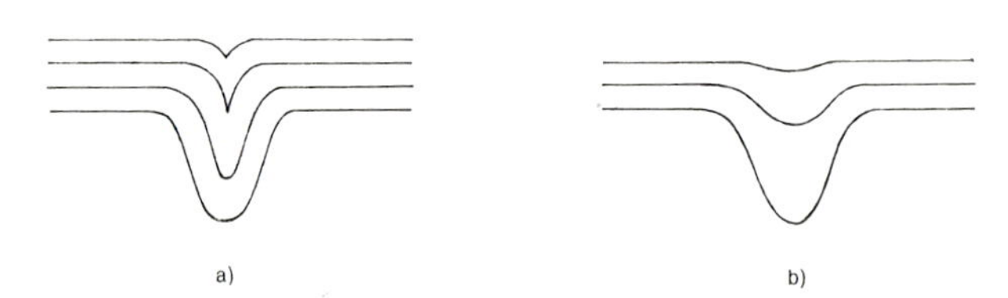

Fig. 1.3. Levelling (a) with uniform metal distribution brought about by good micro throwing power, (After Due Rose et al[9]) (b) with non-uniform metal distribution (more metal in recess than on recess shoulders).

(a) Those which require a low over-voltage for deposition, (e.g. Ag, Sn, Zn, Cd, Th, Pb, Bi).
(b) Those which require a high over-voltage for deposition, (e.g., Fe, Ni, Co, Cr, Mn, Pt), and
(c) Those which cannot be deposited from aqueous solution as single metals (e.g. Mo, W, Zr, Nb, U, Ta).

The rate of reaction of metals with the surrounding medium is related to the surface energy of their atoms. In the body of a metal, all atoms are completely surrounded by others. At a metal surface, however, they are partly surrounded by other atoms and partly exposed to the surrounding medium. Consequently, their energy is greater than that of atoms within the metal. This excess energy is termed surface energy. An indication of the magnitude of the surface energy of a metal is its melting point. The stronger the bond forces, the greater the energy required to disrupt them. The melting point of a metal is the temperature at which the disrupting forces (thermal vibrations) exceed the bond strength between the atoms. Hence, the higher the melting point of a metal, the greater its surface energy.

In an effort to satisfy its available surface energy, a metal tends to attract any substances (e.g. hydrogen, impurities, basic salts, etc.) available in its immediate vicinity within the surrounding medium. The greater the surface energy, the more strongly will the metal attract and adsorb these substances to its surface. The latter will thereby tend to be rendered passive.

Metals of Group A have a relatively low surface energy, and there is therefore little tendency for them to become passive when electrodeposited from their simple salt solutions. Metal ions in solution have limited competition from other substances for the active sites on which growth occurs. Hence, the resistance to be overcome during the discharge and accommodation process is low, and this is reflected by the local metal over-voltage accompanying deposition.

Metals of Group B have a much higher surface energy than those of Group A and hence have a greater tendency to passivity. This greatly reduces the number of active sites for metal atom accommodation. Nucleation will take place from a great many sites and crystal growth will be restricted, which results in a much finer crystal size than in the case of Group A metals. Electrodeposits of Group B will contain a relatively high proportion of impurities. Since their adsorptive capacity for foreign substances is high, their surfaces will be impure. This will result in a considerable deviation from equilibrium of the potential at the metal/electrolyte interface. The tendency of these metals to passivity will affect the rate of cathodic reduction of their ions, and is one of the main reasons for the high over-voltage. Before an ion can enter the crystal lattice it must displace any adsorbed substances from the electrode surface, and this retards its rate of discharge. In addition, since the affinity of these metals for hydrogen is greater than that of Group A metals, their deposition is always accompanied by a greater hydrogen evolution, which is another contributory factor to the retardation of metal ion reduction and incorporation in the lattice.

Metals of Group C have a very high surface energy, i.e. they absorb specific constituents from the surrounding medium, forming films which normally render the metal passive. Therefore such metals cannot be deposited from aqueous solutions singly, but only in the presence of such other metals as are capable of exerting a depolarising action, i.e. as alloys, and then only as very thin films.

Although foreign substances are present in many deposits, particularly in Group B metal deposits, the lattice constant of the deposit is not affected by this codeposition, even if the foreign substances are present in substantial amounts. The metal lattice is, however, highly stressed and this has an important bearing on the chemical and physical properties of the deposit. The electrode potential of a coating containing impurities is usually less noble than that of the pure metal, and the deposit has a lower resistance against corrosion.

Additions of various substances, particularly certain specific organic compounds, are often made to plating baths intentionally in order to modify, or completely change, the properties of the deposits. Very often these additives are decomposed at the cathode and their decomposition products are codeposited.

1.7. Deposition of bright metal coatings

Deposits obtained from electrolytes free from brightening additives are normally matt. Their surface is uneven on a microscale, so that a high proportion of light reflected from them is scattered. In order to ensure maximum specular reflection of light (i.e. a minimum loss due to scattering), the micro-roughness of a surface must be less than the shortest wave length of light. Roughness on a macroscale has little effect on the brightness of a surface.

Matt deposits can be made bright by mechanical polishing provided that the substrate is smooth. However, mechanical polishing is costly in terms of labour, time and metal lost due to thinning of the deposit.

Bright deposits can be obtained directly from the plating bath by incorporating specific additives in the solution. These include surface-active substances (surfactants) and colloids. The mechanism of their action is not yet fully understood, although some of the basic principles involved are beginning to be appreciated. Among the theories proposed, two are of importance. One is based on complex formation and the other on adsorption.

According to the *complex formation theory,* colloid compounds form complexes with the metal cations. Owing to the strong adsorption bond between organic colloids and metal cations, discharge of the complex ions is retarded, and hence the discharge of metal ions at the cathode in the presence of colloid additives is accompanied by a high polarisation. Surfactants can be adsorbed either on the entire cathode or on preferential sites. In the former case, discharge of cations is affected across the entire film formed by the surfactant, whilst in the latter case this occurs only on the free sites of the cathode surface.

The increase in cathode polarisation in the case of *adsorption* of surfactants occurs as a results of either a sharp reduction in size of the active cathode surface, and hence local increase in current density, or an increase in activation energy for the penetration of cations to the cathode surface across the adsorbed layer.

The metal surface charge has an important bearing on the adsorption of both ions and neutral molecules. If the surface charge is negative, the electrostatic repulsion will oppose adsorption of anions, whilst favouring adsorption of cations. However, a positive surface charge will promote adsorption of anions. The potential at which there is no charge on a metal surface is called the zero charge potential.

Whether or not adsorption will occur on a metal surface at a given potential depends not only on the position of the potential relative to the zero charge potential but also on the chemisorption forces operating between the particle and metal surface.

Adsorption of neutral organic molecules usually occurs only within a specific potential range on each side of the zero charge potential of the cathode surface. Beyond that range (in either direction from the null point), they are displaced by ionic adsorption.

If it is known how electrostatic adsorption of specific surfactants occurs on any one metal, e.g. mercury, the potential range over which electrostatic adsorption of these substances on the surface of any other metal is likely to occur can be estimated with the help of the " reduced or ϕ-scale " proposed by Antropov [12, 13].

The way in which electrocrystallisation proceeds is also influenced by colloidal suspensions of metal hydroxides which form in the catholyte during electrolysis in weakly acid or neutral media under conditions in which hydrogen and metal ions are reduced simultaneously. Hydrolysis is then likely to occur in the catholyte and the colloidal suspensions thus formed will possess properties characteristic of jellies. They are readily absorbed on the metal surface and have a strong bearing on the structure and properties of the deposit.

Deposition of bright deposits requires a higher energy than that of matt ones since the inhibition due to adsorption of foreign substances at the cathode must be overcome. Therefore deposition of bright deposits generally takes place at higher over-potentials (i.e. cathode polarisation is higher).

In the deposition of chromium from chromic acid solutions bright deposits are obtained in the absence of brightening agents. Chromium ions deposit through a complex surface film which ensures that deposit growth is randomised.

1.8. Effect of dissolved hydrogen

Dissolved hydrogen has a marked effect on the structure of the deposit. The mechanism of inclusion of this element in the deposit can be diverse. One is by adsorption of atomic hydrogen at the surface during metal deposition. The adsorbed hydrogen partially re-combines and is converted to its molecular form; partly it also enters the crystal lattice of the metal where it may occupy atomic positions or be accommodated interstitially, forming solid solutions. Direct incorporation of hydrogen ions in the crystal lattice in the form of protons is also possible. Yet another method of hydrogen entry can be by formation of chemical compounds with the metal (hydrides).

The specific mode of hydrogen entry into a metal deposit depends on the metal involved and on the parameters of deposition. Although the total hydrogen content of an electrodeposited coating is low, the crystal lattice suffers considerable deformation. Large internal stresses are set up in the deposit, which often results in it becoming brittle.

In addition atomic hydrogen may diffuse into the basis metal, followed by the migration of these atoms to, and their retention in, imperfections in the crystal lattice itself or in much larger mechanical discontinuities, e.g. those associated with non-metallic inclusions. In these "collectors" the atoms unite to form molecules, and it is the significant increase in volume associated with this which leads to the development of very high stresses in the vicinity of the gas pockets.

In ductile metals this change can sometimes be accommodated by plastic deformation, but when, as in high-tensile steels, the ductility is low, cracking may occur. Cracks thus formed can range from tiny fissures to complete and often sudden cleavage through a heavy section, according to circumstances. Baking at the highest temperature that will not of itself injure the component immediately after the plating operation will cause most of the hydrogen to diffuse out. This treatment may be effective provided that cracks, however small, have not already formed during plating, previous pickling treatments or cathodic cleaning.

1.9. Internal stresses in electrodeposits

Internal stress appears to be unrelated to any other physical property of a metal, such as strength, ductility, brightness, etc. In a deposit it manifests itself by the latter being either stretched or compressed as compared with its normal state. When stretched, the deposit is in tension and tends to contract, whilst when compressed (i.e. in the presence of compressive stresses), it tends to expand. The nature of internal stresses and their magnitude depend on the particular metal being deposited, on the operating parameters and on the composition of the electrolyte.

High internal tensile stresses may markedly reduce the corrosion resistance of the deposit, and in extreme cases may even cause it to crack and flake off. Such stresses will also reduce the fatigue life of a plated component, in some instances very seriously.

Internal stresses may arise from two basic mechanisms: (1) lattice misfit between the initial atomic layers of the deposit and substrate; and (2) the manner in which the metal is deposited from the electrolyte. In the first mechanism, the basis metal plays a dominant role, whereas in the second, one or more of a number of factors may be involved. For example, foreign matter may become occluded in the deposit, or hydrogen may be co-deposited. Hoar and Arrowsmith have explained stress formation in nickel deposits on the basis of the formation and behaviour of dislocations[(14)]. The fundamental principles underlying methods of measuring stress in electrodeposits have been reviewed by Walker[(15)].

1.9.1. MEASUREMENT OF INTERNAL STRESS

Although numerous methods have been developed for measuring internal stress in electrodeposits, only a few have found general application. Those used are based on strain measurement which is subsequently converted to stress. Practically all are modifications of Stoney's method[16], introduced in 1909 and itself still widely used. It involves plating one side only of a metal strip which is rigidly held at its upper end. The other side is stopped-off with a suitable lacquer. Tensile stresses cause the strip to curve in the direction of the plated side, whilst compressive stresses cause it to deflect in the opposite direction. The radius of curvature of the plated strip is a measure of the stress.

In the Spiral Contractometer developed by Brenner and Senderoff[17] in 1948, the strip is in the form of a helix, one end of which is fixed to the sleeve of the housing of a calibrated dial, whilst the other is attached to the end of a co-axial torque rod free to move within the sleeve and actuating a pointer traversing the dial. The helical form allows a very long strip to be used within a given compass, thereby greatly increasing the accuracy of measurement.

In 1957, Hoar and Arrowsmith[18] published a method which, although incorporating a new principle in measuring stress, is basically a variation of the bent strip test. However, instead of being allowed to deflect freely, the strip is held in its original position electromagnetically during plating. The advantage of this instrument, which is shown in Fig. 1.4, is that stress can be recorded continuously during electroplating without allowing the strip to strain.

A portable internal stress meter which is available commercially and is suitable for research and the direct control of electroplating processes has been described by Dvorak and Vrobel[19]. The instrument incorporates a precision gauge which accurately measures the changes which occur in the length of a steel strip when it is plated on

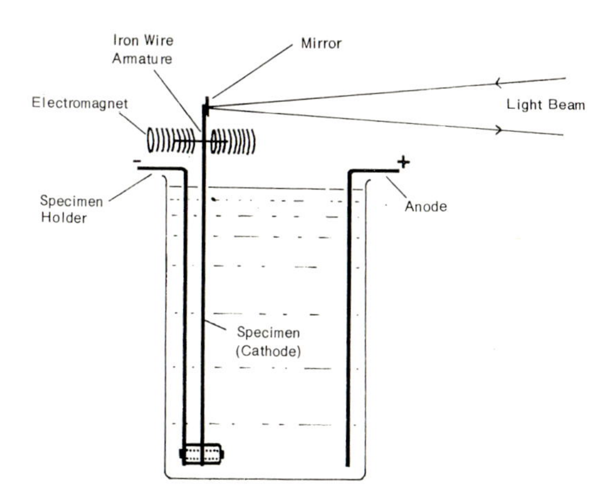

Fig. 1.4. Method of measuring the internal stress in electrodeposits developed by Hoar Arrowsmith.[18]

both sides. The method has very high sensitivity, and enables stress differences as low as 0.2 kp/mm^2 to be detected in nickel deposits; the reproducibility of the results is equally good.

1.10. Hardness and wear resistance

Hardness and wear resistance are not basic physical characteristics but are the result of the interaction of various properties. Although the wear resistance generally increases with hardness, this is not always the case, and hard deposits may exhibit a lower resistance than softer ones.

It often happens that a metal is harder when electrodeposited than after having been severely cold-worked. Since an increase in hardness by cold work is due to lattice distortion, it is evident that the lattices of electrodeposited metals are severely distorted. Impurity incorporation strongly contributes to such lattice distortion.

1.11. Porosity

Porosity in electrodeposits can be due to numerous causes, the most common of which are non-conducting inclusions in, and roughness of, the basis metal surface. Metallic inclusions upon which the hydrogen evolution overpotential is low also represent an important source of coating porosity. Other factors affecting the proneness of the deposit to this type of defect include the structural characteristics of the metal coating (size, shape and orientation of crystals), the composition and pH of the electrolyte, operating parameters, dissolved impurities and brighteners in the electrolyte, anode behaviour and internal stress in the deposit.

1.12. Adhesion

If a coating is to be effective, it must, first and foremost, adhere firmly to the substrate.

It has already been stated that electrodeposited metal atoms are accommodated on the lattice of the substrate in such a way that they either reproduce its crystal structure (epitaxial growth) or form an alloy with it at the interface. The intimate contact between the coating and basis metal required for this to occur precludes the presence of any sort of barrier layer on the surface of the latter. Such layers may be films of grease, oil, or even scale or thin oxide films, and these must be completely removed by means of degreasing, descaling and other operations prior to electrodeposition.

Given a clean substrate surface, the following factors affect the degree of adhesion of the deposit: (i) structure and properties of the substrate metal, (ii) composition of the electrolyte, and (iii) parameters of electrodeposition. Overpickling of steels, and particularly of high carbon steels, may expose extensive regions of carbide, to which a metal deposit adheres poorly.

A large difference of lattice parameter between substrate and deposit metals will preclude epitaxial growth and make alloy formation difficult. An example is the electrodeposition of lead on copper. Adhesion can however, be greatly improved by subsequent heat treatment, which promotes the formation of an intermediate layer of alloy.

The relative coefficients of linear expansion of substrate and deposit metals will have a bearing on adhesion. If the two coefficients differ greatly from each other, flaking-off of the deposit may occur during mechanical polishing, particularly if the work is overheated.

In the electrodeposition of metals such as nickel and chromium, the current should not be interrupted since passivation of the surface is likely to occur and the adhesion of the subsequently-deposited metal film will be weakened. In the case of chromium, interrupting the current is also likely to prevent a bright deposit being obtained.

Some organic compounds and colloids in the electrolyte may lower the adhesion of certain metal deposits drastically, whereas others have no adverse effects.

Internal stress in the deposit, if tensile, tends to reduce adhesion whilst compressive stresses are without effect.

Extensive evolution of hydrogen during electrodeposition and inclusion of this element in the deposit can impair adhesion and lead to blistering. The degree of adhesion between deposit and substrate can be measured by mechanical or physical methods (cf. Chapter 15). In the case of noble metals, an electrochemical method can sometimes be used. Mechanical methods involve measuring the force required to detach the deposit from the substrate (quantitative method) or the simultaneous deformation of deposit and substrate (qualitative method). Physical methods are based on the difference in the coefficient of linear expansion between substrate and deposit; poor adhesion will be shown up by blister or crack formation on heating. Burnishing the coating with a metal strip will also indicate poor adhesion by the formation of a blister.

1.13. Designing metal articles for finishing

The importance of design in relation to the ultimate performance of an electroplated article is all too often neglected. This applies also to other finishing operations such as polishing, anodising and painting. Good design leads to more attractive products, higher sales, and usually lower costs. Difficulties in finishing resulting from unsuitable design create problems for the supplier of plant and processes as well as for the actual finisher and, as often as not, also for the final consumer. Service problems are multiplied, and production delays may occur whilst finishing troubles are sorted out. It sometimes happens that an unsatisfactory design may make an article virtually unplateable except at a prohibitive cost; re-tooling may even be needed, with all the expense and delay that this entails. There is, therefore, no doubt that proper attention to the subject of good design for finishing is a matter of major importance to all branches of the industry[20]. Particularly, much has been written on the influence of design on the plateability of zinc base die castings[21].

Designing articles for finishing—by which is meant chiefly polishing, electroplating and painting—is to a large extent a matter of compromise. Clearly, if suitability of design for finishing were to become a over-riding consideration the world would become a much duller place to live in. Designers would be unduly inhibited and we should lose the wide variety of forms and aspects of finished articles which we now see around us. Moreover, articles which are not absolutely straightforward from the finishing point of view provide a challenge to the metal finisher to improve his processes; it is therefore not altogether a bad thing for him to be presented witn the occasional problem, provided that the designer is aware of the difficulties he is creating and is prepared to face up to them. However, the likelihood that good design for finishing will become so universal that the finisher will be left with no problems to resolve is remote in the extreme. It is desirable that the designer should have some knowledge of the basic principles of finishing processes so as to avoid making them difficult to carry out satisfactorily.

Good design for finishing takes into account the function of the article, its method of manufacture and the requirements of the finishing processes to be applied. By attention to all these aspects, not only will a more serviceable product be obtained, but it will at the same time be less expensive to manufacture. Early discussions between designers and specialists in the finishing process ultimately to be applied are prime requisites. Finishing processes are generally labour intensive, and if they are rendered unnecessarily difficult and result in a high proportion of rejected work, savings in material and fabrication may well be dissipated in high finishing costs. No metal finisher who is proud of his skill would wish to inhibit the designer's freedom of action. He is concerned only with the avoidance of inadvertent errors which create unnecessary finishing difficulties and may be irrelevant to the success of the design.

1.13.1. POLISHING: ARTICLE DESIGN

Many metal articles need some kind of polishing operation before being electroplated. This can be carried out on polishing wheels, by hand, or on automatic machines. Small articles can be barrel polished, which is much less expensive. In this process they are tumbled in a rotating barrel with a suitable abrasive medium. Chemical or electrochemical methods of polishing or brightening are also employed, particularly for stainless steel articles. Beside this, there is a growing trend towards the finishing of metals in sheet and strip form, fabrication being carried out afterwards. This eliminates many of the difficulties of dealing with complex shapes.

In polishing on rotating buffs the following are some of the main points to consider:

(a) Abrupt changes in shape, sharp edges and blind holes are to be avoided.

(b) Large flat areas should be broken up in some way so as to make them less liable to scratches and minor damage. It is also difficult to produce uniformly flat reflecting surfaces.

(c) Projections and protuberances should be avoided since they are difficult to polish and may damage the polishing buffs.

(d) Deep recesses and re-entrant areas should be avoided as far as possible, and all surfaces to be polished should be readily accessible to wheels of normal diameter.

(e) Articles for barrel polishing should be sufficiently strong to withstand the tumbling action, and should not interlock. A dimple or a raised edge is useful on flat surfaces to prevent them sticking together.

1.13.2. PLATING: ARTICLE DESIGN

(a) Deep recesses and abrupt re-entrant areas should be avoided as far as possible, particularly where the deposit has to resist wear or corrosion. Recesses and their edges should be smoothed out; the minimum radius at the edge and base of an indentation should be one-quarter of the depth.
It should be borne in mind that where minimum deposit thicknesses are specified, they are usually not applied to areas which cannot be touched by a ball of 20 mm diameter.

(b) Tubular articles should have provision for complete drainage to prevent carry-over of process solutions during plating. Alternatively tubes should be completely sealed.

(c) Blind holes, cavities, seams and rolled edges which can trap process solutions should be avoided. The space between adjacent holes should preferably be at least twice their diameters; holes and slots should be free from sharp edges. Articles should be assembled *after* plating, particularly if they are made of dissimilar metals.

(d) The design should take into account the need for providing a suspension point giving good electrical contact on the article to be plated when suspended by its own weight. This may be a hole, a thread or a lug located where the surface appearance is relatively unimportant.

(e) Protuberances, sharp corners, edges, fins and ribs draw current preferentially, and will tend to build up with metal during plating. They should therefore be rounded with a radius of at least 1 mm, whilst edges of holes can be countersunk or chamfered.

(f) The electroplating of screw threads presents special problems. Plate distribution tends to be non-uniform, which can result in assembly difficulties. Guidance on this subject can be found in BS 3382, " Electroplated Coatings on Threaded Components ".

(g) Anodising (Chapter 13) is subject to similar considerations to those governing electroplating.

REFERENCES

1. E. Raub and B. Wallhorst, *Arch. Metallk.* 1949, 3, 323.
2. Fundamentals of Metal Deposition, E. Raub and K. Müller, p.65. Elsevier Pub. Co., Amsterdam-London-New York, 1967.
3. M. Genisher, *Z. Elektrochem.* 1953, **57,** 604.
4. G. I. Finch, H. William and L. Young, *Trans. Faraday Soc.*, Discussions No. 1, 1947, 144.
5. K. M. Gorbunova and P. D. Dankor, *Zhur. Fiz Khim.* 1953, **27,** 1725.
6. J. Dash and W. W. King, *J. Electrochem Soc.*, 1972, **119,** 51.
7. A. H. Meathley, *Trans. Electrochem. Soc.*, 1923, **44,** 283.
8. S. Field, *Metal Ind. (London)*, 1934, **44,** 614; *J. Electrodepos. Tech. Soc.*, 1932, **7,** 83.
9. A. H. Durose, W. P. Karash and K. S. Willson, *Proc. Am. Electroplaters' Soc.* 1950, **37,** 193.
10. O. Kardos, *ibid.* 1956, **43,** 181.
11. S. A. Watson and J. Edwards, *Trans. Inst. Metal Finishing*, 1957, **34.**
12. L. I. Antropov, Inhibitors of Metallic Corrosion and the ø-scale of Potentials (paper presented at the 1st International Congress of Metallic Corrosion, London, 1961.
13. L. I. Antropov, Teoreticheskaya Elektro Khimiya, Izd. ' Vysshaya Shkola ' (Theoretical Electrochemistry, Publ. in ' Post-graduate School ').
14. T. P. Hoar and D. J. Arrowsmith, *Trans. Inst. Met. Finishing*, 1958, **36,** 1.
15. R. Walker, ' Internal Stress in Electrodeposited Metallic Coatings '. Metal Finishing Journal Monograph, 1968.
16. G. G. Stoney, *Proc. Roy. Soc.*, 1909, **82**(A), 172.
17. A. Brenner and S. J. Senderoff. *J. Res. Nat. Bur. Standards*, 1949, **42,** 105.
18. T. P. Hoar and D. J. Arrowsmith, *Trans. Inst. Metal Finishing*, 1957, **34,** 354.
19. A. Dvorak and L. Vrobel, *Trans. Inst. Metal Finishing*, 1971, **49,** 153-5.
20. ' Recommendations for the Design of Metal Articles that are to be Coated '. B.S.4779: 1969.
21. W. H. Safranek and E. W. Brooman, ' Finishing and Electroplating Die Cast and Wrought Zinc ', 1973, Zinc Inst. Inc., New York, N.Y. 10017.

BOOKS FOR GENERAL READING

1. ' Electrodeposition and Corrosion Processes' (2nd ed.), J. M. West, Van Nostrand, N.Y., 1971.
2. ' Technology of Electrodeposition ', A. T. Vagramyan and Z. A. Solov'eva, Robert Draper, Teddington, 1962.
3. ' Principles of Electroplating and Electroforming ', W. Blum and G. B. Hogaboom, Mc-Graw-Hill, 1949.
4. ' Corrosion ' Vol. 2, Ed. L. Schreir, George Newnes, 1963.
5. ' Modern Aspects of Electrochemistry ' No. 3, J. O'M. Bockris and B. E. Conway, Butterworth, 1964.
6. ' Fundamentals of Metal Deposition ', E. Raub and K. Muller, Elsevier, 1967.
7. ' Applied Electrochemistry ', N. P. Fedotev, A. F. Alabysher, A. P. Rotinyan, P. M. Vyachestavov, P. B. Zhivotinskii and A. A. Gal'nbek, ' Khimya ' Leningrad Branch 1967.
8. S. Field, *Metal Ind. (London)*, 1934, 44, 614; *J. Electrodeps. Tech. Soc.*, 1932, 7, 83.
9. ' Principles of Metal Surface Treatment and Protection ', D. R. Gabe, Pergamon Press, Oxford, 1972.
10. ' Basic Metal Finishing ', J. A. von Fraunhofer, Elek Science, London, 1976.
11. ' Electrochemical Science ', chapter 6, J. O'M. Bockris and D. M. Drazic, Taylor and Francis Ltd., London, 1972.
12. ' Modern Electroplating ' (3rd ed.), Ed. F. A. Lowenheim, Wiley, 1974.
13. ' Nickel and Chromium Plating ', chapters 2 and 3, J. K. Dennis and T. E. Such, Newnes Butterworths, London, 1972.

CHAPTER 2

Polishing

THE polishing and grinding of metals often constitute an integral part of the finishing operation . In many instances the normal machined surface or the finish of sheet metals as they come from the rolls is satisfactory for the application of electrodeposited coatings or organic finishes. Thus, steel stampings are not commonly polished before zinc plating or spraying with paint, although acid pickling to remove scale may be necessary if the metal has been subjected to heat-treatment at any stage. For decorative finishes, such as bright chromium plating and in most classes of work where a well-finished article is required, some degree of polishing is, however, necessary.

The polishing process is a very involved and complicated one and is as yet not entirely explained[1]. According to one view polishing is simply grinding or wearing away of metal beyond microscopic limits. The generally accepted theory is that, in general, polishing in its later stages does not simply remove further material, but that under the pressure of the polishing abrasives and the temperature rise created by the polishing buff melting of the top layer of molecules takes place, and an amorphous surface described as the Beilby layer, after the formulation of this theory, is built up. To support this view there is evidence that such amorphous surfaces have, amongst other properties, a lower corrosion resistance and a different electrical potential from those of surfaces having an oriented, crystalline structure.

The electron microscope has established that apart from the displacement of crystals, they become gradually fewer and smaller the nearer they are to the metal surface, and may be as small as a few atoms only (50-100 nM). In other words, as a result of the polishing process, the hitherto undisturbed crystal structure becomes more and more dislocated the nearer it is to the surface. It has actually been shown by thermocouple measurements that steel can reach temperatures of 150°—200°C during polishing and brass and aluminium can exceed 200°C. Local temperatures of 500°—1,000°C and even higher may occur on metal surfaces. Particles melted by polishing flow as fast as possible to cooler zones where they are instantly solidified and recrystallised. The crystals in the top layer are evidently broken down to such an extent that so far as size is concerned, there is a cautious state of change from the crystalline to the amorphous state, or vice versa.

It is thus seen that the polishing process would be ideal if the peak areas left after grinding were sufficient to fill in the neighbouring hollows. There are a variety of methods of achieving this :

(a) Burnishing.
(b) Polishing with abrasives using flexible wheels, belts, discs, etc. (mechanical polishing).
(c) Barrel or vibratory polishing.
(d) Electrolytic or chemical polishing.

Nowadays, burnishing is comparatively rarely employed, whilst barrel polishing is a method of producing a sufficiently bright surface especially on the cheaper type of article. The most general method of polishing currently employed involves the use of flexible abrasive wheels in some form or another.

2.1. Mechanical polishing

Burnishing

By burnishing is meant a method of smoothing metal surfaces by flattening out irregularities by the application of pressure; the operation is carried out by means of a hard tool usually made of steel or sometimes of hard stone, such as agate or bloodstone. A considerable variety of tools are used in burnishing which is especially suited to the softer metals, notably brass. The parts to be burnished are most often centred on a lathe and the pressure is applied by means of the tool; flat articles can be burnished by hand. The tool must be kept highly polished, and for this purpose a suitable buff is employed dressed with burnishing putty powder.

During burnishing, the part is kept wet with a weak solution of acetic acid. Periodically, the articles are dipped into cream of tartar solution to prevent tarnishing. They are left in the latter solution until they are ready for final drying off, which is accomplished by first dipping them into a solution of weak nitric acid or sodium cyanide to brighten the surface, after which they are washed and dried in hot water or sawdust.

Grinding

Where a good finish is required, it is necessary to polish the articles by means of a suitable abrasive. If the metal is very coarse and much surplus metal has to be removed, a preliminary grinding operation must first be applied. A large variety of grinding machines are available, making use of solid grinding wheels, horizontal or vertical discs, etc. These may be used either wet or dry. Typical of grinding abrasives are *emery* (consisting essentially of alumina, iron oxide, silicates, and various impurities), which may be bonded together in the form of wheels; *artificial alumina,* which has largely replaced the dangerous silica abrasives such as sandstone; and *silicon carbide,* a very hard abrasive made by fusing silica and carbon at high temperatures in the electric furnace. The grinding operation can have a significant effect on the corrosion resistance of the plated article, especially in the case of zinc-base alloys[2].

After grinding has been carried out, the metal is polished, if necessary, by means of wheels carrying abrasives held on their surface in various ways, as will be described later.

Polishing

For the final stages of smoothing a metal surface, polishing buffs or wheels are used. These wheels are made of a variety of materials such as felt, leather, canvas, calico, etc., and are built up of discs which are clamped and sewn together in a variety of ways. The design of the buff and the selection of the correct type for the job are of prime importance.

For the rougher operations where an appreciable amount of metal must be removed, set up or dressed heads (bobs) are employed, an abrasive being applied to the periphery of the wheel and held there by means of an adhesive. The mixture of the adhesive (which may be a cold one or hot glue) and the appropriate grade of abrasive is brushed on to the head of the wheel and the whole is dried in an oven at a carefully controlled temperature. It is important that the degree of drying should not be excessive as otherwise the head becomes too brittle. For the subsequent polishing operation where a fine finish is required set-up heads are not employed, but a polishing composition is applied to the rotating wheel (buffs) from time to time. This method is almost universally employed for obtaining a high degree of finish on metals.

Polishing broadly covers all the intermediate operations between preliminary grinding and final colouring. There are differences between American and European terminology, but these are tending to disappear. The term ' dressed ' refers to the application of a hard head consisting of a glued abrasive on to the polishing wheel, as has already been mentioned. The hard polishing wheels are made of a variety of materials such as felt or sometimes leather, and are often specially shaped for specific purposes. Mops are wheels made of cotton or calico and are softer than bobs, although they may also utilise a glued abrasive head.

2.1.1. POLISHING LATHES

Polishing wheels must be mounted on to a suitable lathe. The usual lathe for handpolishing normally carries two wheels driven by a single motor; the motor is preferably directly coupled to the spindle and, as a rule, carries a bearing at the end of the casing to provide extra support for the spindle. The polishing wheel proper is attached to the spindle either by means of a clamping plate passing over its threaded end, or similar device. Hand polishing lathes often have a taper-threaded spindle end on to which the wheel fits directly. This latter device is very rapid in use, and there is no need for any further fixing attachment since the speed of the rotating spindle tends to force the wheel on to it. Changing of the polishing wheel is easily carried out.

The thread at the end of the spindle ends are a very useful accessory, enabling the worn thread to be replaced quickly and cheaply. The method of attaching the wheel can also be altered from a taper spindle to a clamping plate very simply by removing the tapered spindle end.

The driving motor should be totally enclosed for protection against the ingress of dust, and the bearings should also be efficiently lubricated in such a way that polishing dust cannot find its way into them.

The speed of the lathe is of importance inasmuch as it affects the peripheral speed of the polishing wheel. Typical speeds are within the range of 1500 to 2,700m per minute. For the rougher polishing operations, it has been found that the lower range of speeds (of the order of 1,500m/min) can be used; these prove more economical than the higher speeds on grounds of wear and tear of polishing wheels and frequency of setting-up. On the other hand, for fine polishing, speeds up to 2,500m or more per minute can be used with advantage. The relationship between peripheral speeds and revolutions per minute for some typical wheel sizes is shown in Table 2.1.

For the higher speeds it is desirable to use more concentrated glue mixes as the abrasive granules will need to be more strongly held. The linear speed at which different metals are most efficiently polished varies, the softer metals in general requiring higher speeds. Thus, for stainless steel or mould metal 2,700 surface

TABLE 2.1

Relationship Between Peripheral Speeds and Revolutions Per Minute For Typical Polishing Wheel Sizes

Spindle r.p.m.	Wheel diameter 15 cm	20 cm	25 cm	30 cm	35 cm
	Peripheral speed : m per min				
1,000	470	630	785	940	1,100
1,200	565	750	940	1,130	1,320
1,400	660	880	1,100	1,320	1,540
1,600	750	1,000	1,260	1,510	1,760
1,800	850	1,130	1,410	1,700	1,980
2,000	940	1,280	1,570	1,880	2,200
2,200	1,040	1,380	1,730	2,070	2,420
2,400	1,130	1,510	1,880	2,260	2,640
2,600	1,220	1,630	2,040	2,450	2,860
2,800	1,320	1,760	2,200	2,640	3,080
3,000	1,410	1,880	2,360	2,830	3,300
3,200	1,500	2,010	2,510	3,010	3,520
3,400	1,600	2,135	2,670	3,200	3,740
3,600	1,700	2,260	2,830	3,390	3,960

m per minute has been recommended, 2,000m for brass, 1,800 to 2,300m for aluminium, and 1,500 to 3,000m per minute for zinc alloy die-castings. The procedures for the polishing of the latter have been summarised by Safranek[3], and those for polishing copper and its alloys reviewed recently by Dewar[4].

In this country polishing lathes are not commonly provided with any means of altering their speed, although a two-speed countershaft or a three- or four-speed gearbox is occasionally fitted. In the U.S.A. variable-speed lathes are more widely used and present many advantages over the fixed-speed type. As the wheel wears down in diameter the peripheral speed decreases if the number of revolutions of the shaft per minute is unaltered, so that poor polishing may result owing to the drag of the slow-moving wheel.

In many of the larger polishing shops lathes are provided with different fixed speeds, so that the correct peripheral wheel velocity can be more readily maintained for each size of wheel.

2.1.2 POLISHING WHEELS

Polishing wheels are made of a variety of materials such as fabric, felt, leather, etc. Fabric wheels are built up of discs of canvas, calico, etc., clamped and sewn together. For rough polishing and where any appreciable amount of metal is to be removed, the wheels employ set-up heads in which an abrasive is applied to the peripheral surface of the polishing wheel and held there by means of glue. A series of such wheels or bobs is employed for each successive operation, making use of a finer grade of abrasive on the head until a high enough degree of finish is obtained. The art of metal polishing involves the skilled use of a suitable cross-polishing technique whereby the scratches left by the coarser polishing media are removed at each stage.

The next operation is carried out with buffing wheels (or mops) without a set-up head on to which the polishing abrasive is applied by holding a stick of a suitable polishing composition against the buff whilst the latter is rotating at speed. Finally, for the highest finish, a colouring operation, using soft mops, is carried out with fine compositions, such as rouge, lime, etc.

The following list is indicative of the many types of polishing wheels used in the industry, although some are only employed for very special purposes.

Felt wheels, which are highly resilient and are fast-cutting, are still used for some applications. Their springiness makes them less liable to cause deep scratches than the harder wheels, so that they are very useful in colouring operations where only a small amount of metal is to be removed leaving a lustrous finish. Felt wheels retain their shape very well and do not lose their contour. Solid felt wheels are the most satisfactory type because of their uniformity, for specialised parts, such as turbine blades.

Stitched canvas or calico wheels are very widely used for all types of polishing operations because of their relatively low cost and high durability; they are exceptionally suited to the coarser grades of abrasive used for rough polishing. They may be used either with set-up heads (when they are known as 'scurf-mops'), or for finer finishing with compositions. Grease is often applied to set-up polishing bobs where a finer cutting action is required; this acts as a lubricant and tends to prevent tearing of the wheel surface.

Canvas disc wheels are used for rough polishing castings and forgings for parts for agricultural implements, stoves plumbing fixtures, etc. The open weave of the canvas enables them to hold glued coarse abrasive heads very strongly. By using different types of stitching a widely varied range of flexibilities can be obtained.

Soft buffs or mops do not make use of glued heads, the polishing composition being applied to them from a bar of the latter whilst the wheel is rotating. The manufacture of mops and buffs is a very specialised industry, and the performance of an individual buff depends on detailed attention to the materials of which it is made and on the method of construction.

Disc Buffs are made from a number of discs of cotton, leather, etc., either held together in the centre to form a loose open wheel, or stitched together to give a harder type of buff. The former is used in the 'colouring' of nickel, brass steel, etc., whilst the stitched wheels are used for the 'cutting-down' of the softer metals. Fig. 2.1 shows a calico mop for manual polishing, fast cutting, finishing and colouring.

Stitched piece buffs are made from off-cuts of material and are used for heavy duty polishing. They are considerably cheaper than the full disc buffs. Various types of

Fig. 2.1. Stitched polishing buff with spiral sewing.
[Courtesy Cooper & Co. (Birmingham) Ltd.]

Fig. 2.2. Continuous strip type ventilated polishing buff.
[Courtesy Cooper & Co. (Birmingham) Ltd.]

sewing are used on stitched buffs, including circular, spiral, tangential, and square sewing, to stiffen the fabric and prevent distortion under pressure. The first two types are most common, and the closer the rows of stitches the harder is the wheel.

The bias buff is the most extensively used type. It makes use of a single bias-cut strip of fabric not more than half the diameter of the buff wound in a helix until the necessary thickness has been built up. As it is wound, the fabric becomes puckered at the outer edge to compensate for the difference between the inner and outer circumferences of the wheel. The result is that the edge of the buff is wavy. This has the advantage of minimising the appearance of polishing marks on the work. This wheel, like the last, is of the ventilated type, so that it can be used with higher pressures and at greater speeds than the conventional type of buff without burning or charring. Polishing is carried out more quickly, and the ventilated buffs are worthy of wider use. Moreover, there is no need for the wheels to be ' rested ' from time to time to allow them to cool, and they do not become overheated under normal working conditions. They are particularly suitable for heavy duty in conjunction with automatic plant. The advantage of bias cutting is that it reduces any tendency towards fraying or unravelling of the material, whilst the pleated face holds composition better so that cutting is faster and colouring is better. As bias buffs wear they tend to become denser which helps to offset the loss of cutting power, which normally accompanies the loss of peripheral speed which occurs. Fig. 2.2 shows a ventilated air-cooled wheel. These wheels can be made of calico for soft metals, or of sisal for mild and stainless steels.

Ventilated buffs are made from fabric treated in such a way that a permanent crimped finish is applied to the fibres to prevent unravelling and untwisting and to improve compound adhesion. Crimping reduces internal friction between threads, thus increasing buff life considerably.

Cotton buffs should be made of long woven staple cotton and of tightly twisted yarn. The more closely woven the fabric, the harder will be the resulting wheel.

In the modern technique of ' crush ' polishing, which demands greater mop resiliency to enable a fuller coverage for the most intricate as well as simpler shaped articles, coupled with a reduction in spindle speeds down to as low as 600 to 900 rpm, buffs with diameters to 50-60 cm, built up to widths of 250 cm can be used[5].

Sisal buffs are of increasing importance today and are used most widely for the finishing of stainless steel. They can be made up from alternate layers of sisal and heavy calico tightly stitched together. Sisal wheels have high cutting properties and are useful for removing draw marks from steel pressings. These buffs are best made from bias cut material. A sisal buff built up from a number of sections is shown in Fig. 2.3, whilst Fig. 2.4 shows the more flexible finger buff. Sisal brushes are often interleaved with calico discs to improve their composition-holding properties and to help them to run cooler.

Fig. 2.3. Stitched sisal mop built up from a number of sections.
[Courtesy W. Canning Ltd.]

Fig. 2.4. Sisal finger buff for complex shaped articles. [Courtesy Cooper & Co. (Birmingham) Ltd.]

Tampico brushes are also employed for removing surface imperfections, tool marks, etc., from complex shaped articles. They are not as powerful as sisal buffs but are rather more flexible.

This by no means exhausts the available types of polishing wheels. Many other kinds of mops, such as swansdown, chamois, and so on, are used for special finishing operations. The selection and use of polishing buffs has been discussed by Rees[6].

2.1.3. ABRASIVES AND THEIR APPLICATION

The abrasives employed for rough polishing or buffing are of various types. Natural abrasives, such as emery, are used, whilst artificial aluminium oxide is being increasingly employed because of its hardness and because of the extent to which it is possible to control not only the grain size but also the crystal form of the abrasive granules. The artificial abrasive powders are manufacured in such a way that they have a high surface capillarity in order that they may adhere well to the polishing wheel when the latter is glued and dressed with them. The range of particle sizes used in general polishing operations are around 60 to 90 mesh for ordinary roughing down to 120 for successive operations, after which the finest grades are employed for finishing operations. The latter are obtainable in grades down to 600 mesh. Uniformity of particle size is of the utmost importance if satisfactory polishing free from scratches is to be obtained.

Modern synthetic abrasives are also specially treated to ensure that they have the best capillarity to enable them to be firmly bonded by the glue. Capillarity is reduced by exposure to dirt or grease, and even exposure of the abrasive to the atmosphere for any length of time will noticeably reduce the strength of a polishing wheel head set up with it. Wheels dressed with a high-grade synthetic abrasive have an open type of coating with a clearance between the grains; the result is that the wheels do not fill so easily with metal particles from the articles being polished. For finish polishing such materials as tripoli, rouge, and chromium oxide find application; these are used in the form of blocks of compositions which are applied to the buffs in the dry state, whilst they are rotating. Siliceous polishing materials are to be avoided wherever possible because of the health dangers arising from the dust which they generate.

Setting-up of Wheels

In setting up the polishing wheel, the latter is first painted with a glue solution or a cold-setting cement and rolled in a trough containing the abrasive. The wheel is then dried out under suitable conditions.

Hide glue was at one time used exclusively for the dressing of wheels, but is now tending to be replaced by cold-setting cements, which are simpler to use. If slab glue is used, it must be soaked overnight; hence ground or flake glue is to be preferred since it only needs to be soaked for a much shorter time. It is then placed in a water-jacketed gluepot, which should preferably be thermostatically controlled at 70°C. If the wheel and the abrasive are heated before the application of the glue (as is very desirable) a lower glue temperature of about 60°C is favoured. Glue loses strength when kept at temperatures above 70°C, and even at lower temperatures fresh quantities should be made up every three or four hours. The strength of the glue is determined by the amount of water used; with high-speed wheels a stronger glue is needed than is the case with lower speeds. Table 2.2 gives the glue strength commonly used under average conditions with various grades of alumina abrasive.

Before applying the glue to the periphery of the wheel the nap of the fabric is raised by running the wheel at high speed against a sharpened steel bar. This operation is also used for 'contouring' the section of the wheel if it is to be used for the polishing of articles of special shape. Gauges are often used to facilitate the

TABLE 2.2
Glue Strength for Alumina Abrasive

Grain size of Alundum aluminium oxide abrasive	Glue %	Water %
24 — 36	50	50
36 — 54	45	55
60 — 70	40	60
80 — 90	35	65
100 — 120	33	67
150 — 180	30	70
220 — 240	25	75

contouring operation. A short nap on the wheel surface helps to bond the abrasive, but if the nap is excessively long it will have an undesirably sharp cutting action. Finally the wheel is dressed with a carborundum stone to smooth the surface. The wheel is then ready for the application of glue or cement.

The hot glue is brushed on to the wheels in a warm room (about 25°C) in order to prevent too-rapid setting; the wheels are then rolled quickly and with even pressure in a trough containing the abrasive until the glue has picked up as much as it can hold. A second coat of glue and more abrasive can then be applied on top of the first if required. To avoid too early gelling of the glue, both the wheel and the abrasive grains are heated to 45 — 50°C. Too early setting of the glue lowers the adhesion qualities and, moreover, the grains do not penetrate into the glue film so that the abrasive content of the latter is too low; the result is a slow-cutting wheel which will tend to overheat.

Uniformity of grain size is important, as is also the shape of the abrasive granules. They must be solid and sharp-cornered to have the maximum cutting qualities. When wheels are being set up with different grades of abrasive the troughs should be placed sufficiently far apart to ensure that larger grains do not fall into the trough carrying the finer granules, or score-marks may result when the wheels are used in polishing operations. The usual types of grain troughs are relatively short and the abrasive must be rolled into the wheel a section at a time with a pounding movement. A recent innovation is the use of a long grain trough in which the entire wheel can be rolled in one continuous motion, using even pressure. The pressure used to pound the wheel into the grain must be uniform, and the actual pressure of application has a marked effect on the polishing action of the wheel. When the wheel appears dry and no glue can be seen, the process is complete.

Cements

The use of cold cements, usually based on silicates, has become of increasing importance in recent years as a replacement for glue. Glue tends to soften with heat, but modern cements are not subject to this effect so that the abrasive is more firmly held at high speeds. The drying of cemented wheels can also be carried out more expeditiously. They will dry in air in about 12 hours, but better results are obtained when they are force dried at about 65°C. The cold adhesives do not lose strength on keeping, as does glue when kept hot, whilst their holding power at elevated temperatures is better. Generally speaking, for the coarser abrasives the cements are more satisfactory than glue, and it is here that they find their chief application. For fine polishing, glue is generally preferred.

Before redressing, wheels are trued by pressing a file against the surface whilst they are running at normal speed. Residual glue is then removed by means of an abrasive brick whilst the wheel is still running, and this also serves to raise a nap on the surface, to which the glue will adhere.

Polishing wheels should be run in the same direction at all times, with the lie of the nap.

Drying

The drying of the glued head must be carried out under controlled conditions if the best results are to be obtained. Hide glue attains its maximum strength not when it has been completely dried, as is sometimes erroneously believed, but when it has a water content of 10 to 12 per cent. The drying room should therefore be controlled both as regards temperature and relative humidity, so that the final moisture content of the glue approximates to the condition at which its strength is at a maximum. The ideal drying conditions usually recommended are 30°C and a relative humidity of 50 per cent.

The drying time depends on the thickness of the glue film containing the abrasive and the strength of the glue used (which in turn depends on the grain size of the abrasive, as shown in Table 2.2). It may also vary to some extent with the size and nature of the material of which the polishing wheel is constructed. Under-drying weakens the glue, so that the wheel will tend to glaze and will have poor cutting qualities. It is also likely to overheat in working, and may even catch fire. If the wheel is over-dried, on the other hand, the glue becomes unduly brittle so that the abrasive granules tear out of the adhesive film easily, and the wheel has to be re-dressed after a short time.

Drying is generally complete in twenty-four to forty-eight hours under the conditions specified, the thicker-coated wheels naturally taking a somewhat longer time. The air in the drying-room should be circulated, and if additional humidification is required this can be obtained by means of a small electronically heated water boiler placed in the room; the heater is connected through a relay to an air hygrometer with contact points attached to the needle, so that the heater is switched on and off between a range of, say, 45 to 50 per cent relative humidity.

In Great Britain it is, however, generally not necessary to humidify the air artificially, and sufficient control can be obtained by varying the temperature of the drying-room to obtain the required humidity. The graph (Fig. 2.5) shows how it is possible to obtain the optimum drying conditions within the drying-room from the temperature and relative humidity of the incoming air. This chart is based on the assumption that good drying conditions are obtained when the vapour pressure of water in the room is 15mm less than the saturation vapour pressure at the same temperature. It is also possible to use infra-red or radiant heating units whereby it is claimed that the heads can be dried completely satisfactorily in an hour or two. The wheel is then balanced statically on a spindle and small dished counterweights can be nailed at a point on its side, if necessary.

Set-up fabric disc wheels must have their surfaces broken up to give them the necessary flexibility. The breaking of the head is done by striking the wheel flatly on the face with a round bar at an angle of about 45 degrees to its axis, going round the whole wheel two or three times. The process is then repeated in the opposite direction, so that the bar falls on to the face of the wheel at right angles to the direction of the first blows.

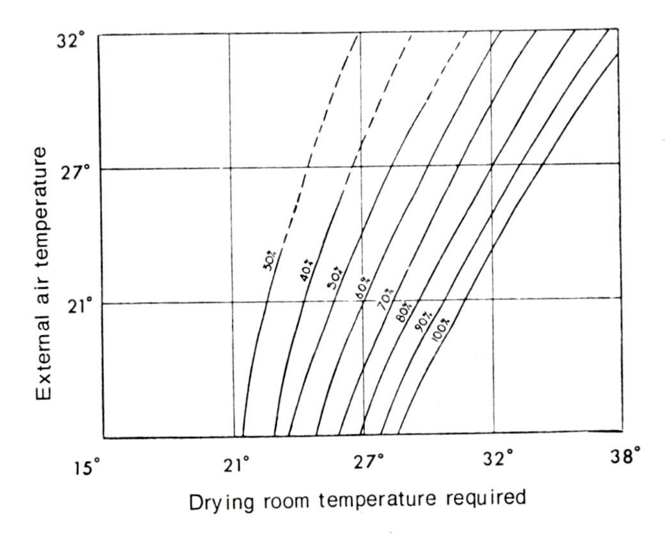

Fig. 2.5. Drying of polishing wheels. Graph showing relation between external air temperature and drying-room temperature necessary for obtaining any required relative humidity. Temperatures are in degrees C.

Satin Finishes

The demand for matt and satin finishes has resulted in the development of non-woven nylon materials impregnated with alumina or silicon carbide. Abrasive impregnated nylon monofilaments which can be made into brushes are also coming into use.

These may greatly reduce the cost of satin finishing as compared with the traditional greaseless compositions used for this purpose. The materials can be made up into buffs for use on manual or automatic polishing machines.

Abrasive belts

Abrasive belts of either canvas or leather are set-up by coating them with glue in a manner similar to that described above, sprinkling the abrasive on to the surface and finally pressing it in with a wooden roller. An alternative method is to add the warmed abrasive powder to the liquid glue of appropriate strength until a paste is formed, which can then be spread on to the belt with a palette knife. The belts are finally dried under conditions similar to those employed in the case of polishing wheels. Occasionally, polishing wheels are dressed in the same way.

The usual methods employed for the setting up of polishing wheel heads described above are slow and laborious and some attempts have been made to improve on them. One method is to incorporate the abrasive into a cold adhesive and apply the mixture directly to the polishing wheel by brush, or better, by means of a spray gun. Owing to the protecting effect of the adhesive it is sometimes advisable to employ a rather larger grain size than usual for the same degree of cut. Much thinner heads are used than in the normal procedure but it is claimed that wheels set up in this way last about five times as long as those headed in the common manner. These heads are especially suitable for obtaining satin or butler finishes.

Coated abrasive wheels

A type of polishing wheel which offers the advantage of being a standardised factory-made product consists of hundreds of pieces of cloth-backed coated abrasive formed into a wheel, spoke-fashion. The abrasive and the cloth wear away simultaneously, so that the wheel takes the shape of the piece being polished and retains this shape throughout the life of the wheel. No dressing is required, and a smooth fast cut can be obtained. Pressures of 0.5 to 1.0 kg/cm^2 are recommended with a peripheral speed of about 3000m per minute.

2.1.4. POLISHING COMPOSITIONS

A considerable variety of polishing compositions is available to the industry, most of them being proprietary products developed as a result of considerable experience by the makers. Partridge[(7)] has comprehensively reviewed the patent literature and data concerning polishing compositions over the period 1960-1971. Polishing compositions consist of abrasives bonded together with a grease base which melts when the composition in bar form is held against the rotating buff; in this way a small amount of abrasive is held by the surface of the wheel.

Polishing compositions are supplied in bars and are manufactured by melting a mixture of selected greases and the abrasive in a jacketed pan, casting the products into moulds. A good composition should be designed to adhere to the polishing wheel just so long as the abrasive retains its cutting qualities and leave no residues to reduce the efficacy of fresh composition which is subsequently applied.

Polishing compositions have a complex action, the function of the harder types of abrasives being to remove the coarser irregularities in the base metal and to have a cutting action. Medium abrasives serve to burnish or flatten out the surface rather than to remove metal while the final ' colour ' is given to the metal by means of the finest abrasives. This last operation is facilitated by the local fusion of the micro-crystals of the surface and results in the formation of the non-crystalline Beilby Layer, which is characteristic of a polished surface.

The formation of the polishing composition involves careful selection of the types of grease and abrasives used, a great deal depending on the absorptive properties of the abrasive itself for the grease. A high degree of fat absorption is desirable since this facilitates the adhesion of the composition to the polishing mop when it is applied and also makes for a mechanically firmer bar which will not fracture when pressed against a polishing wheel. Drier compositions are made by using high-melting stearine-type greases while softer ones utilise the lower melting tallows and waxes. Usually compositions required for cutting down are rather greasier than those employed in the final colouring operations which tend to be rather on the dry side. However, when manual polishing operations are carried out different polishers have their own ideas as to the types of compositions they prefer, while other factors such as wheel speeds and applied pressures are not without influence.

Greaseless compositions are mixtures of glue and abrasive and are easily applied to all types of wheels. They are especially useful in dealing with articles such as stampings from which all imperfections cannot readily be buffed out with grease compositions.

A good composition should be economical in use, clean off readily in operation and be easily removed in the cleaning solutions used prior to plating. Nowadays, the two main types of abrasive used in both bar and liquid compositions are tripoli and alumina, the former being principally used for bright polishing non-ferrous metals, and the latter for mild and stainless steels.

Other abrasives used in polishing compositions are materials such as lime, diatomaceous earth, green chromium oxide, emery, etc. The requirements are rather critical since they must not only be hard but have a very uniform particle size which normally ranges from 50 to 180 mesh. Uniformity is important since the presence of even a few large grains may cause severe scratching of the surface being polished. Abrasives normally have to be calcined in a furnace to make them suitable for incorporation in polishing compositions.

Tripoli compositions are very fast cutting and are specially suitable for producing a high lustre on non-ferrous metals. Tripoli is a hydrated silica and generally comes from either North Africa or the U.S.A. American tripoli has a smaller particle size and greater grease-absorptive qualities (around 50%) than the African material and is, in many respects, more suitable as a polishing abrasive. It is available in a variety of colours, such as grey, yellow or pink, depending on the impurities which are present. The amount and type of grease employed in the bar determines the type of composition necessary for specific purposes. Dry grades of tripoli compound are recommended for lighter work and the more greasy ones for heavier cutting since the grease holds the abrasive on the wheel longer.

Lime compositions are made from calcined dolomite and are widely used as a last operation prior to plating as they help to remove grease residues, the best known grades being Sheffield and Vienna lime. Sheffield lime is a partially slaked material and is commonly applied in lump form directly on to the polishing wheel. Vienna lime has a high magnesium content and has, therefore, greater fat absorptive properties and for this reason is widely used in what are known as 'white finish' compositions. It is the presence of the magnesium which permits this type of lime to be employed in polishing compositions, since the high calcium limes will saponify the fats, and are therefore unsuitable. Lime compositions must be carefully stored to exclude air and moisture since otherwise the lime will react with the atmosphere to form calcium carbonate and the composition will disintegrate.

Emery is a natural mineral consisting mainly of a mixture of equal parts of alumina and magnetite. It is a very useful abrasive and can be obtained down to extremely fine grades known as flour emery prepared by elutriation and sedimentation.

Green oxide compositions are employed for putting a high finish on stainless steel and also for finishing chromium plate. Synthetic alumina has particularly good cutting properties and, like emery, is employed where the highest standard of finish is required.

Alumina is available in numerous grades, and the porosity can be controlled during its manufacture. Thus it is possible to obtain materials with oil absorptions of between 15 and 65 per cent. This gives considerable scope to manufacturers of alumina compositions in producing materials to meet a wide range of requirements.

Liquid compositions

Liquid polishing compositions are of prime importance where automatic polishing is used. In fact, it can be said categorically that the optimum efficiency of an automatic polishing installation is more dependent on the use of liquid polishing materials than on any other single factor, and a high proportion of all automatic machines now use liquid compositions. Liquid compositions have been known for a considerable number of years and originally they consisted of oil or kerosene bases in which an abrasive was suspended. Experiments proceeded desultorily with these with limited success, the chief objections being (*a*) the tendency for the abrasive to settle in the medium and (*b*) the fire hazard resulting from the inflammable nature of the liquid phase.

Progress in the field of emulsion technology and the use of the newer surface-active materials completely revolutionized the formulation of liquid polishing compositions, which are now always low-viscosity water emulsions of oils and greases in which the abrasives are suspended[(8)]. The media themselves are relatively viscous and the suspensions are so stable, due to the close correlation between the specific gravities and viscosity of the media and the abrasive, that little or no settling out occurs even after prolonged standing. A very important feature of such compositions is that they can be formulated precisely to meet the requirements of the job, unlike bar compounds where the formulation is governed by the need to produce a bar which is physically strong enough to handle. The abrasives are applied to the rotating head by means of a spray gun with a large aperture, using an automatic timer operated by a solenoid valve. The compound is sprayed on to the wheel every seven to ten seconds, on the average, the duration of the application being from $\frac{1}{2}$ second to 2 seconds. Low pressure systems using fluid pressures of about 20-30 kg/cm^2 and atomising pressures of 40-60 kg/cm^2 are generally used, but there is a growing interest in high pressure airless equipment using pressures of 70-85 kg/cm^2 which has the advantage of reducing composition losses by overspray, although there is a greater wear of the spray equipment. Higher viscosity components are used in this instance.

Of particular interest is the high velocity spray gun used in conjunction with a low pressure system. Here the high pressures are produced by an intensifier on the gun itself, which has few moving parts.

In hand-operated machines the spray can be put into action by means of a foot valve by the operator but it is not considered that liquid compositions are at present generally suitable for hand-polishing techniques; the reason is that not only may there be excessive throw off from the rotating wheel but the need for frequent changing of the grade of compound complicates the problem of utilizing liquid compositions too much. Liquid compositions are, therefore, almost entirely employed in automatic machines. In automatic machines the valves which actuate the spray are cam operated. The cams can be located at any convenient point, such as the table chains on rotary machines or between the workholders, or on a slowly rotating shaft for small work. In the case of a constant-speed table-type or straight-line machine, cams should be very short. On straight-line machines they may be placed on work carriers, conveyor chains or belts and for long straight work the compound should be sprayed on every 30-40cm which keeps the cutting rate of the mop constant and makes for a high degree of efficiency. The cam also serves to adjust the amount of compound which is sprayed on. It is better to control the amount of compound used by frequent short applications rather than to utilize excessively small gun apertures which may lead to plugging. The rate of metal removal is increased in this way. The angle and the distance of the jet from the rotating wheel are also rather critical. The quantity of compound used affects the rate of cut, but if this is increased excessively the efficiency of the process is reduced in relation to the amount of composition used. To avoid wastage of composition through throwing-off it is desirable to run the mops at as low a speed as it is practicable and to apply composition while they are in contact with the work being polished. With modern compositions, gravity feed cannot be employed, pump operation from the container in which the material is delivered, or from a pressure tank, being normally used. Several guns are quite commonly served by a single pressurized container. Pump operation is simple and safe but the problem of designing a pump to stand up to the abrasive is not an easy one. Where wide wheels are used, such as in the polishing of sheet, traversing or oscillating guns are employed.

One ancillary advantage of liquid polishing compositions is that as they contain emulsifying agents they are much more readily removed during subsequent cleaning operations (prior to electroplating, for example), than the solid bar type. This applies also to the machine itself which is much more easily kept clean. Also, their water content results in better cooling of polishing wheels so that the life of these can be extended by up to 50 per cent or more and the fire hazard is reduced at the same time. The wastage in the form of unused ends is entirely eliminated so that the cost advantage is substantially in favour of these compositions, which are remarkably economical in use.

2.1.5. ABRASIVE BELT POLISHING

A modern development is the method of abrasive belt polishing in which the polishing wheel is used simply as a base for an abrasive belt which runs around it and is supported at its remote end by a backstand idler. The wheel itself may be of any suitable resilient material. Sometimes an ordinary canvas mop is used, or one made of a suitable grade of rubber or synthetic rubber will often prove very satisfactory.

This system has the advantage that factory-produced belts can be used which are uniform and cool running, whilst the skilled work of setting up and drying polishing wheels is entirely eliminated. It is, however, a method which is best used for rough polishing as the belts lack the flexibility for dealing with finer polishing operations, on intricate parts especially.

An important factor in performance is the contact wheel which is employed. Fig. 2.6 shows the arrangement of two different contact wheels and belts. The angle and depth of serration profoundly affect the cutting action of the abrasive belt, When the belt is new and sharp, a low pressure is desirable, and this should be increased gradually as the belt wears. For most metals belt speeds should be between 1,500 and 2,400 m/minute, 1,800 to 2,000 m/minute being generally employed for glue bonded belts and 2,000 to 2,400 m/minute for the resin-bonded types. For work with severe contours the belt speeds should be reduced.

Resin-bonded belts have the highest heat resistance and toughness and are, therefore, used chiefly for coarse cutting, final polishing being done on glue-bonded belts where

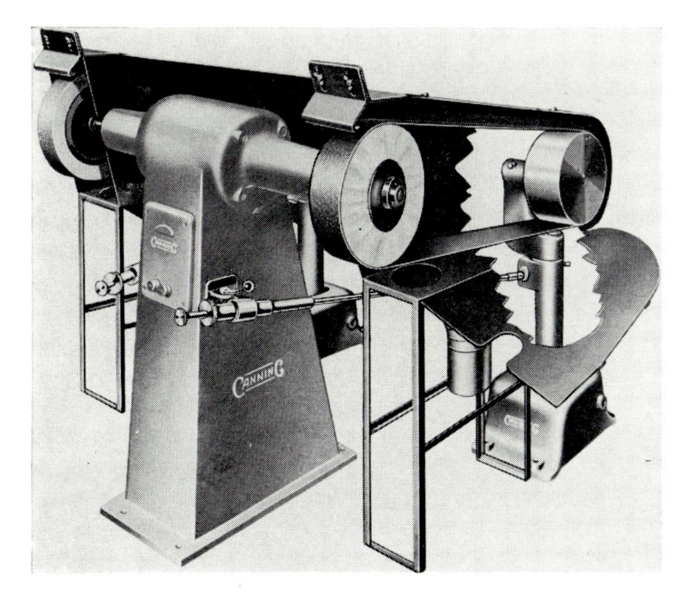

Fig. 2.6. Abrasive belt polishing unit with two contact wheels and belts.
[Courtesy W. Canning Ltd.]

the smoothest finish is required. An intermediate type of belt consisting of a resin over glue bond is available and is used for general polishing operations. Aluminium oxide is used for harder steel and silicon carbide for the softer metals as a general rule, but this need not necessarily be the case, as aluminium oxide has been used successfully on aluminium and zinc-base alloy die-castings[9].

Lubricants are commonly used on belts, the types employed being similar to those used when the metals are being machined. Surface-active agents in the oils are useful in promoting wetting of the belts. The lubricant is dripped onto the belt some 7-10 cm ahead of the idler wheel or is sprayed on. Sometimes grease bar compounds are used.

Coated abrasive belts must be properly stored if maximum life is to be obtained, particularly where glue-bonded belts are employed. Temperatures of 15° to 20°C and 40 to 50 per cent relative humidity are desirable and a storage cabinet to maintain these conditions should be part of the equipment of every polishing shop.

2.1.6. THE ART OF POLISHING

Metal polishing is still in many respects an art, and it is difficult to lay down general rules. Morgan[10] has discussed the principles of polishing and the polisher's skill. Peripheral wheel speeds must be selected to give optimum results, not only from the point of view of the appearance of the finish, but also to ensure that due economy in labour and materials is being obtained. The appropriate wheel of the correct flexibility must be selected for each job. Care should be taken in the handling of the article to prevent sharp edges meeting the polishing head and causing the latter to wear unduly quickly.

The different grades of polishing are termed ' roughing ', ' greasing ', ' finishing ', etc., depending on the cycle of operations used. For roughing a fairly coarse abrasive is used on a dry wheel, but for the finer polishes grease or compositions are applied to the wheel. The lubricant not only helps in the production of a smoother and brighter finish, but also prevents loading or glazing of the abrasive head and reduces discoloration by preventing the wheel from overheating. The polishing wheel life is therefore also considerably increased. Lubrication is particularly helpful in the case of the softer metals such as aluminium and zinc-base alloys.

The observations of Pinner[11] on the influence of the direction of rotation of the polishing wheel with respect to the direction of movement of the work are significant. Grinding and deep cutting for removing surface imperfections and polishing scratches are carried out by moving the work in the opposite direction to that in which the wheel is rotating. It is important, however, that the final polishing operation just prior to plating should be accomplished with the work moving in the same direction as the rotation of the wheel when polishing steel prior to nickel plating. In this manner, microscopic steel slivers are removed, and the absence of such particles definitely favours the production of smooth deposits which are more easily buffed and will possess improved corrosion resistance. The final wheel should also be a partly used one to reduce the tendency for scratches to appear to a minimum.

The following are some typical polishing cycles for a number of metals. The initial grade of abrasive used in each case depends, of course, on the condition of the metal surface. If the surface is good, the coarser polishing operations can be omitted. Where emery is recommended it can with advantage be substituted by the corresponding grade of synthetic alumina.

Aluminium

Polish with a greased felt wheel and 90 emery, following by 120. Next use a tripoli composition on a stitched mop and finish with an open cotton buff with lime composition.

Steel

If the surface is very coarse, roughly polish with emery of 60 mesh on canvas wheels, followed by 90 and then 120 on greased cotton bobs. For subsequent nickel plating an improved surface can be obtained by finishing with levigated alumina on an open mop.

Stainless Steel

Polish with 90 alumina, cross-polishing to remove scratches. Follow with a white composition on a cotton buff, and then finish with a chromium oxide composition. Emery should not be used on stainless steel, as the particles of iron oxide present may become ingrained in the metal surface and promote corrosion.

Speeds of 2,000 to 2,400 m/minute are recommended for grinding and polishing. Stainless steels have low thermal conductivity so that they are easily overheated. For this reason excessive pressures during the polishing or buffing operation must be avoided. The types of surface finish that can be obtained by polishing stainless steel and the equipment necessary has recently been reviewed[12].

Zinc-base alloy die-castings can be polished with 120 emery on the parting ' flash ' of the casting only, then buffed with a stitched mop using tripoli composition. It is desirable to have as high a finish on the die as possible. This reduces the amount of polishing to a minimum and so avoids the danger of polishing through the surface ' skin ' of the casting. The polishing of zinc base die-castings has recently been discussed by Safranek and Brooman[13].

Nickel plate is polished directly on cotton mops with lime composition.

Chromium plate can be ' coloured ' on a basil mop, either with a white composition or, if the deposit is very dull, with green oxide composition.

2.1.7. DUST EXTRACTION

Dust extractors are essential on polishing equipment. Careful design of extractor cowls and duct-work is needed to ensure that there is a minimum of bends and constrictions in the latter, and that the positioning of the former does not interfere unduly with the handling of the work. Automatic machines present special problems; in one installation where a very large amount of dust was generated a successful solution was found by putting each machine in a compartment of its own, and extracting air at high speed through a water curtain at the back of the compartment. A running water floor also served to prevent dust accumulation. The same water was pumped around continuously, the accumulations of sludge in the water being filtered out on wire mesh screens.

2.1.8. AUTOMATIC POLISHING

Nowadays a great deal of metal polishing is carried out on automatic machines, of which there is a great variety[14] designed for the polishing on mass-production lines of specific articles, such as motor-car handles, hubcaps, tubes, or metal sheet. Whilst there are a limited number of basic designs in principle, which are described below, they are generally adapted to the polishing of a limited range of products by the use of suitable jigs and finishes.

Rotary table machines

The rotary table machine can be either of the continuous type, where the table is rotated at a slow speed, or an indexing type where the table is indexed every few seconds. The former machine is used for high production rate on such compounds as door knobs, lipstick cases, fountain pen caps and barrels, etc., while the latter is normally used when a lower production is required on larger articles such as saucepans, hub caps and other components when there are solids of revolution. The indexing type of machine can also be used for out-of-round shapes by adopting special camming devices. One of the latest developments on the indexing type is the introduction of independent drive on every work spindle, thus enabling the direction of rotation of any component to be clockwise or anti-clockwise on different spindles as required, while the speed of rotation can also be varied from spindle to spindle.

Straight-line machines

The straight-line type of machine is used primarily for the polishing of long articles such as bumper bars, mouldings, strip, sheet, and so on, while the rotary-type machine is capable of polishing smaller articles at very high rates.

Reciprocating-type machines

In this machine the work is reciprocated back and forth under one or more adjustable polishing heads. The machine has a table on which a work-holding platen is mounted, this being actuated by a motor-driven worm gear and a rack and pinion arrangement. It has an adjustable stroke mechanism operated by a reversing control and two limit switches and, if necessary, a solenoid-operated timer control can also be installed to permit the work to reciprocate under the polishing wheel for a predetermined time cycle.

This type of machine is normally used for long extrusions, architectural mouldings, nameplates and larger components such as stainless-steel stove tops, large radiator grilles, etc. The length of the machine is controlled by the component length and the number of polishing heads required, with a minimum of twice the length of the component, while platen speed can be varied from about 3 to 12m per minute. Loading and unloading is carried out at the same end.

Fixture-return-type machine

In this type loose fixtures are mounted on platens which are carried through the polishing sequence by means of dogs mounted on a conveyor chain powered by a variable-speed drive (Fig. 2.7). The platens are automatically disengaged from the driving dogs and slide on to an unloading platform after the polishing operation, the fixtures being returned to the loading end either by a motorised belt conveyor or by a roller track. Usual fixture speeds are approximately 25 to 10m per minute, and many different components can be handled on this machine.

Horizontal-return-type machine

The horizontal-return-type machines are useful for the polishing of windscreen frames, toaster bodies, refrigerator shelves and similar articles. Loading and unloading

Fig. 2.7. Fixture-return-type machine.

are at one end and the components travel round the machine, which may be up to 20m long, at speeds from 3.5 to 14m per minute. These machines can utilise special camming arrangements in conjunction with fixtures on the platens so that articles can move and swivel into position for each successive polishing operation. A chain driven sprocket is mounted on a gear that is pinion driven through a worm gearbox coupled to an adjustable-speed drive, the motor being connected to the latter by means of a fluid coupling. Platens are provided within the frame of the machine to act as fixture holders. The platens are driven by special platen carriers, each having two sealed ball bearings and assembled to the sprocket chain. A variation on this type of machine is the' over-and-under ' type in which the fixtures return beneath the machine to the load and unload stations at each end. Such machines can be up to 25m long and are suitable for polishing articles, such as bumper bar ends, cigarette lighter cases, powder compacts, and so on. Fig. 2.8 shows a machine of this type which uses both abrasive belts and ventilated mops.

Fig. 2.8. Horizontal-return-type machine showing simultaneous use of abrasive belts and mops.

Rectangular-type machine

Rectangular-type machines are of the horizontal design, the conveyor line being open in the centre to allow free-standing polishing lathes to be used on both sides of it to reduce the overall length. A typical application for this type of line is for polishing steam iron bodies and articles of intricate shape such as tail-lamp bezels.

The most recent development of the rectangular machine is the universal type, shown in Fig. 2.9. The standard machine has a 13m perimeter, a variable-speed drive unit in one corner and take-ups in the two opposite corners, 30cm long cast iron platens, supported at top, bottom and sides by rollers, are driven by a special drive chain. These platens travel around the side of the machine and have a dovetail mounting surface for the spindle assemblies. Separate motorised roller chain arrangements with back-up guide rails can be placed at any position under the drive chain. This chain engages sprockets on the ends of the spindles and rotates the part for efficient polishing or buffing operations. This machine can be used both for solids of revolution and also for intricate shaped articles by means of camming arrangements.

Rotary-conveyor-type machine

In the rotary machine design a totally enclosed base is employed on which a platen or top table is fixed. This is made in 12 sections and is supported by heavy-duty roller bearings. Large machines up to 10m in diameter have been built and, by totally enclosing the entire machine, many of the dust problems normally associated with polishing can be completely eliminated.

Fig. 2.9. A 13 m perimeter open centre automatic polishing machine set up for the buffing of automobile tail light castings.

[Courtesy Acme-Murray Way International Ltd.]

Fig. 2.10. Automatic machine for the intermediate grinding of stainless steel strip. [Courtesy Acme-Murray Way International Ltd.]

Special-purpose machines

In certain cases there are components which are not suitable for the types of of machine described above and in these instances special-purpose machines have to be used, particularly where large production is required. Typical machines are those for polishing tubes, in which the latter are passed between polishing wheels by means of feed devices and suitable chucks. These machines can be obtained with either one, two or three heads to give the standard of finish required and can normally be operated by one man. Feed speed can be varied from 1.5 to 10m per minute.

Sheet and strip can be polished on one side by what are known as flat stock polishing machines. Abrasive belts or normal polishing mops can be used in these machines, depending again on the type of finish required, and continuous processing can be obtained by means of a bank of two to six machines which are fed by rollers. Normally the spindle-carrying wheels are slowly oscillated to reduce the tendency for score marks to occur. A modern machine for the intermediate grinding of stainless steel strip is shown in Fig. 2.10.

An adaption of the continuous strip machine is one designed for the polishing of of flat-irons. Here the parts are fed one after another under a series of polishing wheels. These are mounted independently, and are capable of movement in a vertical plane to enable them to keep in contact with the different levels as the flat-irons pass under them, while a lateral movement aids in the production of a uniform polish. This type of machine is capable of dealing with many types of components of relatively flat section.

Large special-purpose machines for the complete finishing of bumper bars are also available. Fig. 2.11 shows the latest concept in bumper machines. The conveyor track is mounted above the machine, leaving the polishing residues to fall to the floor

Fig. 2.11. Bumper polishing machine with overhead conveyor track.

and not on the track. This makes for greater cleanliness and easier head and setting manipulation. Camming facilities are mounted on the overhead structure .

The automatic polishing of tubes or rods can be accomplished in two types of machine. In the centreless polisher the tubes are passed between the polishing wheels by means of profiled feed wheels and suitable chucks; the rate and angle of feed and also the pressure applied during polishing are capable of fine adjustment. An alternative method is to make use of an abrasive-coated continuous band travelling over an inner cushioning band to control the pressure during polishing.

Adjustable lathes

In conjunction with the machines described above it is normal practice to use floor-standing lathes of special design. These are individually motorized and allow for adjustment in all planes; they can be fitted with spring-loading devices giving a vertical float of up to 25cm. They are normally powered by 5- to 20-h.p. motors and can also be fitted with air-lift devices which enable them to be lifted out of action during the indexing of the components. For the polishing of bumper bars a special lathe has been designed to give a float of up to 30cm by means of a counter-balancing device. Another modern development is the use of very wide mops and in one or two universal straight line machines, mops of up to 60cm wide are now in use. These mops are usually of the ventilated type and can be used either as solid or spaced mops. In the latter case spacers are inserted between the discs to give a very soft cushion mop, thus enabling them to conform even to the most complex shaped articles.

These machines have also been designed to give an oscillating movement, as referred to in the description of the sheet polishing machine, so that score marks and polishing lines can be reduced to a minimum.

Semi-automatic machines

Although automatic polishing machines are undergoing rapid improvement and are becoming increasingly versatile there are, however, certain limitations which should be appreciated when their installation is being considered. In the first place, with fully automatic polishing it is not possible to allow for periods of dwell to eliminate particularly large imperfections, such as can be done when manual polishing is carried out. Some form of rectification operation may, therefore, be required at times. Another point is that they may not always be capable of polishing all areas of complex shaped articles without the introduction of a disproportionate number of stages. A certain amount of hand finishing may, therefore, be needed on certain components, but this need not be a major item where the bulk of the polishing is carried out on the automatic plant. These limitations may, however, be reduced by the use of semi-automatic machines where the time of polishing is more closely under the control of the operator of the plant.

Semi-automatic machines have been designed to cover all work that can be carried out on fully automatic machines; where the output is limited these semi-automatic machines can be a big factor in the saving of floor space and polishing time. Examples include an oscillating machine which can be used with a conventional hand lathe for polishing and buffing out-of-round work such as oval cooking utensils, cake pans, tureens and similar items. A sprocket attachment allows the work to oscillate across the face of the rim, and vertical and horizontal adjustments facilitate quick set-ups.

Another useful machine is what is known as the roller feed unit, which is also used for out-round work such as automobile trim, ventilator surrounds and general automobile mouldings. The work is mounted on a sprocket-driven fixture and the work-table rotates, maintaining a constant point of contact of the work in relation to the wheel. This machine can either be controlled manually or fitted with an air lift and time control for automatic cycle operation. This unit may also be furnished with an individual adjustable floating head and a separate fabricated table to handle larger types of mouldings, not suitable for a standard machine.

Endless bands and portable wheels

For the polishing of difficult shapes, such as cycle forks, agricultural implements, hooks, etc., endless bands are used on suitable attachments. These are dressed with emery or other suitable abrasives in a similar manner to that employed in the case of polishing wheels. For other purposes portable wheels are used, and these can be attached by means of flexible shafting either to the end of a polishing spindle or directly to an electric motor.

Composition applicators

One of the problems in the design and operation of automatic polishing machines is the application of the polishing composition. Using the conventional type of composition bar, it is necessary to have a motorised or air-actuated reciprocating carrier which will slowly advance the bar on to the wheel and withdraw it at periodic intervals. The design of a device of this kind to accommodate different lengths of bar and to inch it forward as it wears, always keeping the pressure of application constant, is not easy and applicators require a considerable amount of maintenance, apart from their initial high cost. Wide bars such as may be needed for polishing sheet, are especially difficult to handle by means of applicators. In addition, their presence interferes with the efficient operation of dust extraction hoods which should conform as nearly as possible to the polishing head to reduce the dust problem. In fact, one of the advantages of automatic polishing is that it is possible to take dust away more effectively than with hand-operated spindles.

Another problem with composition applicators is the fact that as the bar wears down the machine must be stopped from time to time to enable a new one to be inserted. As the bars at each spindle wear down at different rates and have to be replaced at irregular intervals, a constant watch must be kept to ensure that they are replaced in time; the stopping and starting of machines on each occasion can result in a loss of operating time of as much as 25 per cent. Not only is this a serious matter, so far as production is concerned, but there is a substantial waste of bar ends which cannot be utilised. Sometimes they are remelted but this is not an altogether satisfactory procedure.

Attempts have been made to develop continuous bar compounds in the form of coils to reduce the need for stopping machines so frequently, but there are considerable problems in formulating a composition which has the required properties and at the same time is sufficiently flexible to be coiled without cracking. It is better to use straight bars which are as long as is practicable.

The use of slow speed wide buffs on automatic machines in recent years has enabled a very wide range of articles to be polished in this way without the use of a multiplicity of heads. Two heads, for example, can do the work hitherto done by six or eight heads. The articles are buried in the faces of the buffs during polishing. These may be up to as much as 200cm on wheels running as slowly as 600 to 1,000 r.p.m. The polishing heads may be parallel to the direction of feed, or may be angled.

Advantages and disadvantages of automatic polishing

Machines of this type are used for the polishing of metals prior to electroplating and also for the buffing of the plated deposits. They are rapid in action, sometimes reducing the polishing time by as much as 90 per cent as compared with that taken for hand-polishing. There are, however, certain features about automatic polishing machines which need careful attention if the best results are to be obtained.

In the first place, they may not be capable of polishing all areas of parts of difficult shape, so that an additional hand-finishing operation may have to be applied subsequently. This will often represent a relatively costly proportion of the entire finishing process owing to the additional handling involved. Secondly, in order that polishing may be effective, and to avoid jumping or chattering of the polishing wheels, there is a tendency for the pressures applied to be made substantially higher than those used by the hand-polisher. This, together with other effects, may result in a rather high rate of wear of polishing wheels in some cases. A third disadvantage of fully automatic polishing is that the machine is not able to 'dwell' on local imperfections for the purpose of polishing them out. Semi-automatic machines enable such defects to be dealt with, on the other hand.

High pressures are especially to be avoided in the finishing of nickel-plating where the removal of 25 per cent or even more of the nickel deposit in the finishing operation is by no means uncommon. Zinc-base die-castings should also not be polished too heavily prior to plating to avoid wearing through the dense surface skin and exposing the relatively porous interior of the casting. Furthermore, plated deposits will be disproportionately thinned at the point where the wheel first makes contact with the article unless very great care is taken at the commencement of the polishing stroke to ensure that the initial contact is made smoothly.

2.2. Barrel polishing

Barrel polishing provides a very cheap method of finishing metal parts, such as castings, stampings, forgings, and many kinds of small parts, which require the removal of burrs, etc. Accurately machined components and parts threaded externally are not suitable for barrelling.

The operation is usually carried out in two stages; in the first stage the parts are scoured to remove roughness, protuberances or scale, after which they are given the final polish in a second barrel.

It must be stressed that barrel finishing is an operation which demands special consideration in the case of each type of article being polished; the best conditions in every instance can often only be determined by actual experiment.

The barrels used are of various types, including horizontal closed barrels, or open-inclined barrels which are usually employed for final brightening and for drying in sawdust. The barrels may be circular in section or polygonal and made of wood, steel, cast-iron, plastics, or rubber-lined metal. Wood barrels should be of beech or oak, and are often reinforced with metal. Resin-containing softwoods are undesirable. Hexagonal or polygonal barrels give a better cutting action than circular barrels because of the accelerated fall of the parts as they drop successively from each face of the barrel as it revolves. They are, however, best avoided for the initial scouring of softer metals. The barrel speed varies, the higher speeds (of the order of 60m per minute) being used for light articles, while heavier components are barrelled at speeds down to half this rate.

Sometimes the container is given an oscillatory motion in addition to the rotary movement. This helps polishing and prevents nesting, but the wear on the bearings of this type of barrel is very heavy.

2.2.1. SCOURING

For scouring, the parts are put into the barrel together with materials such as scrap pieces of steel of suitable form (e.g. tube ends, stars, etc.), steel balls, and an abrasive which may be aluminium oxide, sand, granite chippings, rottenstone, etc. For the final polishing, less aggressive abrasives, such as diatomaceous earth and putty powder, are employed. Either paraffin (or mixtures of paraffin and oil) or water may be added to the polishing barrel as a carrier. The former is extremely useful, especially on steel; but if water is employed the presence of a small amount of alkali is desirable in order to prevent rusting.

Two sizes of steel balls are best used together when hollow parts are being treated; the larger size is then able to deal with the outside surfaces, while the smaller balls reach the interiors. Steel balls must be protected from rust when not in use by storing in soapy water.

Although it is practicable to remove scale in the scouring process, the hardness of the oxide deposit makes this somewhat difficult. It is therefore advisable where possible to degrease and pickle scaled articles prior to barrel polishing. This applies even more strongly to parts which have oil burnt on to them, as a result of annealing (notably those made of brass). Such components should be degreased prior to annealing if the best results are to be obtained. Barrels used for scouring iron or steel parts should not be used for softer metals, to avoid damage to the latter by any harder metal particles that may be left in the barrel. If it is desired, however, to use a barrel which has been employed for steel parts to finish zinc, for example, it should be thoroughly washed out with paraffin to remove any iron particles that may remain inside.

As an example a suitable mixture for the barrel polishing of small steel or brass parts would consist of: (i) powdered rottenstone; (ii) a mixture of equal parts of oil and paraffin; and (iii) steel scrap or balls if necessary. The parts being polished should not constitute more than about 50 per cent of the total mixture.

As a general rule, barrels are run about two thirds full. Soap is included in the mixture to maintain a foam when non-ferrous metals are being treated, as it is found that this is conducive to a bright finish. Excessive soap concentrations are best avoided as they reduce the cutting action, although they may, at the same time, improve the lustre of the finish at the cost of a longer barrelling time.

On the question of barrel speeds, the barrels can run up to 30 revolutions per minute; speeds in excess of this are impracticable as undue damage to the articles being treated will result. For parts weighing up to 1 kg, 24 r.p.m. is a useful general speed; higher speeds result in a centrifugal action, causing excessive impingement and damage or complete locking of the mass so that no polishing action occurs at all.

The components are packed in the barrel in such a way as to avoid nesting. The barrel is filled to within a few inches of the top, since if too much space is left they will be damaged during the polishing operation. The barrel is closed and run for a few minutes, after which it is reopened. It will then be found that some settling has taken place, and more parts are then put into the barrel to fill up the space. The time of polishing may be of the order of two to six days, depending on the nature of the parts being treated and on the finish required.

When the operation is complete, the barrel is emptied into a sieve, where the oil drains off; the parts are then separated by hand. When non-ferrous metals are being polished, the steel scrap, balls, etc., used as polishing assisters can be removed by means of a magnet. Ferrous parts can, with advantage, be dipped in a rust inhibiting, or de-watering oil after treatment.

The purpose of the detergent in barrel burnishing is to keep the metal clean and remove metal particles from pores in the abrasive which tend to clog it. Such clogging substantially reduces the cutting power of the abrasive. The components should remain in the polishing medium as far as possible without rubbing against one another. The volume ratio of components to abrasive chips is between 1:2 and 1:10, and the barrels should be filled to about 50—70 per cent of their volume. The ratio needs to be smaller with larger or more delicate articles.

The abrasive must be changed at intervals, as it loses its sharpness. The oil-paraffin mixture can be used for longer periods, being thinned with paraffin from time to time.

2.2.2. FINAL POLISHING (BARREL BURNISHING)

After scouring, when a higher degree of polish is required, the parts are removed from the barrel, degreased or washed in paraffin and put into a finishing barrel. If any tarnishing has occurred, it may be necessary to pickle lightly for a few minutes in 5 per cent sulphuric acid or, if the parts are of a copper alloy, to clean them in a dilute cyanide solution.

The finishing barrels must be kept clean and free from grease or oil. The barrel is filled with a solution containing soap-flakes, preferably with the addition of a small amount of sodium cyanide to act as a brightener. Trisodium phosphate is also a useful brightening addition agent. In any event, the solution must be kept well on the alkaline side if the best results are to be obtained. Into the solution, together with the parts being polished, are put small steel balls together with material such as steel scrap and leather cuttings. Small wood blocks are also useful (as a filling material, rather than as a polishing agent), whilst rosin-free sawdust (boxwood and beech dust) are sometimes added. Limestone chips are a specially useful constituent when a high lustre is required. The parts should constitute some 30 to 50 per cent of the contents of the barrel. The barrel is filled completely, and is run for at few minutes, after which it is opened and more filling material (such as wood blocks) added to prevent excessive damage to the parts when the barrel is rotated. The polishing operation should be complete in six to twelve hours, after which the parts are removed and transferred to heated open sawdust barrels, in which the final drying and polishing is completed in about five minutes; they are then ready for the subsequent plating or other finishing operation.

TABLE 2.3

Recommended ratio of parts to abrasive for barrel polishing

' Box ' Volume of Parts (cm³)	Stone to Part Ratio	
	Ferrous	Non-Ferrous
up to 4	1 : 1	may be self-tumbling
4 to 8	2 : 1	1 : 1
8 to 16	5 : 1	2 : 1
16 to 32	8 : 1	3 : 1
32 to 50	10 : 1	5 : 1
50 to 65	13 : 1	7 : 1
65 to 80	15 : 1	8 : 1
80 to 100	18 : 1	10 : 1
100 to 200	25 : 1	12 : 1
200 to 325	35 : 1	15 : 1
325 to 500	45 : 1	17 : 1
500 to 800	55 : 1	20 : 1
800 to 1,600	100 : 1	25 : 1

A special problem in barrel polishing is presented by soft metal parts with a harder metal insert. Thus, motor-car handles are made of zinc-base alloy, into which a steel shaft is often cast. Such parts cannot be barrelled in the conventional manner, as the steel shafts would damage the soft zinc. By holding the handles in the interior of the barrel in suitable fixtures they can, however, be successfully polished. The loading of such a barrel is a slow process, whilst its capacity is naturally somewhat limited. An alternative method which has been proposed is to protect the steel shafts by some sort of soft covering such as a rubber sleeve.

2.3. Vibratory finishing

Vibratory finishing has overtaken the older barrelling and tumbling techniques because of its greater speed and effectiveness. A huge variety of plants of this type is now available. The operation of one plant used for vibratory finishing of castings in the U.K. has recently been described(15). The limitations of the conventional rotating burnishing or tumbling barrel is that practically all the polishing action occurs as the articles slide from the upper to the lower part of the barrel during rotation. Only a small part of the contents is being polished at any given time, whilst shielded areas may receive no treatment at all. Thin and long components are also liable to be damaged during the operation. An important development has been the introduction of the toroidal-shaped barrel which allows parts to circulate freely whilst retaining their relative positions, minimising or avoiding part impingement[16].

Vibratory machines are used for descaling, deburring, removing flash, grinding marks, and casting defects, and for polishing[17]. The contents of the machine consisting of the components together with abrasives, polishing compound and a certain proportion of water ,are subjected to a continuous oscillating movement which produces scarring and abrasion. Increasing the amount of water reduces the abrasive action. The machines are normally loaded to about 90 per cent of capacity, which compares very favourably with the 50—60 per cent loading to which conventional barrels are limited to allow for the necessary sliding movement to occur during rotation.

Vibratory polishing machines can be broadly divided into centrifugal and spindle types.

The *centrifugal machines* have two rotational movements which greatly multiply the pressure of the abrasive medium against the work load. The barrels are cylindrical and may be mounted around a rotating turret plate, so that they rotate in the reverse direction to the plate at a somewhat lower speed.

In the *spindle machines* the articles to be polished are held in fixtures which are lowered into a tub containing the abrasive compound and water. The rate of rotation of the tub is about 100—200m per minute, and it is simultaneously vibrated at a constant velocity. The spindles themselves rotate slowly in the opposite direction at 8-12 r.p.m. The rotary and vibratory actions are controllable individually, and the spindle speed can also be varied. Processing time is about 10—20 minutes. The spindles are finally raised automatically for unloading.

The spindle-type of plant is mostly used for high finishing on precision components because of the need for attaching each article to the spindle separately.

Vibratory machines operate at up to 3,500 oscillations per minute, and can process parts ten times or more rapidly than the usual tumbling barrel. The larger machines run more slowly, around 1,800 vibrations per minute[18]. Open top machines are frequently used; these enable the progress of the process to be inspected.

Combined rotary and vibratory machines have been designed which rotate more slowly and have the advantage of not being subject to the stratification of the work to which vibratory machines are subject.

Rotational machines are less aggressive than the vibratory types, so that they can be used in conjunction with softer abrasives, although these wear more rapidly.

A spiral vibratory finishing machine with ' in-tub ' mechanical separation and discharge of components is shown in Fig. 2.12. End-unload machines are the most versatile types, and are capable of processing a larger workload per litre of machine capacity than others. They can be used for many standards of finish on a wide range of metals. A ' long-tub ' vibrator 335cm long is shown in Fig. 2.13. Delicate parts can be fixtured or free loaded.

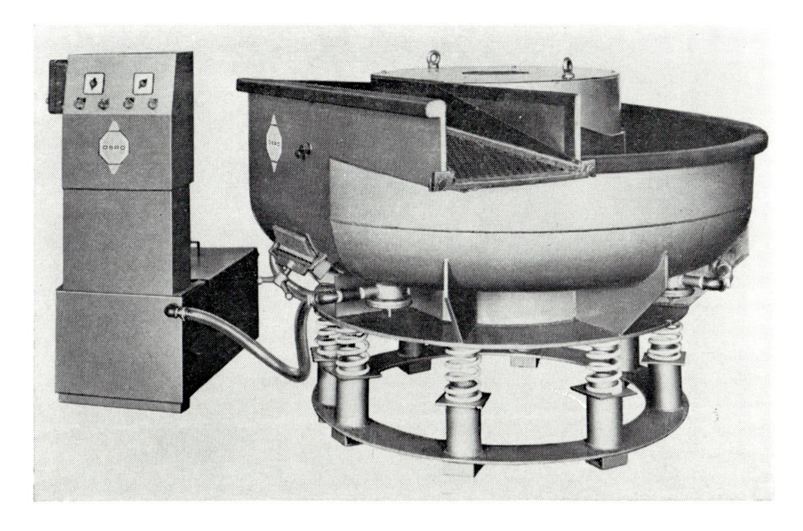

Fig. 2.12. Vibratory finishing machine with 'in-tub' mechanical separation controlled from one console. [Courtesy Osro Ltd.]

Fig. 2.14 shows a high energy finishing machine. Such high speed, high energy machines can process work up to 30 times as fast as conventional vibratory machines. They are designed for mass production finishing of small and medium sized components including forgings, pressings, castings and complex machined parts in most metals.

Fig. 2.13. Long-tub end-unload vibratory polishing machine. [Courtesy Osro Ltd.]

Fig. 2.14. High energy, high speed finishing machine with compound re-circulation system and mechanical separation of components and compounds. [Courtesy Osro Ltd.]

Abrasives

The abrasives used in vibratory finishing play a vital role in the process, and can be either natural or manufactured. Two to six parts of media are usually used to each part of work. The amount of abrasive should be sufficient to prevent articles from coming into contact with one another. The size and shape of the abrasive chips is important, since they should not jam into recesses in the components being processed. Larger chips have a more vigorous action, especially on edges and protuberances, than smaller ones but produce rougher surfaces. Mixing a proportion of large chips with smaller ones reduces impact during the treatment of large-size articles made of soft metals such as zinc.

Natural abrasives are random-shaped 'chips' of granite, limestone, flint or similar materials and are used for deburring and polishing in conjunction with an abrasive cutting compound. Granite acts mainly as a carrier for the compound, but limestone gives smooth finishes and wears much more rapidly. Another abrasive is quartzite which is made up of chips of a mass of small particles bounded with a softer material. It has a long life, and is used for light finishing.

Manufactured Abrasives consist of alumina or silicon carbide, and are made in random or pre-formed shapes, such as cones, pins and spheres[19]. Some typical media and compounds are shown in Fig. 2.15. Alumina bonded with ceramic is used because

Fig. 2.15. Typical media and compounds for vibratory finishing.
[Courtesy Osro Ltd.]

of its rapid cutting action, toughness and long life; it may be used on its own without an additional cutting compound.

These media can be mixed with steel balls, cones or other shapes to act as burnishing agents, and produce a high finish.

Polishing compounds

Compounds are usually employed in conjunction with the abrasives and can be acid, alkaline or neutral. The acid types are used mainly for diecasting. They are mixed with water, and should cover the work completely to avoid pitting and etching.

Alkaline compounds contain phosphates, silicates, sequestering and wetting agents and give solutions having a pH of about 9. They assist in cutting and prevent rusting of steel components. They are also useful where oil or lubricants are present on the work. Foaming should be avoided as it reduces the cutting action.

The neutral compounds are usually soap based and are used for final finishing.

Cutting compositions are also often added to speed up the abrasive action. These are usually complex mixtures designed for specific applications and may contain detergents, inhibitors and buffering agents as well as abrasive powders.

Chemical accelerators

A ten-fold acceleration in the salvaging or pre-treatment of zinc alloy die-castings for electroplating is claimed for the use of a solution of 10—40 g/l of sodium bisulphate and 8—15 g/l of sodium dichromate at pH 1.1—1.6 in the vibratory finishing process by Safranek and Miller[20]. The surface is also improved and the abrasive wear is reduced.

Application

Vibratory polishing is finding increasing uses in a large variety of industries, such as jewellery, door handles, hinges, plumbing and motor car parts. The process of vibratory finishing can be employed for deburring as well as for polishing before plating. Some representative examples are discussed by Kittredge[21].

2.4. Electrolytic polishing

Electrolytic methods of polishing metals have aroused considerable interest, both on account of the reduction in cost and labour which they are capable of bringing about and also because of the fact that in some instances it is possible to produce a higher degree of finish than is obtainable by mechanical methods of polishing. This applies particularly to the case of aluminium.

The modern techniques are based on the work of Jacquet[22] who investigated methods for polishing metals for metallographic examination which would be capable of producing a smooth surface without distorting the crystal lattice, which is an inevitable concomitant of the formation of a Beilby amorphous layer. He found that it was possible to obtain a high degree of polish on steel specimens by making them anodic in various electrolytes, such as a mixture of perchloric acid and acetic anhydride. Thus, it was found that copper could be polished to produce surfaces more highly reflecting than those resulting from mechanical polishing methods. The copper was made the anode in a solution containing about 400 g/l of phosphoric acid or pyrophosphoric acid at a temperature of 15° to 20°C under carefully controlled conditions of current density, location of anodes, etc.

The perchloric acid-acetic anhydride mixtures give better results on steel, but the critical influence of difference in the type of steel and even the method of heat-treatment, coupled with the nature of the chemicals used (the mixture can become explosive under some conditions), makes it very difficult to apply the process to commercial articles.

The action which takes place in the electropolishing bath occurs as a result of the highly polarised conditions prevailing at the anode. During treatment a film is produced which may be gaseous or liquid, in the latter case usually a layer of solution containing a high concentration of salts of the anode material. If this film has a high resistance to the passage of current, and the surface consists of depressed and elevated areas, it is clear that the thickness of such a film will be greater in the depressions so that the current density will tend to be high on projections and low in the protected depressions. The high current density areas will thus be dissolved away so that the entire surface of the metal will tend to be flattened and evened out.

Polarisation effects may also come into play which tend to make the depressions relatively cathodically passive and the elevations anodic due to differences in the concentration of the anodic film. Neufield and Southall[23] recently investigated the electropolishing of aluminium in a variety of electrolytes and found that the surface films produced do not have a common structure or morphology.

Parts to be treated must be separately racked; the time of treatment is about 5 to 20 minutes. Slight agitation is helpful but must not be excessive. The main object of movement is to prevent streaks due to the passage of oxygen bubbles over the parts being treated. Racks are best protected by means of an insulating lacquer or coating, but pure copper racks can be successfully used in the uninsulated state with many of the stainless steel electrolytes as the metal becomes passive after a few seconds and does not dissolve. Stainless steel and aluminium are the most commonly electropolished metals, since they are especially difficult to polish mechanically.

2.4.1. STAINLESS STEEL

Stainless steels (especially the 18 : 8 austenitic type) provide a useful and practical application for electrolytic polishing methods. Stainless steel is most difficult to polish mechanically; high pressures must be applied, and the low thermal conductivity of the metal makes overheating of the polishing wheels liable to occur. Good polishing is essential in the case of the austenitic steels, and overheating, a poor finish or surface contamination are likely to result in a surface of low corrosion resistance which is liable to pitting or rusting in service.

It has been found that stainless steel can be polished anodically in a variety of acids, apart from those already mentioned, of which sulphuric-phosphoric and sulphuric-chromic acid mixtures with the addition of polyglycols are the most effective and simplest to apply in a commercial sense, being relatively free from difficulties of operation and from obnoxious fumes.

A sulphuric-citric acid bath consists of 55 to 60 per cent of citric acid, 15 per cent of sulphuric acid, and the rest water. It is maintained at a temperature of 85° to 95°C. The solution has a relatively long life, and low current densities can be employed. During electropolishing the metals removed are converted to sulphates and precipitate out of the bath to some extent as a sludge. This is removed from time to time. The use of a small amount of methyl alcohol in this bath has also been advocated. A minimum current density must be applied before the polishing action is initiated. The temperature must also be raised as the bath ages.

Organic materials, such as glycerin or polyglycols, which have a favourable effect on anodic polishing, lower the critical current density below which the insulating viscous film no longer exists. They also serve to minimise the chemical etching of the metal surface by the acid.

The critical current density is important in electrolytic polishing since, if it falls below the minimum, the diffusion of the anodic film proceeds at a faster rate than that at which it can be maintained. For this reason it is important to avoid excessive agitation or movement of the electrolyte and also to avoid the setting-up of convection currents as far as possible. Current densities in excess of the minimum have no adverse effect even when they are relatively high. This is useful, since it makes the process applicable to parts of involved shape where wide variations in anodic current densities are to be expected. The addition of sulphuric acid to the phosphoric acids tends to reduce the chemical action on the steel, and also increases the rate of anodic dissolution, thus facilitating the polishing action.

2.4.2. CARBON STEELS

Generally for satisfactory results carbon steels must be free from local defects or pits as these tend to be increased in depth and emphasised by electropolishing methods. For smoothing ground or polished surfaces electropolishing can be helpful. Thus, a 100μm finish can be brought down to 75μm or less, whilst by starting even with a 15μm finish some improvement can be effected. The sulphuric-phosphoric acid electrolyte is applicable to carbon steels, and is normally operated at current densities of 100 to 50 A/dm^2, and at temperatures as high as 85°C. Slow movement of the work in the bath is desirable.

De-burring can also be carried out by this method, but the cost of racking, etc., may be prohibitive. The amount of metal to be removed may also lead to excessively frequent renewal of the solution.

A sulphuric-phosphoric acid bath containing a certain amount of chromic acid can be used for the polishing of carbon steels and also some alloy steels. Very high current densities are employed (up to 100 A/dm^2) together with high temperatures. The chromic acid is reduced to the trivalent state during working, so that the solution becomes greenish in colour; when the trivalent chromium content becomes too high (in excess of about 3 per cent), the operation of the bath is adversely affected. The solution is generally controlled by means of a hydrometer, and is discarded when the density becomes too high.

The use of this type of solution involves extraction equipment, for chromic acid spray is evolved during its operation. A potential of 12 to 15 volts is required.

2.4.3. COPPER AND BRASS

Electrolytic polishing of copper and brass can be carried out in a 53 per cent solution of pyrophosphoric acid as the electrolyte. The current density is 8 to 10 A/dm^2 with an applied voltage of 1.6 to 2 and a solution temperature of 15°C to 22°C.

Orthophosphoric acid solutions have also been employed, solutions of 25 to 60 per cent concentration giving good results. Similar current densities can be employed at 2 volts, the bath being operated at room temperature. The solutions seem to work better when some ageing of the electrolyte has occurred. The presence of a certain amount of copper in the bath stabilises the voltage and enables more uniform polishing to be obtained. The articles should be removed from the bath before the current is cut off, since copper may otherwise be re-deposited on the metal surface as a result of a back e.m.f. being set up.

Sometimes a white deposit is left on the polished surface; this can be removed by immersion in a dilute solution of orthophosphoric acid.

Good results can be obtained on brass and copper from phosphoric acid-chromic acid electrolytes; a typical solution contains:

Chromic acid	50 per cent
Phosphoric acid	10 per cent
Water	40 per cent

The usual current density is about 50 A/dm^2 and the temperature is 40°C. The polishing time varies naturally with the state of the surface, but 5 to 10 minutes is generally sufficient.

Baths containing chromic acid deteriorate rapidly, however, as trivalent chromium and copper accumulate in the solution, and when 2 to 4 per cent of these materials are present the solutions must be discarded. The solutions are used in plastic or in lead-lined tanks with wire-reinforced glass insulating shields. A loading of 0.5—1.0 amps per litre is usual. Air or anode rod agitation should be provided.

2.4.4. SILVER

Silver can be electrolytically polished by anodic treatment in a bath of the same general composition as is used for plating. One solution which has been recommended consists of:

Silver (as $KAg(CN)_2$)	30 to 45 g/l
Potassium cyanide	30 to 40 g/l
Potassium carbonate	40 to 60 g/l

A voltage of about 5 volts is applied, the current density being of the order of 1 to 2 A/dm^2. The polishing time is fifteen to twenty seconds.

2.4.5. ALUMINIUM

Electrolytic polishing is of outstanding importance in the case of aluminium, although it has been replaced by chemical polishing to a considerable extent. Mechanical polishing is difficult when fully bright finishes are required on aluminium and its alloys, and the electrolytic and chemical polishing methods enable brighter finishes to be obtained at lower cost.

The Brytal process[24] was one of the earliest in the field, and is still employed where mirror-bright finishes are required on large flat areas. It is especially suitable for super-purity aluminium and for the magnesium and magnesium-silicon-containing aluminium alloys. The treatment consists in making the article anodic in a solution containing about 15 per cent Na_2CO_3 and 5 per cent Na_3PO_4 for 10-15 minutes at a current density of 3-5 A/dm^2 at 80°C.

The Alzak[25] process uses a fluoborate electrolyte consisting of a 2.5 per cent solution of fluoboric acid. The treatment time is 5-10 minutes at 1-2 A/dm^2.

The Battelle Process[26] is based on phosphoric acid and sulphuric acid, with or without the addition of chromic acid. It is more rapid than the above-mentioned processes, but gives a less specular finish.

Applications

Electrolytic polishing is used for stainless steel decorative items such as trim on cookers and automobiles, and for wire goods, screen wiper arms and bathroom fittings. Copper and brass costume jewellery is electropolished before plating or lacquering. Electrolytic polishing is also employed for superfinishing and deburring.

General remarks

There can be little doubt that electrolytic polishing methods have a considerable future. The electropolished surface is usually free from stain, whilst the technique enables a variety of surface finishes ranging from a fully bright to a satin or matt finish to be obtained. Owing to the good ' throw ' of the solutions it is also possible to polish in recesses and at depths not normally accessible to mechanical polishing wheels. A surface does not need to be plane for a mirror finish to be obtained on it by electrolytic methods.

On the other hand the use of electropolishing methods under commercial conditions introduces complications which have not yet been fully surmounted. The high current densities usually employed necessitate large sources of current; racks and conductors must also be of generous dimensions. The racks themselves must be insulated effectively owing to the destructive action of the solutions used under anodic conditions. These solutions are, as a rule, themselves highly corrosive to tanks and equipment. The heating and cooling equipment often needed are also subject to very severe conditions.

Adequate surface finish, smoothness and the cleanliness of the surface have their effects on the polish obtained and on the economics of the process. Deep scratches, dirt and metallic or non-metallic inclusions should be removed prior to treatment by mechanical means for the most satisfactory results. Attractive finishes can also be produced by shot-blasting prior to electropolishing, whilst two-tone effects can be obtained by partial stopping-off during treatment. A reasonable degree of cleanliness of the metal is desirable before immersion in the electrolyte, although a chemically clean surface is not essential.

Electrolytic polishing brings about a definite improvement in corrosion resistance. This is ascribed to:

(1) Improvement in the micro-profile of the surface layer by the removal of the deformed structure and foreign inclusions (which favour corrosion), and a reduction in the size of the active surface.

(2) The formation of a thin passive film.

(3) The removal by dissolution of the chromium impregnated layer which occurs in the case of stainless iron alloys if a sufficient depth is etched away.

2.5. Chemical polishing

In this process no current is applied, the articles being immersed in solutions of acids such as phosphoric, nitric and acetic. The temperatures used range from room temperature to as high as 95 °C for the fastest action. It is used in the main for the brightening of aluminium. Phosphoric-nitric acid mixtures are contained in stainless stainless steel tanks; heating is by means of stainless steel steam coils. Efficient fume exhaust plant is essential. This should cover the rinsing and neutralising tanks.

The advantages of a non-electrolytic process are obvious inasmuch as it can be used on intricately shaped articles or in barrels, whilst the difficult problem of racking is overcome. Indira and Shenoi[(27)] have recently dealt with the advantages and applications of chemical polishing and have tabulated the solution compositions for polishing Al, Mg, Zn, Cd, Cu, Fe and steels, Pb, Ag, stainless steels, Ta and Ti.

A considerable number of processes are in use for aluminium automobile trim and many other applications. They are based on phosphoric and nitric acids with the addition of heavy metal nitrates or sulphates (Cu, Ni, Fe, Co, Ag, Cd, etc.) and weak acids such as acetic, boric, citric or molybdic acids. The solutions are used at 90°-140°C and require agitation and efficient extraction. A typical solution contains 80 per cent H_3PO_4 and 4 per cent HNO_3, the balance being water; it is operated at 90°C, the treatment time being 30 seconds to 5 minutes.

Amongst the more important patented processes are Alupol[(28)], which makes use, for example, of a bath containing phosphoric acid 50%, nitric acid 6.5%, sulphuric acid 25%, acetic acid 6%, water 1.5% and nickel nitrate 3%. The widely used Erftwerk Process[(29)] developed in Germany employs a solution containing nitric acid 13%, ammonium bifluoride 16% and lead nitrate 0.02%. The bath is maintained at 55°-75°C, the treatment time being 15-20 seconds.

Probably the most widely used process in the United Kingdom is Phosbrite 159[(30)], which is basically similar to the Alupol process. It gives a high degree of smoothness on a wide range of alloys, as well as excellent specular reflectivity. It can be used on the magnesium and heavy-metal-containing alloys provided there is not a high proportion of silicon present.

Chemical polishes for treating electrodeposited nickel[(31)] and for the rapid removal of grinding and lapping scratches from iron and steel[(32)] have been reported.

REFERENCES

1. Walter Burkhart, H. Silman and C. R. Draper, 'Mechanical Polishing', Robert Draper, 1960.
2. M. R. Caldwell, *Proc. Am. Electroplaters' Soc* 1951, **38**, 83.
3. W. H. Safranek, *Metal Progress* 1967, **91**, No.5, 95.

4. J. P. Dewar, *Metal Finishing J.*, 1973, **19**, No. 218, 70.
5. J. P. Dewar, *Product Finishing*, 1974, **27**, No. 9, 20.
6. J. L. Rees, *Metal Finishing J.*, 1969, **11**, 388.
7. J. Partridge, Noyes Data Corp., Park Ridge, NJ 07656, U.S.A., 1972, 204pp.
8. H. Silman, *Electroplating and Metal Finishing*, 1960, **13**, No. 5, 159.
9. E. T. Candie and S. L. Doughty, *Proc. Am. Electroplaters' Soc.*, 1950, **37**, 257.
10. W. L. Morgan, *Plating*, 1972, **59**, No. 634, 636.
11. W. L. Pinner, *Plating*, 1955, **42**, 1039; 1956, **43**, 50; 1957, **44**, 763.
12. Stainless Steel, 1972, No. 23, 14.
13. W. H. Safranek and E. W. Brooman, ' Finishing and Electroplating Die Cast and Wrought Zinc ', 1973, Zinc Inst. Inc., New York, NY 10017.
14. *Die Casting and Metal Moulding*, 1972, **4**, No. 6, 14.
15. *Metal Finishing J.*, 1973, **19**, No. 225, 288.
16. J. R. Strom and D. R. LaValley, *Plating*, 1972, **59**, No. 3, 236.
17. F. Sautter, *Electroplating and Metal Finishing*, 1968, **19**, No. 7, 248.
18. W. E. Brandt, Trans. 3rd Natl. Diecasting Congress 1964, Paper 22.
19. K. K. Radarmer, Trans. 4th Natl. Diecasting Congress 1966, Paper 1402.
20. W. H. Safranek and H. R. Miller, *Plating*, 1972, **59**, 38.
21. J. B. Kittredge, *Plating*, 1971, **58**, No. 12, 1203.
22. P. A. Jacquet, ' Le Polissage Electrolytique des Surfaces Metalliques et ses Applications ', Editions Métaux, 1948.
23. P. Neufield and D. M. Southall, *Electrodeposition and Surface Treatment*, 1975, **3**, 159, 68.
24. N. D. Pullen, *J. Inst. Metals*, 1936, **59**, 151.
25. L. B. Mason, U.S. Pat. 2,108,603 (1938); 2,040,617-8 (1936).
26. C. L. Faust, U.S. Pat. 2,550,544 (1951).
27. K. S. Indira and B. A. Shenoi, *Products Finishing*, 1973, **37**, No. 11, 80.
28. A. Venet, Fr. Pat. 967,207 (1950).
29. Brit. Pats. 693,776; 693,876 (1950).
30. W. K. Bates and C. D. Coppard, *Met. Finishing J.*, 1958, **4**, No. 37, 5.
31. V. O. Nwoko, *Electrodeposition and Surface Treatment*, 1975, **3**, 219.
32. B. W. Chrust and L. C. Smith. *J. Iron & Steel Inst.*, Tech. Note, 1973, Feb., 155.

CHAPTER 3

Degreasing and cleaning

THE removal of oil, grease, and other contaminating foreign matter is particularly important before almost any industrial finish can be successfully applied to metals. Indeed, it is probably true to say that the greatest single cause of finishing defects is poor cleaning of the metal surface. Early in the history of electroplating, it was found that silver electrodeposits would not adhere to a basis metal unless the latter was thoroughly cleaned, and such operations as boiling in potash and hand scouring began to be commonly used. Nickel plating was even more critical, but similar methods were reasonably satisfactory until the advent of chromium plating. It was soon found that chromium was deposited in such a highly stressed state that practically no nickel plate as produced at that time could be chromium plated without the nickel peeling off. Entirely new methods, such as electrocleaning and acid etching, had to be introduced in order to make nickel plus chromium plating the reliable operation it is today.

Virtually all metal components will require cleaning at some stage, even where no subsequent finishing operation is involved and in some cases the standard of cleanliness required is as high as, or even higher than, that required as a preparation for electroplating, e.g. high purity cleaning of inertial navigation equipment[(1)] using chlorothan VG. organic solvent. Amongst the most important items which require cleaning are hydraulic apparatus, instruments, and electrical and electronic components in the modern motor-car.

The type of cleaning process employed is important. For example, the use of strong electrolytes such as caustic soda on steel prior to phosphating can be deleterious. Solvent degreasing can in certain cases make articles highly prone to rusting in storage, whilst lubrication may be made difficult by some cleaning methods. An increasing amount of attention is being paid to metal cleaning processes with respect to the economics aspects of the subject, including the important question of water consumption. Although the degree of cleaning required prior to electroplating is known to be critical it is obviously wasteful to use an expensive method of obtaining an unnecessarily high standard of cleanliness on a metal surface, and in fact it may even be undesirable to do so. British Standards Institute have published a code of practise for the cleaning and preparation of metal surfaces[2].

Automatic electroplating plants usually incorporate elaborate cleaning cycles which must be designed to remove the worst kind of contamination likely to be encountered, yet at the same time the methods employed must be carefully controlled so as not to result in overcleaning. This is a hazard, particularly in the case of sensitive metals such as aluminium or zinc die-castings and can lead to defective electrodeposits. In automatic plants the plating operations proper occupy a relatively small part of the total length of the plant as compared with the cleaning, intermediate rinsing and drying operations.

Mechanisation[3] is of special interest in the field of metal cleaning, and methods which depend on manual operations or skilled handling have tended to be superseded by those capable of being made fully, or at least partially automated. In the case of engineering products where the presence of any dirt particles could result in failure in service, cleaning would, in the past, often be carried out by individual hand operations coupled with careful scrutiny of the parts concerned, but nowadays this is wholly impracticable. It is therefore necessary to provide equipment through which the parts concerned can be conveyed or passed with the completed assurance that the required degree of cleaning will be obtained with as near complete certainty as is humanly possible, whatever degree and type of contamination is likely to be encountered.

Often a cleaning problem arises not because of any inherent difficulty in removing the contaminant as such, but by virtue of the fact that design considerations[4] lead to shapes which may entrap cleaning solutions or which are difficult for a cleaning medium to reach. The result is that swarf and grit particles, traces of grease or machining compounds, trapped cleaning solution and so on, can be left behind, so that troubles in service are subsequently encountered.

3.1. Types of Contaminants

The types of surface contamination which have to be removed can be broadly classified as follows:

(a) Oils, greases, soluble oils, drawing compounds, lubricants and the like left on the metal surfaces subsequent to mechanical operations such as rolling and pressing.
(b) Polishing residues, which may include greases, soaps, abrasives, abraded metal and textile material from polishing buffs.
(c) Metal swarf, grit and dust deposited from the atmosphere.
(d) Tarnish and rust films developing during storage or handling.
(e) Solder flux residues.

Broadly speaking, the oily contaminants on metal surfaces are of two groups, viz. non-saponifiable and saponifiable oils. The former are mainly mineral oils whilst the latter are found chiefly in the animal and vegetable fats which consist essentially of organic soaps formed by the combination of a long-chain fatty acid and an alcohol, generally a polyhydric alcohol such as glycerol.

The removal of solder flux residues has recently been reviewed by Creed[5] and the cleaning of printed circuits soldered with tin by Imbert[6].

3.1.1. MINERAL LUBRICANTS

The first group of materials mentioned above constitutes a major source of contamination. It is essential to use lubricants in most metal-forming and fabricating processes, and these lubricants contain a large variety of constituents. Besides mineral oils, vegetable and animal oils and fats are often present, as are greases, which may be of either the lime- or soda-based types. Sometimes, for special purposes where very high pressures are employed, fillers such as zinc oxide are included in the compositions, whilst insoluble metallic soaps, including metallic oleates, stearates and naphthenates, may be present in the form of extreme pressure lubricants. Ordinary alkali soaps have also been employed to some extent. In addition, there are numerous synthetic lubricating constituents which are stable under high pressures and which are especially difficult to remove. The presence of silicones can cause serious cleaning problems, and their use in these compositions is to be avoided. When it is borne in mind that all these materials may exist on metals that must be subjected to a cleaning operation and that many lubricants are proprietary and of undisclosed composition, the magnitude of the problem can be realised.

The mineral oils which are very widely used in engineering processes are substantially non-saponifiable. These can be removed by means of organic solvents or by treatment in alkaline solutions, making use of the detergent, as distinct from the saponifying characteristics of the materials.

The composition of a lubricant is governed almost entirely by effectiveness for the particular mechanical operation for which it is required. Generally speaking, scant attention is given to the problem of removing it before any subsequent finishing processes which may have to be carried out, although it must be said that manufacturers of pressing and rolling compounds are giving increasing thought to this aspect of the application of their products, and often make special claims regarding the ease with which they may be removed. Too much must not be expected in this respect, however, since a good press lubricant must be inherently tenacious and bond to the metal surface; hence, to a certain extent, good lubricating qualities and ease of removal are conflicting requirements.

3.1.2. ANIMAL AND VEGETABLE OILS

The animal fats are most commonly solid at room temperatures and consist largely of the glycerides and palmitic and stearic acid; on the other hand, the liquid vegetable oils contain a much higher proportion of oleic acid glycerides. Liquid fats are converted to the more valuable hard fats by hydrogenation in the presence of catalysts. On a large scale these materials are also extensively used in lubricants for metal forming operations, and are also important constituents of metal polishing compositions.

The chief constituents of the hard fats are tristearin and tripalmitin, which are formed by the reaction of glycerin, which acts as a try-hydric base, with the corresponding fatty acid. Triolein is formed similarly by the combination of oleic acid with glycerin. Some of the properties of the fatty acid constituents in natural fats and oils are shown in Table 3.1.

TABLE 3.1

Some fatty acid constituents of natural fats and oils

Fatty Acid	Composition	Melting point °C	Sp. gr. at melting point
Lauric acid	$C_{11}H_{23}.COOH$	43.6	0.875
Myristic acid	$C_{13}H_{27}.COOH$	54.0	0.862
Palmitic acid	$C_{15}H_{31}.COOH$	62.0	0.853
Stearic acid	$C_{17}H_{35}.COOH$	69.0	0.845
Oleic acid	$C_{17}H_{33}.COOH$	14.0	0.854

When fats are heated with water under pressure or with sulphuric acid, hydrolysis takes place, so that the fat is broken down into glycerin and the constituent fatty acid:

$$\begin{array}{l} CH_2.O.CO.C_{17}H_{35} \\ CH.O.CO.C_{17}H_{35} \\ CH_2.O.CO.C_{17}H_{35} \end{array} + 3H_2O \rightarrow \begin{array}{l} CH_2.OH \\ | \\ CH.OH \\ | \\ CH_2OH \end{array} + 3C_{17}H_{35}.COOH$$

Stearin (3 molecules) Glycerin (1 mol) Stearic acid (3 mol)

In practice, when natural fats are hydrolised in this way, mixtures of the fatty acids present are obtained.

When alkalis are used in place of water, hydrolysis occurs very much more rapidly, and alkali salts of the fatty acid are produced together with glycerin. These salts constitute the common soaps and the process of boiling fats with alkalis to produce such soaps is termed 'saponification'. So far as cleaning processes are concerned, the main aspect of interest is the fact that, whereas the oils and greases are insoluble in water, the sodium and potassium of the fatty acids formed by the action of alkalis on fats are entirely soluble. Thus, saponification by alkalis of such oils and greases effects their speedy removal. It is to be noted, however, that the soaps of alkaline earth metals, e.g. calcium (or lime) soaps, and the magnesium soaps are not soluble to any extent in water. They are precipitated from the soluble soaps when hard water containing calcium and magnesium salts is used, and the removal of these insoluble salts from the metal surface onto which they may be precipitated can present difficulties.

3.1.3. SOLID MATTER

Usually, finely divided particles of buffing, lapping and drawing compounds, etc., are held fast to the metal surface by the polymerised oily matter which is itself difficult to remove. Treatment of the work in a vapour degreasing plant will remove the oily matter from the soil but may leave the solid particles adhering tenaciously directly to the metal surface. The particles are bonded to the surface by physical and chemical adsorption forces, the degree of adhesion depending on the size of the particles. It is evident that coarser particles will have fewer points of contact with the surface than finer ones, thus explaining why finely divided contamination is in general the more difficult to remove. Spring[7] has concluded in his discussion of the removal of particulate matter, that when solid particles are encased in an oily matrix a very efficient solvent cleaner may cause problems by rapidly dissolving the oil but leaving the solid matter even more strongly attached to the surface. A cleaner which is less efficient, slower acting and requires agitation will be more efficient with regard to removing the solid contamination by virtue of the mechanical action of the agitation process.

3.2. Cleaning processes

The cleaning procedures used in the preparation of metals for electroplating have been described by Linford[8]. The principal systems can be tabulated as follows: There are three groups of substances used in metal cleaning and degreasing—viz.: (a) organic solvents; (b) detergents consisting of aqueous solutions of various materials, usually alkalis, with various additions; (c) emulsified cleaners.

3.2.1. SOLVENT WASHING

Greases and oils are soluble in organic liquids such as petroleum fractions, carbon tetrachloride and trichlorethylene, and grease removal can be effected by washing in these solvents. In choosing a solvent for a particular application many factors need to be considered[9], but an ideal solvent for metal cleaning would meet the following requirements. It should be: (a) stable, cheap and readily available; (b) non-flammable; (c) an effective solvent for all types of oils, greases, waxes and tars; (d) non-toxic; (e) of low viscosity and surface tension to facilitate effective penetration of greasy deposits; (f) readily separable from the extracted matter with a high percentage of recovery; (g) non-corrosive to metals even at elevated temperatures. It is difficult to meet all these requirements, but some solvents approach them much more nearly than others.

The procedure of washing parts in a bath of cold liquid solvent has several disadvantages. Firstly, a film of the solvent containing a varying amount of oil in solution is left on the work; when this evaporates, a certain amount of the contaminant remains on the parts being cleaned. Secondly, the action of the solvent used in a cold still bath is apt to be relatively slow, necessitating the use of large equipment. Thirdly, the operation involves much handling and manipulation of the parts being cleaned.

Although a certain amount of rough cleaning is still carried out by washing in paraffin, solvent naphtha, or other more-or-less inflammable solvents, this procedure is nowadays largely confined to the removal of accumulations of grease mixed with considerable amounts of grime on a small scale. Under conditions where efficient modern industrial degreasing plants using high-powered non-flammable solvents are available there is little to be gained by the use of methods of this kind.

3.2.2. EMULSION CLEANING

The difficulty of removing organic solvents from a metal surface when they have accumulated any quantity of oil has resulted in the use of mixtures which are capable of dissolving the oils and then readily emulsifying them so that both the cleaning agent and the grease can be removed by washing with water. Numerous other advantages are claimed[10] for emulsion cleaning processes. These include increased efficiency with regard to the physical cleanliness of the treated surface, safer working conditions and easier control. The degreasing agents are either brushed over the articles to be cleaned, or the articles may be immersed in the emulsion, or sprayed with it, after which the residue is washed off with water. The solvent and the grease which it has taken up are emulsified by the water, leaving the work in a clean condition. Emulsion cleaners are usually based on a non-halogenated solvent/water mixture containing a wetting agent to establish a stable emulsion. An organic rust inhibitor may be added. When dried, such an emulsion cleaner leaves a very thin film of rust preventative. The cleaners contain between 1 and 5% of the organic solvent, and the pH is normally around 8.

3.2.2.1. Di-phase cleaners

The di-phase cleaners have some advantages over the homogeneous solvent emulsions. They are variable emulsions in which the solvent phase consists essentially of a hydrocarbon with a high flash point, whilst the emulsion phase is an emulsion of the solvent in water with surface active agents. In use, the cleaner breaks down into two separate immiscible phases, a 2 to 4 per cent concentration of the solvent being usually present. As the solvent floats on the top of the waterphase, these cleaners are generally used in spray washing machines at temperatures of 70° to 80°C. The process is highly suitable for incorporation into an automatic plating line. The cleaning action of a di-phase solution of this kind is two-fold; first the oleophilic contaminants are wetted and dissolved by the solvent phase, and secondly hydrophilic materials including inorganic salts and soluble oils are taken up by the aqueous emulsion phase. The surface-active constituents of the unstable emulsion also help to facilitate the removal of contaminants by concentrating at the metal-liquid interface, thus assisting the cleaning action.

The mechanism of the action of a di-phase cleaner depends on the differential wetting and dispersing properties of the system. Thus is it found that powdered metals such as zinc, for instance, are rapidly wetted by the solvent phase and not by the aqueous phase, whilst on the other hand non-metallic mineral abrasives such as alumina, lime or pumice are affected conversely. It is thus seen that the behaviour of a di-phase emulsion is quite different from that of a stable emulsion of a solvent in water. It has been found that such a system can also suspend between four and six times as much insoluble material as a stable emulsion without re-depositing it on the work.

In a spray cleaning machine, the di-phase solution is taken off from just below the surface of the reserve tank so that both phases are sprayed onto the articles being cleaned. It is clearly of little use to have a means of removing grease and dirt from articles put through a plant unless it is also possible to remove the contaminants from the cleaning medium in turn. This is taken care of in the di-phase system, because oils and greases concentrate in the solvent layer at the top of the tank, whilst solid particles and dirt either fall to the bottom or concentrate at the interface between the aqueous and solvent layers. Di-phase cleaners are especially suitable for the removal of grinding, drawing and polishing compounds, abrasives and grease from practically all metals; as the solutions are chemically neutral, this includes those sensitive to alkalis, such as zinc and light alloys. Di-phase cleaning is especially suitable for small parts since traces of adsorbed fatty acids are left behind after cleaning and these act as a mild corrosion preventative. If plating has to be carried out after cleaning, however, further washing stages are required, and these can be incorporated in the same plant.

Emulsion cleaning is considerably cheaper than vapour degreasing (see Section 3.2.3) in operation but has the disadvantage that the work is wet after treatment so that some kind of film must be left after drying. On some kinds of articles rinsing and drying is sufficient, but on very small and precision parts the presence of even a thin film may be deleterious and trichloroethylene vapour degreasing is to be preferred. On the other hand, vapour degreasing leaves the surface so grease-free that it is peculiarly susceptible to corrosion, and steel articles are therefore liable to rust after this operation has been carried out.

The essential advantages of the di-phase system compared with organic solvent degreasing have been summarised as follows:

(1) The work can be treated on the plating racks.

(2) The process can be incorporated in an automatic plating line.

(3) Fewer people need handle the work — this saves on labour and rejects due to handling.

(4) The system will remove all polishing composition.

(5) Running costs are lower.

(6) The health hazards are much lower than with solvent degreasing.

As has already been indicated, mechanical agitation is particularly useful in cleaning operations with the di-phase system, and Fig. 3.1 shows a machine designed for this purpose. The work is loaded into a container which is moved rapidly up and down pneumatically through the heated di-phase solution. A compressed air supply of about 550 KN/m^2 is needed to operate the unit. Provision can also be made for the work to be spray-rinsed with filtered solution after the immersion sequence, which takes about 2 minutes. A series of these units can be coupled together where a number of cleaning and rinsing operations, for example, have to be carried out.

The exact cleaning cycle or plant to be employed with the di-phase system depends on the articles being cleaned and on the type of soil to be removed. For light and medium soils a single stage spray cleaning operation followed by two rinsing stages

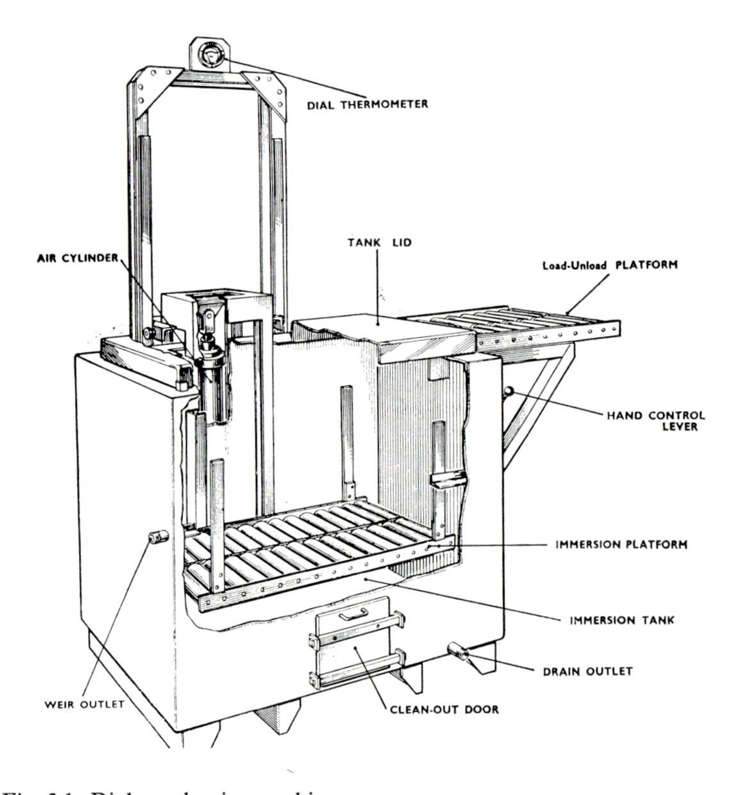

Fig. 3.1. Diphase cleaning machine.
[Courtesy Oxy Metal Industries (G.B.) Ltd.]

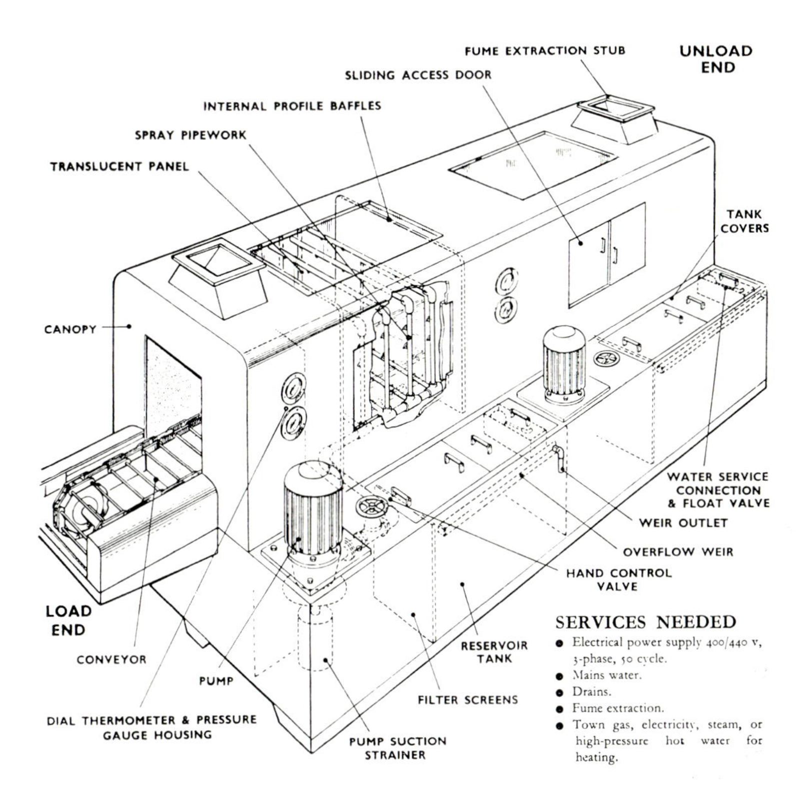

Fig. 3.2 Two stage conveyor-belt spray washing machine for diphase cleaning.
[Courtesy Oxy Metal Industries (G.B.) Ltd.]

are usually sufficient, but for extremely heavy contamination, a soak stage of 3 to 5 minutes in the di-phase cleaner tank at 80°C, a spray wash at 50° to 55°C for one minute, and a spray rinse in a mild alkali followed by a final spray rinse in hot water may be needed. It is thus seen that a considerable degree of flexibility is possible, depending on the result it is desired to achieve.

3.2.2.2. Field of application of emulsion cleaning

Emulsion cleaners are used because of their low operating costs and efficiency. They produce a physically clean surface which is, however, adequate only for subsequent phosphate treatment, painting or similar processes. If plating is to be carried out, a chemically clean metal surface is needed, so that an alkaline treatment is essential after emulsion cleaning and before the plating operation. Emulsion cleaning is therefore a substitute for organic solvent degreasing, which likewise only effects a physical cleaning of the metal surface.

3.2.3. CHLORINATED HYDROCARBON DEGREASING

In modern metal finishing practice, practically all solvent degreasing where it is carried out, makes use of a chlorinated hydrocarbon as the solvent medium, and it is generally employed in specially-designed plant. There are several stabilised chlorinated hydrocarbons manufactured commercially for a variety of degreasing applications; all of them are very powerful solvents for oils, greases and waxes, and have the great advantage of non-flammability. They confer their non-flammable characteristics on other solvents when mixed with them provided the proportion of the flammable constituent is not excessive. Some of the characteristics of a number of chlorinated hydrocarbons are shown in Table 3.2.

TABLE 3.2
Characteristics of principal chlorinated hydrocarbon solvents

Solvent	Composition	Sp. Gr.	Boiling-point °C
Methylene dichloride	CH_2Cl_2	1.346	42.0
Chloroform	$CHCl_3$	1.50	61.2
Carbon Tetrachloride	CCl_4	1.62	76.8
Dichloroethane	$CH_2Cl.CH_2Cl$	1.25	83.7
Tetrachloroethane	$CHCl_2.CHCl_2$	1.601	145.0
Pentachloroethane	$CHCl_2.CCl_3$	1.690	159.0
Trichloroethylene	$CHCl.CCl_2$	1.47	86.7
Perchloroethylene	$CCl_2.CCl_2$	1.624	120.0
Trichlorofluoroethane	$CF_2Cl.CFCl_2$	1.58	47.6
Trichloroethane	$CCl_3.CH_3$	1.314	70.75

All these solvents are to a varying extent narcotic and toxic (e.g. carbon tetrachloride is very toxic whereas trichloroethane is comparatively safe). Carbon tetrachloride attacks copper and lead slightly; trichloroethylene on the other hand does not affect these metals, but iron and steel are very liable to rust after degreasing by means of this solvent. This is probably not due to any attack by the solvent on the metal itself, but because the grease-free surface which it leaves is especially liable to be corroded by the atmosphere. On the other hand, some reaction takes place between aluminium and magnesium and hot trichloroethylene or its vapour, and when these metals are in a finely divided state, the reaction may be quite rapid, resulting in the partial decomposition of the solvent.

The use of chloro and chlorofluorinated hydrocarbon solvents for cleaning has recently been reviewed[11].

3.2.3.1. Trichloroethylene

On account of its extremely high solvent properties for oils and its low toxicity as compared with many other compounds of this group, trichloroethylene is the most extensively used organic metal degreasing solvent.

Trichloroethylene is a clear, colourless liquid, readily volatile and boiling at 86.7°C. The low specific heat and latent heat of vaporisation make it especially useful in hot solvent and vapour degreasing plants, since it is readily boiled with a small input of power. Thus, the heat required to convert 1 kg of water into steam is sufficient to vaporise $8\frac{1}{2}$ kg of trichloroethylene. Owing to its low surface tension trichloroethylene penetrates readily into crevices and wets metal surfaces quickly. Trichloroethylene is non flammable, neither does is form combustible or explosive mixtures with air. It dissolves most greases, oils, tars and gums very readily, and also rubber and many resins. It is not affected by water and, in fact, the contaminated solvent may be recovered by steam distillation. The presence of water can, however, result in corrosion of metal if the solvent has undergone any degree of decomposition with the liberation of hydrochloric acid. Water dissolves in trichloroethylene to the extent of 0.025 per cent by weight at 18°C.

Trichloroethylene decomposes at temperatures in excess of 130°C, so that local overheating of the solvent is to be avoided, whichever means of heating is adopted. In the case of heating by steam coils, the pressure should not exceed 2.1 kgf/cm^2 (30 lb/sq in) pressure, in order to avoid such overheating.

3.2.3.1.1. Stabilisers for trichloroethylene

Apart from high temperatures, trichloroethylene is affected by light, exposure to which results in slow decomposition with the liberation of hydrochoric acid. Such decomposition is clearly undesirable, not only because of wastage and deterioration of solvent, but because the liberated acid may severely attack the metal being degreased and also the galvanised steel of which degreasing plants are generally constructed. For this reason all commercial trichloroethylene used for degreasing purposes contains added stabilisers. The materials used are generally basic organic compounds, such as amines. Table 3.3 shows the ratio of the degree of decomposition of trichloroethylene to which 0.1 per cent of various stabilisers were added when the samples were simultaneously exposed to a mercury vapour lamp for a given period.

3.2.3.1.2. Distillation of trichloroethylene

As the solvent builds up in oil content, it must be distilled in order to free it from contamination. The amount of oil and grease contained by the solvent at any stage can be gauged with sufficient accuracy from a determination of its specific gravity. Table 3.4 shows the specific gravity of trichloroethylene liquor at 15°C when containing various amounts of dissolved oil. This Table is based on contamination by oil of specific gravity 0.926, but is approximately correct for the general types of oils removed in the degreasing of metal parts.

When it becomes heavily contaminated with oil-containing soaps, trichloroethylene is liable to foam, so that redistillation for the cleaning and recovery of the solvent should be carried out with care.

TABLE 3.3

Ratio of Degree of Decomposition of trichloroethylene containing various additions, when exposed to ultra-violet light

Stabiliser (0.1 gm per 100 cm^3 solvent)	Decomposition ratio
Dibutylamine	220
Diphenylguanidine	160
Diphenylamine	132
Monoethylamine	126
Cresol	88
Diethylamine	73
Aniline	69
Triethylamine	48
n-Amylamine	45
Pyridine	10
Cyclohexanol	2
Benzol	1

3.2.3.2. Perchloroethylene

This solvent, although more expensive, is preferred to trichloroethylene for certain applications. For articles of light gauge, the use of perchloroethylene permits a longer condensing and draining time to be obtained because of the higher boiling point of the solvent. On the other hand, the higher boiling point involves the use of greater steam pressures for heating; this is offset to a small extent by the somewhat lower heat of vaporisation of perchloroethylene (50 cal/g as compared with 57.2 cal/g for trichloroethylene[12].

TABLE 3.4

Trichloroethylene content of oil-contaminated solvent

Density °Tw	Specific gravity	Percentage of trichloroethylene
94	1.47	100
83	1.41	90
72	1.36	80
61	1.30	70
50	1.25	60
39	1.19	50
28	1.14	40
17	1.08	30
6	1.03	20
–	0.99	10
–	0.95	5

The greater density of perchloroethylene vapour is also helpful in reducing solvent losses by keeping the vapour within the plant, whilst the high boiling point serves to eliminate water vapour from the solvent—this reduces any tendency for hydrolysis to occur with the formation of hydrochloric acid. It has also been claimed that perchloroethylene is less toxic than trichloroethylene, but this view has not been substantiated. The maximum permissible concentration of either solvent in the atmosphere is 100 p.p.m. v/v.

Perchloroethylene is more stable in the presence of aluminium than trichloroethylene and is worth employing where large quantities of this metal are being handled.

Under conditions where water-cooling coils are not able to control the level of trichloroethylene vapour in a degreasing plant, perchloroethylene proves more satisfactory. Such conditions prevail in hot climates and also at high altitudes.

The physical and chemical properties of trichloroethylene and perchloroethylene are shown in Table 3.5 in comparison with those of water.

TABLE 3.5

Characteristics of trichloroethylene and perchloroethylene compared with those of water

	Trichloro-ethylene	Perchloro-ethylene	Water
Formula	C_2HCl_3	C_2Cl_4	H_2O
Molecular weight	131.4	165.85	18
Molecular point (°C)	−87.1	−22.35	0
Boiling point (°C)	86.7	120.7	100
Density of liquid (at 15°C) g/ml	1.47	1.63	1.0
Density of vapour (air = 1)	4.6	5.7	0.63
Specific heat cal/g	0.241	0.215	1.0
Latent heat of vaporisation (cal/g at boiling point)	57.2	50.0	539
Surface tension (dynes/cm^2) at 20°C	30.0	32.0	72.7
Vapour pressure (mm Hg at 20°C)	59	17	17.5
Boiling point of azeotrope with water (°C)	72	87.7	—
Solvent: water ratio (w/w) of azeotrope	13.5 : 1	5.3 : 1	—

3.2.3.3. Trichlorotrifluoroethane

This highly stable, almost non-toxic solvent is a moderately good solvent for oils and greases but does not attack a wide range of resins and plastics. It is therefore of particular value for cleaning composite metal/plastic/resin components such as printed circuit assemblies. Radio and television sets have been operated whilst completely submerged in the solvent, illustrating that the solvent does not in any way affect electrical components. Dismantling precision electrical instruments in order to clean isolated parts is therefore unnecessary. Materials which have been cleaned without damage in trichlorotrifluoroethane include PVC, magnetic tape computer memory units, photographic films, camera parts and ball bearings.

The solvent can be used for cold cleaning or in a vapour degreasing plant. It is much more expensive than either trichloroethylene or perchloroethylene but the special advantages outlined above will in some circumstances justify its use.

3.2.3.4. Trichloroethane

Another chlorinated hydrocarbon which is finding application is 1,1-trichloroethane ($CH_3.CCl_3$). This product has solvent properties very similar to those of carbon tetrachloride and is one of the least toxic of the chlorinated hydrocarbons (M.P.C. 1000 p.p.m.). An inhibited grade is commercially available. It is completely non-flammable and does not attack sensitive metals, such as zinc and aluminium, rubber and many plastics. It is mainly used cold for cleaning by dipping, spraying or wiping but is not suitable for aqueous solvent recovery as it is liable to decomposition in the presence of water. However, where oil redeposition must be avoided, this solvent is to be recommended(13).

3.3.3.5. Degreasing plants

In order to obtain the maximum advantage from the grease-dissolving properties of trichloroethylene or other chlorinated hydrocarbon cleaners, the solvent must be used in a properly designed plant; this will also ensure that the degreasing operation is carried out with due economy and with the maximum of safety.

The three main types of plant in use are: (a) vapour plants; (b) liquor-vapour plants; (c) multi-liquor plants. The plants themselves are generally constructed of galvanised steel, which stands up very well to the action of the trichloroethylene liquid and vapour provided decomposition of the solvent does not take place; in the latter event, rapid attack of the galvanised coating will occur and, in the presence of water, stainless steel construction is recommended. The parts to be cleaned may be lifted into and out of the plant by hoist or by hand, suspended from hooks or contained in baskets. Various types of mechanically operated plants have been designed in which the parts are loaded on to conveyors and pass through automatically, being discharged at the other end in the degreased condition.

3.2.3.5.1. Vapour plants

The vapour-type plant consists essentially of a tank in which a small amount of solvent is boiled. Near the top of the tank is a bank of copper condensing coils through which cold water circulates, so that the vapour is prevented from rising to an excessively high level and overflowing the tank. The arrangement is shown diagrammatically in Fig. 3.3. The cooling coils do not, of course, prevent the

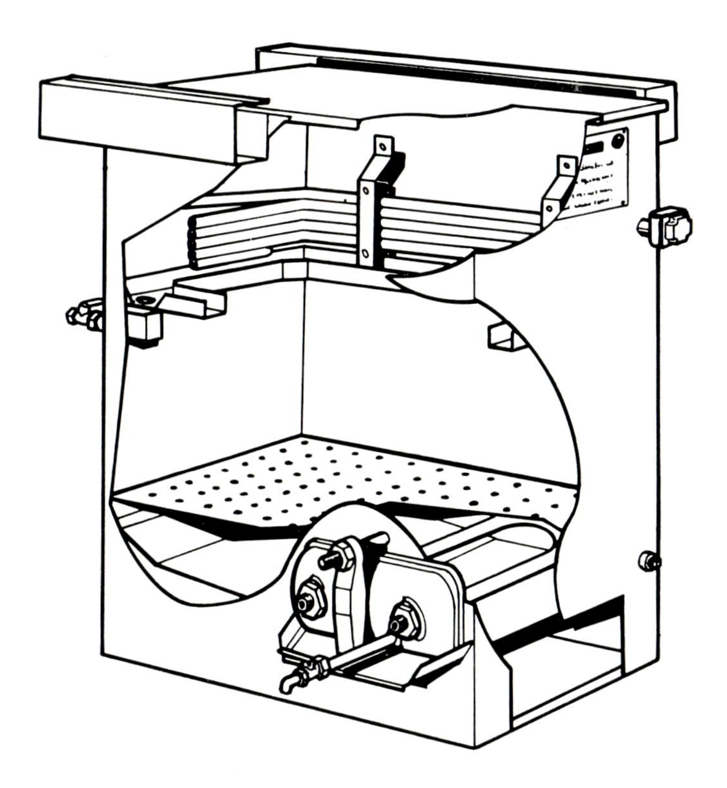

Fig. 3.3 Vapour plant, popular model, steam heated.

[Courtesy I.C.I. Ltd.]

conduction of heat through the walls of the plant to the region above the coils; this results in convection currents being induced which tend to assist the release of trichloroethylene vapour into the atmosphere. For this reason a channel is welded to the exterior of the plant just above the vapour level to reduce conduction of heat up the walls of the plant and thus to lessen convection losses.

In operation, the metal parts to be degreased are suspended in the vapour. Since they are cold to start with, vapour condenses onto them, dissolving the oil and running off in the sump. This process continues until the articles reach the temperature of the vapour, when no further condensation can take place. Clearly, with articles of thin section, the condensation-cleaning action will continue for a shorter time than with more massive articles, which take considerably longer to heat up to the vapour temperature.

It is to be noted that, as only pure condensate reaches the surface being cleaned, there is no possibility of contamination taking place as a result of dirty solvent residues being left on the metal surface. Also, the condensation and running-off of the solvent has a mechanical cleaning action, so that not only is oil removed but also,

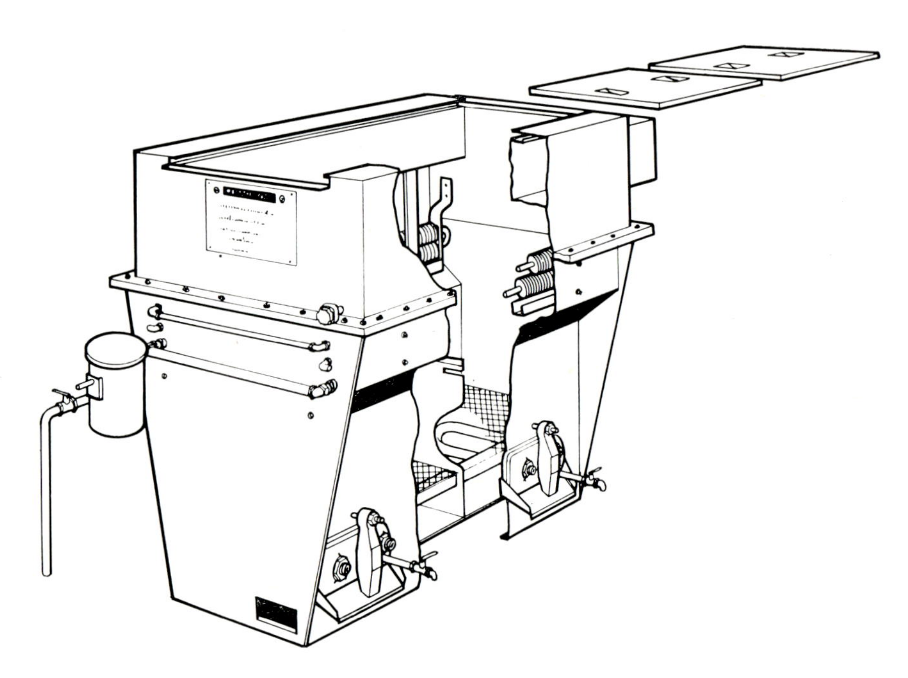

Fig. 3.4. Multi-liquor (two-stage) steam heated plant providing boiling liquid and vapour treatments. [Courtesy I.C.I. Ltd.]

to a certain extent, abrasives, metal swarf and soaps which may not themselves be actually soluble in the solvent.

Recently, air cooled degreasing plants have evolved from the need to save water and energy[14]. It is essential, however, to use water cooling where large amounts of condensate are involved.

3.2.3.5.2. Liquor plants

In the multi-liquor types of plant (see, for example, Fig. 3.4) the bottom of the tank consists of two or more compartments, each containing boiling solvent. The articles to be cleaned are immersed successively in the various sections, the cleanest solvent being in the last compartment. This is achieved by arranging for the condensate from the cooling coils at the top of the tank to be delivered into the final compartment. This type of unit is, however, less in demand than the simpler vapour or liquor-vapour type in which most work can be satisfactorily degreased.

3.2.3.5.3. Liquor-vapour plants

Broadly speaking, vapour degreasing adequately removes oils and greases, even when these contain a small amount of non-soluble contaminants. When such contamination is present to any considerable extent, however, liquor-vapour plants are to be preferred. Thus, articles which have been polished and retain a certain amount of polishing composition in recesses need to be immersed in the boiling solvent to remove the abrasive and soap residues. As the grease dissolves, the other materials are released and are washed out by the movement of the boiling solvent.

Liquor-vapour plants are also preferred where material of thin section is being treated. Such work heats up very quickly so that vapour degreasing is not really effective unless a second immersion is given, the parts being allowed to cool off in between. This is a laborious procedure, and it is therefore better to treat such parts in a liquor-vapour plant.

Care should be taken to lift parts out of the degreasing plant slowly and at a uniform speed to prevent drag-out of vapour, and of liquid in cupped parts. Various types of racks and baskets have been designed to handle articles so as to effect rapid degreasing with a minimum of solvent loss.

In operating the liquor-vapour type of plant, two methods are available: (i) where the work is greasy and free from solids such as polishing abrasive, it is desirable to remove the bulk of the grease by vapour treatment and follow this by a rinse in the liquor compartment; in this way little grease or dirt is introduced into the liquor compartment, so that the solvent is kept very clean by the automatic flow of condensate from the cooling coil into it; (ii) where the work is covered with polishing dirt, it is essential that the solid particles should be removed as fast as they are released by the dissolution of the grease-bond; therefore, the work is washed directly in the liquor compartment and raised out of it steadily to avoid excessive drag-out. The purpose of the vapour compartment in this case is to receive the dirty solvent displaced from the liquor compartment by the continuous flow of clean distillate into it. The successful action of the plant in this case depends on keeping the dirty liquor in the vapour compartment boiling freely so as to produce the maximum amount of condensate to provide automatic self-cleaning of the solvent in the liquor compartment.

3.2.3.5.4. Heating degreasing plants

The heating of degreasing plants can be carried out by gas, steam, electricity, high-pressure hot water, or oil. In the case of gas-heated plants, burners are located below the plant and should give a uniform degree of heating over the whole bottom, with an adequate flue to remove the products of combustion. A remote control valve is connected to two thermostats, one being located at the top of the cooling coil, where its action is to cut off the gas supply to the main burner if the flow of cooling water is restricted, thus preventing the vapour level from rising unduly high. The second thermostat is in the sump, and cuts off the gas supply if a temperature of 130°C is exceeded in the liquor when distillation is nearing completion. Heating trichloroethylene to temperatures above this figure leads to decomposition and the

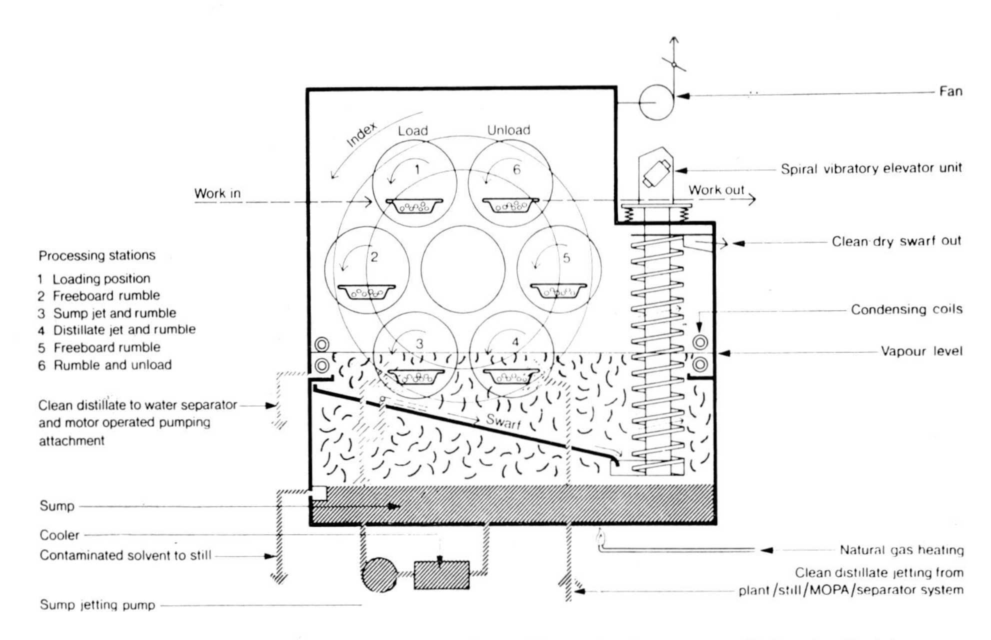

Fig. 3.5. Fully automatic vapour type plant with two jetting stages utilizing 6 cylindrical rumbling baskets for cleaning small parts. [Courtesy I.C.I. Ltd.]

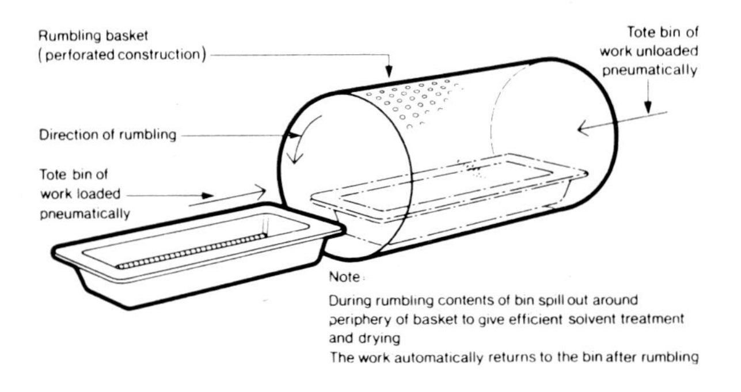

formation of acid, which will attack the plant. If the temperature should become extremely high as, for example, when the solvent comes into contact with an open flame or red-hot surface, breakdown of the compound takes place with the formation of toxic phosgene and hydrochloric acid. Exposure of the solvent to such conditions must therefore be avoided at all costs.

Steam heating is especially suitable for use with degreasers. The steam pressure should not exceed 2.1 kgf/cm^2 (30 lb/sq. in.) with clean trichloroethylene in the plant. A safety device is fitted to such plants to shut off the steam if the water supply to the cooling coils fails or becomes inadequate.

High-pressure hot water can also be employed successfully for heating. It is difficult to arrange for the thermostatic control of hot water plants, and for this reason an orifice plate is provided in the hot water system to regulate the rate of heat transfer by restricting the flow. The size of the orifice must be determined by trial and error. It is also desirable to fit a warning light and a bell operating in conjunction with thermostats, to notify the operator if either the sump temperature becomes too high or if the vapour level overflows the cooling coils due to failure of the water supply.

Electrical heating is readily carried out by means of thermostatically controlled heaters, fitted below the sump to avoid local overheating. This is a flexible method of heating, although it is inclined to be expensive.

Table 3.6 shows the operating characteristics of five typical standard manually operated plants. The figures quoted for output and solvent consumption are only approximate but are representative of those that can be obtained in the degreasing of heavy metal components.

3.2.3.5.5. Automatic plants

Automatic plants for solvent degreasing are preferable to the open-top degreasing plants described above, although they are rather more expensive initially. They have the advantages of low labour costs, the avoidance of high solvent losses due to bad handling of the work into and out of the plant, and the elimination of practically all health hazard to the operator.

In the simpler plants, baskets of work are loaded by hand, but the sequence of operation is carried out automatically; Fig. 3.6 shows an automatic machine of this type for vapour degreasing only. The components to be cleaned are placed on a platform either individually or in baskets, and the operator presses a button when the doors close and the platform is lowered into the vapour. Only when the vapour level has risen above the level of the work does the platform rise through the drying zone to the unload position. The doors then open with a bell signal and the work can be removed. Whilst the doors are open a suction fan prevents solvent vapour leaving the machine. It is thus seen that the time cycle is automatically governed by the quantity and type of work in the plant and is not left to the discretion of the operator.

Fully automatic machines, such as the one shown schematically in Fig. 3.7, are available for multi-stage degreasing operations; here the work is transferred through liquid, vapour and spray cleaning compartments. In many machines the work baskets can be made to rotate by the operator when required; this is particularly helpful in dealing with many types of contaminants and with flat articles and blanks which are liable to stick together. Solvent loss due to removal by ‘ cupping ’ in hollow articles is also largely eliminated by this method. The lower part of this type of machine is divided into two or more solvent baths, as required. Each of these has its own thermostatically controlled heating system, and can be used as a cold or hot emulsion

TABLE 3.6

Dimensions and Operating Characteristics of Standard Solvent Degreasing Plants

Type of Plant	Effective size of compartments			Overall size of plant* (approx)			Solvent charge (litres)	Fuel consumption max		Cooling water (litres/h)		Time to start from cold (min) (approx)		Maximum output (kg/h) (approx)		Solvent consumption (kg degreased per litre) (approx)
	Length (cm)	Breadth (cm)	Depth (cm)	Length (m) (cm)	Breadth (m) (cm)	Depth (m) (cm)		Gas† (m^5/h)	Steam (350 kn/m^2) (kg/h)	Gas	Steam	Gas	Steam	Gas	Steam	
Vapour	50	23	38	1 15	0 68	1 6	18	2.1	17	112	290	12	6	270	450	165
Vapour	92	60	60	1 60	1 22	1 45	45	6.8	45	340	630	15	6	820	300	190
Liquor-vapour	50	23	38	1 15	1 22	1 30	100	4.7	32	250	540	40	15	225	450	83
Liquor-vapour	60	46	46	1 50	1 38	1 45	245	8.5	47	450	800	50	25	570	900	83
Multi-liquor (two-stage)	33	33	33	1 30	1 22	1 52	230	5.7	32	360	450	45	25	200	410	100

* On steam heated plants the safety device (measuring approx. 75cm × 25cm × 85cm) is not included in these dimensions.

† Figures are based on an average calorific value of 18,500 J/m^3.

Fig. 3.6. Simple automatic 4 compartment machine for degreasing only with panels removed to show lift and transfer mechanism. An ultrasonic stage can be added if desired. [Courtesy I.C.I. Ltd.]

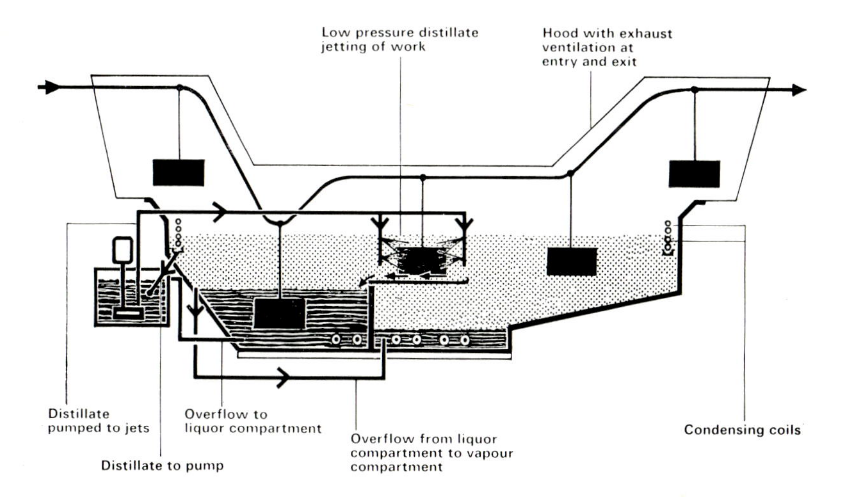

Fig. 3.7. Fully-automatic multi-stage liquor-vapour degreasing plant. [Courtesy I.C.I. Ltd.]

cleaner, a vapour state cleaner, or as a container for preservative dip.

In the operation of the plant shown in Fig. 3.8, a button is pressed which brings a completed work basket to the door opening around which an exhaust duct is fitted to deal with traces of vapour which might escape. The basket is then removed and replaced with a further basket of work to be cleaned. At the same time each of the other baskets in the plant has moved one position forward, with the final basket in the drying position. Rotation of the baskets is under the control of the operator. For fragile parts which cannot safely be rotated, rectangular baskets which rock up and down in the baths can be provided.

Another special purpose mechanised plant designed to degrease parts in boiling solvent liquor is shown in Fig. 3.9. The conveyor consists of a single chain to which is attached a series of compartmented carriers suitable for holding deep drawn hollow articles which, in this particular instance, are of the order of 2.5 cm diameter by 7 cm deep. The drawn parts are placed mouth downwards in the carriers; these pass up over a sprocket wheel and then take the work down into the boiling liquor mouth upwards. Here the chain again rises upwards, inverting the components so that the liquor drains out freely. After travelling through the cooling turret, the conveyor chain passes under the plant to the starting point. It is important to note that in this final stage the work no longer rests against the back of the carrier but is tilted towards the open front, whence it falls into baskets or on to a conveyor belt.

To prevent the liquor in which the parts are washed from becoming unduly charged with grease, a concentrator forms part of the plant. The greasy liquor overflows into this as it is displaced by clean condensate from the cooling coil. Another measure to maintain the cleanliness of the liquor consists in the provision of

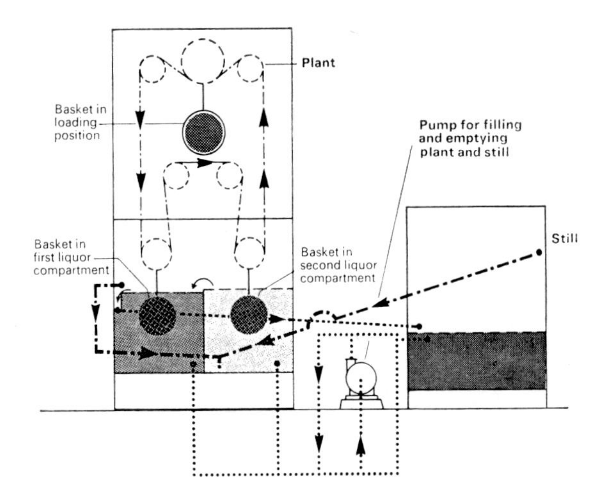

Fig. 3.8. Semi automatic basket type cleaning plant with two-liquor stages with interconnected still. [Courtesy I.C.I. Ltd.]

Fig. 3.9. Special-purpose mechanised plant for degreasing parts in boiling solvent liquor.

a baffle plate which diverts into the concentrator the initial condensate which contains the bulk of the grease dissolved as soon as the work enters the vapour. With a conveyor speed of 120 cm per minute, an output of 5,000 articles per hour is possible.

3.2.3.5.6. Solvent economy

In order to obtain the maximum economy in solvent consumption and to ensure that the best results are obtained, solvent degreasing plants must be operated correctly. In the open-topped plant of the hand-operated type, work should be lifted from the plant slowly and steadily in such a way as to avoid disturbance of the vapour surface as far as possible. The speed of the lift in the drying-off zone of the plant should be of the order of $1\frac{3}{8}$ metres per minute[15]; where the conveyor passes into and out of hooded turrets at the end of the totally enclosed plants, which substantially reduce solvent loss, the conveyor speed can be up to about 3m per minute.

Lids are fitted to open-topped plants, and these should be kept closed as far as possible during idle periods. Roller or sliding types of lids are to be preferred to loose or hinged lids since they result in less draught and disturbance of the vapour-air interface when the lids are open or closed. In practice, of course, there is a constant diffusion between the air and the vapour phases, and the use of the lids results in a considerable reduction in the rate at which diffusion can take place. Gas-heated plants are especially liable to solvent losses by diffusion, due to the convection currents brought about by the gas flames and the heat from the flue pipe.

Solvent degreasing plants should be located where their is a minimum of draught, but in cases where high solvent losses exist for one reason or another, it sometimes pays to install a solvent recovery unit. This consists essentially of a narrow duct placed round the top of the plant through which a low-power fan draws a stream of air. The solvent-laden air extracted in this way is passed through a container filled with activated carbon which absorbs practically all the solvent carried out of the plant. This can then be recovered by periodic steam distillation of the solvent-impregnated carbon.

Pressings and cup- or thimble-shaped articles generally cause the withdrawal of considerable amounts of solvent from the plant. Such articles should be placed horizontally in baskets with the open ends facing in one direction. The baskets can be lowered into the plant with the open ends tilted slightly upwards so that air will not be trapped; when the baskets are removed, they are tilted in the opposite direction so that both solvent and vapour will run out. It is desirable to have some kind of rack which is placed just above the vapour level on which the basket can be rested before withdrawal from the plant, in order that the bulk of the solvent may have an opportunity to condense. This tilting operation can be carried out mechanically in the case of hoist-operated plants carrying heavy loads. If a drying rack cannot be fitted due to lack of room in the plant itself, it is useful to have a water-cooled tank closely fitting the basket nearby in which the baskets can be left for a few minutes. The solvent condenses in this tank and sinks to the bottom, from where it can be collected at the end of each day.

Gray[(16)] has recently discussed methods for minimising solvent losses in vapour degreasing plants.

3.2.3.5.7. Solvent purification and distillation

In automated plants it is clearly undesirable to have to stop the plant, perhaps several times a day, to manually remove sludge and swarf from its sump. However, all mechanised cleaning plants are nowadays designed with an attached solvent purification plant[(17)]. Such purification equipment usually comprises:

(1) Pumping unit.
(2) Container for pure solvent.
(3) Container for contaminated solvent.

(4) Distillation plant.
(5) Filter for the removal of coarse sludge.
(6) Filter for the removal of fine sludge.

In any plant, when a considerable amount of oil and dirt has accumulated in the solvent, the latter must be distilled off. As grease builds up in it, the rate of production of vapour decreases, while the boiling-point of the liquor rises. Both these effects are undesirable, excessive boiling temperatures in particular leading to solvent decomposition. Thus, a liquor composed of a 50 : 50 mixture of trichloroethylene and oil distilled at a rate of approximately 73 per cent of that of clean solvent in a gas heated plant; in a steam-heated unit this rate was reduced to only 23 per cent under the same conditions, while the presence of 35 per cent of oil reduced the rate of vaporisation of solvent by 50 per cent. A liquor composed of 20 per cent trichloroethylene and 80 per cent grease boils at about 130°C, and this temperature cannot safely be exceeded; it is, however, strongly advisable to clean the plant well before this stage is reached.

Special facilities are provided for carrying this out, the condensed solvent being collected in a suitable vessel placed below the distillate pipe leading from the plant. Distillation is continued until the sump temperature reaches 130°C, when the sump thermostat automatically comes into operation and no more solvent will distil. At this stage the plant is allowed to cool, and the residues are removed from the sump through the mud doors by means of a rake. Distillation should not be carried out if the level drops below the steam coils. In this case the oily residue is run off and added to the contaminated liquor when the next occasion for distillation arises.

Frequent cleaning is desirable to avoid decomposition of the solvent, which can be brought about not only by local overheating but also by the presence of such materials as light alloy swarf. The fact that the solvent has become acid is generally shown by the presence of a haze of hydrochloric acid over the tank together with an acid odour and, later, by the formation of green corrosion products on the copper condensing coils and corrosion of the tank walls. The plant must be stopped at once in these circumstances.

3.2.3.5.7.1. Aqueous distillation

Normally, and especially where there are a number of plants in operation, all plant residues are distilled in a separate distillation unit which can be an ordinary vapour plant assigned for this purpose. One method is to use live steam if available, or wet distillation, whereby water is added to the oily liquor so that, in effect, steam distillation takes place. There is no risk of overheating the solvent towards the end of the distillation process as is the case when heavily oil-contaminated solvent is being distilled alone.

The addition of water to clean trichloroethylene results in the formation of an azeotropic mixture boiling at 73°C (instead of 87°C, the boiling-point of the pure solvent), so that practically all the solvent can be recovered by the use of temperatures little above 100°C. The amount of water required is not critical, about 10 litres being sufficient for the most commonly-used plant sizes.

When aqueous distillation is employed, a water separator must be fitted. This prevents water being carried into the drum into which the distillate flows, whilst it also allows the stabiliser to be retained in the trichloroethylene if the water flowing from the separator is constantly returned to the greasy liquor.

A further advantage of this method is that the time for cleaning a plant is much reduced, it being only necessary to drain or pump out the dirty solvent before sending it to the still. Meanwhile, the plant can be immediately filled with a fresh charge of solvent and is ready for use again.

3.2.4. ULTRASONIC CLEANING

Cleaning by ultrasonic methods is one of the most important newer developments for removing contamination from intricate parts and a great deal of literature concerning many aspects of the technique has been published. It has been claimed that[18] cleaning with ultrasonic methods can reduce the tendency for hydrogen embrittlement under some circumstances. The cost of ultrasonic plant is somewhat high but the speed and certainty of the process make it worth while, particularly for expensive articles or where the consequences of failure can be serious. Essentially an ultrasonic cleaning plant consists of a tunable ultrasonic oscillator, a power amplifier, a transducer which converts the electrical energy into accoustic or mechanical energy, and a container in which the articles are immersed in the liquid cleaning medium. Chlorinated hydrocarbon solvents are normally used. Trichloroethylene, perchlorethylene and trichlorotrifluorethane are especially suitable[19] for ultrasonic cleaning. The surface tension of the latter is lower than that of trichloroethylene, which enables it to enter and wet recesses more easily. The addition of a surface active agent to stabilised trichloroethylene together with a small volume of water produces a two-phase mixture which is said to be capable of efficiently removing wax- or soap-based polishing compounds.

There are three classes of transducer materials available to convert the electrical into sonic energy. These are:

(1) Piezo-electric crystals such as Rochelle salt or quartz.

(2) Ceramics based on lead zirconate or barium titanate.

(3) Magnetostrictive materials, usually nickel or cobalt alloys.

The most important factor in obtaining satisfactory service over long periods is the stress limit which the materials can withstand without failure due to stress or fatigue, the latter being usually the more important. Quartz crystals may be used in transducers for small installations because of their good mechanical strength and

and stability, but for larger units ceramics are used. As the size of the crystal determines its frequency, and in view of the fact that it is not easy to obtain large quartz crystals, such transducers have been used principally in the high frequency range, i.e. 250 kilocycles to 2 megacycles per second. The ceramic materials can be used at lower frequencies and require lower voltages than quartz crystals. Nowadays lead zirconate-titanate transducers are widely used. These have the advantage of very high energy conversion efficiencies and can withstand temperatures up to 150°C.

Magnetostrictive devices depend for their operation on the dimensional changes produced in some ferromagnetic metals and alloys when an external magnetic field is applied. All metal transducers comprising closely stacked laminations are capable of delivering high ultrasonic energy outputs. It is usual to use frequencies of the order of 100,000 cycles/second. The electrical energy to be fed to the transducer is produced by transistorised oscillators built in sizes of from 100 watts up to several kilowatts. The incorporation of a feed-back circuit from the transducer has eliminated tuning problems by sensing the resonance point of the transducer and automatically stabilizing the generator frequency at this value.

The transducer is mounted or inserted into the tank containing the solvent where it vibrates at its own natural frequency when energised by the oscillator. The shock waves thus produced cause vibrations in the liquid which exert the cleaning action.

During one half cycle of the wave a vacuum is developed which helps to extract soil from the pores in the metal surface. If the wave amplitude is high enough a degree of cavitation is produced[(20)] on account of the solution being temporarily detached from the surface. It is this cavitation effect which is mainly responsible for the cleaning action[(21)]. The other half cycle of the wave results in the development of a very high pressure which forces cleaning solvent into the pores of the metal surface. The development of cavitation is favoured by low frequencies and high amplitudes. At low frequencies the erosion effect of the cavitation may be very pronounced.

For the cleaning of components such as watch parts, higher frequencies are best since this avoids violent cavitation action which is not particularly good for dealing with the minute particles of lapping compound and the extremely thin surface films which are difficult to remove. High frequency oscillation reduces cavitation and results in an effective scrubbing action which will not erode or damage the metal parts concerned. At higher frequencies energy losses during transfer to the cleaning medium become extremely high.

The influence of variables in ultrasonic cleaning with an aqueous detergent solution have been investigated by Crawford[(22)]. The test results indicated that at low intensities it may not be possible to remove some soils completely, irrespective of immersion time. With pulsed inputs the average rather than the peak power is the factor which determines the cleaning efficiency. It was further found that the transducer must operate at its resonant frequency for best results (Fig. 3.10) and that the liquid must be kept free from suspended soils for maximum cleaning efficiency (Fig. 3.11).

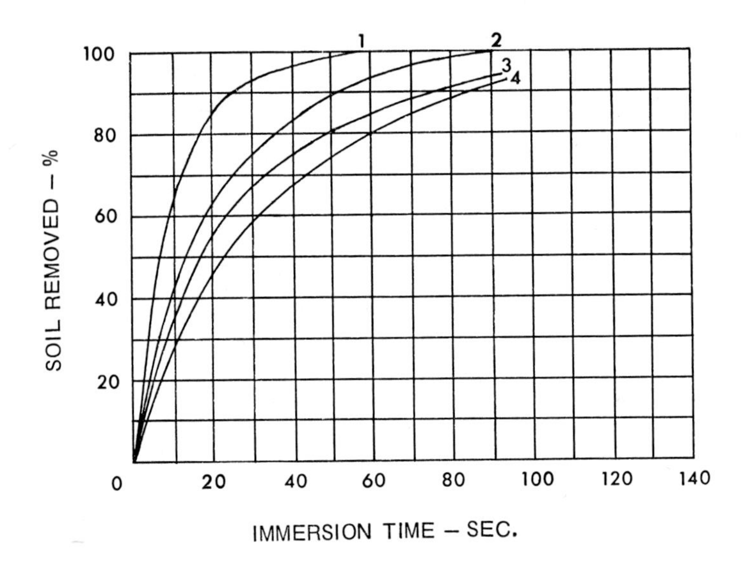

Fig. 3.10. Effect of frequency shift on rates of cleaning.[22]

Frequency:

1. 22.2kHz (resonance) 2. 21.2kHz
3. 20.0kHz 4. 19.0kHz

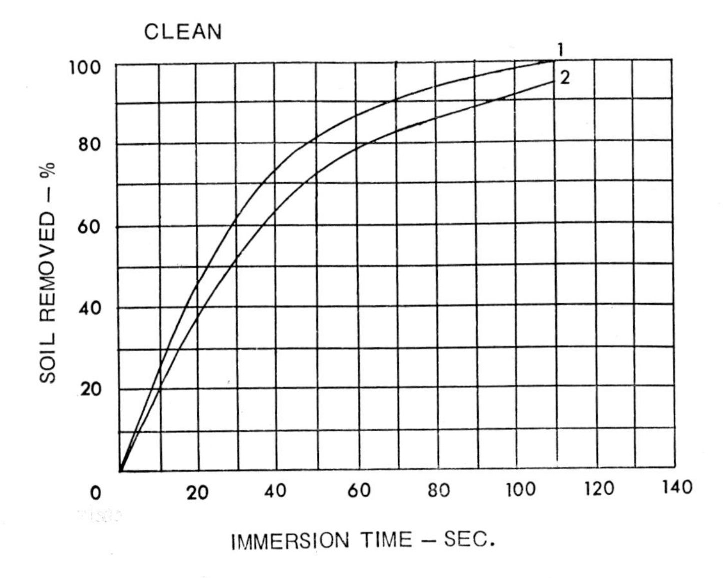

Fig. 3.11. Effect of suspended soil on rate of cleaning[22]

1. Clean solution.
2. Suspended soil. Liquid at 100°F with added detergent. Ultrasonic intensity 1.5 W/cm².

3.2.4.1. Applications of ultrasonic cleaning

The engineering and electronics industries provide a prolific field of application for ultrasonic cleaning and it is only possible here to deal with a limited number of examples. Geckle[23] has recently reviewed some examples of ultrasonic cleaning. The earlier attempts at ultrasonic cleaning utilised a simple tank into which the transducer head was placed, but it was soon found that such a procedure was not adequate. Modern ultrasonic cleaning plants are necessarily much more elaborate and provide a complete sequence of operations to enable very rapid, high-quality cleaning to be carried out continuously; this applies to both manual and automatically-operated plants.

In the manufacture of ball bearings, for example, clearances of 0.0005 cm are common, and it is normal to grind ball bearings to the nearest 0.00002 cm so that even the most minute particles of dirt on such components can cause an increase in the frictional torque of the bearing. Before the development of ultrasonic cleaning, the best available cleaning method was immersion in a petroleum solvent. It took as long as two minutes to clean each bearing, and even then the results were still variable. By using ultrasonic energy, cleaning can be carried out automatically much more rapidly, and it is possible to design plants which will clean every recess in completely assembled bearings. Provision is made in these machines for the removal of contaminated solvents and their replacement by clean liquid by continuous circulation and filtration through two-stage filters. These will remove particles as small as 1 micron in diameter. Similar plants have been designed for cleaning hydraulic components, particularly cast iron cylinders for automobile brakes, servo-motor laminations and commutators, fine-pitched screw threads and worms, zero-backlash gears, electric shaver heads, needle valves, valve seats, fuel injection nozzles, carburettors, instrument parts and electronics components.

As the maximum energy of the ultrasonic vibration is liberated in a straight path from the transducer through the cleaning liquid, it is evident that the solvent should have free access to all surfaces which are to be cleaned, and that air pocketing and shielding should be avoided; hence if the work is of a complicated shape or contains blind holes (which is often the case) a simple static immersion in a tank of trichloroethylene would not be adequate. Two techniques are generally used in such cases: (1) the articles are revolved either on racks or in a barrel in a horizontal plane above the sonic head; or (2) the solvent container may be made airtight and a partial vacuum applied to the system to ensure that there is no air pocketing. The first method is usually adequate when the recesses or holes are not less than 0.5 cm diameter, and it has the advantage of reducing shadowing to a minimum. The vacuum method is specially suitable for dealing with the smallest holes and has in fact been developed to deal with the special problems of the watch industry.

3.2.4.2. Cleaning of watch and electrical components

Watch parts have provided an important field of application for ultrasonic cleaning because of the high standard of cleanliness required, the large quantity of components involved, and their high intrinsic value. Also, if the visual inspection, which is essential when manual cleaning methods are employed, can be eliminated, the actual cost of the plant becomes a matter of secondary importance.

Watch parts, as is well known, are extremely complex in shape and often include pinions and teeth with blind holes up to about 2 mm in depth and 0.5 mm in diameter. The magnitude of the problem is indicated by the fact that hard-packed polishing compound has often to be removed from such holes and from the teeth of extremely large numbers of articles of this description.

In one type of ultrasonic cleaning plant which has been designed for this purpose, which is of the manual type giving a high degree of flexibility in operation, five separate compartments are used. The articles are placed in small nickel baskets which can be vibrated mechanically at each stage. The first tank contains solvent in which pre-cleaning is carried out. The second compartment contains the ultrasonic head and can be partially evacuated, whilst the third and fourth compartments are for rinsing. In the fifth compartment, drying is carried out by means of dust-free air, the components being heated by an infra-red unit the air temperature can be regulated and provision is made for extracting the air from the compartment to avoid turbulence.

The amplitude of the ultrasonic energy can be regulated, whilst filtration and pumping equipment for the solvent are also provided, as well as cooling and exhaust equipment. For fully automatic operation, a rotary machine, in which a central shaft carrying eight radial arms to each of which a work basket is attached, is available. Each arm is lowered hydraulically into a separate tank in turn, an indexing system transferring the baskets from one tank to the next, using an umbrella principle for raising and lowering the arm. The rate of cleaning is exceedingly fast and in actual practice the time cycle of the plant is governed principally by the time required for the circulation and filtering of the cleaning fluid, since if traces of contaminated solvent are left on the work, the effect of the cleaning operation will be nullified.

A peculiar aspect of the effect of ultrasonic cleaning is that when watch parts have been treated in this way it becomes almost impossible to oil them satisfactorily. The reason is that a slightly contaminated surface promotes an increase in the interfacial tension between the oil and the metal, so that a droplet applied to it will remain in place; if this very thin film (which is often of only monomolecular thickness) is removed, the oil spreads rapidly and does not remain at the points where it is needed for effective lubrication. Such spreading also results in an increased rate of evaporation and ageing of the oil, which is clearly undesirable. Watch components which have been cleaned ultrasonically are therefore treated in such a way as to increase this interfacial tension in order that it should be possible to lubricate them effectively. In the ' Epilame ' process, a thin film of stearic acid is applied to the metal either by dipping or brushing of the component with a dilution of the acid in a solvent or by exposure to its vapour. The vapour treatment is considered to be superior, giving coatings only about 0.00001 mm thick which remain effective for a sufficiently long time.

The use of ultrasonic cleaning techniques in the electrical and electronics industry is widespread[24, 25]. Applications include cleaning individual components such as capacitors, semiconductors and potentiometers, printed circuit boards and assemblies, television and radio chassis, electric motors and other complex electrical and electronics components. Ultrasonic cleaning is particularly valuable for cleaning apparatus which has been in service for many years. The cleaning of some such equipment by manual methods is a very long and tedious process.

A guide to commercially available ultrasonic cleaning equipment has recently been published[26]. Also a review[27] covering cavitation, cleaning chemicals and applications of ultrasonic degreasing.

3.2.5. SOLVENT DRYING

A method of using solvent vapour for the stain-free drying of electroplated articles has also been developed. This involves immersing the wet article in a boiling, specially-formulated trichloroethylene-based mixture so that the water is evaporated off as an azeotrope. The trichloroethylene contains about 1 per cent of a surface-active additive which causes the solvent to wet the metal preferentially, thus displacing water together with any dissolved salts from the surface. The additive is rinsed off in clean solvent, which then evaporates. The method is suitable for all kinds of plated metal pressings and stampings, as well as for tubular and wire work. It can be used on finishes such as nickel, silver, gold and chromium plate, as well as on phosphated, anodised and electrolytically polished articles.

A review dealing with stain free drying using dewatering fluids has recently been published[28].

It is carried out in a more-or-less conventional stainless steel or otherwise protected two-compartment degreasing plant. A water separator and rim ventilation are incorporated. The trichloroethylene containing the additive is in one compartment of the plant, whilst the second (rinse) compartment is filled with plain trichloroethylene. The water-wet articles are immersed in the first section for, say, 15-30 seconds, after which they are transferred to the rinsing section for a further 15 seconds or so. They are then slowly withdrawn and, after remaining in the vapour zone above the solvent for a short time, removed from the plant.

In the case of large articles, where a vapour rinse may be inconvenient, a three-compartment plant can be used; one tank is used for drying, the other two being used for rinsing. Since the azeotropic mixture of trichloroethylene and water, which contains 7% water, boils at 73°C as against 87°C for the solvent alone, water in the drying compartment which has been removed from the work with the aid of the additive is rapidly volatilised. The solvent and the water are condensed on the coils of the plant simultaneously, the latter being removed in the water separator whilst the former is returned to the plant. The arrangement is shown in Fig. 3.12.

Perchloroethylene can be used for this purpose with advantage in place of trichloroethylene, since its azeotope contains 16 per cent water. It is also more stable, less toxic, and has a lower heat of vaporisation. Perchloroethylene has, however, a

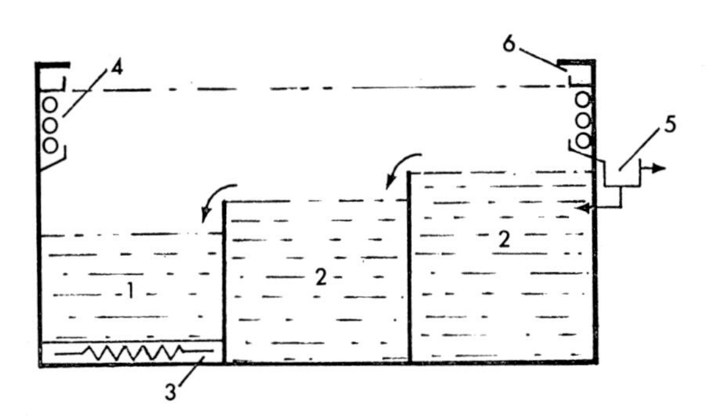

Fig. 3.12. 3-Stage solvent drying bath.

1. Drying bath.
2. Rinse bath.
3. Heater (with thermostat)
4. Condensing coil.
5. Water separator.
6. Ventilation.

rather high boiling point (121 °C), and is therefore to be used with caution on articles which may be affected by higher temperatures.

The solvent method of drying has the advantage of being able to deal with intricately shaped components readily, whilst the labour and space requirements are low. It is also relatively rapid in operation. On the other hand, it often involves the need for removing racked articles in the wet state and transferring them to baskets for treatment as the trichloroethylene attacks many types of rack coating. The cost of the solvent used is also not inconsiderable, whilst the process will only operate successfully if the articles are of sufficient mass to retain adequate heat to evaporate the residual solvent after rinsing. It is also unsuitable for chromated zinc- or cadmium-plated articles, as the solvent has the effect of dehydrating the conversion coating and reducing its effectiveness. Parts dried in this way are also exceptionally free from grease and oil, so that they may be more liable to develop corrosion, particularly in recessed areas where the plating is liable to be thin.

3.2.6. ALKALINE CLEANING

Alkaline cleaning methods are used either alone, for removing surface contamination of an oily nature from metals, or after solvent degreasing to produce a chemically clean surface suitable for subsequent electroplating or other finishing operations. The alkalis which are employed in present-day practice are relatively limited in number; each has its own specific characteristics, however, and, as a rule, a combination of several alkalis is used in the formulation of cleaners. In addition, small quantities of other materials may be added to facilitate wetting, deflocculation or peptisation of colloidal surface contaminants, together with synthetic wetting agents (or surfactants) to lower the surface tension of the solutions if desired. A review and classification of alkaline cleaners in the plating shop has recently been given[29].

Saponification consists essentially in the formation of water-soluble soaps by the interaction of the alkali present in the cleaner with the oil or grease which it is desired to remove. In the case of present-day alkali cleaner formulations which are fast-acting and powerful, however, saponification plays a relatively small part in the cleaning process, as the reaction is a relatively slow one. Mineral oils are not capable of being saponified by alkalis to any appreciable extent, so that other mechanisms must be brought into play if an effective cleaning action is to be effected when these are present in quantity. The most important of these are emulsification and deflocculation.

3.2.6.1. Emulsification

The efficiency of a detergent as an emulsifying agent is determined by its ability to promote the spontaneous dispersion of the oil in the form of fine droplets (which may be of the order of 10^{-5} mm in diameter or less), coupled with its ability to prevent their re-coalescence. Emulsifying agents are characterised by their tendency to concentrate in the interface between the oil and the water. Their molecular structure consist usually of a long-chain hydrocarbon portion, which has an affinity for the oily part of the dispersion, attached to a polar or water-soluble atom or group of atoms, e.g. a sodium atom, sulphonic acid derivative, etc. This type of structure is typical of the soaps and the synthetic wetting agents, which will be examined in greater detail later.

It was at one time assumed that the emulsification of an oil takes place by the penetration of the detergent solution through the oil film by virtue of its low surface tension, with consequent wetting of the underlying metal. Hence, the oil would be split up into a multitude of droplets and thus be separated from the basis metal. It has however, been shown that, on immersion in a detergent solution, the oil film contracts, so that its surface area is not increased[30]. In contracting, it forms globules which are then readily separated from the metal surface by the alkaline solution. This indicates that the wetting of the metal by the alkaline solution is of greater importance than the wetting of the actual oil film.

Emulsifying agents are also always, to a degree, wetting agents, and it is difficult to distinguish clearly between these phenomena.

3.2.6.2. Deflocculation

Deflocculation or peptisation, as it is often termed, will also take place as a result of detergent action. By deflocculation is meant the disintegration of the contaminant prior to its emulsification and suspension in the aqueous phase. In this sense deflocculation is the opposite to agglomeration. The deflocculating agent acts by separating the particles so that the residual forces tending to hold them together, or to cause flocculation, are prevented from acting. The effect may be the result of adsorption on the surfaces of the oily particles, so that the formation of stable agglomerates is prevented. When a dirty metal surface is immersed in an alkaline solution, the first process which takes place is probably some degree of saponification of any saponifiable material which may be present by the alkali. This is followed by a wetting action in which the detergent concentrates at the interface between the contaminant and the detergent solution, so that deflocculation and emulsification are brought about.

The material removed must be suspended in the cleaning solution in such a way that rinsing will leave a clean metal surface. An unstable dispersion is not satisfactory as it may lead to the precipitation of the grease on the metal. The stability of the dispersion depends on the adsorption of the detergent at the oil-water interface. Micelles carrying similar electrical charges can also facilitate stabilisation; hence, the addition of very small quantities of colloidal material to a detergent may often give a very marked improvement in its cleaning properties.

The ability of a particular solution to clean a metal surface is thus a combination of a considerable number of factors, including the lowering of the interfacial tension between the water and the oily phases, the dispersing and deflocculating powers of the solution, and foaming properties which promote suspension.

It is of interest to note that scouring and mechanical action may in some cases actually give inferior cleaning[(31)]. This can probably be ascribed to the breakdown of the emulsion of the oil in the cleaner as a result of the oil particles being forcibly brought into close contact with one another, thus causing the protective surface film to break down; precipitation of the oil then takes place on the metal surface.

Where a great deal of grease has to be removed, it is sometimes an advantage to have two cleaners. The first is used as a soaking bath, the alkali employed not having especially good emulsifying properties. The grease is removed here and floats to the surface, whence it can be skimmed off. Thus, the bulk of the solution does not become heavily contaminated with emulsified oil. The partly-cleaned work is then transferred to the second cleaner, which should have good emulsifying and peptising properties.

3.2.6.3. Constituents of alkaline cleaners

Caustic soda, NaOH, is one of the most useful of the alkaline cleaning agents. Its solutions have the highest pH value and saponifying properties of any of the alkalis, although its emulsifying properties, so far as solid dirt and mineral oils are concerned, are not very good. Its high solubility enables strongly alkaline solutions to be produced whilst, unlike sodium carbonate, its use will not result in the formation of insoluble compounds. Owing to its causticity it is, however, more difficult to handle than sodium carbonate, whilst it is considerably more expensive than the latter alkali. Caustic soda is available in the fused or flake form. The latter is to be preferred for ease of handling in the making up of cleaning solutions. The commercial material may contain up to 2 per cent of sodium chloride.

Caustic soda appears to allow the re-deposition of oily soils on metals after its cleaning action has been completed as its concentration is reduced, for example, by rinsing.

Another disadvantage of caustic soda is that it is most difficult to rinse off from metal surfaces and residues may form carbonates with atmospheric carbon dioxide. A substantial improvement can be effected by the incorporation of a small amount of sodium hexametaphosphate in the solution. As little as 0.001 per cent will have a marked effect on improving the rinsability characteristics.

Caustic potash, KOH, has been used to some extent as an alkaline cleaner in electroplating practice in preference to caustic soda because of its higher electrical conductivity, despite the fact that it is considerably more expensive. The commercial product contains about 90 per cent KOH.

Sodium carbonate, Na_2CO_3, is the cheapest and most readily available alkali suitable for use as a constituent of alkaline cleaners and is widely used. The anhydrous salt is known in the trade as soda ash. This is the most economical form in which it can be purchased. The decahydrate, $N_2CO_310H_2O$, washing soda, is efflorescent and contains a high proportion of water. Soda ash dissolves, however, rather slowly and for this reason the monohydrate, $Na_2CO_3H_2O$, produced by crystallising the solution at temperatures above 35°C, has found some favour. This material is stable, readily dissolved, and has only a low water content. It is still, however, relatively expensive, and is therefore not used extensively. Sodium carbonate solutions wet some kinds of contaminants more readily than do caustic soda solutions, whilst they also tend to rinse soils away more readily.

Mixtures of caustic soda and sodium carbonate are known as causticised ash, and these are available for detergent purposes whilst a double salt, sodium sesquicarbonate, $Na_2CO_3.NaHCO_3.2H_2O$, is also marketed.

Sodium silicates are also useful detergents. The term includes materials ranging from ' water-glass ', a viscous liquid, to granular sodium metasilicate in various states of hydration. The compositions of the alkali silicates as supplied vary considerably, and they do not always correspond to the requirements of a precise chemical formula. Broadly speaking, so far as metal cleaning requirements are concerned, the important factor is the $Na_2O:SiO_2$ ratio of the material. The product having a content of 9.5 per cent Na_2O and 3.05 per cent SiO_2 is the most economical material to use. Sodium metasilicates having the approximate compositions $Na_2SiO_2.5H_2O$ and $Na_2SiO_3.9H_2O$ as well as others, are commercially available. Silicates are useful in the cleaning of metals which are readily attacked by alkalis, such as aluminium and zinc, as they have an inhibiting effect on the action of the more aggressive alkalis on these metals[32]. Their emulsifying properties are also good and their presence does not unduly depress the emulsifying characteristics of soap-containing solutions. The metasilicate is preferable if a considerable proportion of saponifiable oils is likely to be present.

It is likely that one of the essential functions of alkalis in cleaning is to react with and remove adsorbed layers of fatty acids which cause oily soils to adhere to hard surfaces. Silicates are adsorbed by such surfaces, and this helps to displace films of polar fatty material and prevent their subsequent redeposition.

Trisodium phosphate is widely employed in metal cleaning on account of its free-rinsing properties, and the fact that its action on sensitive metals such as zinc and aluminium is small. It is, however, a relatively mild detergent, and its peptising action is not as good as that of sodium metasilicate. Trisodium phosphate is commercially available as a free-flowing material in the form of needle-shaped crystals consisting substantially of the dodecahydrate, $Na_3PO_4.12H_2O$. It is readily soluble in water, the solubility increasing rapidly with increase in temperature. Thus a saturated solution at 20°C contains approximately 10 per cent by weight of Na_3PO_4, whilst at 100°C it contains 45 per cent. A solution of 0.2 per cent by weight of the crystalline salt has a pH of 11.7 at 20°C. The crystals contain impurities in the form of sodium hydroxide and sodium carbonate, but these constituents do not usually exceed 2.5 per cent. The anhydrous product is also marketed as a white powder.

Phosphates are very useful in formulations containing silicates, the surface tension being appreciably lowered when more than 35 g/l are present. This results in an improved cleaning action. Trisodium phosphate also counteracts the tendency of silicate constituents of alkaline cleaners to produce films on zinc, particularly in anodic cleaning (section 3.2.6.7).

Tetrasodium pyrophosphate, $Na_4P_2O_7$, finds rather more limited application in cleaning. It is less soluble than the tribasic phosphate, but the solution is very stable and hydration to the orthophosphate takes place only very slowly. A 1 per cent solution has a pH of 10.2, equivalent to that of a neutral soap. It is an effective water-softener, and its use in combination with soda ash and metasilicate has been recommended for the cleaning of aluminium. A number of other phosphates have also been developed for use in cleaning mixtures.

Borax, $Na_2B_4O_7.10H_2O$, is sometimes used as a very mild alkali in cleaners. Its solutions undergo hydrolysis with the formation of boric acid, which is a very weak acid, leaving an alkaline solution.

$$Na_2B_4O_7 + 3H_2O \rightarrow 2NaBO_2 + 2H_3BO_3 \quad \text{Concentrated solutions}$$

$$NaBO_2 + 2H_2O \rightarrow NaOH + H_3BO_3 \quad \text{Dilute solutions}$$

Soaps, such as rosin soap or sodium oleate, have been included in cleaners in small amounts to decrease the surface tension of the solutions and promote wetting and peptisation. The soaps suffer from the disadvantage of having relatively low solubility in cold water, and so are liable to leave deposits on articles on rinsing. This solubility is decreased still further if any considerable proportion of caustic soda is present in the cleaner. They are also precipitated by hard water.

Sodium aluminate is sometimes found in proprietary cleaner mixtures since, like sodium silicate, it reduces the attack of alkalis on metals such as zinc and aluminium. It is prone, however, to leave films on the latter metals which are somewhat difficult to rinse away.

Cyanides may also be employed in cleaners because of their ability to remove tarnish films from such polished metals as brass. It is preferable to omit cyanides from cleaner formulations and to use a weak solution of sodium cyanide as a post-cleaning dip. This will have the effect of removing any tarnishing which may occur during the cleaning operation and help to reduce the problems of effluent treatment.

Chelating or complexing agents[33] such as EDTA or gluconic acid have found use in metal cleaning compositions because of their ability to complex or take into solution water-insoluble compounds such as oxides, sulphates and carbonates,

$$\text{e.g.}\quad Na_2H_2Y + MO \rightarrow Na_2MY + H_2O$$

where Na_2H_2Y is the disodium EDTA Salt, MO is metal oxide and Na_2MY the water-soluble EDTA complex, being the ethylene diamine tetra-acetic acid radical. In practice it is found that it is also possible to remove various surface compounds such as burnishing compositions, phosphate and chromate coatings, and welding fluxes. EDTA is produced on an industrial scale and is sufficiently inexpensive to be used as an additive for increasing the efficiency of conventional alkaline cleaners. Such cleaning solutions are normally used hot and are especially beneficial where cathodic cleaning (see Section 3.2.6.7) is being carried out.

Because chelating agents are able to complex water-insoluble compounds, their presence in effluents will make the removal of normally water-insoluble toxic compounds more difficult.

Sodium gluconate is a useful constituent of alkaline cleaners, especially of the strong type, because it forms complexes with many metal ions, thus prolonging the life of the solution[31]. The gluconates confer good rinsability, saponifying and emulsifying properties on the cleaning bath, and also have a softening action on the water used. They also prevent the formation of deposits on tanks and heating elements.

The following is the composition of a typical gluconate bath:

Sodium hydroxide NaOH	10 g/l
Trisodium phosphate $Na_3PO_4.12H_2O$	10 g/l
Sodium metasilicate $Na_2SiO_3.9H_2O$	5 g/l
Sodium carbonate Na_2CO_3	15 g/l
Tensagex DP24 (surfacant—see Wetting agents below)	0.3 g/l

The bath is operated at 50-60°C with air agitation or forced circulation.

Wetting Agents

Wetting agents, or surfactants (surface active agents) as they are more accurately designated are frequently used in alkaline cleaners on account of their valuable properties in promoting emulsification and reducing the surface tension of solutions of alkalis which, as we have seen, is not appreciably lower, and sometimes even higher, than that of water.

Most synthetic wetting agents are not, in general, seriously affected by the presence of heavy metal or alkaline earth compounds, and so are stable even in acid solution. A very large number of wetting agents is now available under different commercial names, but it is to be borne in mind that not all are suitable for use in alkaline cleaners. Some tend to be unstable and are precipitated out, whilst others may undergo decomposition under the prevailing pH conditions. The fact that such a material lowers the interfacial tension between water and oil does not automatically ensure that it will increase the tendency of the aqueous phase to displace oil from a solid surface; the interfacial tension between the solid and liquid phases must also be considered. Thus, distilled water may be effective in removing a mineral oil-umber suspension from steel under controlled conditions, whilst neutral soap solutions would not do so in the absence of alkali. This can be explained as being due to hydrolysis of the soap, fatty acids causing the latter to become attached to the metal. Approximately 0.2 per cent of oleic acid needs to be present in the oily part of the contaminating fluid to increase the adhesion markedly.

A typical wetting agent consists of a long hydrocarbon chain molecule to which is attached a water-attracting group, such as an $—OSO_2ON_2$, $—COONa$, or an $—SO_3Na$ group. A proper balance in the molecule must exist between the water-attracting or polar group and the hydrocarbon group since if the former predominates excessively, the material will lack surface-active properties, and if the latter so predominates, the compound will not be sufficiently water-soluble to be of value. Generally, surfactants have a molecular weight of the order of 300 to 350 and contain a hydrocarbon chain of at least eight carbon atoms. For good detergent properties a fairly long chain containing at least twelve carbon atoms is necessary, although wetting properties develop with considerably shorter chains.

When a surface-active agent of this description exercises its function in promoting the wetting of oil, the molecules of the wetting agent orientate themselves in such a way that the polar groups enter the aqueous portion of the mixture and the hydrocarbon chains attach themselves to the non-aqueous constituent. In the case of solid particles, the hydrocarbon group is adsorbed at the surface of the solid in a similar manner (this phenomenon finds larger scale practical application in the flotation processes for the concentration of mineral ores).

The sodium and potassium salts of the long-chain fatty acids having eight or more carbon atoms all have surface-active properties and constitute the soaps whose detergent properties have been known from very early times. Their solubility in cold water varies. Thus, sodium stearate, having a C_{18} carbon chain, is practically insoluble in cold water. The introduction of a double bond in the molecule, as is the case with sodium oleate, increases the solubility. Solubility increases rapidly as the temperature is raised, however, and the soaps then exhibit excellent detergent qualities. On the other hand, soaps may have poor detergent properties if the molecule is not suitably balanced; thus, the soaps of linoleic acid and ricinoleic acid (both C_{18} unsaturated fatty acids) are good wetting agents but poor detergents.

Surface-active materials can be conveniently divided into three main groups; (a) anion-active, (b) cation-active, and (c) non-ionic.

Anion-active compounds

In the case of the anion-active compounds, the molecule ionises in solution in such a way that the large organic ion carries a negative charge, and hence is discharged at the anode when the solution is electrolysed. Soap is an anionic-active agent which dissociates in water to give the simple metal cation Na^+ and an anion of the general formula $CH_3(CH_2)_2$ COO^-. At an oil/water interface the hydrophobic group COO– will be attracted to the oil phase and the anions will orientate themselves as shown in Fig. 3.13. The net result is the production of an electrical double layer surrounding the oil droplet which causes the oil droplets to repel each other.

This type of surface-active material is simpler to manufacture and less expensive than either the cation-active or non-ionic products.

Perhaps the earliest and best known of the anionic wetting agents is Turkey Red Oil, which is sulphonated castor oil. This product has its limitations, as it is difficult to manufacture a uniform product whilst it is inferior to soap both as regards wetting and detergent properties. The next group to be used was the sulphated fatty alcohols, such as lauryl and cetyl alcohols. The alcohols themselves are produced by the catalytic reduction of the corresponding fatty acids, which can be obtained from coconut oil or ethylenic petroleum hydrocarbons, after which sulphation is carried out. The reactions concerned may be represented as follows:

$$R.OH + HO.SO_2OH \rightarrow R.OSO_2.OH + H_2O.$$

On treatment with caustic soda the sodium salt is formed:

$$R.O.SO_2.OH + NaOH \rightarrow R.O.SO_2.ONa + H_2O.$$

R may be any type of fatty alcohol radical such as lauryl, myristyl, etc.

These alkyl sulphates are stable in neutral or even acid solutions and are essentially similar to soap in chemical structure, the —COONa group of the soap being replaced by the —$O.SO_3Na$ group.

The sulphonated alkyl-aryl compounds are worthy of mention. These have a sodium sulphonate group attached to any alkyl-substituted aryl group. These compounds are characterised by their high surface tension-depressing properties; their detergent value is, however, poor, and they do not foam readily.

Sulphonated amides, of the general structure of $R.CO.NH.C_2H_4.SO_3Na$ (where R is an alkyl group), are also good wetting agents but are not very stable to strong alkalis.

Amongst the most effective of the surface tension-depressants are the esters of the sulphonated di-basic acids such as succinic acid. However, they do not give stable foams and their detergent properties are not high.

The ethyl hexyl phosphates are another efficient group of anionic surfactants. Thus, the sodium salt of di (2-ethylhexyl) phosphate gives good results in aqueous solution containing from 1 to 2 per cent of alkalis, or from 2 to 5 per cent of neutral salts.

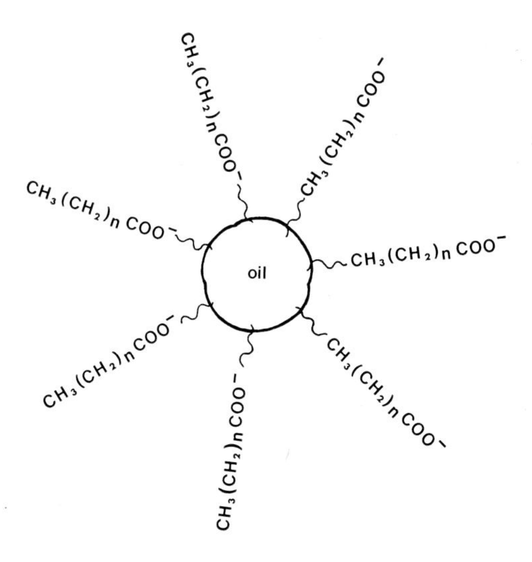

Fig. 3.13. Attraction of an anionic compound to an oil droplet.

Cation-active compounds

In the case of the cation-active compounds, the molecule ionises in such a way that the greater part of it carries a positive charge, the negative ion (which may be a halide, for example) being relatively small. Typical of the cation-active substances are the quaternary ammonium compounds and the primary and secondary amine acid salts. These cation-active materials are especially useful in neutral or in acid solutions where they possess excellent detergent, peptising, and emulsifying properties. One advantage of the cation-active materials is the fact that they will not react with compounds of the heavy metals as do anionic agents. Typical chemicals of this group are:

```
              CH—CH
            //     \\
C16H33—N            CH
       / \         /
     Br   CH=CH
```

Cetyl pyridinium bromide

```
 ┌  CH3      CH3 ┐ +
 │     \    /    │
 │       N       │ Cl-
 │     /    \    │
 └ C16H33    CH3 ┘
```

Trimethyl cetyl ammonium chloride

Non-ionic materials

The non-ionic surface-active compounds are extremely important in that they do not ionise in solution to any substantial extent, and they are therefore very stable in acid, neutral and alkaline solutions. They are also unaffected by hard water. They may, however, be decomposed in the presence of strong oxidising acids. The non-ionic compounds can be used in the presence of materials which would inactivate anionic surfactants. They are capable of emulsifying greasy soils, oils and waxes and hold them in stable suspension, and are therefore particularly useful in detergent formulations.

The solubility characteristics of these substances are a most important consideration in selecting a product for a particular use. The hydrophilic properties of the molecule are contributed by the ether-oxygen groups, the greater the number of ether linkages the greater the water solubility. It is possible to obtain a complete range of combinations, from complete oil solubility to complete water solubility. Generally, where a surfactant is required to emulsify a material, it must be adequately soluble in it.

Applications of surface-active agents in alkaline cleaners

It is important to note that anion-active and cation-active materials should not be mixed, as they may mutually interact with the precipitation of an insoluble compound. Furthermore, in the case of electrolytic cleaning (see Section 3.2.6.7), an anionic compound would prove more effective when anodic cleaning is used than would a cation-active material, and *vice versa*. Another factor that has to be considered is whether a cation-active material, for example, might be precipitated at the cathode due to the neutralisation of the charge on the non-polar ion, especially if the acid is insoluble (only the sodium salt actually used being soluble), as is not infrequently the case.

Surfactants can be conveniently included in appropriate proportion in compounded cleaners supplied in powder form. Alternatively, when cleaners are being made up by the user the wetting agent can be added to the solution before use, and in such cases liquid materials can be conveniently employed. It is found in practice that there is an optimum concentration of surface-active agents beyond which further addition has no appreciable effect in bringing about greater reduction in surface tension.

For most materials 0.1 to 0.3 per cent of wetting agent is the maximum that need be added. Fig. 3.14 shows the relation between surface tension and wetting agent concentration in 4 per cent caustic soda solution at 20°C for five typical wetting agents. Tergitol 7 is an alkyl compound in which the active $—SO_4Na$ group is attached to the middle of the molecule; it is represented by the formula:

$$C_4H_9.CH(C_2H_5)C_2H_4.CH(SO_4Na)C_2H_4CH(C_2H_5)_2.$$

Calsolene oil HS and Lissapol LS are of the sulphated fatty alcohol type; Perminal WA is a sodium naphthalene sulphonate, whilst Alphasol OT is the dioctyl ester of sulphosuccinic acid.

The concentration of wetting agent should be kept as low as possible consistent with good wetting properties; it is often found that increasing the concentration results in the formation of residual films on the metal surface which may interfere with subsequent processes. These films are especially liable to form if wetting agents which are not sufficiently stable in the alkali are used; in such cases decomposition products are formed which may adhere to the metal surface very tenaciously. The rinsability

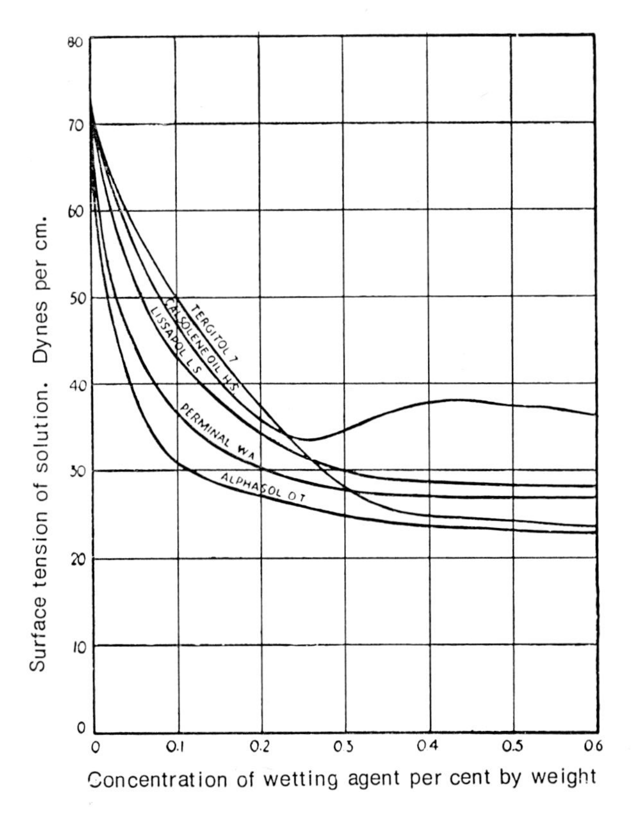

Fig. 3.14. Effect of wetting agents on surface tension of 4% NaOH solution.

of a wetting agent depends both on its solubility and its molecular structure. Clearly, the more strongly surface-active the compound is, the more difficult it will be to rinse off.

Biodegradability of surfactants

An important aspect of the use of surfactants is biodegradability since it has been found that certain types of surfactants are not destroyed in sewage treatment plants. They can then enter rivers and streams, where they produce foaming whilst residues may also remain in potable water, which is clearly undesirable.

Biological resistance is a function of the branching of the structure of the alkyl group of the chain; resistance to degradation increases with increase in the degree of branching as well as with increase in the length of the ethylene oxidising chain and of the main hydrocarbon-derived chain. The chemical nature of the hydrophilic group is relatively unimportant in this respect. Nowadays, therefore, the use of 'soft' or early biodegradable materials, such as those based on ethoxylated straight chain alcohols, is preferred.

3.2.6.4. Balance of cleaner composition

The composition of an alkaline cleaner must be carefully balanced so that the optimum cleaning efficiency is obtained with the least attack on the metal. Some metals (e.g. zinc and light alloys) are very readily attacked by alkalis, so that the pH of solutions used on such metals must be controlled. The formulation of metal cleaners has been studied by Horikawe *et al.*[35].

3.2.6.5. Properties of alkaline cleaners

3.2.6.5.1. Activity

The activity of a cleaner may be measured by its active Na_2O content, this being defined as the amount of Na_2O available at a pH in excess of 8.3, since alkaline solutions have little cleaning action at lower pH values. The total Na_2O content is determined by titration with acid, using methyl orange as an indicator; the active Na_2O is found by titration in the presence of phenolphthalein, which changes colour at the appropriate pH value. The active and total Na_2O contents of some of the commoner alkalis are given in Table 3.7.

TABLE 3.7
Active and Total Na_2O Contents of Some Alkalis

Alkali	Active Na_2O	Total Na_2O
NaOH	75.5	76.0
$NaOH.Na_2SiO_3.5H_2O$ (50:50)	51.5	52.6
$Na_2Sio_3.5H_2O$	28.0	29.2
$Na_3PO_4.12H_2O$	10.0	18.0
Na_2CO_3	29.0	58.0
$Na_4P_2O_7$	8.1	23.3
$Na_2B_4O_710H_2O$	8.4	16.3

3.2.6.5.2. pH value

The pH value of the solution is often accepted as a partial measure of the effectiveness of a cleaner. It is also a useful guide in estimating the relative conductivities of alkalies, as the conductivity of an electrolytic cleaner is an important factor in its efficiency. The pH values of solutions of different concentrations of a number of alkalis are shown in Table 3.8.

The pH of a cleaner should be about 9 to 10 for the more sensitive metals such as zinc, aluminium and tin-lead alloys, increasing to about pH 10 to 11.5 for other metals where only light grease has to be removed. For heavier contamination, pH values in excess of this may be found necessary. The general run of cleaners used in electroplating have pH values of between 11.5 and 13 for copper, brass and

TABLE 3.8
pH Values of Various Concentrations of Representative Alkalis at 20°C

Alkali	Concentration						
	%	%	%	%	%	%	%
	0.01	0.05	0.1	0.15	1	5	10
NaOH	11.6	12.1	12.6	13.0	13.5	14.0	14.0
Na_2CO_3	10.4	10.7	11.0	11.2	11.4	11.5	11.6
$Na_2SiO_3.5H_2O$	10.6	11.2	11.6	12.1	12.5	12.9	13.4
$Na_3PO_4.12H_2O$	10.5	10.9	11.3	11.6	12.0	12.3	12.6
$Na_2CO_3.NaHCO_3.2H_20$	10.2	10.1	10.1	10.0	9.9	9.7	9.4
$Na_2B_4O_7.10H_2O$	9.1	9.1	9.1	9.2	9.2	9.3	9.4

steel, but where substantial soil deposits have to be removed, cleaners having pH values as high as 13 to 14 are often employed.

The pH of a cleaner may be determined either by means of a potentiometric technique or more simply by the use of chemical indicators which undergo colour changes within the pH range concerned. The colour changes of some of the indicators which prove useful in the measurement of the pH of alkaline solutions are given in Table 3.9.

TABLE 3.9
Colour Changes of Alkaline pH Indicators

Indicator		
Thymol blue	Yellow to blue	8.0— 9.6
o-Cresolphthalein	Colourless to purple	8.2— 9.8
Phenolphthalein	Colourless to red	8.3—10.0
Thymolphthalein	Colourless to blue	9.3—10.5
Alizarin yellow G	Yellow to orange	10.1—12.1
Brilliant cresyl blue	Blue to yellow	12.0—12.4
Tropaeolin O	Yellow to orange	11.1—12.7

3.2.6.5.3. Surface tension

An important point in connection with soaps and alkaline solutions is that the surface tension of such solutions decreases as the pH rises. The surface tensions of solutions of sodium oleate at different pH values have been determined by Powney and Addison[36] and are shown in Fig. 3.15. The oil globules in suspension are themselves electrically charged; this facilitates the formation of a stable emulsion since the mutual electrostatic repulsion of the globules prevents the reprecipitation of grease on the parts being cleaned. The maximum charge on the particles is found at a comparatively low concentration of alkali (approximately corresponding to N/1000 NaOH), so that increasing the concentration of the latter results in two opposing

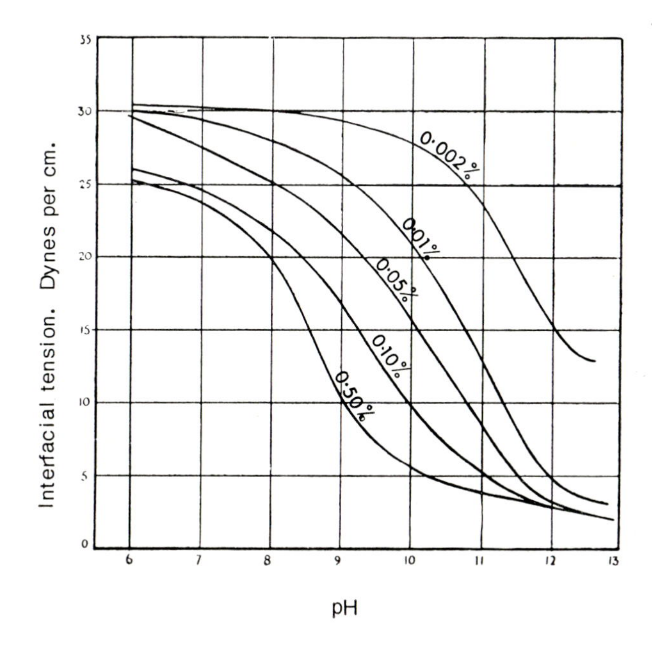

Fig. 3.15. Effect of pH on interfacial tension of various concentrations of sodium oleate against xylene.

effects, viz. reduction of surface tension and a simultaneous decrease in the charge of the emulsified globules. This may account for the good emulsifying properties of some of the milder alkalis and for their effectiveness at low concentrations. Excessive alkali concentrations and high pH may result in breakdown or instability of the emulsified grease due to a reduction in the electrostatic charge on the particles.

In order to promote the peptisation and emulsification of oils and greases, cleaning solutions should have a relatively low surface tension; unfortunately, however, solutions of the common alkalis have relatively high surface tensions (see Table 3.10).

TABLE 3.10
Surface Tensions of some Alkaline Solutions

Alkali (2 per cent solution)	Surface Tension (20°C) (dynes/cm)
Water	72.8
Sodium hydroxide	77.0
Sodium carbonate	77.0
Trisodium phosphate	76.2
Sodium silicate	76.3
Sodium aluminate	75.6
Soap	28.0

It will be seen that soap solutions have a considerably lower surface tension than solutions of alkalis; this is the case even when the soap is present in very small concentration. For this reason soap is a very useful constituent of an alkaline detergent, but even when it is not deliberately added a certain amount of soap gradually forms due to saponification of the vegetable oils and fats.

It is likely that the presence of a soap or a wetting agent not only assists in the wetting of the oil film directly, but also facilitates peptisation of the contaminant and the stabilisation of the resulting emulsion. It was found by Miller[37] that the addition of alkalis to soap solutions decreases their surface tension when measured against benzene; the effect of caustic soda is the most pronounced, trisodium phosphate and sodium silicate having comparatively little influence (see Table 3.11).

TABLE 3.11
Surface Tension of Alkalis in Soap Solution Against Benzene (40°C)

Alkali	Percentage alkali on 0.03 per cent solution								
	0	0.006	0.012	0.024	0.03	0.048	0.06	0.09	0.096
NaOH	26.8	19.1	14.1	7.1	—	—	—	—	—
Na_2CO_3	26.8	23.4	20.0	18.8	—	14.6	—	—	7.9
Na_3PO_4	26.8	—	—	—	20.0	—	16.8	13.7	—
*Sod. silicate	26.8	—	25.8	25.3	—	24.3	—	—	22.9
**Sod. silicate	26.8	—	25.0	23.0	—	20.9	—	—	18.6

* Containing 10.5 per cent Na_2O and 26.70 per cent SiO_2
** Containing 20 per cent Na_2O and 30 per cent SiO_2

3.2.6.6. Plant for alkaline cleaning

The plant used for the simplest types of alkaline cleaning consists of tanks heated by steam, gas or electricity, in which the parts to be cleaned are immersed in the solution either attached to suitable racks or packed in baskets. Arrangements may be made for circulating the solution, and an overflow at one end will help to remove surface grease accumulations. Where high temperature solutions are employed, exhaust ducting running along one or both sides of the tank (depending on its width) is very desirable to remove steam and maintain good working conditions.

Normally alkaline cleaning solutions do not evolve noxious fumes, but where caustic soda is used in any proportion as a constituent of an electrocleaner, the gas liberated at the electrodes results in some caustic spray being carried into the atmosphere, which can be unpleasant. Alkaline cleaners (both of the electrolytic and immersion types) are almost invariably incorporated as essential features of automatic plating plants.

When steam coils are used for heating it is advisable to have them arranged vertically against one wall of the tank. The convection currents result in good circulation of the cleaning liquor which, as has been seen, greatly facilitates the

Fig. 3.16. Conveyorised machine for spray washing/rinsing/drying of components and sub-assemblies. [Courtesy A. N. Marr Ltd.]

cleaning efficiency of the solution. Plate-type heat exchangers are especially effective and convenient.

A variety of conveyorised pressure-spray washing machines are available, which have arrangements for projecting the hot alkali solution from a series of jets on to the parts to be cleaned. The usual types of alkaline solutions are employed with pressures of up to 350 kgf/m^2 at the jets. The temperatures used are 80°C or more, and the solution is fed to the nozzles via a recirculation pump. The parts to be cleaned travel through the plant on a conveyor or may be transported by walking beams or skid rails. The two-compartment types of unit are the most satisfactory, the second chamber using clean hot water only to rinse away alkali. Careful design and location of the jets are necessary to ensure that every part of the article being washed is reached by the solution. Fig. 3.16 shows a machine for washing, rinsing and drying of components and sub-assemblies, the general mechanical arrangement being similar to that illustrated in Fig. 3.17.

Plants have also been produced for the cleaning and pickling (see Chapter 4) of components in the same unit. The highly corrosive nature of the conditions demands the most careful attention in the design of this type of plant.

A wide range of mechanical cleaning plants of all kinds is now in regular use in industry, and provided steps are taken to ensure that they are kept in good working order and that suitable alkaline solutions are employed for the work in hand, excellent results are obtained.

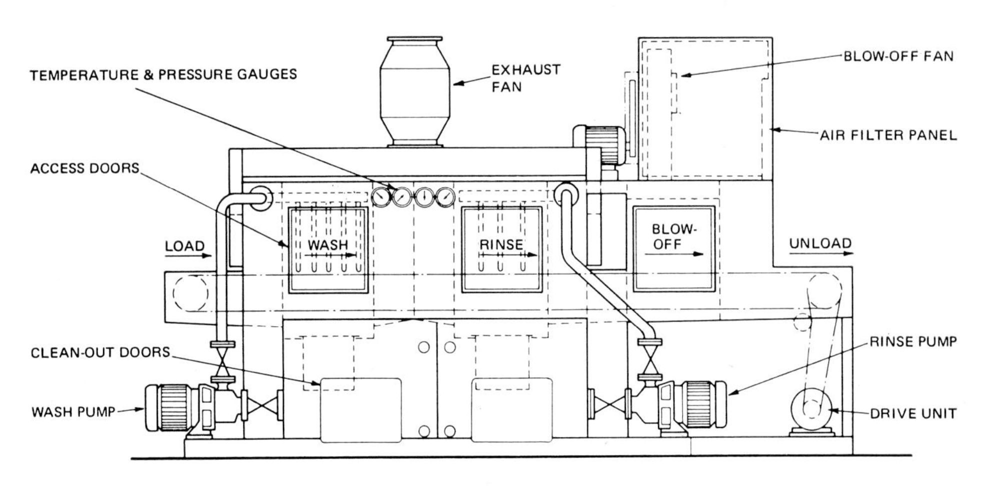

Fig. 3.17 General arrangement of spray washing machine similar to that illustrated in fig. 3.16. [Courtesy A. N. Marr Ltd.]

3.2.6.7. Electro-cleaning

The use of electrolytic alkaline cleaners is common practice in electroplating processes. Fundamentally, electrolytic cleaning consists in suspending the articles to be cleaned in an alkaline solution and making them one electrode in a direct current circuit. An applied potential of 6 to 12 volts may be used, and the articles are made anodic[38] or, more usually, cathodic. The cathodic current densities employed range from 3 to 10 A/dm^2. A considerable volume of hydrogen is liberated by the current at the surface of the work being cleaned, and this gas formation effectively cleans the metal surface by its mechanical scouring action, which expedites the work of the alkaline solution. Some free alkali is formed at the cathode during electrolysis. Some peptising action probably also occurs so that the grease is not re-deposited on the work. Where a steel tank is used this may be made the anode, but it is preferable to use separate anodes in order that the current densities should be more readily controllable. Nickel or stainless steel are suitable metals for anodes as they are neither filmed nor attacked by the solutions used.

Cathodic cleaning is preferred because twice as much gas (hydrogen) is liberated at the cathode as at the anode (where oxygen is evolved), so that a more rapid cleaning action might be expected. In actual practice, however, this difference in the effectiveness of cathodic, as opposed to anodic, alkaline cleaning is by no means as marked as might at first sight appear. Anodic cleaning may encourage the deposition of smut due to electrophoretic effects[39] or due to undesirable etching of some metal surfaces[40]. On the other hand, cathodic cleaning has the disadvantage that should any traces of soft metals become dissolved in the solution, these will be deposited on to the work by the current; there is also the likelihood that hydrogen absorption will take place[41] with the possible embrittlement of steel parts.

If anodic cleaning is to be used, even for a very short time, great care must be taken to ensure that attack of the metal by alkali does not take place.

The plating-out of dissolved metallic impurities on the cathodes is overcome in some plants by the use of current reversal, whereby the current is reversed momentarily just before the cleaned articles are removed from the tank. Thus, any metallic films which may have been deposited are re-dissolved. This procedure is not essential, however, and a great deal of large-scale cleaning on sensitive metals, such as zinc alloys, is carried out without the use of a reversing current. Any films which may be deposited are very thin and can be removed by subsequent operations after cleaning, such as a brief immersion in cyanide solution or in weak hydrochloric or sulphuric acid.

Where sensitive metals are to be cleaned prior to plating in bright nickel solutions, the final electrolytic alkaline cleaning should, in preference, be anodic. Zinc alloys and brass must not be subjected to a final cathodic alkaline treatment if good nickel deposits are to be obtained.

Electrolytic cleaners are commonly used hot, and the tanks can be arranged with a suitable overflow to enable surface grease accumulations to be readily skimmed off. They are also equipped with cathode bars from which the work is suspended. The solutions do not differ substantially in composition from those employed in non-electrolytic cleaning processes. The desirability of a high electrolytic conductivity must, however, be borne in mind in formulating them. Elevated temperatures, relatively high concentrations and solutions of high pH value all favour increased electrical conductivity. Thus, between 20°C and 100°C the conductivities of 3 per cent solutions of most of the common alkalis increase between two and a half and four times.

Caustic potash is sometimes favoured in electrocleaners in preference to caustic soda on account of the higher conductivity of its solutions at all concentrations. The maxium conductivity of a solution of caustic potash occurs at a concentration of approximately 6 g equivalents per litre whilst with caustic soda the maximum is at about 4 g equivalents.

Neutral electrolytic cleaning baths (pH 6 to 8) have found some application. They are based on neutral salts such as sodium sulphate and surface-active agents. They can be used anodically or cathodically and will not attack even the most sensitive metals.

Wetting agents have to be used with great discretion in electrolytic cleaners; some of the difficulties that may arise with them under the influence of electrolysis have already been discussed. They should be of the non-foaming type to avoid the accumulation of liberated foam, which can result in minor explosions when the gas is ignited by sparks when the work is removed from the cathode bars. Addition of insoluble polar materials in emulsion form (e.g. hexyl or octyl alcohol) has been recommended for inclusion in the solutions to reduce foaming, but their use presents some disadvantages as they are relatively volatile and vaporise off when used in hot solutions.

3.2.6.8. Alkaline cleaning of specific items

3.2.6.8.1. Steel

Cold-rolled steel presents a formidable problem in metal cleaning on account of the tenacious nature of the films which may form on the metal surface under some conditions of rolling. Palm oil lubricants which are used in rolling mill practice are decomposed under the influence of heat and pressure with the liberation of free fatty acids which attack the surface. Palmitic acid can form gummy, insoluble deposits of iron palmitate under such conditions. When soluble oils are employed as coolants, they become overheated and may tend to char and oxidise; oxidation is especially prone to take place with such oils owing to the large surface area of the minute globules which is exposed to the air. The decomposition products are then liable to be very strongly adsorbed at the metal surface. The steel surface can act as a catalyst in such oxidation processes. Similar contamination can occur with the lubricants used in deep drawing and pressing operations. The trouble is often enhanced if the metal is bright annealed in an oxygen-free atmosphere.

The alkaline cleaning of stainless steel has recently been dealt with in an A.S.T.M. technical publication[42].

3.2.6.8.2. Automatic strip cleaning

Steel and non-ferrous metal strip may be cleaned at high speed, and special plants are available for this purpose.

In a plant for the cleaning of steel strip prior to annealing, the strip is first subjected to heavy jets of warm alkaline solution to remove the main bulk of the oil present. The oil which is removed is skimmed from the surface of the spray tank from time to time. Next, the strip passes to the electrocleaner and then through water sprays and revolving brushes to remove alkali and smut. It finally enters a hot air-blast drier, delivering air at a temperature of 120 to 170°C. The clean, dry steel is then ready for bright annealing.

The strip passes through the cathodic electrocleaner at speeds of more than 500 m per minute, treatment time being less than three seconds. The current densities used are of the order of 5 to 10 A/dm^2. The anodes are steel grids. The volume of solution drag-out from the cleaning tanks is very high under such conditions and rollers and air jets are employed to keep the losses low. The solution used can consist of either sodium metasilicate or phosphate in concentrations of 10 to 30 grams per litre, although proprietary cleaners are also used. Silicates are also considered especially suitable for this application, with or without fortification by the addition of caustic soda.

3.2.6.8.3. Aluminium strip

The continuous cleaning of cold-rolled aluminium strip is carried out on a large scale to ensure that it is free from oil and grease. This is essential if the aluminium is to be lacquered or lithographically printed, as is often necessary when the material is used for packaging.

In one plant (Fig. 3.18), strip 120 cm wide passes over rollers and is sprayed first with a hot alkaline solution, after which it is rinsed with hot water, and then sprayed with weak (0.5 per cent) nitric acid solution. It is finally dried by hot air. Nylon brushes are provided after the hot alkali to remove any traces of contaminants which have not been dealt with by the alkali solution. The type of solution can be modified to produce an etched or a smooth surface, as required. The acid section is constructed of stainless steel. The stainless steel or plastic spray nozzles are fed by totally immersed pumps in the solution storage tanks beneath the plant. The strip travels at a speed of up to 100 m per minute, whilst the overall length of the plant is 14 metres.

After drying, the strip is directly lacquered or printed by rollers. For some purposes, the plant can be adapted to produce a thin anodised film on the strip after cleaning. This improves the adhesion of the lacquer to the strip.

Fig. 3.18. Plant for continuous cleaning of cold-rolled aluminium strip.
[Courtesy Wean Engineering Co.]

3.2.6.9. Testing degree of cleanliness; cleaner control and evaluation

The evaluation and testing of cleaning solutions and compositions are not easily carried out. Perhaps the simplest method consists in suspending a plate as uniformly coated as possible with the oil or contaminant which it is desired to remove in the detergent solution. The time which elapses before it can be lifted out without showing evidence of a ' water-break ' on the surface is a measure of the effectiveness of the cleaner. When cleaning is complete, the solution will run uniformly from the metal surface without the formation of globules or rivulets. With a little practice, reasonably accurate comparisons of cleaning times can be made.

The ' water-break ' method has, however, been criticised on the grounds that not only is the break dependent on the thickness of the water film, but false water-breakless surfaces can also be obtained as a result of preferential adsorption of hydrophilic material[43]. This adsorbed material can be removed by a dip in a 5 per cent solution of sulphuric acid. Alternatively, an acidified copper sulphate solution can be used for the test piece dip after cleaning, as copper will then be deposited only on the chemically-clean areas. The latter technique enables a quantitative estimate of cleaning properties to be made by measuring the areas covered by the immersion deposit of copper. By determining the time required for cleaning with different solutions, the relative merits of detergents can be evaluated.

The use of oil carrying a fluorescent dyestuff to indicate when cleaning is complete has been described. After immersion in the cleaner for a given time the plates are examined or photographed under ultra-violet light, and since the amount of fluorescence is proportional to the quantity of oil adhering to the metal surface, a rapid and convenient method of judging the efficiency of a metal-cleaning compound is provided.

As a result of tests carried out using the fluorescent method for evaluating cleaners, it was found by Morgan and Lankler[44] that a proprietary wetting agent of the long-chain alkyl-aryl sulphonate type reduced the time required by various alkaline cleaners to remove oily contamination from metal surfaces. Thus it was found that by using a 10 per cent solution of a cleaner consisting of :

Caustic soda	10 per cent
Trisodium phosphate	55 „ „
Soda ash	35 „ „

at a temperature of 60°C, fluorescence was still evident after ten minutes, showing incomplete removal of the oil. In the presence of 5 per cent of the wetting agent, however, a 2 per cent concentration of this cleaner effectively removed the oil in ten minutes.

A useful method of testing has been termed the ' residual soil ' method which can be applied both for the evaluation and routine testing of cleaners. The test pieces are first subjected to the cleaning solution, carefully rinsed and dried and then weighed. The residual contaminants are then washed off by treatment with petroleum ether after which the specimen is rinsed with alcohol, dried and re-weighed.

The amount of contaminant found in this way is a direct measure of the cleaning efficiency of the detergent solution. It is claimed that this method will indicate residual contamination of the order of 0.1 to 0.2 mg for a test panel of 85 cm^2 total area and will give no change in weight for such a test panel if it has been chemically cleaned[45].

In another method[46] steel test panels which have been subjected to a standard contaminating and cleaning cycle are rinsed in running water at 50°C for five minutes to eliminate false water-breaks and then sprayed with clean water by a fine atomiser. Any residual oily areas are shown up as a spray pattern by this method, and the area of this pattern can be estimated. Again, this gives a quantitative measure of the efficiency of the cleaner which has been termed the ' cleaning efficiency index '.

Further work on this method of testing shows that the oil film, which is initially continuous, shrinks and yields many discontinuities in which no oil is evident[47]. Those areas that are still covered with oil, however, have thicker films than originally. Finally, the oil assumes a spherical form with minimum attachment and is then removed. The water spray pattern apparently shows the area from which the oil has been removed either temporarily by shrinkage or permanently by physical removal.

In applying these tests, uniform metal surfaces should be used and care should be taken to rinse off last traces of surface-active agents. Also, loosely adherent oil picked up from the surface of rinse waters should not be allowed to dry onto the metal.

The control of alkaline cleaners in actual use is most difficult, if not impossible, to carry out. pH and active Na_2O determinations are of some value, but solutions must be discarded when they begin to contain any substantial amount of emulsified oil and dirt. There is no ready means of determining this apart from the actual performance of the cleaner, and a fresh solution must be made up when its effectiveness appears to be reduced.

REFERENCES

1. *Product Finishing*, 1972, 25, No. 9, 28.
2. Code of Practise for Cleaning and Preparing Metal Surfaces B.S.I. CP.3012:1972.
3. E. Flaminio, *Trattamenti e Finitura (Milano)*, 1975, 15, No. 9, 31.
4. C. M. Postins, *Met. Finishing J.*, 1969, 15, No. 180, 426.
5. K. E. Creed, *Nucl. Sci. Abs.*, 3004, No. 10, 476. GEPP 123. N.T.I.S.
6. B. Imbert, *Galvano-Organo*, 1974, 43, No. 442, 243.
7. S. Spring, *Plating*, 1965, 52, 1297.
8. H. B. Lindford, *Plating*, 1965, 52, 1262-6.
9. *Prod. Finishing*, 1965, 18, No. 5, 60.
10. J. L. Bish and A. M. Tucker, *Prod. Finishing*, 1967, 20, No. 2.
11. *Ind. Finishing*, 1973, 25, No. 303, 4.
12. Non-Flammable Solvents for Industry, p. 88, I.C.I., Millbank, London.
13. J. A. McIntyre and R. A. McDonald, *Plating*, 1973, 60, No. 6, 633-5.
14. B. Barsy, *Metall*, 1975, 29, No. 6, 636.

15. R. E. Shaw, *Trans. Inst. Met. Finishing,* 1959, **36**, 107.
16. D. P. Gray and W. E. Pegram, *Prod. Finishing,* 1974, **27**, No. 5, 12.
17. G. Kern, *Metal Finishing J.,* 1969, **14**, No. 158, 39.
18. J. W. Mee, *Trans. Inst. Met. Fin.,* 1963, **40**, 242.
19. L. K. Griffin, *Metal Finishing J.,* 1968, **14**, No. 161, 162.
20. K. Tesser, *Galvanotechnik,* 1956, **47**, 553.
21. E. A. Neppiras, *Soviet Physics Accoustics,* 1962, **8**, 1, July.
22. A. E. Crawford, *Prod. Finishing,* 1967, 20, No. 7, 42.
23. R. A. Geckle, *Metal Progress,* 1975, **108**, No. 2, 35.
24. P. Knaggs, *Industrial Finishing,* 1971, **23**, No. 278, 12.
25. A. E. Crawford, Some New Developments in Printed Circuitry, I.M.F. Publication (1967).
26. *Metal Finishing Plant & Processes,* 1974, **10**, No. 5, 147.
27. R. D. Stafford, *Ind. Finishing,* 1973, **25**, No. 303, 10, 12, 43.
28. *Galvano, Teknisk Tids (Oslo),* 1972, **15**, No. 4, 6.
29. *Metal Finishing J.,* 1974, **20**, 232, 106.
30. S. Spring and L. F. Peale, *Met. Progress,* 1947, **51**, 1102.
31. Foster Dee Snell, *Ind. Eng. Chem.,* 1943, **35**, 107.
32. Ch. Rossmann, *Galvanotechnik + Oberflächenschutz,* 1964, **5**, 166-7.
33. J. K. Aiken and C. Garnett, *Electroplating and Metal Finishing,* 1957, **10**, No. 2, 31.
34. D. C. Horner, *Metal Finishing,* 1966, **12**, 99-102, 115.
35. N. R. Horikawa, K. R. Lange and A. B. Middleton, *Proc. Am. Electroplaters' Soc.,* 1964, **51**, 86-91.
36. J. Powney and C. C. Addison, *Trans. Farad. Soc.,* 1938, **34**, 635.
37. E. B. Millard, *Ind. Eng. Chem.,* 1923, **15**, 810.
38. D. W. Bloor, *Electroplating and Metal Finishing,* 1971, **24**, No. 6, 7.
39. P. D. Liddiard, *Metal Industry,* 1944, **65**, 14201.
40. S. S. Frey, 48th Ann. Proc. A.E.S., 128 (1961).
41. C. A. Sapffe and C. L. Faust, *Proc. Am. Electroplaters' Soc.,* 1940, **28**, No. 1, 23.
42. 'Cleaning Stainless Steel' A.S.T.M. Spec. Techn. Publ. 538 Amer. Soc. Testing & Materials, Philadelphia, Pa., U.S.A., 1973, pp. 239.
43. C. Harris, *A.S.T.M. Bull.,* 1945, **136**, 31.
44. D. M. Morgan and J. C. Lankler, *Ind. Eng. Chem.,* 1942, **34**, 1158.
45. A. Mankowich, *Met. Finishing,* 1947, **45**, 77.
46. S. Spring, H. I. Forman and L. F. Peale, *Met. Finishing,* 1946, **44**, 297.
47. S. Spring, *Met. Finishing,* 1950, **50**, No. 2, 65.

BOOKS FOR GENERAL READING

1. 'Industrial Cleaning', Prism Press, Melbourne, 1974.
2. 'An Introduction to Metal Degreasing and Cleaning', A. Pollack and P. Westphal, Robert Draper, Teddington, 1963.

CHAPTER 4

Descaling, pickling and drying processes

THE general term of ' scale ' is applied to the oxide coatings which develop on metals when they are exposed to elevated temperatures. Scale can form during a variety of operations, e.g. in casting, forging, hot rolling and annealing.

The composition and physical structure of the scale depends on the conditions under which it is formed. Thus, such factors as the temperature and duration of heating, the mechanical effects of rolling, forging, etc., the composition of the furnace atmosphere and the duration of the cooling cycle all influence the type of scale which is produced. Before almost any finish can be applied to metals, this scale must be removed. The removal of scale is also carried out between pressing and drawing operations to eliminate the heavy wear of tools and dies which would take place were an attempt made to work scaled metal.

Much investigatory work has been carried out on the structure of scale on metals; this will be referred to later since the structure of the scale has a profound effect on the process of descaling by pickling.

4.1. Blast and centrifugal cleaning

The removal of scale from surfaces by projecting abrasive particles at high speed on them is a convenient method, and is commonly used for a variety of articles. Moreover, the successful use of this method does not depend on the chemical nature of the scale, and it can be applied equally well to scale formed under all conditions; naturally, however, harder or more ingrained types of scale take somewhat longer to remove and may involve the use of different sizes and types of abrasive. The method is also useful for removing burrs (if they are small), directional grinding lines and for producing a matt finish or a keying surface where this is required.

The process is also especially suitable for the removal of siliceous scales produced in welding or sandcasting ;these scales are strongly resistant to acids, and blast-cleaning presents a satisfactory method of dealing with them. In the case of rolling scales, however, care must be exercised to avoid driving particles into the metal surface as these may then act as centres for corrosion when subsequent finishing coatings have been applied. In practice the blast cleaning operation is carried out in two main types of plant: (a) pressure-operated and (b) centrifugal.

4.1.1. PRESSURE PLANT

Here the abrasive is passed from a hopper to a closed chamber into which air is blown at high pressure. The air picks up the abrasive grains and carries them through a hose and thence to a nozzle. The abrasive particles are directed onto the surface to be cleaned by a nozzle which may be fixed or moveable, as shown in Fig. 4.1, and the operation is continued until the scale is removed. This is usually accomplished very quickly.

To enable the plant to function continuously, an intermediate chamber is provided between the hopper and the pressure chamber. While the plant is in operation the valve between the pressure and the intermediate chamber is left closed, whilst that between the latter and the hopper is open; thus, the abrasive is collected in the intermediate chamber. When the supply in the pressure chamber is exhausted the valve connecting the two latter compartments is opened. The air is then admitted to the intermediate chamber, automatically closing the hopper valve. The abrasive passes into the pressure chamber, and after the closure of the valve leading to the latter, blasting can be continued. The abrasive is returned to the hopper and must be sieved from time to time to remove dust derived from both worn abrasive and the surface being cleaned.

For the cleaning of small castings, forgings, etc., a barrel type of unit is satisfactory; in this, the parts are turned slowly in a cylindrical perforated barrel rotating about a horizontal axis, the blast of abrasive being directed onto them. The abrasive falls through the perforations and is then sifted through a dust separator and returned for re-use. Barrel plants are among the cheapest to operate but they are, generally, not suitable for very fragile parts. These are sometimes dealt with by putting small lengths of rubber hose into the barrel to prevent impact of the pieces on one another.

Another type of plant makes use of rotating tables onto which the parts to be cleaned are placed within a cabinet. This method is especially suitable for flat parts; for longer components, reciprocating tables are sometimes used.

In the gravity method of induction air pressure blasting, the abrasive is stored overhead. An induction type of gun is used into the nozzle of which air is introduced through a jet. This creates a partial vacuum in the gap between the jet and the nozzle so that the abrasive is sucked into it. The air then expands at the nozzle tip and is directed towards the article to be treated. The effect is less powerful than direct pressure blasting, but the plant is somewhat cheaper initially and maintenance costs are rather lower.

Fig. 4.1. High pressure grit blasting.
[Courtesy Tilghman Wheelabrator Ltd.]

4.1.1.1. Nozzles

In abrasive blasting, the efficiency of the nozzle is the key factor affecting cleaning costs and speed of treatment. Nozzle efficiency depends on the distance of the nozzle from the work and its size. As wear occurs, the nozzle becomes less efficient. The nozzle orifice should be kept as small as possible in relation to the work, as compressed air is expensive. The length of the nozzle is important—a longer nozzle produces a more powerful jet, whereas a short nozzle covers a larger area but shows a slightly lower abrasive velocity.

Discharge nozzles and air jets should be used in the ratio of approximately 2 to 1, i.e. the discharge nozzle should have about twice the internal diameter of the air jet. Nozzles are subject to wear, the abrasive causing them to wear internally and to become cone-shaped from the inlet along their entire length. This causes the abrasive stream to spread sometimes as much as four times the extent of a new nozzle. Suction or gravity feed equipment does not produce wear in the internal diameter of the nozzle. Since air only passes through the air jet, wear occurs only at the discharge nozzle through which the combined air and abrasive are discharged.

Nozzles are generally from 6 mm to 15 mm in diameter and are made of either cast iron, alloy steel, or special wear-resisting materials such as sintered tungsten carbide or boron carbide. The rate of wear of iron or steel nozzles is very high indeed, and as the bore increases, the air pressure must be correspondingly raised if the same velocity is to be imparted to the abrasive. For this reason, it is worth while using carbide nozzles, despite their higher initial cost, since increased efficiency is attainable with them. The life of carbide nozzles may be of the order of 1,500 hours, which is many times longer than that of the much cheaper cast iron nozzles.

4.1.1.2. Air supply

The air consumption of a blasting plant of this kind is considerable, the pressure used varying from 1.5 to 7.5 kg/cm². Higher pressures give better rates of

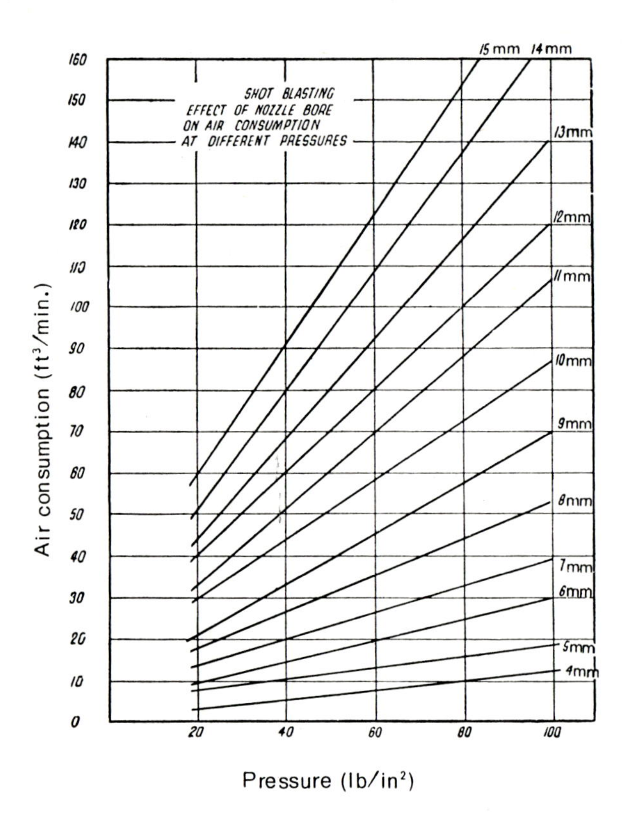

Fig. 4.2. Effect of nozzle bore on air consumption.

cleaning, but there is a corresponding increase in the rate of wear and tear of the plant, abrasive and nozzles. Fig. 4.2 shows the air requirements for different diameters of nozzles under varying conditions of air pressure.

An inevitable consequence of compressing air for blasting is the condensation of moisture which can result in severe corrosion problems[1]. Although condensation can be controlled by the use of filters and traps these do not remove water vapour from the air stream. If this moisture is not removed by means of a drying unit (utilising a drying agent such as silica gel), clogging of the grit feed tube may result. Should this occur the plant must be shut down and the feed tube and orifices cleared.

4.1.2. CENTRIFUGAL BLAST-CLEANING PLANT

In this type of plant the need for compressed air is eliminated. The abrasive is allowed to fall on to a rotating wheel or series of impeller blades from which it is projected centrifugally at high speed on to the parts being cleaned. By careful design it is possible to control the direction of the abrasive grit very accurately, and better coverage can be obtained than by the use of nozzles and compressed air. The elimination of the air system simplifies the plant considerably, and it is claimed that the power consumption is also lower owing to the fact that much of the energy expended on the compression of the air in the pressure type of plant is lost and is not conveyed to the abrasive.

Centrifugal abrasive cleaning plants can be readily conveyorised, the parts being passed through the chamber continuously on a moving belt, as shown in Fig. 4.3. To minimise wear of the belt and the plant it is essential to keep the conveyor fully loaded with work, otherwise maintenance costs may become unduly high. The chief source of trouble in centrifugal cleaners is wear of the impeller blades; under unfavourable conditions these may require frequent replacement. A hard alloy should be used for these blades; it has been found that high-silicon iron withstands the abrading action particularly well. Generally speaking, this type of equipment presents many advantages, and the operating costs are often lower than when compressed air is employed, especially when smaller articles are being treated.

4.1.3. VACUUM BLASTING PROCESSES

The shot blasting of large articles which cannot conveniently be put into a cabinet can be carried out by the Vacu-Blast method[2], which can be operated anywhere in the factory or on site. Essentially, the cleaning gun itself acts as a cabinet, the blast nozzle being surrounded by a vacuum cone which picks up the abrasive as soon as it has done its work, and returns it to a reclaimer. Here it is separated from the debris and returned for re-use. The method can be used for removing mill scale and rust, for cleaning metals before welding, and for the removal of weld scale and slag after welding. It is also employed for the cleaning of the interiors of narrow tubes.

The abrasive action of the particles roughens the surface, giving rise to numerous indentations. If the surface is too smooth, subsequently-applied organic coatings may not adhere satisfactorily. If on the other hand the surface is coarsened too much, bridging may occur[4] which is the failure of the coating to cover between peaks or prevent high peaks protruding through the coating. In order to obtain the desired anchor pattern or surface roughening, it is important to use accurately size-graded abrasives. These are intially more expensive but over a period will prove cheaper in use.

Three standards of cleanliness for blasted surfaces are given by British Standard Specification 4232-1968:

(1) First Quality—completely free of contamination or discoloration.
(2) Second Quality—clean. At least 95% completely clean, and in any sq in (6.25 sq cm) 90% shall be clean.
(3) Third Quality—clean. At least 80% completely clean, and in any single sq in (6.25 sq cm) 60% shall be clean.

For the descaling of stainless steel by blast methods, stainless steel shot or alumina abrasives are required. Iron-containing materials becoming embedded in the steel surface markedly reduce the corrosion-resistance of stainless steels. The shot must also be kept free from fine particles or fragmentary material of foreign origin if the best results are to be obtained.

4.1.5. PLANT MAINTENANCE

Blasting equipment needs careful maintenance if the best results are to be obtained. Thus, small leaks in abrasive hose connections result in loss of power and efficiency. Similar considerations apply to worn nozzles, which should be replaced in good time. Static charges sometimes build up on rubber hose, which are liable to give the operator a shock and may result in perforation of the hosing. The use of conducting rubber (loaded with graphite) eliminates this source of trouble. As mentioned earlier, water in the compressed air is undesirable and steps should be taken to eliminate it either by a good after-cooler on the compressor, by means of a suitable water trap, or by the use of a drying plant containing a drying agent. A most efficient centrifugal automatic air dehumidifier is available which can be directly installed in the pipeline.

The centrifugal type of plant demands careful attention to the balance of the impeller blades. If one blade wears prematurely, the whole set-up is thrown out of balance, so that severe vibration and damage to the machine and bearings may result. Special abrasive cleaning equipment is a desirable adjunct to every type of plant, as an abrasive containing a high proportion of dust will give poor results. Proper ventilation is also needed to prevent abrasive accumulating within the machine, making for inferior results and poor working conditions. Good maintenance also eliminates delays due to blockages in the pipe system.

4.1.6. VAPOUR BLASTING

In vapour blasting, an abrasive is suspended in a liquid which is delivered to the blasting nozzle by means of a circulating pump. A suction feed method is generally used in which the abrasive is fed to the gun at low pressure. The abrasive, which is non-metallic, is suspended in a liquid (usually water plus a corrosion inhibitor), and ranges in size from 60 mesh to as fine as 5000 mesh. The plant for vapour blasting consists of a water-tight cabinet where work is cleaned, a hopper tank for mixing, storing and collecting the suspension, and a gun equipped with two lengths of flexible hose, one for the abrasive feed and the other for the air supply. After blasting, parts are usually dipped into a rinse tank containing water and a corrosion inhibitor. When rapid drying is desired, the solution should be kept at about 60°C; compressed air can be used to blow-off excess water. The process is a flexible one as it is possible to vary the particle size and hardness of the abrasive, the air pressure, the distance from gun to work, and the liquid-to-abrasive ratio.

The abrasives are relatively inexpensive and have a useful life of 25 to 40 operating hours, which makes the process an economical one. Nozzles also last well, since the abrasive has little tendency to erode metal when suspended in liquid.

Vapour blasting is primarily used for fine finishing, and with suitable abrasives it is possible to hold tolerances of 0.0001 cm, where required. The process offers a simple method of de-burring, both externally and on the interiors of small parts. In particular, it is widely used for the cleaning of moulds and dies.

A typical application of wet blasting is in the production of cover plates for sewing machines[5]. These steel plates are processed in an automatic blasting machine utilising fine silicon carbide abrasive applied by two centrifugal wheels. The plant is capable of producing 500 plates per hour, the surface being suitable for a final satin finish with nickel and chromium plating.

Articles which have to be hard chromium plated are also improved by liquid blast cleaning the surface of the part to be plated before the chromium is applied.

The type of abrasive and the pressure at which it is used affect the finish of a part being cleaned. For ferrous metals, the pressure ranges from 4 to 7 kg/cm^2, depending on the degree of cleaning required and condition of the work; for non-ferrous metals the pressure is from 0.7 to 4 kg/cm^2. For producing a satin finish, the air pressure is usually lower, ranging from 0.2 to 1.5 kg/cm^2.

4.2. Pickling

The term 'pickling' is applied to the removal from metals of scale and oxide coatings, usually resulting from heat-treatment, forging or casting operations. The thickness, composition and structure of the scale depend on the conditions under which it is formed. The temperature and duration of heating and the nature of the atmosphere (whether oxidising or reducing) have a profound influence on the scale and hence on the ease or otherwise with which it can be removed. Pickling is generally carried out by immersion in acid or, less frequently, by spraying with acid for varying periods of time, depending on the metal and type of scale to be removed, sulphuric and hydrochloric acids being the principal acids used for ferrous metals. The former is cheaper and can be used hot, whilst the latter must be used at relatively low temperatures because of its volatility. Other acids, such as nitric, hydrofluoric, and chromic acids or combinations of these and other acids, are employed for the pickling of non-ferrous metals and special alloys; sometimes neutral salts are added to these. Superimposed current has sometimes been used to accelerate pickling, the work generally being anodic. The very obstinate oxides, present particularly on alloys resistant to high temperatures, can be removed successfully in certain molten salt mixtures; this is an expensive method, however.

Commercial sulphuric acid is purchased at 95 per cent strength and must be diluted for use, the concentrations generally used for pickling being between 5 and 25 per cent. As it is practically non-volatile, it can be used hot; in practice, the temperature is raised gradually as the amount of dissolved metal in the bath increases to counteract the reduction in the speed of descaling which takes place with increasing concentration of dissolved metal.

Hydrochloric acid is marketed in the form of an aqueous solution of about 30 per cent strength. As the chlorides produced by pickling in this acid are more soluble than the corresponding sulphates, hydrochloric acid does not have to be discarded so frequently; in addition, its action is more rapid than that of sulphuric acid. The lower temperature at which it is used also means that heating costs are less, so that its use is tending to increase.

The hydrogen liberated during pickling generates a fine acid spray which is unpleasant, so that the use of exhaust equipment is recommended.

For a brief review of acid dips used in plating sequences the reader is referred to the recent article by Mohler[6].

4.2.1. ACID PICKLING OF FERROUS METALS

In the case of steel and ferrous metals generally (where pickling finds its principal large scale application), the superficial oxide layer is, to quite a considerable extent, insoluble in the acids commonly employed for this purpose. The scale is actually removed largely by spalling off as a result of the attack by the acid on an intermediate lower oxide layer lying between the basis metal and the scale itself. This helps to make the process an economical one, since only a relatively limited proportion of the acid is employed in dissolving the iron, with the result that the pickle lasts much longer than it would otherwise do. All steel sheet must be pickled before tinning or galvanising, and vast tonnages of steel are pickled annually in the course of these processes alone.

4.2.1.1. Composition of scale

The composition of the scale formed on steel under conditions of oxidation in air at high temperatures has been very thoroughly investigated and the subject reviewed by Pfeil and Winterbottom[7], and more recently by von Fraunhofer and Pickup[8].

It is generally considered that the oxide formed when iron is heated is made up of three distinct layers (Fig. 4.5):

(a) The outermost layer, which is relatively thin and contains the highest proportion of oxygen; this consists of ferric oxide, Fe_2O_3.

(b) An intermediate layer, which is rather thicker and consists of magnetic oxide of iron, Fe_3O_4.

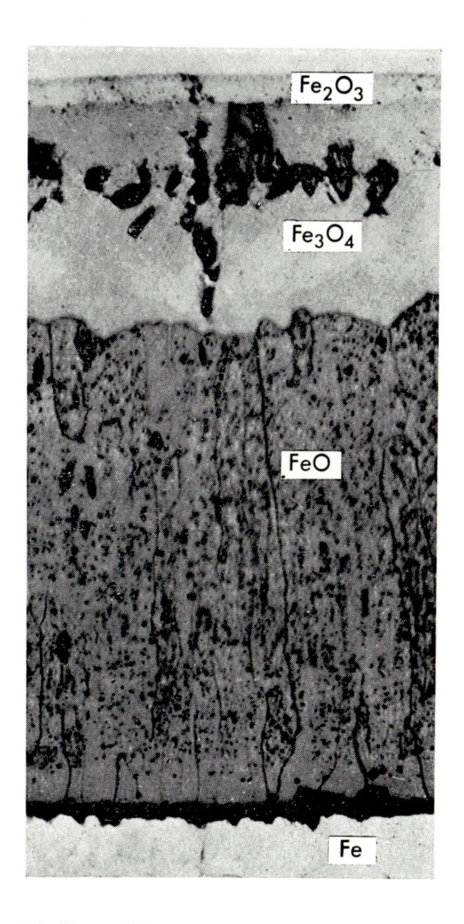

Fig. 4.5. Iron oxidised in air at 700°C for 24 h. ×500[7]

(c) A relatively thick layer in proximity to the iron, containing the highest proportion of the latter compound and having a total composition approximating to the formula FeO. This layer, termed wüstite, decomposes below 570°C into a eutectoid of iron and magnetite, Fe_3O_4. In the case of scale formed below this transition temperature the innermost layer of FeO is usually absent and the inner layer consists of Fe_2O_3 or Fe_3O_4, or a mixture of both.

The precise composition and the structure of the scale will be modified in practice by the conditions under which heating is carried out[8, 9]. Fig. 4.6 indicates the rate of scale formation at various temperatures in an air current flowing over a heated iron surface at the rate of 1 cm/second. However, the rate of oxidation is affected markedly by surface preparation and pre-treatment.

In the case of hot rolled steel, the actual rolling temperature plays an important part in the amount and composition of the scale formed. Thus, higher temperatures result in heavier scales, which, moreover, contain a higher proportion of ferrous oxide.

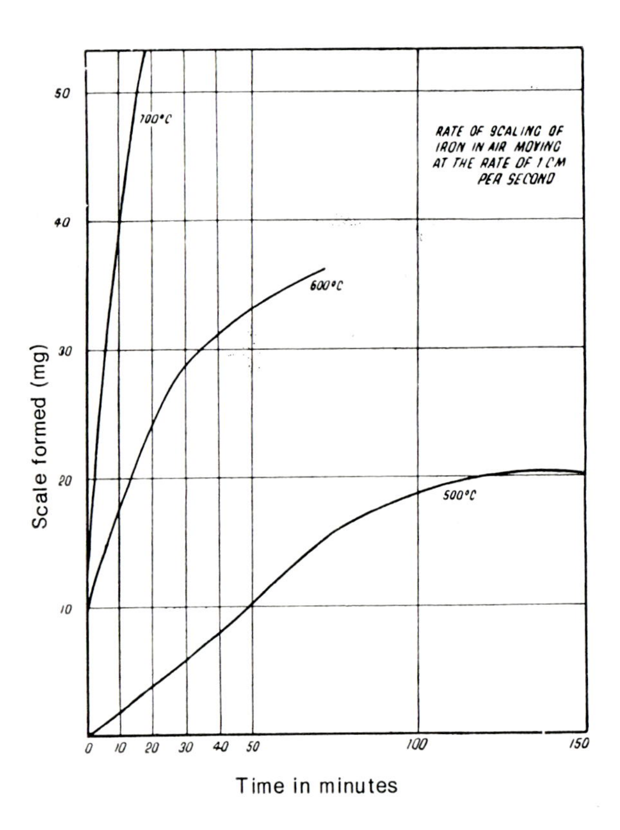

Fig. 4.6. Rate of scaling of iron in air moving at 1 cm/sec.

4.2.1.2. Preliminary degreasing

The pickling acids do not readily remove grease, and the presence of even a thin film of oil anywhere on the surface of the metal will very considerably reduce the rate of pickling over the area concerned. In sheet mills, where pickling is carried out soon after heat treatment, no preliminary degreasing is generally necessary, but in engineering works, where there is a strong likelihood of oil contamination of the metal surface, it is essential to degrease the metal before pickling. This can be carried out satisfactorily in hot alkaline cleaners (see Section 4.2.3) (with or without the application of current (Section 4.2.4)) or, by degreasing in perchloroethylene or trichloroethylene vapour, although the latter process is much more costly. The use of wetting agents in the pickling acid can help in some instances in the pickling of slightly greasy work.

4.2.1.3. Pickling reactions

With sulphuric acid, the reactions which take place during the pickling of iron and steel are largely confined to the two lower oxides and the metallic iron, viz:

$$Fe_2O_3 + 3H_2SO_4 = Fe_2(SO_4)_3 + 3H_2O$$
$$FeO + H_2SO_4 = FeSO_4 + H_2O$$
$$Fe + H_2SO_4 = FeSO_4 + H_2$$

Reduction of any ferric sulphate that forms occurs as a result of the presence of nascent hydrogen:

$$Fe_2(SO_4)_3 + 2H = 2FeSO_4 + H_2SO_4$$

The corresponding reactions with hydrochloric acid are:

$$Fe_2O_3 + 6HCl = 2FeCl_3 + 3H_2O$$
$$FeO + 2HCl = FeCl_2 + H_2O$$
$$Fe + 2HCl = FeCl_2 + H_2$$

Reduction of ferric chloride by nascent hydrogen:

$$FeCl_3 + H = FeCl_2 + HCl$$

The dissolution of magnetite, Fe_3O_4, in either acid is slow, this oxide being soluble in acids only with great difficulty.

The solubilities of iron, FeO and Fe_2O_3 in these acids have been determined by Bablik (Table 4.1)[10]. It will be seen that in the case of the lower concentrations of hydrochloric acid a substantial proportion of the total attack takes place on the iron base; with acid of 10 per cent concentration, however, considerable dissolution of the oxides occurs. In highly concentrated hydrochloric acid the pickling reaction is more in the nature of a chemical dissolution of the scale than mechanical removal.

TABLE 4.1

Solubilities of iron and its oxides in pickling acids

Temp. °C	Acid concentration per cent		Dissolved from 100 (g) sample in 1 hour Fe	FeO	Fe_2O_3
20°	H_2SO_4 :	1	6.0	3.9	0.14
20°		5	15.0	4.8	0.56
20°		10	35.0	6.4	0.98
40°		10	97.7	9.0	1.4
20°	HCl :	1	20.8	0.112	0.48
20°		3	31.6	0.36	0.76
20°		5	40.7	0.71	0.83
20°		7	50.1	1.6	1.8
20°		10	72.0	10.6	7.5

With sulphuric acid at lower concentrations the weight of oxide dissolved as compared with that of metallic iron is considerably greater than is the case with hydrochloric acid at corresponding concentrations. The dissolution of the oxide in sulphuric acid at all concentrations plays a very important part in the pickling reactions. In the case of sulphuric acid at all concentrations the ratio of metallic iron dissolved to that of the oxides is much greater at elevated temperatures than at room temperature. The ferrous oxide (FeO) layer in contact with the steel is the most soluble of the oxides present in scale.

During pickling of high temperature scale[11] the acid penetrates to the FeO layer through pores and fissures in the scale. Since the FeO layer also contains dispersed particles of Fe and Fe_3O_4 formed by partial decomposition on cooling, microscopic electrochemical cells (i.e. cells of Fe/acid/Fe_3O_4) are set up and result in the outer part of the scale peeling off as shown in Fig. 4.7. With scale formed at temperatures of less than 570°C the inner layer of FeO with its associated decomposition products is often absent so that scale removal occurs by acid penetrating the scale and attacking the Fe_3O_4 and the basis metal (Fig. 4.8). This is a much slower process than the removal of oxide scales formed above 570°C.

In sulphuric acid, the proportion of the scale which is removed by being split off, as opposed to that removed by being chemically dissolved, is greater than with hydrochloric acid. This is a factor which makes for greater economy in acid consumption when the former acid is used.

Raising the temperature of sulphuric acid is more effective in increasing the rate of dissolution of metal and oxide than is the case with hydrochloric acid. Metallic iron is also much more soluble in the latter acid than is the scale, and this difference is greater than with sulphuric acid.

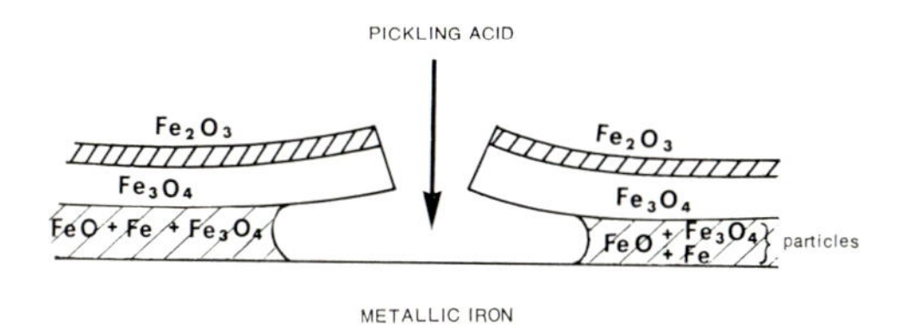

Fig. 4.7. Removal of scale formed on steel at high temperatures (>570°C)[11].

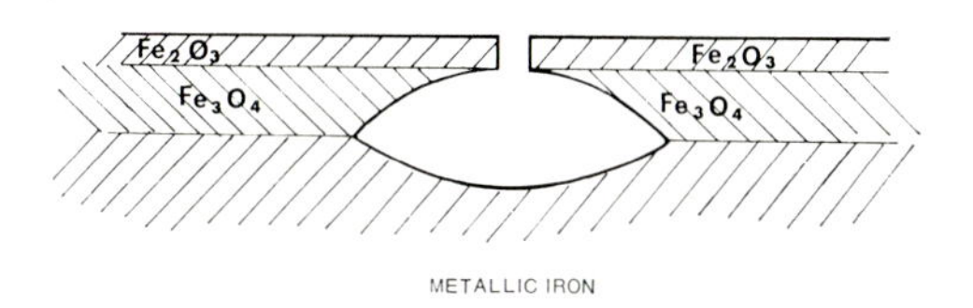

Fig. 4.8. Removal of scale formed on steel at low temperatures (<570°C)[11].

4.2.1.4. Sulphuric acid

The action of concentrated (95%) sulphuric acid on iron and steel is very small. Starting from a dilute acid solution it is found that, as the concentration of the acid increases, the pickling time decreases until it reaches an optimum at about 25 per cent strength; the pickling rate then falls rapidly. On the other hand, the effect of temperature is very marked; it is seen from Table 4.2 that increasing the temperature of 5 per cent sulphuric acid from 18°C to 60°C reduces the pickling time to about one-tenth. For this reason, sulphuric acid pickling should be carried out in the hot acid for maximum efficiency.

As iron sulphate builds up in the solution, however, the rate of pickling is reduced. Fig. 4.9 shows the speed of pickling of rolling scale in the presence of ferrous sulphate[12]. The scale contained 47.3 per cent FeO and 51 per cent Fe_2O_3. The

TABLE 4.2

Effect of temperature on pickling time

Acid	18°C	40°C	60°C
5 per cent H_2SO_4	135	45	13
10 per cent H_2SO_4	120	32	8
5 per cent HCl	55	15	5
10 per cent HCl	18	6	2

restraining effect of ferrous iron salts in pickling solutions is also less at elevated than at low temperatures due to the greater solubility of the salts at higher temperatures. At low concentrations of sulphuric acid the effect is less marked than with higher concentrations. Ferric salts accelerate the rate of pickling, but this is only of academic interest since the trivalent iron content of a pickling solution is normally negligible.

4.2.1.5. Hydrochloric acid

Commercial hydrochloric acid consists of a solution in water of 30 to 35 per cent of hydrochloric acid gas. The volatility of the acid makes its use impracticable at elevated temperatures; above 40°C the rate of loss of acid is very rapid indeed. In actual practice the pickling speed is increased by using higher concentrations of acid rather than elevated temperatures. In this respect hydrochloric acid differs from sulphuric acid, concentration being of little influence and the chief method of increasing pickling rate being by elevating the temperature (cf. Figs. 4.10 and 4.11). Hydrochloric acid tends to give a smoother and whiter pickled surface than does sulphuric acid, and it is therefore favoured in some operations despite its higher cost and slower action.

The addition of hydrochloric acid to sulphuric acid is sometimes practised.

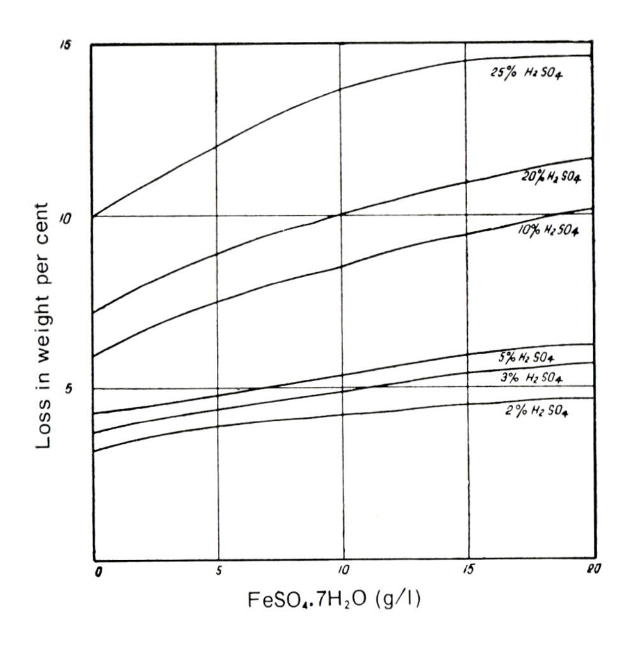

Fig. 4.9. Effect of ferrous sulphate on rate of solution of rolling scale in sulphuric acid.

Pickling is accelerated by such additions and the higher solubility of ferrous chloride enables the solution to be used rather longer before it is discarded.

4.2.1.6. Hydrofluoric acid

Although the removal of scale from ferrous metals is generally carried out by pickling in simple solutions of sulphuric or hydrochloric acid, where castings are concerned, the presence of hydrofluoric acid is often necessary to dissolve away the siliceous material present on the surface. Welding scale is also apt to contain fused silica or iron silicates, due to the electrode coatings. Hydrofluoric-sulphuric acid mixtures at temperatures of 70 to 75 °C are useful in such cases. The removal of the

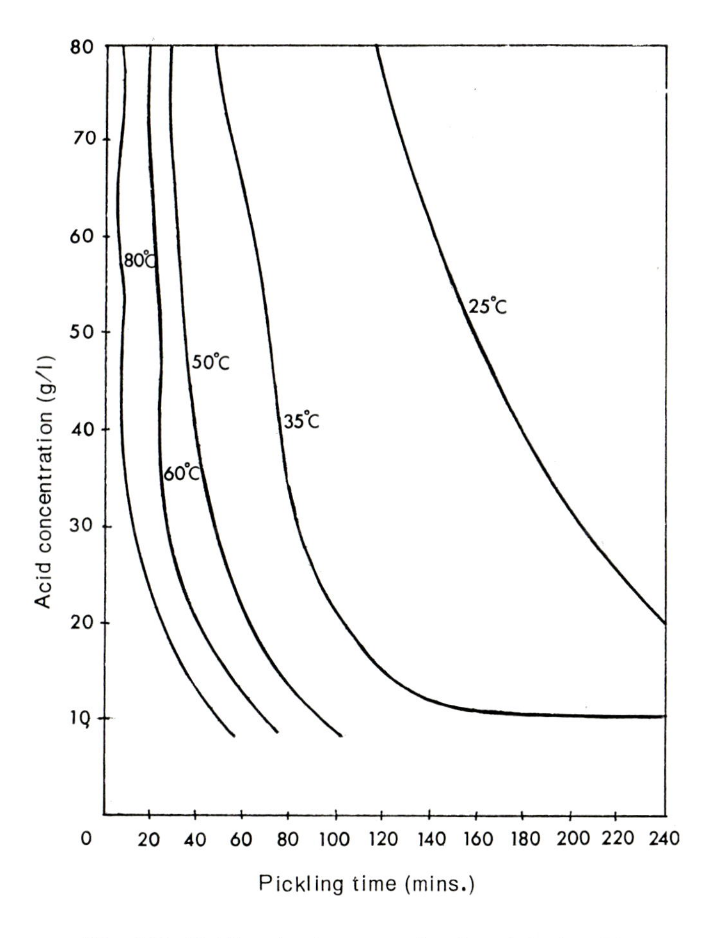

Fig. 4.10. Pickling time/concentration for sulphuric acid.

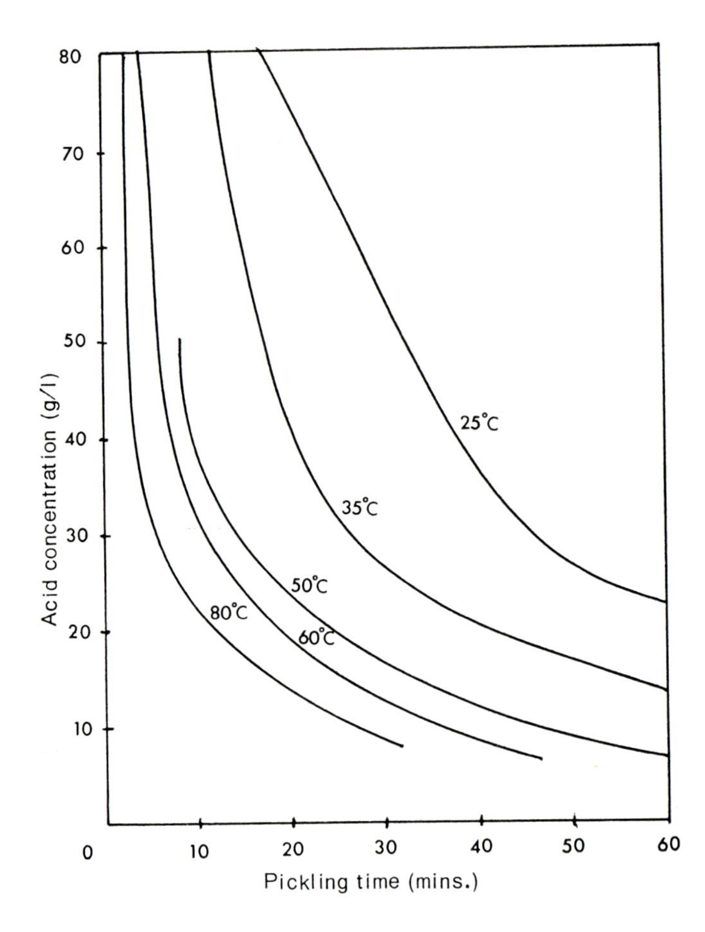

Fig. 4.11. Pickling time/concentration for hydrochloric acid.

scale occurs as a result of: (a) the chemical dissolution of the scale coating, and (b) ' splitting off ' of the scale, as has already been stated.

4.2.1.7. Sludge formation

After pickling, a layer of adherent black sludge remains on the metal surface; much of this sludge also accumulates on the bottom of the tank. In sulphuric acid pickling, the deposit contains up to 0.6 per cent of iron dissolved in the acid; with hydrochloric acid the amount of sludge produced is substantially less. This sludge also contains a high proportion of the nobler metals present in traces in the steel being pickled, and these are precipitated onto the steel surface. Hence, thorough washing of pickled metal to remove these deposits is important.

4.2.1.8. Inhibitors

In the pickling of ferrous metals, the scale is very often partially lifted in some areas before the general surface has been descaled. Various materials have, therefore, been developed for addition to pickling baths to reduce attack on the basis metal without appreciably affecting the rate of descaling. These inhibitors also reduce hydrogen evolution and hence the likelihood of hydrogen embrittlement and the formation of pickling blisters on sheet metals due to the pressure of occluded hydrogen below the metal surface (see Section 4.2.5). They also reduce roughening and pitting due to excessive metal dissolution. The excess acid consumption and loss of weight of metal can be considerable in the absence of inhibitors. Thus, it was found that in pickling 100 m^2 of steel sheet (both sides) for fifteen minutes in a bath containing 3.5 per cent of sulphuric acid at 95°C without an inhibitor, 3 kg of scale were removed accompanied by a loss of 33 kg of metal. In other words over 90 per cent of the acid was used in dissolving excess iron over and above that required for the descaling process proper.

An inhibitor should be completely stable in the pickling acid even at elevated temperatures, be easily and completely soluble in the acid, and retain its efficiency on prolonged storage. In other words, the first essential in an inhibitor, apart from its efficacy in this direction, is that it should be a highly stable compound. It must also not be prone to form films on the metal surface which will be difficult to remove and so interfere with subsequent processes such as galvanising, enamelling or electroplating. It should, of course, effectively reduce attack by the acid on the bare metal surface, and minimise hydrogen evolution.

Inhibitors are only added in very small quantity, e.g. of the order of 0.05 to 0.10 per cent of the pickling acid, and are sometimes mixed with diluents to aid in stabilising them or to increase their solubility. In any case, there is a definite optimum concentration above which their effectiveness does not increase.

The subject of inhibitors for hydrochloric and sulphuric acid pickling solutions has been reviewed[13, 14] and the theory of corrosion inhibition in pickling solutions discussed by Wilmotte and Benezech[15]. A large number of such inhibitors are now known; from the chemical point of view nearly all of them fall into one of the following four classes[16]:

(1) reducing agents (aldehydes)
(2) nitrogen-containing organic compounds
(3) sulphur-containing organic compounds
(4) nitrogen- and sulphur-containing organic compounds

The action of pickling inhibitors appears also to depend on the presence of certain chemical groupings; typical of compounds in general use are heterocyclic nitrogen compounds such as pyridine and quinoline, which are especially effective. Other substances, such as substituted ureas and thioureas, mercaptans, aldehydes, ketones and organic acids find application. The properties of the inhibitor are determined by the length and characteristics of the hydrocarbon and polar portions of the molecules and their stereo-chemical arrangement.

Pickling inhibitors are available as foaming and non-foaming types. The purpose of the former is to produce a blanket on the surface of the solution which prevents the liberation of unpleasant acid spray. Unfortunately, excessive foam formation may cause the inhibitor to concentrate in the surface layer, which is undesirable.

The relative efficiencies of some compounds as inhibitors are shown graphically in Fig 4.12.

The mechanism by which inhibitors act has been the subject of much speculation and the reader is referred to the text by West[17] for a relevant account.

4.2.1.9. Wetting agents

When inhibitors are added to pickling acids the pickling time is usually increased [19] since the dissolution rate of the oxide as well as of the metal is decreased (although to a much lesser extent). This undesirable effect of pickling inhibition may often be

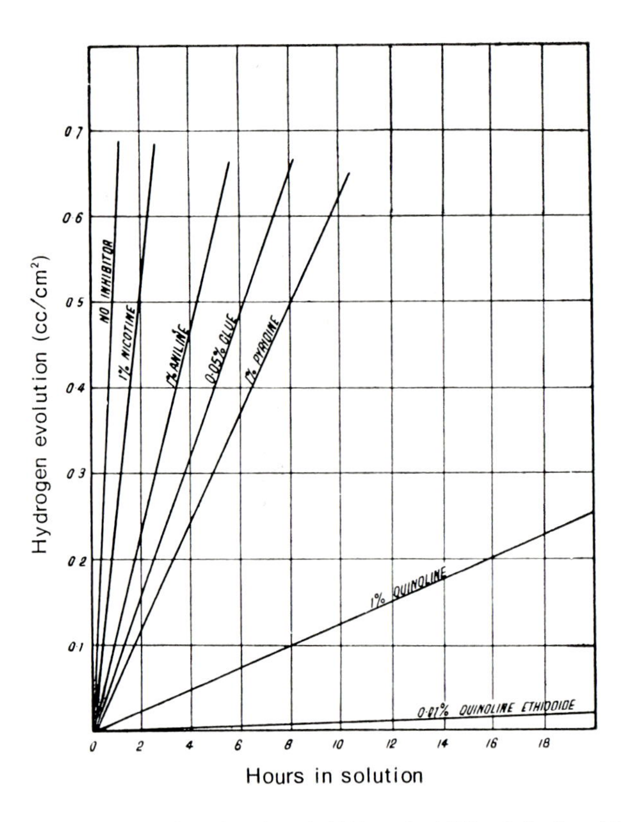

Fig. 4.12. Relative efficiencies of various inhibitors in 34% sulphuric acid at 250°C.

prevented by the addition of a suitable wetting agent. In some cases wetting agents which also have inhibiting properties are used alone. In general, wetting agents have a number of functions in a pickling process[16]:

(1) To accelerate the pickling process (pickling activators and accelerators).
(2) To support the action of the inhibitor and to stabilise the distribution of dissolved constituents in the pickling solution.
(3) To render the pickling process uniform.
(4) To reduce the drag-out losses of pickling acid.

The wetting agent lowers the surface tension of the pickling acid, enabling better penetration of the scale by the acid to be obtained. In addition, the removal of hydrogen bubbles is assisted, so that the acid reaches the metal surface more rapidly and uniformly.

Many suitable wetting agents are available for use in pickling acid solutions. Typical examples are the alkyl naphthalene sulphonates, the quaternary ammonium compounds with long hydrocarbon side chains, and the aliphatic sulphonates. Such wetting agents are available under commercial names.

Recently a pickling solution containing detergents has been reported[18] for the combined cleaning and pickling of steels.

4.2.1.10. Washing after pickling

Thorough washing after pickling is essential, as considerable amounts of sludge and acid-containing iron salts adhere to the metal when it is lifted from the pickling solution. If imperfectly washed off, these salts set up corrosion and can contaminate subsequent processing tanks (galvanising baths, plating solutions, etc.) into which they may enter. Residual salts can also accumulate in pores in the metal and can subsequently ooze out, damaging any finishing coat that may have been applied. Thus, in one case it was found that rapid corrosion of nickel-plated brass lamp holders occurred after only a few hours' service. The corrosion products contained a very large proportion of nitrates, which resulted from the retention of nitric acid or nitrates in the pores of the brass during pickling, and these found their way through the nickel plate when the lamp holders became heated during use.

4.2.1.11. Phosphoric acid

Phosphoric acid can be used alone as a pickling acid for steel; in this case it is employed at a concentration of approximately 15 per cent and at a temperature of about 70°C, but the relatively high cost of the acid is against its general use. It has the advantage that any traces of acid which may be left behind will not, in general, promote corrosion in view of the insolubility and non-hygroscopic nature of the heavy metal phosphates. If, however, the articles are to be plated subsequently, special care must be taken to avoid drying-out of the phosphoric acid before the rinsing operation, as insoluble films may form which may create adhesion problems. To prevent this, the temperature of the pickling bath must be kept sufficiently low having regard to the heat retention and thickness of the articles being pickled[20].

On the other hand, phosphate surface films are advantageous where painting is to be carried out since they provide a good undercoating for the paint film to adhere to[21]. The use of phosphoric acid rust removers without subsequent rinsing will be satisfactory only if the products resulting from the reaction between the rust remover and rust (or iron) are virtually insoluble and show neither an acid nor alkaline reaction. From the ternary equilibrium diagram for the system Fe_2O_3-P_2O_5-H_2O shown in Fig 4.13, it can be seen that at concentrations of up to 35% phosphoric acid the only phases in equilibrium with the solution are magnetite and $Fe_2O_3.P_2O_5.5H_2O$, which is the solid form of the tertiary ferric phosphate $FePO_4.2.5H_2O$. At higher concentrations of phosphoric acid, other phosphate species are also in equilibrium with the solution. These other phosphates are acidic and undesirable from the point of view of rust control. Svoboda and Knapek[22] have, however, pointed out that conditions of derusting are much more complex than those involved in the determination of equilibrium diagrams.

In practice the actual concentrations of acid, when applying the phosphoric acid solution to a rusty surface, will vary because (a) water evaporation occurs which tends

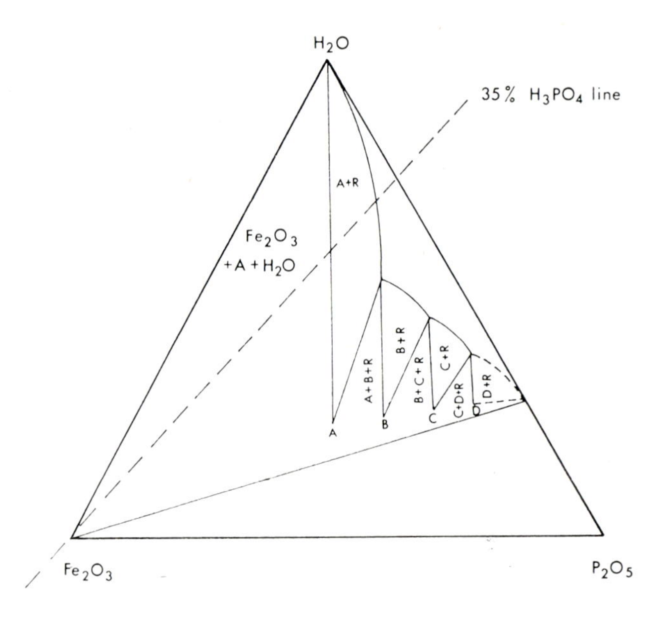

Fig. 4.13. Equilibrium diagram for the ternary system Fe_2O_3-P_2O_5-H_2O at 25°C.[22]

A = $Fe_2O_3.P_2O_5.5H_2O$ B = $Fe_2O_3.SP_2O_5.8H_2O$

C = $Fe_2O_3.3P_2O_5.10H_2O$ D = $Fe_2O_3.3P_2O_5.6H_2O$

R = solution.

to concentrate the acid and (b) reaction of the acid with the metal will reduce its concentration. By studying the reaction velocities of iron and phosphoric acid it was found that this reaction alters the phosphoric acid concentration to a lesser extent than does evaporation of water. Hence, if strong phosphoric acid solutions are used, the danger exists that the effective concentration of acid will exceed 35%, thus leading to the formation of undesirable acid phosphates.

Where subsequent rinsing is employed a suitable concentration of phosphoric acid is in the range of 32-35%. To obtain good penetration of the rust film it is recommended that 2% n-butyl alcohol be added to function as a wetting agent.

4.2.1.12. Stainless steel

The pickling of stainless steels bears little relation to the technique used with other ferrous metals because the nature of the scale on these metals is essentially different. The various solutions used have recently been reviewed by Wilmotte and Benezech[23].

4.2.1.12.1. Nitric acid baths

The austenitic stainless steels, containing 18 per cent chromium and 8 per cent nickel, are commonly pickled in nitric-hydrochloric or nitric-hydrofluoric acid mixtures. One typical solution consists of equal volumes of water and commercial hydrochloric acid containing about 5 per cent of nitric acid, used at 50° to 60°C; another solution contains 6 per cent nitric acid and 0.6 per cent hydrofluoric acid and is used at similar temperatures.

An alternative process makes use of two successive baths into which the metal is immersed:

Solution I

Sulphuric acid	15 per cent
Hydrochloric acid	5 per cent
Water	Balance

(Temperature 60°C)

Solution II

Nitric acid	20 per cent
Hydrofluoric acid	0.5 per cent
Water	Balance

(Temperature 30° to 40°C)

The metal should be rinsed between the two pickles.

Inhibitors have been recommended for use in these pickling acids, but they generally lack stability in the presence of nitric acid, so that they are decomposed in solution relatively quickly.

It will be observed that these solutions consist of an acid, together with an oxidising agent (i.e. nitric acid). The latter tends to oxidise the lower oxides of the metals present in the alloys to the comparatively soluble higher oxides, which are then more readily dissolved by the acids present. Other oxidising agents may be introduced in place of nitric acid so as to avoid the liberation of dangerous and unpleasant fumes of oxides of nitrogen. Dichromates, for example, have been employed to some extent in sulphuric acid pickling baths for non-ferrous alloys such as brass.

4.2.1.12.2. Ferric sulphate baths

Ferric sulphate plus hydrofluoric acid can also replace the nitric acid in pickling solutions for stainless steel, a concentration of three to six parts of ferric sulphate to one part of hydrofluoric acid being used. The best results are obtained when the composition of the bath corresponds to the theoretical formation of ferric fluoride in the solution according to the equation:

$$Fe_2(SO_4)_3 + 6HF = 3H_2SO_4 + 2FeF_3$$

The solution is then colourless and practically fumeless owing to the fact that little free hydrofluoric acid is present. Excessively high hydrofluoric acid concentrations cause too much attack on the steel whilst if the ferric sulphate concentration is too low, the rate of pickling will be much reduced. The ferric ion probably holds the fluoride in a complex by a buffering action so that the actual concentration of fluoride ions is maintained at a sufficiently high level to enable descaling to take place, but in insufficient concentration for attack on the steel to occur.

A solution capable of removing tenacious scales from austenitic stainless steel sheet contains 12 per cent of ferric sulphate and 3 per cent of hydrofluoric acid, the bath temperature being 65 to 70°C. The pickling time required is ten to fifteen minutes. The advantages presented by this method of pickling are as follows: (i) the pickling operation is practically free from fumes and spray when the solution is in balance; (ii) the attack by the solution on those areas where the scale has been removed is negligible due to the passivating effect of the ferric sulphate on the metal. This is very important in view of the difficulty in obtaining an inhibitor which will be stable in pickling solutions of conventional type which attack the descaled stainless steel surface relatively quickly (see Fig. 4.14).

When scales have been formed in a strongly reducing atmosphere it may be desirable to increase the ferric sulphate concentration beyond the proportions already mentioned, as such scales will contain much lower oxides and will therefore require the presence of a relatively higher concentration of oxidising agent. On the other hand, scales formed in highly oxidising atmospheres will require a greater relative proportion of hydrofluoric acid. Increasing the overall concentration of the pickling solution and the temperature increases the rate of pickling.

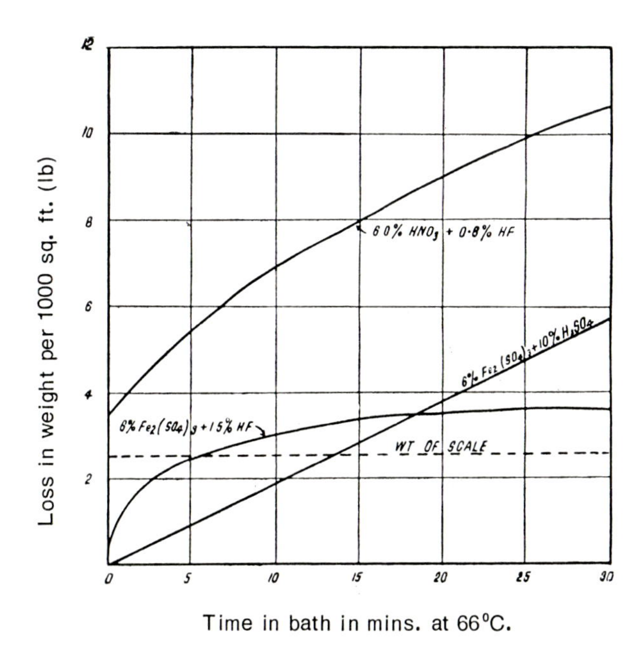

Fig. 4.14. Rate of attack on cold rolled annealed 18:8 stainless steel sheet in typical pickling solutions.

The high chromium irons do not pickle as satisfactorily in this type of solution as do the austenitic stainless steels. Ostrofsky[24] has described a process for pickling high chromium iron involving an electrolytic treatment at 2.5 A/dm^2 in fused caustic soda.

4.2.1.13. Rinsing

Descaled metal should first be washed in running cold water, preferably using high-pressure spray-jets. It should then be washed in another running water tank, this second tank being kept practically uncontaminated with incoming acid residues. The metal is then immersed again in very hot water to dissolve residual salts, being left in the tank for a sufficiently long time to acquire the temperature of the water. This will effect the removal of most of the acid from the metal pores. Sufficient heat will also be retained for the metal to dry off when removed from the tank unless it is of very thin section.

On removal from the water, pickled steel is liable to rust very quickly, so that some sort of temporary corrosion protection is usually employed[25, 26]. Temporary protection may be obtained by immersing the work in a dilute solution of nitrates, nitrites, chromates, pickling inhibitors or phosphoric acid (1% solution). A different approach is to use a thin, easily removable oil film which can be brushed or sprayed onto the metal surface.

4.2.2. SALT-BATH DESCALING

Salt-bath descaling provides a convenient method of dealing with difficult scale-removal problems, particularly where mixed metals are involved. Consideration[27] must, however, be given to the following safety aspects: fumes, chemical hazards, spattering, heating safety, fume extraction and mill operation.

One of the first of these methods was the sodium hydride process[28]. The hydride is generated within the pickling bath as a result of the interaction of sodium and cracked ammonia. A generator is employed consisting of a welded mild steel box open at the bottom and with a hole at the top large enough to admit blocks of sodium, and covered with a lid. This box is immersed in molten caustic soda to a depth of 30-35 cm, and projects about 15-20 cm above the liquid. When the cracked ammonia is introduced into the box below the liquid level, the gas bubbles through the molten caustic soda, meets the layer of molten sodium floating on the surface within the generator and produces sodium hydride in the bath in a concentration of $1\frac{1}{2}$ to $2\frac{1}{2}$ per cent.

Being a powerful reducing agent, the hydride reduces the oxides present on the metal to a finely divided powder or flaky form; there is no attack whatsoever on the basis metal. Most metal oxides are fully reduced by sodium hydride. For example, magnetite Fe_3O_4 and cupric oxide CuO react as follows:

$$4NaH + Fe_3O_4 \rightleftharpoons 3FE + 4NaOH$$
$$NaH + CuO \rightleftharpoons Cu + NaOH$$

Chromium oxide is not reduced to the metal but converted to an easily removed, loosely adherent lower oxide:

$$NaH + Cr_2O_3 \rightleftharpoons 2CrO + NaOH$$

The operating temperature of the bath is 350° to 370°C, and the action is extremely rapid; the same bath can be used simultaneously for almost all types of scale. The alkali hydroxide formed during scale removal replenishes the caustic soda content of the bath and reduces drag-out losses.

Small parts are treated in baskets and larger parts are suspended from hooks. They must be dry and preferably preheated before immersion, drained for a few minutes after treatment, and quenched in cold water; this enables the reduced scale to be rinsed away readily. Finally, the articles are washed in hot water to remove residual caustic soda.

The complication of the sodium addition and the need for an ammonia cracking plant can be avoided by the use of molten caustic soda in combination with salts that have a reducing action on the scale. The salts used have the property of reducing the refractory oxide to one of a flocculant nature capable of solution in a weak acid. A commercial process based on this system is available which can be used for descaling any metal that is not affected by molten caustic soda at 500° to 520°C; the bath may be heated externally or, more satisfactorily, by means of immersed electrodes. It can be successfully employed on most non-ferrous metals including nickel and nickel-chromium alloys, titanium, and copper-silicon alloys, as well as the stainless steels for which it was originally introduced.

In another patented process[29] the descaling of steel is carried out by entering the metal into a molten bath of alkali metal hydride in alkali metal hydroxide. Oxidation of the melt surface is prevented by covering it with a layer of paraffin oil. The bath is operated at a relatively low temperature since a eutectic mixture of sodium and potassium hydroxide is used as alkali metal hydroxide. The treatment takes 1 to 15 minutes, depending on the nature and thickness of the scale. The metal is quenched in water and then lightly pickled for a minute or two in dilute hydrochloric or sulphuric acid to remove the scale, most of which has, in any case, been spalled off by the water quench. A final dip in 10 per cent nitric acid solution is helpful in producing a bright finish.

The salt bath processes result in minimum losses of metal and are widely applicable; admittedly, however, they have the disadvantage of high costs in maintaining the bath temperature. Working with molten caustic soda also means that the operator must wear goggles and protective clothing.

4.2.3. ALKALINE DERUSTING SOLUTIONS

An alternative to acid pickling solutions for rust removal is the use of alkaline solutions which remove rust without attacking the underlying steel surface. The alkali derusting solutions contain a suitable chelating agent, such as E.D.T.A. or sodium gluconate, and have the advantage of being capable of removing surface soils, oils and many paints, so that cleaning and derusting are effectively combined. Sodium gluconate[30, 31] is a valuable additive to alkaline derusting solutions because it sequesters metal ions and prevents the precipitation of insoluble ferric salts. This results in a very clean surface ready for electroplating or other finishing operations. The process may be operated either by simply immersing the metal parts in the solution, or electrolytically which results in a much more rapid removal of rust and scale. Whichever method is adopted, the composition of the solution is determined by firstly fixing the ratio of caustic soda to gluconate on the basis of whether the contamination to be removed is predominantly rust or surface soil. If rust is the major component of the contamination, then a high proportion of gluconate to hydroxide is used. The actual concentrations of the constituents depend on the amount of contamination to be removed. Wetting agents added to the solution improve the derusting action, whereas other additives function as dispersants assisting in the prevention of sedimentation.

The alkaline derusting solutions are used at temperatures in excess of 80°C and preferably in excess of 90°C. In the electrolytic method the work to be derusted is made the cathode, current densities of between 2 and 10 A/dm^2 being employed. A suitable solution contains[30] sodium hydroxide and gluconate in the ratio of between

1:1 to 9:1 depending on, as already mentioned, the type of major contamination to be removed.

4.2.4. ELECTROLYTIC PICKLING

In electrolytic pickling the metal is made either the anode or the cathode in an acid or alkaline solution or in a solution of a neutral salt. Anodic pickling is often preferred since, of course, hydrogen is not liberated at the anode and therefore hydrogen embrittlement does not occur, as may be the case when cathodic pickling is employed. Cathodic pickling does however, have the advantage that the surface of the metal is 'cathodically' protected against corrosion. In both anodic and cathodic pickling the scale is removed chiefly as a result of the mechanical effect of the gas liberated by electrolysis at the metal surface. Dilute sulphuric acid is often used as the electrolyte, current densities of 2 to 10 A/dm^2 being employed. The difficulties of applying the current and uncertainty as to whether the results are worth the relatively elaborate equipment needed have limited the general application of the process.

4.2.4.1. Bullard-Dunn process

The Bullard-Dunn method[32] of electrolytic pickling has, however, had some measure of success, a number of plants having been installed for dealing with difficult pickling problems. The electrolyte consists of a hot dilute sulphuric acid pickle of the normal type containing about 0.1 per cent of stannous sulphate. The current density employed is high—of the order of 6 to 8 A/dm^2, the articles to be descaled being made the cathodes. The anodes are mostly of high-silicon iron, a small percentage of them being of tin in order to maintain the tin content of the solution.

As pickling proceeds, immediately an area becomes descaled, metallic tin is deposited on to it, and this inhibits further attack by the acid on that area and helps reduce hydrogen embrittlement of the steel. When all the scale has been removed, the surface is left covered with a thin film of tin. This may be left on the surface, forming a good undercoat for subsequent painting or enamelling; alternatively, it can be removed rapidly by anodic treatment in caustic soda using silicon-iron cathodes on to which the tin is deposited, and these can then be put into the pickling bath, so that the tin is recovered.

Corrosion tests in salt water on rolled steel window-frame sections descaled by the Bullard-Dunn process and painted showed that the amount of corrosion produced was only 50 per cent of that obtained by painting directly on to shot-blasted and phosphate-coated material. When samples from which the tin had been removed were exposed, the performance of these finishes was about equal.

Fig. 4.15. Plant for pickling cooker components at Stoves Ltd., Liverpool.
[Courtesy W. Canning & Co. Ltd.]

4.2.4.2. Anodic pickling

In this case the increased pickling rate is due to the mechanical scrubbing action of the oxygen gas evolved. The mechanical cleaning action for a given electrolysis current will be less efficient than in cathodic cleaning since the volume of oxygen gas produced is only half that of the corresponding volume of hydrogen which would be produced cathodically.

Anodic pickling is usually carried out in sulphuric acid pickling solutions. The use of either very strong (>70%) acid solutions[33] or high anodic current densities (>20 A/dm^2)[34] enables the iron to become passivated, so that the pickling action is almost entirely due to the mechanical action of evolved oxygen.

Electrolytic pickling should not be confused with electrolytic polishing or electro-brightening as applied to stainless steel and other metals (see Chapter 2).

4.2.5. HYDROGEN EMBRITTLEMENT

When hydrogen is liberated at a steel surface during pickling, electroplating or electro-alkaline cleaning, a proportion of the hydrogen is absorbed by the metal surface and diffuses into the interior of the steel. Such hydrogen is liberated in the atomic state, and it is in this form that it is absorbed; molecular hydrogen does not diffuse into iron or steel at the temperatures prevailing in these operations.

The absorbed hydrogen is not uniformly distributed throughout the metal, but is concentrated chiefly near the surface. Absorption also appears to be catalysed by the presence of elements such as phosphorus, sulphur and arsenic, which are capable of forming hydrides.

The quantity of hydrogen absorbed may be quite considerable and has two major effects in metal finishinng: (a) while it is present in steel an embrittlement of the metal takes place, (b) on subjecting the metal to high temperature finishing processes, especially galvanising or enamelling, the gas may be liberated with the formation of blisters in the steel surface or within the protective coating.

Edwards[35] in 1924 investigated the rate of diffusion of hydrogen through mild steel sheet of different thicknesses, and found that the amount of hydrogen diffusing at room temperature was roughly inversely proportional to its thickness (Fig. 4.16). As the temperature rises the amount of hydrogen diffused decreases appreciably, although it must be borne in mind that under practical conditions of acid pickling there is simultaneously an increase in the total hydrogen liberated as the temperature of pickling increases.

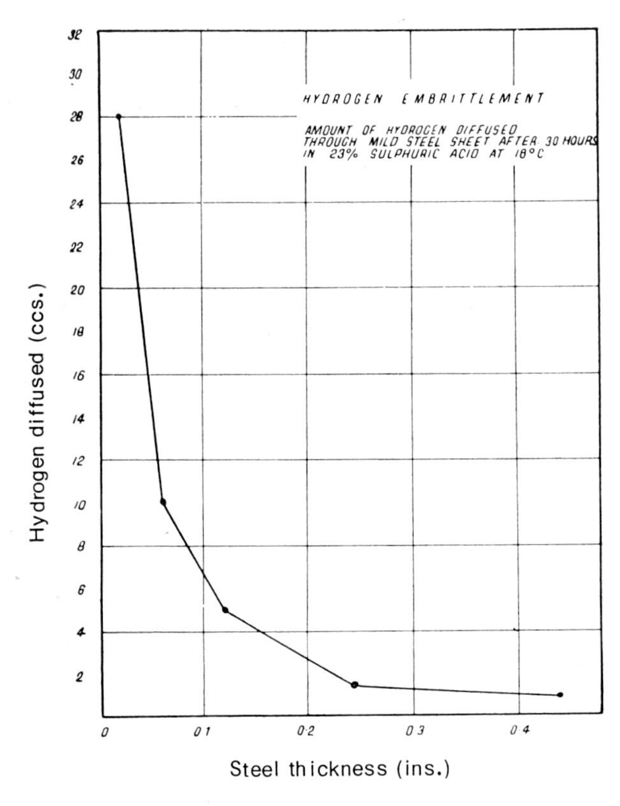

Fig. 4.16. Amount of hydrogen diffused through mild steel sheet after 30 hours in 23% sulphuric acid at 18°C.

Broadly speaking, those factors which are conducive to an increased rate of attack by the acid on the metal, e.g. higher temperature, greater acid concentration or the use of more active acids, all lead to a reduction in the ratio of the amount of hydrogen diffusing through the metal to the total hydrogen liberated. Thus Zapffe and Sims[36] found that increasing the acid pickling temperature from 20 to 80°C increased the hydrogen absorption thirtyfold, but the ratio of liberated to absorbed hydrogen was found to have increased fivefold. On balance, probably, the shorter pickling period resulting from the use of higher pickling temperatures causes a net reduction in the amount of hydrogen absorbed in a given time.

The cathode efficiency of the plating process has little bearing on the amount of hydrogen absorbed or the degree of embrittlement produced[37]. Thus, some processes developing no visible cathodic gassing whatsoever produced the greatest embrittlement, considerably exceeding the effect of straight ' hydrogen plating ', i.e. cathodic pickling. In this respect copper plating is similar to cadmium, zinc and chromium plating processes. It is interesting to note that processes which do not produce adherent coatings also do not produce embrittlement, and reagents added to baths to change the characteristics of the deposit also change the characteristics of the embrittlement; hence each condition requires separate evaluation. Orthotolyl thiourea used as an inhibitor increases the proportion of hydrogen entering the metal lattice even though the actual amount produced is lessened[38].

4.2.5.1. Effects of embrittlement

The effects of hydrogen absorption manifest themselves in an embrittlement of the steel, although it is important to distinguish between loss in mechanical strength due to roughening of the metal surface in pickling and that caused by hydrogen absorption. Noble[39] found that susceptibility to hydrogen embrittlement increases, in general, with hardness. The maximum susceptibility occurs at the temper brittle range of 260° to 288°C except in steels that are not subject to a temper brittle range, e.g. molybdenum-bearing steels, in which case the temperature of maximum susceptibility is much lower. The presence of hardening constituent, e.g. carbon, increases susceptibility to embrittlement, whilst alloying elements such as copper, nickel and chromium also influence the absorption of hydrogen in steel[40].

A method of testing for embrittlement has been developed by Zapffe and Haslem[41]. In this, duplicate wires about 1.5 mm dia. x 10 cm long are carefully cleaned by hand polishing with emery paper, fastened between anodes and immersed to a depth of 5 cm in the electrolyte to be tested. After plating for the required time the wires are withdrawn, rinsed and immersed in water at 0°C to delay changes in the condition of the wire with respect to hydrogen. Thirty seconds after the cessation of plating the first wire is subjected to the bend test, and after 90 seconds the duplicate is tested. This technique suppresses any error which might result from ageing before testing. The test consists in bending the wire specimen at a constant rate of 5 degrees per second around an approximately 1.5 mm dia. radius placed at the centre of the

5 cm plated length. The angle at which fracture occurs is a measure of the degree of embrittlement produced.

Steel, heat-treated to a hardness of under 40 Rockwell C, is very little subject to embrittlement, although cold-worked steel is susceptible at a lower hardness than the same material heat-treated to a corresponding hardness. Hydrogen embrittlement tends to be dissipated on standing at room temperature, and this can be hastened by heating; the precise conditions for complete relief, however, depend on the nature of the embrittling medium and on the thickness of the material. The temperature to which the metal is raised is a more important factor in the elimination of embrittlement than is the duration of heating. Stainless steel is particularly susceptible to embrittlement.

TABLE 4.3
Relief of Hydrogen Embrittlement

Mean temperature, °C	18	19	143	143	177	177	204
Time (min)	10	60	30	60	10	30	30
Crushing stregnth loss, per cent	65	30	30	0	45	10	0

Mean temperature, °C	80	80	80	80	100	100	99	177	232
Time (min)	10	30	60	120	60	120	120	120	120
Crushing strength loss, per cent	60	55	50	35	15	0	52	15	0

Of the plating processes, it is of interest to note that chromium produces the greatest amount of embrittlement and lead the least, while for electrodeposits of equal thickness, tin is the easiest to relieve and silver the most difficult. Hydrogen embrittlement is a particularly serious risk in cadmium plating, as this coating is widely used in the protection of aircraft components (see Chapter 12).

Embrittlement can be removed by heating in water, air or oil, or by ageing at room temperature, the effectiveness of the respective media being in this order. Complete restoration of crushing strength is obtained by heating in air for one hour at 145°C or half an hour at 205°C.

Table 4.3 shows the results obtained by various treatments tested for the relief of hydrogen embrittlement caused by pickling.

4.2.6. PICKLING OF NON-FERROUS METALS

In the pickling of non-ferrous metals, the mechanism is rather different from that involved in the pickling of iron and steel inasmuch as almost the entire oxide layer must be removed by dissolution in the acid itself. By far the most important group of non-ferrous metals that requires pickling consists of copper and its alloys (including brasses and bronzes), together with the heat-resisting alloys based on nickel and chromium.

Pickling is employed either to remove the scale and oxide from non-ferrous metals as a final operation to produce a clean and attractive surface which may subsequently be polished and plated, or at an intermediate stage in fabrication, where subsequent pressing, forming or related operations have to be carried out, to avoid damage to tools. The pickling of non-ferrous metals has been the subject of a review by Silman[42].

4.2.6.1. Copper alloys

The oxide film formed on copper when the metal is heated to a high temperature in air consists of an outer scale of cupric oxide (CuO) under which is an inner scale of cuprous oxide (Cu_2O). The outer scale is normally in a state of stress and may flake off in cup-shaped fragments.

At lower temperatures (up to about 600°C) the rate of oxidation is parabolic, approximately following the law:

$$\text{Rate of oxidation} = Ae^{-Q/RT}$$

where A and Q are constants, R is the gas constant, and T is the absolute temperature.

Dunn[43] studied the oxidation of copper-zinc alloys at different temperatures and found that all alloys with less than 80 per cent of copper oxidise at practically the same rate over the range 580°-880°C. The copper oxide produced at the metal surface is reduced by zinc diffusing from the interior, so that the scale consists of practically pure zinc oxide. With a copper content of 80 to 90 per cent in the alloy, the rate of zinc diffusing is unable to keep pace with the rate of copper oxide formation, so that the scale becomes progressively richer in copper. With a 90: 10 copper-zinc alloy, the oxide layer contains 90 per cent of copper.

The factors influencing the type of scale which will be formed on brass are therefore: (a) the rate of diffusion of oxygen through the oxide, (b) the degree of oxidation of the metal, (c) the diffusing of zinc to the metal surface, (d) the reduction of copper oxide by metallic zinc, and (e) the re-entry of copper into the metal surface. When the rate of diffusion of the zinc is not rapid enough, the mechanism breaks down and normal oxidation takes place. Arsenic (up to 0.2 per cent) and nickel (up to 1 per cent) are without effect on the oxidation of brass, but aluminium strongly inhibits the rate of scaling; thus the presence of only 1.9 per cent of aluminium will lower the rate of oxidation to one-fortieth of its value in the absence of this element. The effect is probably due to the reduction of the zinc and copper oxides by the aluminium.

Copper and its alloys are commonly pickled in hot sulphuric acid under similar conditions to those prevailing in the pickling of ferrous metals, although obviously ferrous and non-ferrous alloys must not be treated in the same bath. The acid concentration is raised from about 5 per cent, when the bath is new, to 10 per cent when the bath has been in use for some time, as a result of the additions that have to be made to regenerate it. The temperature must also be raised from, say, 40°C to about 80°C as the metal concentration builds up, in order to maintain the pickling rate. The oxide is removed largely by chemical dissolution. This type of pickling is

is increased by the dichromate so that a brightened surface is produced. This solution may, however, etch the surface appreciably, especially in areas which may have suffered local dezincification during heat-treatment.

Ferric sulphate may also be employed in place of nitric acid in bright dips to eliminate the toxic fume problem. Ferric iron is capable of oxidising the copper to the soluble cupric state, and is suitable for use on most copper alloys, including brasses, silicon bronzes and aluminium bronzes. It is used in the proportion of some 10 per cent of the sulphuric acid content of the bath. The appearance of the pickled brass is uniform, but the colour is darker than that obtained with the sulphuric acid-dichromate mixtures. It is considerably less bright than that obtained by nitric acid dipping, and this is regarded as a serious disadvantage of the process. When the alloy contains beryllium or silicon, the sulphuric acid may with advantage be replaced by hydrofluoric acid.

As stated, pickling in sulphuric acid containing oxidising agents other than nitric acid does not produce a fully bright finish. Various additions have been recommended from time to time which enable brighter surfaces to be obtained, but although the finish is not as bright as that obtained from nitric-sulphuric acid mixtures, it may be adequate for many purposes. One brightening bath for brass consists of a solution of 9 per cent ferric sulphate to which is added either 0.1 per cent cream of tartar, 3 per cent acetic acid, 0.5 per cent citiric acid, or 0.5 per cent glycollic acid. This pickle is generally used cool, but in some cases a hot solution may be preferred. For nickel silver, 0.5 per cent of adipic acid has been found to be a useful addition agent. By the inclusion of about 10 per cent of sulphuric acid in the solution a combined descaling and brightening bath is said to be obtained.

Fintschenko[45] has recently described a copper pickling process using sulphuric acid-hydrogen peroxide solution with stabiliser.

The rates of attack on brass of some pickling solutions are shown in Fig. 4.17.

4.2.6.1.2. After-treatments

Nitric acid pickling of brass needs special care in washing since there is a tendency for the acid to be retained in pores in the metal; this may ooze out later, especially if the metal is heated during lacquering operations, etc., causing spotting and discoloration.

One technique for preventing the tarnishing of brass after bright dipping is to produce a thin soap film on the brass surface by dipping into a solution of about 5 grams per litre of cyanide in which 5-10 grams of neutral soap have been dissolved. The solution is kept slightly warm and the articles are immersed after thorough rinsing and then rinsed again. If proper rinsing is not obtained prior to immersion in the cyanide-soap mixture, the latter will be precipitated on the work as a gummy fatty acid film.

The soap film protects the articles from tarnishing for several days and does not interfere with subsequent electroplating operations.

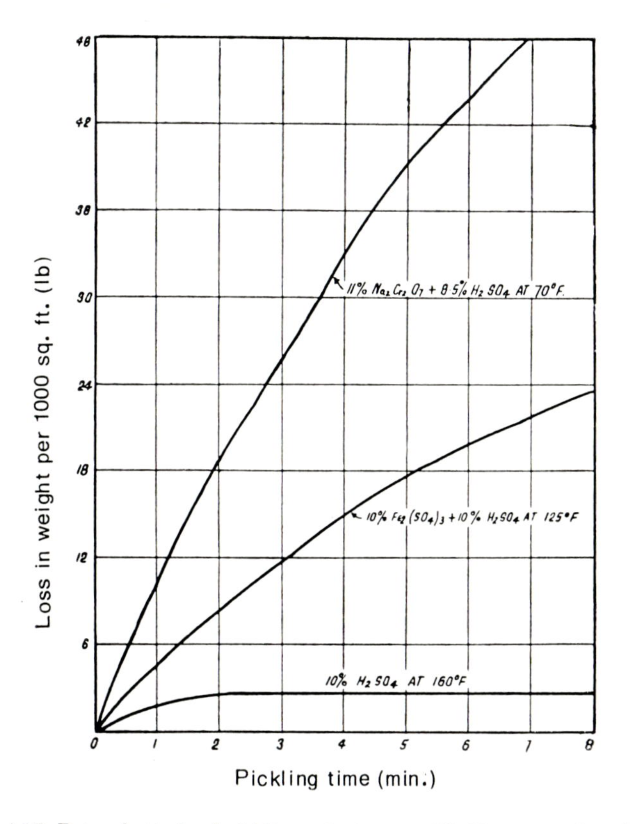

Fig. 4.17. Rate of attack of pickling solutions on 85 : 15 copper-zinc alloy.

4.2.6.1.3. Nickel silvers

The nickel silvers form a group of copper-nickel-zinc alloys which have good corrosion resistance and, being white in colour, are employed in both domestic and industrial applications. They were at one time widely used for decorative and architectural fittings and for such items as bath taps, but now they are most extensively employed for making spoons and forks, which are generally silver plated. Essentially, the nickel silvers are brasses in which part of the zinc has been replaced by nickel to impart a white colour to the material. This limits the composition so that nickel silvers contain 45 to 75 per cent copper, 5 to 18 per cent nickel, the balance being zinc. Certain grades may contain a small amount of lead to facilitate machining. The nickel silvers can be hardened only by cold working, and the mechanical properties of the wrought alloys are determined by the degree of annealing or by the amount of cold working performed after the annealing operation.

Acid pickling is most commonly employed to remove oxides formed during annealing. This is carried out most often in a 10 per cent (v/v) solution of sulphuric acid at 50 to

65 °C, more concentrated solutions being employed where the scale is heavier. After pickling, thorough rinsing is necessary since traces of residual acid in the rinse waters may lead to staining. When the scale has been removed by pickling in sulphuric acid, the surface tends to be matt because of the slight etching that takes place. Brightening is, therefore, usually necessary, and this is carried out by dipping in a solution of sulphuric acid of similar concentration to which 25-50 g/l of sodium dichromate has been added. The bath is used at 25-40 °C. Bright dipping in the usual nitric/sulphuric acid dip for brass is often employed; this gives a more brilliant surface.

4.2.6.1.4. Other special copper alloys

Amongst other copper alloys which require special treatment are the aluminium bronzes, which are pickled in 10 to 20 per cent sulphuric acid to remove the oxide film that forms on the metal when it is annealed; alternatively a short immersion in concentrated hydrochloric acid may achieve the same result.

Beryllium copper, employed for electrical contacts and springs, is very difficult to pickle. A black oxide coating is formed and the usual after-treatment consists in dipping the metal into a 10 per cent solution of sulphuric acid at 50 to 60 °C and, after rinsing, to follow this by a further dip in sulphuric acid of about the same concentration to which a small amount of sodium dichromate has been added. This brightens the work and has some passivating effect. Cleaning can also be carried out in a 10 per cent hydrofluoric acid solution containing about 10 per cent ferric sulphate as the oxidising agent. This is a non-passivating treatment.

4.2.6.1.5. Fume disposal

Nitrous fumes from bright dipping tanks are heavy and are best exhausted from high-velocity slots extending across the dipping tanks. The air velocity across the tank should be not less than 18 metres per minute for every 30 cm from the slot lip to the opposite wall of the tank.

Often the extraction fans exhaust to the air, but if possible the extracted air should be washed to remove the nitrous acid. To ensure scrubbing being effective, the air speed should be reduced to 90 metres per minute during the passage of the fumes through a scrubbing tower packed to three quarters of its length with stoneware tower packing. A high-velocity jet spray ring is fitted to the top of the chamber for spraying soda solution, allowing a contact time of about two minutes to neutralise the acid. Fig. 4.18 shows the lay-out and construction of a plant used for the removal and washing of nitrous fumes from a pickling plant for brass or copper.

4.2.6.1.6. Electrolytic copper recovery

The recovery of copper from pickling solutions is now practised fairly extensively since considerable economies can be effected by this procedure where pickling is carried out on any substantial scale.

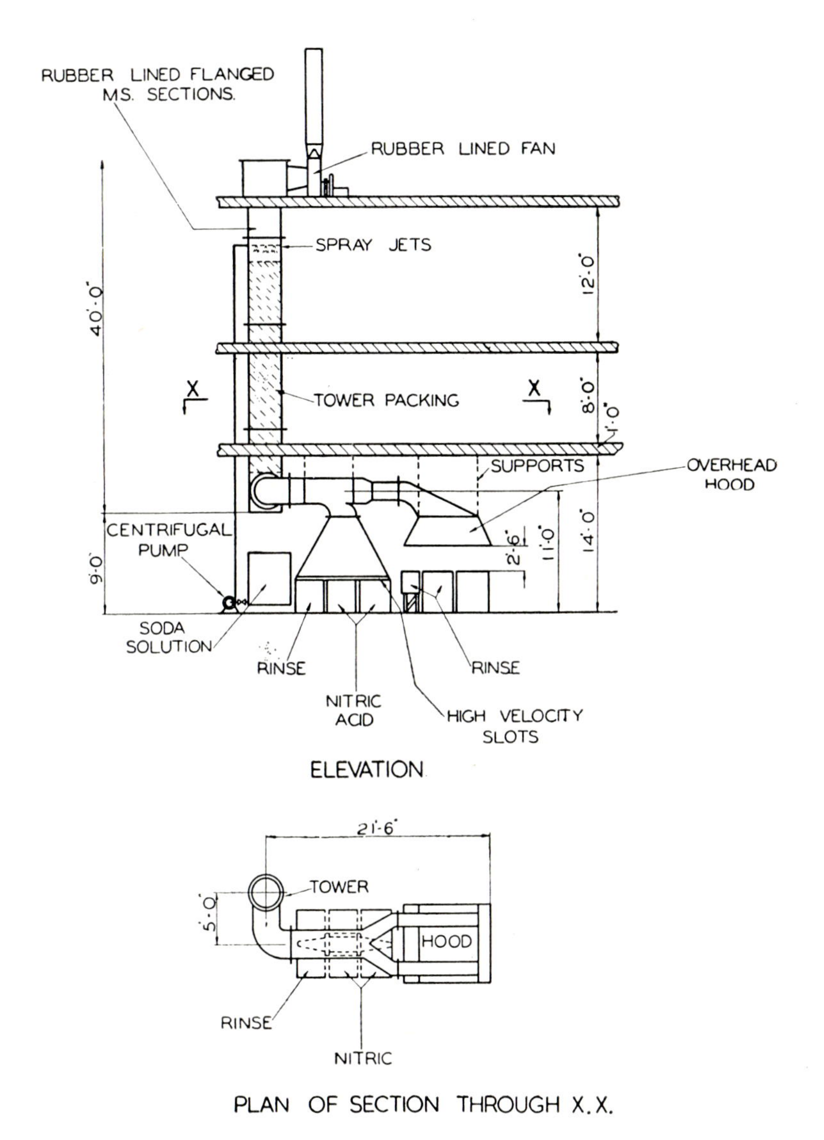

Fig. 4.18. Diagrammatic section of fume dispersal system for bright dipping plant.

The deposition of coherent copper from sulphate electrolytes is dependent on the availability of an adequate supply of copper ions at the cathode surface. These conditions are met basically by relatively high copper and low acid concentrations, and by low current densities and elevated temperatures. Of course other factors enter, and in designing a recovery system these have to be taken into account. Moreover, the composition of the electrolyte is often outside the control of the plant designer.

High conductivity (low resistivity) of the electrolyte is desirable since this reduces power requirements. For a given concentration of copper sulphate the resistivity of the bath is affected by the temperature and acid concentration; for a solution containing 5 per cent copper sulphate the resistivities at 25 °C are as follows:

H_2SO_4, g/l	0	50	100	150	200
Resistivity, ohm-cm	65	4.9	2.58	1.88	1.55

Excessive concentrations of sulphuric acid lead to the formation of a high resistance film at the anode consisting of copper sulphate, but moderately high acid concentrations make for low voltage requirements. High current densities are undesirable as they increase the tank voltage, reduce efficiency, and lead to powdery deposits.

Temperature acts in the opposite direction to current density; too low a temperature of the electrolyte leads to hard or powdery deposits whereas a higher temperature increases the solubility of copper sulphate.

Metallic impurities in the electrolyte which are more noble than copper (e.g. silver, gold, platinum and palladium) are co-deposited with it, whilst those which are less noble remain in solution. The cell voltage, and hence the power consumption, increases with increase in impurity content of the electrolyte.

Addition agents improve the quality of the deposit. Glue is useful in this respect; it is often used in conjunction with sulphate-lignin liquor, hydrolysed casein, or thiourea. The rate of addition is something like 4 parts per million parts of electrolyte per day.

Current density as such does not greatly influence the cathode efficiency. Neither do changes in acid concentration between 15 and 22 per cent or copper concentrations between 1.5 and 4.5 per cent. Nickel does not affect efficiency, but the influence of iron is uncertain as much depends on its state of oxidation in the electrolyte.

As the concentration of copper in the electrolyte falls, the cathode efficiency also falls. The decrease becomes noticeable under commercial conditions when the copper content has dropped to 0.02 per cent; by this time the cathode efficiency has fallen to about 50 per cent.

4.2.6.2. Nickel and nickel alloys

Amongst the most important and, incidentally, the most difficult groups of metals to pickle are the nickel alloys, which include the heat-resistance types; the latter are of much importance in gas-turbine production. The oxides formed on such nickel alloys as Monel, Inconel and Nimonic are highly resistant, and the type of scale produced varies with the heat-treatment conditions to which they are subjected. Sometimes, if these conditions are not favourable, patches of scale may remain which are particularly difficult to remove. Excessive pickling is undesirable as it can lead to serious etching of the surface.

Thorough cleaning and degreasing are necessary before pickling, as otherwise uniform action of the acid will not be obtained and local damage to the metal can result.

Alkaline cleaners are the best for the purpose and these may be based on sodium hydroxide or sodium carbonate, preferably with the addition of trisodium phosphate or sodium metasilicate to aid detergency and to improve rinsing. Solution concentrations of 10 to 20 per cent are recommended, which are operated at fairly elevated temperatures (80° to 90°C). Vapour degreasing in chlorinated solvents is also used on small parts, but this method, although satisfactory, is relatively expensive.

4.2.6.2.1. Flash pickling

Tarnish resulting from very thin oxide films, such as occurs on drawn articles, wire rivets, and similar components, can readily be removed by flash pickling for a minute or two in nitric acid mixtures. Two solutions are generally used. For Monel metals, which consists essentially of nickel and copper in the proportions of about 4:1, a double sequence is recommended. The first bath consists of:

Water	1 litre
Nitric acid (38° Bé)	1 litre
Common salt	60 to 90 grams
Temperature	20° to 40°C
Time	$\leq$5 seconds

The parts should be thoroughly cleaned in this first dip, allowing only a short time for each immersion, then rinsed in hot water at 80°C and followed by a rapid dip in a second solution consisting of:

Water	1 litre
Nitric acid (38° Bé)	1 litre
Temperature	20° to 40°C
Time	$\leq$5 seconds

The second dip should be followed by rapid rinsing and neutralising in 1 or 2 per cent ammonia solution. The parts should then be dried by dipping in boiling water followed, for example, by drying in sawdust.

Only one dip is required for nickel in a sulphuric acid/nitric acid mixture made up from:

Sulphuric acid (66° Bé)	1½ litres
Nitric acid (38° Bé)	2¼ litres
Water	1 litre

After cooling, 30 g of common salt is added. The time of immersion is 5-20 seconds at a temperature of 20-40°C.

The rate of action of this pickle may be retarded by reducing the sulphuric acid to ½ litre and the nitric acid to 1¾ litres.

The procedure is to warm the parts first by dipping in hot water, and then to immerse in the acid bath for 5 to 20 seconds; this is usually sufficient to brighten nickel. The parts are rinsed in hot or cold water, neutralised in dilute ammonia, and dried by dipping them in boiling water or by means of sawdust.

4.2.6.2.2. Hot-formed products

Forgings and hot-rolled components (including wire rod in coil) which are oxidised but have subsequently been heated in a strongly reducing, sulphur-free atmosphere develop spongy, more or less lightly adherent layers. On nickel these consist of metallic nickel, and on Monel of a mixture of metallic nickel and copper. The oxide film of Inconel 600 (an alloy of 80 per cent Ni, 14 per cent Cr, balance Fe) is selectively reduced to a mixture of nickel oxide and chromium oxide, the latter resulting in a characteristic green or greenish-brown colour. A sulphuric acid solution, operated in properly equipped and ventilated plants, is employed for the pickling of this group of metals.

After pickling, the parts should be rinsed in hot water and neutralised in a 1 to 2 per cent ammonia solution. The operation is often facilitated by occasional intermediate scrubbing with pumice on a fibre brush. It is advisable to maintain separate baths for the treatment of Monel and nickel. The solutions work better after having been used for a time; therefore, when making up new solutions, about 2 per cent of spent solution should be added to the fresh bath.

In the case of forgings, hot-headed bolts and hot-rolled or hot-formed products which have been annealed and allowed to cool in air, the oxide formed is extremely thin and tightly adherent. It is also glossy in appearance and very dark in colour Under bad conditions, or when there is much sulphur in the atmosphere, heavier scale is formed. Hydrochloric acid, with additions of either cupric chloride or sodium dichromate, is employed, e.g. to remove a moderately thick oxide scale from Monel and Inconel. A typical solution consists of:

Hydrochloric acid (20° Bé)	½ litre
Cupric chloride	30 grams
Water	1 litre

The time of treatment is 20-40 minutes at 80°C.

4.2.6.2.3. Nimonic alloys

The Nimonics are a series of high-temperature-resistant alloys used extensively in aircraft and gas turbines. Essentially they consist of some 80 per cent nickel, the remainder being mainly chromium. These alloys are annealed at 900° to 1000°C in either oxidising or protective atmospheres. If the metal is not cleaned before annealing, non-uniform oxide films will form, leading to difficulties in pickling. The oxides show considerable variation in ease of removal; acid or alkaline pickles can be employed.

Either of the following solutions may be used for the pickling of these alloys:

(a)	Nitric acid (conc)	20 litres
	Hydrofluoric acid (60 wt.%)	5 litres
	Water	75 litres
	The time of treatment is up to 1 hour at 15°C.	
(b)	Ferric sulphate solution (containing 64 g of salt/100 cc)	4 litres
	Hydrofluoric acid (60 wt.%)	1 litre
	Water	17 litres
	The time of treatment is up to 6 hours at a temperature of 60°C.	

No addition of hydrofluoric acid should be made after the first mixing, but ferric sulphate solution must be added from time to time to maintain the concentration of ferric sulphate (ignoring the ferrous iron) at approximately its original strength, i.e. 10-12 g/100 ml of liquid.

Pickle (b) has a tendency to produce pitting and 'worm-holing' on the metal, especially if components being pickled come into contact with one another or if the time of immersion is unduly long.

On removal of the components from these pickles, rinsing in water followed by light rubbing, or washing in a stream of water under pressure, is necessary in order to remove the loosened scale and free the metal from acid.

Descaling can also be carried out in a caustic alkali solution followed by an acid treatment. The solution consists of:

Caustic soda	200 grams
Potassium permanganate	150 grams
Water	1 litre

It can be contained in a steel tank and heated by a steel steam coil to near boiling point. As the liquor does not attack the metal but only the scale, articles can be immersed for lengthy periods. Normally they require about 8 hours in the liquor; after rinsing with water and immersion in a 10 per cent sulphuric acid solution, the scale is easily removed by gentle rubbing or by water washing under pressure.

A development of this process makes use of the accelerating properties of organic nitro-compounds in the acid. A typical solution of this type for pickling high-chromium alloys consists of sulphuric acid 200 g/l, metanitrobenzene sulphonic acid 40 g/l, and hydrofluoric acid 16 g/l. The bath is operated at 65°C.

These alloys can also be conveniently descaled in fused alkaline salt baths based on caustic soda or by the sodium hydride salt bath process (Section 4.2.2), followed by a 30 second drain and a final water-quench. Any loosely adhering material should be removed by water hosing. The components are then immersed for 30 to 40 minutes in an acid bath containing 7% by volume of nitric acid and 1.7 g of ferric chloride per 100 ml of the total solution; this is used at 60°C. Finally, they are water hosed to remove any smut and acid that remain.

4.2.6.3. Aluminium and its alloys

As aluminium reacts with both acids and alkalies, pickling can be carried out in either type of bath. The action is quite vigorous and only weak solutions are needed. Aluminium may be cleaned in a 5 to 10 per cent solution of hot caustic soda or in other alkalies, such as sodium carbonate or trisodium phosphate. There is a rapid evolution of gas with some liberation of spray, which can be reduced by the addition of a wetting agent. After rinsing, the articles are subjected to a short weak acid treatment, such as immersion in nitric acid or in a nitric/hydrofluoric acid mixture, to remove smut.

The alloys of aluminium containing copper react rather more vigorously with caustic alkalies than pure aluminium; thus, at 80°C the rate of attack is about six times greater than at 50°C in the case of pure aluminium, but with the aluminium-copper-magnesium alloys the rate of attack is ten times greater because of the local electrolytic cells formed by the precipitation of copper intermetallic compounds. In view of the rapid increase in the rate of attack with rise in temperature, close control is necessary. Excessively low temperatures or too concentrated an alkali solution will not affect the rate of attack but will also give rise to irregular attack and penetration, so that the resulting sheet becomes very patchy in appearance.

The most rapid rate of attack on aluminium occurs with hydrofluoric acid, although most other acids are also very aggressive. Concentrated nitric acid, however, does not attack aluminium. Table 4.4 shows the rate of attack of various acids on the metal.

The attack on aluminium by alkalies is considerably affected by the presence of halides, which result in the etching of the metal surface and a reduction in brightness; hence, if maximum reflectivity is to be obtained, the presence of chlorides or fluorides must be closely controlled.

Sulphuric or nitric acid may be used to produce a matt surface on aluminium sheets; the latter acid is used at a concentration of 3 to 5 per cent and the former at

TABLE 4.4
Loss in weight of aluminium in various acids

Acid	Temp. °C	Loss in weight per day g/m^2
1% Hydrochloric acid	20	2.4
	50	43
1% Hydrofluoric acid	20	226
2% Sulphuric acid	20	2.1
3% Nitric acid	20	1.15
	60	18.2
5% Nitric acid	20	1.38
	60	27.1

around 3% at 80°C. The time of immersion is 2 to 10 minutes. Attack is slow, and it also tends to be somewhat non-uniform. These acids should be used only on pure aluminium or on alloys in which the alloying constituents are in solid solution. When etching in alkalies, a 50 per cent concentration of nitric acid is recommended as an after-dip. A fine silvery finish can be obtained by pickling in a 1:500 solution of hydrofluoric acid. This solution is also suitable for sand castings which contain small particles of silica in the surface, as well as for aluminium alloys containing silicon. Hydrochloric acid of similar concentration can also be employed, but the results are not quite so attractive.

4.2.6.3.1. Chemical milling of aluminium

Chemical milling is an increasingly important method of forming metal strips by etching away unwanted areas to different thicknesses in suitable pickling solutions or etching agents. Its main application has been in the aircraft industry on aluminium and its alloys to reduce weight without the cost and complication of mechanical removal of the metal[46], but it can also be used on ferrous metals. Chemical milling is particularly useful for forming complex patterns on curved surfaces without mechanical distortion; it is also valuable for tapering, step-etching and the sizing of plate and sheet.

The process involves four steps, viz., preparation, masking, etching and stripping. For aluminium, hydrochloric acid or caustic soda solutions are used, whilst aqua regia is employed for low- and high-alloy steels. Etchant recovery and regeneration systems are desirable to maintain the uniformity of the rate of attack, to decrease operation costs and to reduce the problem of effluent disposal. Strippable coatings of various types are used for local masking.

The indications are that chemical milling has no adverse effect on the physical properties of aluminium and its alloys.

4.2.6.4. Magnesium

Castings represent the most common form in which magnesium has to be pickled to remove oxides and contaminants generally. A suitable bath for cleaning both sand- and die-castings consists of chromic acid 280 g/l, nitric acid 25 ml/l and hydrofluoric acid 8 ml/l. The object of the hydrofluoric acid is to remove sand particles; where these are not present, as in wrought articles, a dip in a solution of 180 g/l of chromic acid is advantageous. Other recommended solutions contain small amounts of sodium nitrate and calcium fluoride in addition to the chromic acid..

To obtain satisfactory durability of paint coatings on magnesium, it is essential that a suitable conversion coating be applied to the metal (Chapter 13). Before treatment, castings are generally cleaned in a 2 to 5 per cent caustic soda solution and then pickled in a 5 per cent sulphuric or nitric acid bath to remove the casting skin, unless they have been machined. If they have been machined, the acid treatment is omitted.

4.2.6.5. Titanium

During fabrication and heat-treatment titanium develops a complex oxide scale[47] which consists mainly of titanium dioxide (TiO_2) with underlying layers of Ti_2O_3 and TiO. The scale remains unattacked by most acids unless they are highly concentrated; it can, however, be removed by immersion for a short time in a 2 to 5 per cent solution of hydrofluoric acid at room temperature, or more rapidly at higher temperatures; the time of treatment is only 2 to 5 minutes[48]. A further solution which has been recommended is composed of nitric acid 450 ml, fluosilicic acid 100 ml, ammonium bifluoride 100 ml and sulphuric acid 450 ml. The nitric acid passivates the titanium surface and reduces the possibility of hydrogen embrittlement.

Another pickle, which does not react with the scale but attacks the underlying metal and thus spalls off the scale, is a solution of fluoboric acid (5 to 20 per cent) used at 75° to 95°C. Treatment with a hot sulphuric acid/calcium chloride solution followed by a nitric/hydrofluoric acid bath gives a bright surface free from pits, with no evidence of hydrogen absorption; hence the formability of the metal is unimpaired. All these solutions, however, attack the basis metal so that some degree of etching occurs.

4.2.6.6. Mixed metals

A difficult problem in pickling is presented when a component consists of a mixture of metals. The commonest of such cases is in the pickling of articles which have been brazed or soldered. Immersing steel articles which have been silver-soldered in sulphuric acid pickling solutions may result in the accumulation of copper salts in the bath and the metal is then re-deposited on the steel.

There is no simple answer to this problem; every case must be treated on its merits but, broadly speaking, if the acid is kept frequently renewed, assemblies of this kind can usually be pickled without undue difficulty.

4.2.6.7. Noble Metals

Gold is generally too soft for practical use in jewellery and dentistry, and is therefore alloyed mostly with silver or copper, or both of these metals. A certain amount of discolouration therefore occurs as a result of oxidation in the heating of the metal during fabrication. The oxidation is removed by a pickling operation.

In the case of gold-copper alloys, a short treatment in 2-5 per cent hot sulphuric acid in a barrel is usually sufficient to remove discolouration in a few minutes. If silver is also present, either a 50 per cent solution of nitric acid or a mixture of equal parts of nitric acid, sulphuric acid and water give satisfactory results.

Gold which accumulates in the pickles can be recovered by precipitating it out with zinc dust.

Silver tarnishes in the atmosphere at varying rates, depending on the degree of pollution. This is a result of the formation of silver sulphide by reaction with sulphur-containing gases in the atmosphere. In the case of silver-copper alloys, the discolouration is due to a mixture of silver and copper sulphides. After fabrication, during which silver articles are heated to about 600°C, the conventional treatment is to boil them in an aqueous solution which contains 3 per cent tartaric acid and 5-6 per cent sodium chloride for 10-20 minutes after a short preliminary dip for a few seconds in a 50 per cent solution of nitric acid. Alternatively, the operation may be carried out in a 5-10 per cent solution of boiling sulphuric acid, when the copper oxide is dissolved, the silver being only slightly attacked. To accelerate the action of the sulphuric acid, a small amount of an oxidisation agent such as potassium permanganate may be added with advantage.

Sulphide tarnish on silver can also be removed by immersion in a 30 g/l solution of potassium cyanide to which 1 g/l of zinc cyanide has been added. Good results can also be obtained by immersion in a solution of sodium thiosulphate which has the advantage of being non-toxic.

Small silver articles can be rapidly deoxidised by putting them in contact with aluminium foil or zinc granules in a hot 60 per cent sodium carbonate solution. An electrochemical reaction occurs in which the silver sulphide is reduced to metallic silver by the hydrogen evolved.

Platinum metals form an exceedingly thin oxide film, which is generally removed by mechanical polishing after cleaning in an alkaline solution, which does not attack them.

4.2.7. DRYING AND TEMPORARY PROTECTION

The drying of metal components after pickling and washing is very important, and a variety of equipment is available for the purpose. This is described in detail in pp. 213-221. The objectives of drying are to prevent tarnish, corrosion and water spotting.

The simplest method is to rinse and immerse the articles in hot water, preferably deionised, and allow to dry in air. To be effective, however, they should have sufficient mass to retain enough heat to vapourise the water film. Other methods of drying include hot air, infra-red radiation, solvent rinsing and centrifugal removal of water. The oldest system of drying plated articles is perhaps the method of putting them in rotating barrels with resin-free sawdust; the process takes 20-60 minutes, the residual sawdust being removed by blowing with compressed air. It is not suitable for articles with deep holes or recesses.

Wetting agents are useful for drying articles with large surface areas, and dewatering agents are frequently used to displace the water by an organic, mildly protective film by immersion.

Pickling generally renders the metal surface sensitive to further chemical reaction, so that it is necessary that there should be as little delay as possible before the pickled work goes on to the next operation.

If on the other hand for any reason, it is not possible for the next finishing process to be carried out immediately, the work should be protected from corrosion. If the parts are to undergo other processes, the protection should be ' temporary '.

A protective system of this kind should have the following characteristics:

1. The protective coating should be easy to apply and to remove.
2. It should have little or no effect on the appearance, dimensions or properties of the surface.
3. The parts must be capable of being finished rapidly and without undue labour costs.
4. The cost of the protective should not be too high in relation to the cost of the part.

In applying the temporary protective, the parts can be immersed in a dilute solution of the corrosion protective or of compounds which will remove any tarnish films that are formed. For ferrous metals, dilute chromate, nitrate or nitrite solutions are suitable or the pickled work may also be stored in dilute hydrochloric acid before the next operation. Dilute cyanide, tartrate or tartaric acid solutions are usually used for copper.

Another method of protecting the pickled components is to provide a protective

Fig. 4.19. Hot air drying section of a cleaning and phosphating machine.
[Courtesy Oxy Metal Industries (G.B.) Ltd.]

coating which is easily removed. For this purpose the metal is dipped for a short period in a dilute aqueous solution of compounds which act as corrosion inhibitors. After this they are allowed to dry in the air without rinsing.

Temporary protection is also afforded by a thin, continuous film of an anti-corrosive oil or grease. These can be applied by brushing, spraying or dipping the work in the liquid or semi-liquid material. After application the solvent evaporates and, depending on the concentration of the solution, a film remains on the surface consisting of the dissolved grease or oil. The solvents are volatile organic liquids such as petrol or chlorinated hydrocarbons.

In order to save solvent, emulsifiers and corrosion inhibitors are also added to mineral oils. Together with water these give a homogeneous emulsion which is applied to the metal surface where it leaves a thin protective film. Emulsions of this type are very economical in use as they exert their protective action at high dilutions (1 : 10) while the use of water as a diluent does not involve special safety precautions. The film is removed simply by immersing the parts in water or by wiping them with a damp cloth, leaving the surface in its original condition. Materials of this type are also known as water displacing oils or dewatering oils and lacquers.

Fig. 4.19 shows the hot-air drying section of a cleaning and phosphating machine in which the heated air enters the oven through a series of adjustable cast aluminium nozzles. They are arranged to provide a mechanical blow-off effect to accelerate drying, and enable an air-blast of 3,500 m/min velocity to be applied over the entire surface of the work.

4.2.8. PICKLING PLANTS AND THEIR OPERATION

Acid pickling was, for many years, regarded as a very crude process not amenable to close control. Fumes and poor working conditions were considered inseparably associated with pickling operations. To-day, however, it is possible to operate a pickling plant under conditions as good as those prevailing in many other parts of a factory.

4.2.8.1. Construction

The manufacturers of pickling equipment have contributed in no small measure towards achieving this. Early plant for pickling consisted of tanks made of wood planks, bolted together and strengthened with metal reinforcement. Pickling tanks have also been made of slate, stone and even of concrete, covered with bituminous or asphaltic coatings. These proved successful to varying degrees, but there was always the difficulty of leaking joints. Available acid-resisting jointing compositions are seldom capable of withstanding the expansions and contractions of tanks of this description and of resisting acid penetration, especially when hot sulphuric acid is used. Wood tanks lined with chemical lead have been used for sulphuric acid pickling. They are not suitable for hydrochloric acid, but they can be used for hydrofluoric acid

Fig. 4.20. Concrete pickling tank lined with pvc and brick. [Courtesy Nordac Ltd.]

pickles. Steel tanks may be similarly lead-lined, but perforation of the lead soon causes complete failure of the tank.

Pickling tanks may suitably be constructed of acid-resisting brick combined with special cements. Epoxy-resin cements are suitable for this purpose. Non-metallic pickling tanks constructed of laminated synthetic resin-impregnated fibrous materials of various kinds are also used, the resins being usually of the phenolic type.

Steel tanks for pickling are nowadays usually lined with hard rubber or with polyvinyl chloride. The latter can be applied in sheet form and bonded to the steel with suitable cements, or the material may be applied as a plastisol by dipping the entire tank. Plastics have the advantage over rubber of being unaffected by oxidising acids under most conditions of ordinary use.

Where heavy steelwork is to be pickled, an acid-resisting brick lining over the rubber serves to protect the latter from mechanical damage. Such tanks are capable of giving many years of trouble-free service. The linings are also applicable to hydrochloric acid pickling, and to solutions containing hydrofluoric acid at temperatures up to 65 °C. Plants of these types are shown in the accompanying illustrations. Fig. 4.20 illustrates a reinforced concrete tank lined with p.v.c. and protected by blue-bricks in acid-resisting cement. Fig. 4.21 shows a sheet steel pickling installation of similar construction. Armour plate glass tanks may also be employed for pickling in special cases[50] but they are relatively expensive and are limited in size.

Fig. 4.21. Sheet steel pickling plant. [Courtesy Nordac Ltd.]

Fig. 4.22. Rubber lined tube pickling installation. [Courtesy Nordac Ltd.]

4.2.9. PROCESS CONTROL IN PICKLING

Sulphuric acid pickling baths are made from the strong acid, purchased preferably as the purer and more concentrated commercial grade which has a density of 1.84 (66° Bé) and contains 95 to 97 per cent sulphuric acid. Good quality acid is desirable, since the presence of contaminants such as arsenic may have an inhibiting action and thus materially reduce the rate of pickling, as has already been pointed out.

Hydrochloric acid is purchased in the form of a 30 to 35 per cent solution. An approximately 10 per cent solution of the commercial acid can, therefore, be made up by adding one part of the acid to two parts of water by volume. The pickling solutions commonly employed vary from 5 to 10 per cent in acid content, and once a suitable concentration has been decided upon, daily acid additions must be made to maintain the strength.

4.2.9.1. Analytical control

The said concentration can be determined by direct titration of the solution with standard caustic soda using phenolphthalein as an indicator. A practical innovation suitable for use on the plant consists in adding pellets of sodium bicarbonate containing a little phenolphthalein to a sample of the pickle taken in a suitable measuring container. Each pellet can be of such a weight as will correspond with the presence of 1 per cent of acid in the pickle. The number of pellets which have to be added to the sample before the solution turns pink is then a measure of the percentage of the acid present.

The iron content should also be determined at frequent intervals since the accumulation of iron salts in the acid reduces the rate of pickling. After a time it is not feasible to add more acid, and the solution must be discarded and a new one made up. Density measurements alone do not given a true indication of the acid and iron contents of the bath since both these constituents influence the density of the solution stimultaneously.

Winterbottom and Reed[57] have constructed nomograms which are useful and adequate for control purposes. They relate the iron and acid contents with the density of the pickling solution in degrees Beaumé for both sulphuric and hydrochloric acid pickling. For sulphuric acid the nomogram has been based on the relation: $(Fe) = 2.0\ [T - 0.132(H_2SO_4)]$, and for hydrochloric acid, on $(Fe) = 2.56\ (T - 0.098(HC1))$. Analytical methods are, however, generally applicable and easy to carry out. As all the iron in the solution is in the ferrous state, titration with standard potassium permanganate solution enables the iron content to be determined directly. In the case of hydrochloric acid solutions it is necessary to add manganous sulphate and phosphoric acid to prevent oxidation of the acid to chlorine, and to prevent the development of the ferric iron coloration.

Colorimetric papers for the rapid determination of iron in pickling solutions can be prepared by dipping filter paper strips into a 1 per cent solution of ammonium thiocyanate in acetone. By treating these with a series of standard solutions of ferric chloride, red colours corresponding to each concentration of iron can be obtained; these colours can then be compared with that produced by a drop of the pickling solution which is to be tested. The iron in the pickle should have been oxidised previously to the ferric condition by the addition of a few drops of nitric acid.

As has already been stated, when the iron content reaches some 8 per cent in the case of sulphuric acid and 12 per cent in hydrochloric acid, the pickling rate is reduced to such an extent that the bath must be discarded. In actual practice, it is often found not an economic proposition to run sulphuric acid pickling baths with an iron concentration in excess of 6 per cent, or hydrochloric acid baths containing more than 9 per cent of iron.

The operating temperature in sulphuric acid pickling is best kept at about 60°C in the case of new solutions, gradually raising the temperature to 70—75°C as the iron content builds up. The pickling operation itself may take from three to twenty minutes, depending on the thickness and condition of the scale. Excessively long pickling times are to be avoided in order to prevent subsequent processing difficulties. If scale removal is not effected satisfactorily in a relatively short time, changes should be made in the pickling process itself (by altering the acid, temperature, etc), rather than by attempting to obtain the required results by extending the duration of pickling.

4.2.10. WASTE DISPOSAL

The waste acid must be neutralised in suitable units by means of alkali before being discarded. The time-honoured method is to pass the waste acid through a chamber built of acid-resisting brick and containing lime, often mixed with steel swarf to prevent consolidation of the neutraliser. It is more convenient to use a lime slurry for neutralisation, mixing it with the acid at the point where it enters the neutralising pit and agitating the mixture by means of air or a pump. The best arrangement consists of a two-compartment tank; the smaller compartment is the mixing chamber where both milk of lime and waste acid are fed together in appropriate proportions. The mixed lime and acid solution then passes over a weir into the larger compartment where the sludge is kept in suspension by means of perforated air pipes at the bottom. When neutralisation is complete the tank contents are discharged by means of a pump or a steam ejector.

A neutralising plant can consist of an outer wall of reinforced concrete, asphalt or rubber, lined with an inner protective lining of blue brickwork set in siliceous or resinous cement. The weir walls must be liquid-tight and built and lined integrally with the outer walls to ensure this. However, increasingly stringent Water Authority regulations are making it necessary to use more sophisticated methods of acid waste treatment. An HCl regeneration plant for spent pickling liquors has been described by McManus[58].

Ferrous salts can be recovered from spent sulphuric acid pickling solutions by crystallisation in specially designed plants, and this system is being increasingly adopted where large-scale pickling operations are carried out. The liquor containing iron equivalent to 85 per cent of maximum solubility is pumped into a crystalliser filled with coils through which refrigerated or deep well water is circulated. The crystalliser is also fitted with a jacket through which the water circulates, the acid being rapidly stirred meanwhile. By lowering the temperature of a 10 per cent sulphuric acid solution from 60°C to 20°C, the iron content of the solution can be reduced to about 6 per cent in some 2 hours. The ferrous sulphate crystals formed in cooling are precipitated at the bottom of the crystalliser (which is conical in shape) when the stirrer is stopped. The regenerated acid is pumped back for further use, the crystals being run off, drained, washed and dried in a centrifuge.

Acid recovery plants can effect considerable savings besides enabling the pickling process to be carried out under close control and at relatively low temperatures. The ferrous sulphate crystals can be calcined to produce iron oxide.

When ferrous sulphate is heated in air at fairly low temperatures, $FeSO_4.7H_2O$ crystals lose 3 molecules of water of crystallisation until the lower hydrate, $FeSO_4.4H_2O$, is reached. At 167°C this compound is oxidised to basic ferric sulphate; this in turn decomposes at 492°C (in practice, temperatures of 550°C to 600°C or more may be used) into ferric oxide, Fe_2O_3, and sulphur trioxide SO_3, which is readily converted into sulphuric acid by absorption in water, or better, in fairly strong sulphuric acid. The iron oxide finds useful application as a paint pigment and in the manufacture of polishing compositions.

Three patented methods of decomposing ferrous sulphate in practice are worthy of mention. In the first, the ferrous sulphate is mixed with about 30 per cent of its weight of ferric oxide, Fe_2O_3, and 1 per cent of sodium carbonate to neutralise free acid; it is then dried and sintered at 750°C to 850°C, when sulphur trioxide is evolved, leaving Fe_2O_3. Alternatively, the iron sulphate is roasted and treated with ammonia gas and water to produce ammonium sulphate and iron oxide.

In another process[59] the septahydrate is introduced into a fluidised bed roaster where it is converted to a free-flowing monohydrate powder which, in turn, is decomposed at 700°C into sulphur and iron oxide in the proportion of 1 kg of each of these products per 5 kg of monohydrate. The Autoxidation process developed by Ruthner reacts the dissociated monohydrate with hydrochloric acid gas, precipitating ferrous chloride and regenerating sulphuric acid. By heating the chloride, the hydrochloric acid is recovered and used again and the iron oxide is sold to steel mills. In the Zaher process the waste liquor is concentrated in a spray tower, further enriched with strong acid, and the monohydrate which separates is filtered off and roasted to ferric oxide and sulphur dioxide.

Jangg[60] has recently reported a process for separating H_2SO_4 and iron electrolytically from spent pickling liquors using magnesium cathodes.

REFERENCES

1. *Product Finishing,* 1969, **22**, No. 7, 62.
2. *Electroplating and Metal Finishitng,* 1955, **8**, 67.
3. *Product Finishitng,* 1968, **21**, No. 6, 104.
4. P. B. Wharton, *Product Finishing,* 1969, **22**, No. 8, 57.
5. J. B. Hignett ' Symposium on Pretreatment for Metal Finishing Processes ', p.19, d.30, Institute of Metal Finishing, 1965.
6. J. B. Mohler, *Metal Finishing,* 1972, **70**, No. 11, 48.
7. L. B. Pfeil and A. B. Winterbottom, ' A Review of Oxidation and Scaling of Heated Solid Metals ', Dept. of Sci. Inc. Res., 1935.
8. A. J. von Fraunhofer and G. A. Pickup, *Corr. Sci.,* 1970, **10**, 253.
9. G. Garber, *J. Iron and Steel Inst.,* 1959, **192**, 153.
10. H. Bablik, *Iron Age,* 1929, **123**, 879.
11. U. R. Evans, ' Introduction to Metallic Corrosion ', p.142, Arnold, London, 1960.
12. P. Dickens, *Stahl & Eisen,* 1939, **59**, 364.
13. H. Gehman, *Wire and Wire Prods.,* 1966, **41**, 1609, 1612, 1703.
14. H. A. Cork, *Ind. Finishing,* 1973, **25**, No. 303, 14, 16.
15. R. Wilmotte and J. Benezech, *Galvano,* 1973, **42**, No. 434, 667.
16. M. Straschill, ' Modern Practice in the Pickling of Metals and Related Processes ', Chapter 3, Robert Draper Ltd., Teddington, 1963.
17. J. M. West, ' Electrodeposition and Corrosion Processes ', Van Nostrand, Toronto, N.Y. and Princeton, 1971.
18. H. A. Cork, *Ind. Finishing,* 1974, **26**, No. 317, 11.
19. E. Mayer, *Galvanotechnik,* 1959, **50**, 598.
20. M. App, *Galvanotechnik,* 1961, **16**, 520.
21. A. A. B. Harvey, ' Symposium on Pretreatment for Metal Finishing Processes ', p.6 Institute of Metal Finishing, 1965.
22. M. Svoboda and B. Knapek, *Product Finishing,* 1969, **22**, 41.
23. R. Wilmotte and J. Benezech, *Galvano,* 1973, **42**, No. 432, 335.
24. J. N. Ostrofsky, U.S. Pat. 2,261,744.
25. M. Straschill, *Corrosion Prevention and Control,* 1961, **8**, 46,283.
26. M. Straschill, *Electroplating and Metal Finishing,* 1950, **3**, 253.
27. E. Taylor, *Metal Finishing,* 1973, **71**, No. 1, 36.
28. E. L. Cody, *Materials and Methods,* 1946, **23**, 1278.
29. Brit. Pat. 1,139,201.
30. 'The Use of Gluconates in Alkaline Cleaning ', Product Information Bulletin, No. 310. Pfizer Ltd.
31. M. J. Reide, *Galvanotechnik,* 1967, **58**, 176.
32. J. Kronsbein, *J. Electrodepositors Tech. Soc.,* 1940, **16**, 55.
33. Tainton, U.S. Pat. 2,063,529.
34. Ger. Pat. 572,453.
35. C. H. Edwards, *J. Iron and Steel Inst.,* 1924, **110**, 9.
36. C. A. Zapffe and C. E. Sims, *Met. Alloys,* 1941, **13**, 737.
37. C. A. Zapffe and M. E. Haslim, *Plating,* 1949, **36**, 906.
38. L. Felloni and G. Bolognesi, *Ann. Chim. (Rome),* 1957, 47, 985, 996.
39. H. J. Noble, *Iron Age,* 1941, **45**, Nov. 27.
40. S. A. Balezin and N. I. Narushevich, *Zhur. Priklad Khim,* 1960, **38**, 2536.
41. C. A. Zapffe and M. E. Haslem, *Trans. Am. Soc. Metals,* 1947, **39**, 241, 248.
42. H. Silman, *Metallurgical Reviews,* 1959, **4**, No. 16.
43. J. S. Dunn, *J. Inst. Met.,* 1931, **46**, 35.
44. P. H. Margutils, *Proc. Am. Electroplaters' Soc.,* 1955, **42**, 60.

45. P. Fintschenko, *Met. Finishing J.*, 1974, **20**, No. 233, 138.
46. C. C. Shepherd, *Proc. Am. Electroplaters' Soc.*, 1960, **47**, 51.
47. A. D. Mcquillan and M. K. Mcquillan, 'Titanium', Chapter 12, Butterworths, London, 1956.
48. W. B. Stevens, *J. Metal Prog.*, 1958, **73**, 87.
49. S. H. Grinrod, *Electroplating*, 1965, **18**, No. 12, 418.
50. G. L. West, *Metals and Alloys*, 1945, **21**, 413.
51. *Wire Industry*, 1953, **20**, 1081.
52. *Indust. Gas.*, 1953, **17**, 194, 48.
53. W. E. Hoare, 'Symposium on Pretreatment for Metal Finishing Processes', p.31, 36, Institute of Metal Finishing, 1965.
54. *Product Finishing*, 1967, **20**, No. 1, 60.
55. *Product Finishing*, 1967, **20**, No. 6, 45.
56. *Product Finishing*, 1969, **22**, No. 2, 54.
57. A. B. Winterbottom and J. P. Read, *J. Iron & Steel Inst.*, 1932, 2, 182.
58. G. J. McManus, *Iron Age*, 1972, **210**, No. 16, 83.
59. D. W. Becker, *Industrial Chemist*, 1960, **36**, 604.
60. G. Jangg, *Electrodeposition and Surface Treatment*, 1972, **1**, No. 2, 139.

BOOKS FOR GENERAL READING

1. M. Straschill, 'Modern Practice in the Pickling of Metals and Related Processes', Robert Draper Ltd., Teddington, 1963.
2. H. J. Plaster, 'Blast Cleaning and Allied Processes', Vols. I and II, Portcullis Press, London, 1974.
3. E. W. Mulcahy, 'The Pickling of Steel'. Industrial Newspapers, London, 1973. (Deals with both plant and methods of pickling.)

CHAPTER 5

Process plant and equipment

ELECTROPLATING and related metal finishing operations have long left the stage when they were carried out by craftsmen dipping articles into tanks of liquids of arbitrary compositions. Hand operated plating equipment is by no means obsolete, but where large-scale production is concerned automatic plant is nowadays regarded as essential. Fig. 5.1 shows a hand-operated plating line while Fig. 5.2 illustrates a plating barrel made of moulded polypropylene panels. The door clips are of titanium.

Basically, the electroplating operation itself is simple, although the complete process sequence used in electroplating is elaborate since meticulous cleaning is necessary, and this in itself may involve a considerable number of separate operations. The after-treatments, including rinsing and drying, must also be carried out with care, and require numerous stages, with differing times for each. Plant for mechanical treatments (grinding, polishing, etc.) and for cleaning have been discussed in Chapters 2 and 3 respectively.

The lay-out of a plating plant needs the most careful consideration. The entire plant should be designed for flow production, and to reduce unnecessary movement of personnel or work to a minimum. The arrangement of the processes, in the case of manually operated plant, should facilitate the successive operations of polishing, pre-cleaning, racking, plating, unracking and inspection. Stripping of rejected work must also be catered for. Adequate facilities for the maintenance and repair of plating racks, including rack insulation equipment, are also desirable.

Ventilation of the plating shop is an essential part of the installation, and adequate provision must be made for the intake of air into the shop to compensate for that removed by exhaust fans.

Electroplating plant requirements are rather critical as regards choice of materials because of the corrosive nature of the solutions used, and the general conditions of service to which the plant is subjected.

An electro-plating installation would comprise the following main items:

(1) Plating, cleaning, and rinsing tanks made of suitable materials.
(2) A source of direct current supply at a low voltage (usually not exceeding 12 volts, although for anodising up to 60 volts may be required).
(3) Equipment for the heating of various tanks.
(4) Ancillary plant, such as measuring and recording instruments, filtration units, pumps, air compressors or blowers for agitation of solutions, and exhaust ducts and fans to remove noxious fumes.

5.1. Automation

Electroplating processes are very amenable to automation. The articles to be processed are transferred through the various tanks, either on racks or in barrels, and a wide selection of plants is now available to meet almost any requirement. In addition, plating and other finishing operations are carried out as a continuous process on metal in strip form and on wire. Automation has been greatly facilitated as a result of bright plating, which has eliminated intermediate polishing. The economics of automatic plating has been discussed by Cummins Jr.[1]. Chemical dosing equipment which will not be dealt with here has been reviewed by Rubinstein[2].

In general, the principal advantages of automatic plating plant may be summarised as follows:

(1) Consistent adherence to the operation sequence, thus reducing defective work.
(2) Improved working conditions and the elimination of the need for protective clothing for operators.
(3) Uniformity of quality and the maintenance of plating specifications.
(4) Improvement in factory layout and economy of floor space.
(5) Effective plant utilisation, ensuring that a pre-determined output is maintained.
(6) The feasibility of using deep tanks and large rack sizes.
(7) The reduction of damage due to handling.
(8) Reduction in labour costs.

For large-scale production, automatic plant is nowadays regarded as essential to ensure that both quality and output should be maintained. The basic principles of the design and construction of automatic plating plants have been discussed by Silman[3].

5.2. Fully-automatic plant

In modern installations the entire sequence of plating operations is carried out on the plant, from cleaning and plating to rinsing and drying. Where components are being plated they are transferred smoothly on racks or in barrels through the necessary

Fig. 5.1. Hand-operated plating installation.

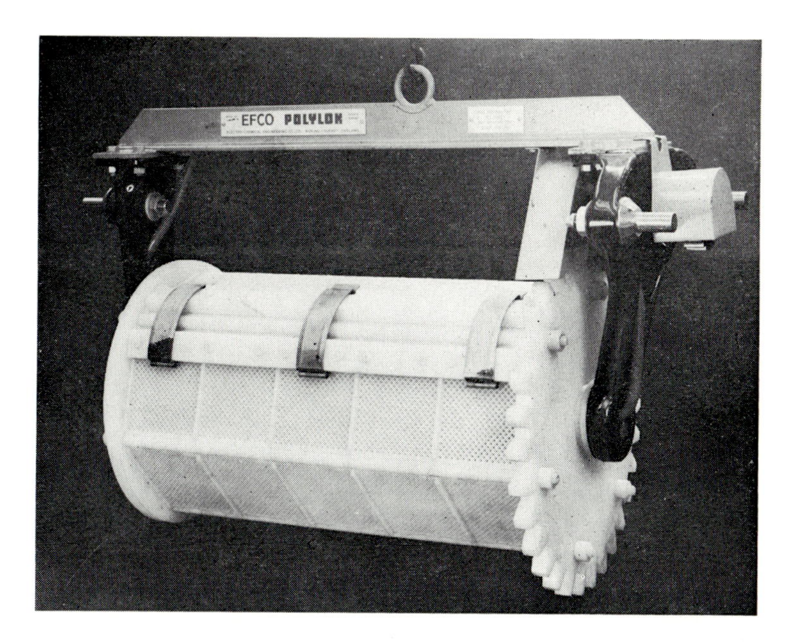

Fig. 5.2. Polypropylene moulded electroplating barrel. The hangers at the ends are of P.V.C.-coated steel. [Courtesy Oxy Metal Industries (GB) Ltd.]

processes successively, the time for each ranging from, say, a minute or two up to hours if need be. Provision is made for the application of controlled currents at appropriate stages, the work being made either the anode or the cathode as required.

Early automatic plants were large and cumbersome and usually designed and built for a particular type of production. Nowadays, there is a high degree of standardisation, and modern plants are more in the nature of machine tools, having a high degree of flexibility in operation. Neither are such plants necessarily confined to the mass production of large quantities of similar articles which do not vary much in design, machines being available which will cope with a very wide variety of work.

This has been made possible by improvements in engineering and design, and by increased demand which has favoured plant standardisation. An equally important factor, however, has been the improvement of the electroplating solutions themselves. They are less critical with respect to operating conditions and also enable deposition to be carried out at higher speeds. Hence plants can be smaller and lighter in construction as well as lower in cost. In the case of bright nickel, which is by far the most extensively used of all plating processes, modern solutions enable articles of varying shapes and dimensions to be treated without the special racking or the very close current control hitherto considered essential.

5.3. Types of automatic plant

Automatic plants can be broadly classified as being of the ' through ' or of the ' return ' type. In the former, the articles are loaded and unloaded at opposite ends of the machine, but in the latter they travel through a series of tanks arranged in the form of a closed loop and are loaded and unloaded at the same point.

Return-type plants have been widely favoured over the years; they have the advantage that as the work returns to the loading position, the machine can be controlled by a single operator. The need for sending plating racks back to the loading end of the plant, which is a problem with straight-through types of machines, is also eliminated. However, with the more recent introduction of the programmed automatic plant, which will be referred to later, there has been a renewed interest in the through-type plant. These machines consist essentially of a line of tanks along each side of which runs a conveyor carrying a series of lateral bars from which the plating racks are suspended. Four or five lines of work can pass through such machines side by side, the suspender bars travelling through the tanks and being lifted when necessary to enable the racks to be transferred from one tank to the next. The output of such machines is high, whilst the width of the tanks enables large articles to be plated if required. Transfer may be carried out by a variety of mechanisms which are designed to lift the bars carrying the racks into successive tanks and to move them forward progressively within the process tanks themselves. Linear motors in automatic machines have considerable advantages over conventional motors in that they require little maintenance and are relatively free of wear[4].

5.3.1. RETURN-TYPE RACK PLANTS

The return-type rack machines are ideally suited to production where the output requirement is high, and the cycle time is fixed. In one widely used design of plant, a hydraulically operated lifting frame is used for transferring the racks of work from one tank to the next; this runs the length of the machine and is alternatively raised and lowered, carrying sections of the cathode track, where partitions between tanks occur, with it. In the main, a single cathode track is used, but where the articles are not too large, a double track plant is entirely practicable. Exceptionally, triple track plants have been built, but difficulties arise with the movement of the work racks around the ends of the machine, whilst the rate of loading and unloading becomes too high for ease of handling. The racks are moved forward intermittently by rachetted pushers attached to reciprocating T-bars acting on sliding suspenders which carry the racks along the cathode track. When the main plating operation is a long one, the tank in which this is carried out can be extended beyond the carriage at one end of the plant, the movement within it being taken over by a separate mechanism.

The pushers move the racks in the tanks whilst they are immersed intermittently at approximately one minute intervals, the racks being stationary except when a transfer is being effected somewhere in the plant. For exceptionally short immersion times devices for delayed set-down of racks at certain stages can be incorporated in the plant. At the ends of the machine, pushers attached to segments oscillating in a horizontal plane guide the work suspenders around the machine on the curved cathode track. A single-track return-type plant of this description will deliver from forty to eighty racks (equivalent to about 20 sq m) of work per hour; two-track machines deliver up to 160 racks per hour on average. The cycle is, however, variable to some extent, being controllable by the central selector switch box. The arrangement is shown diagrammatically in Fig. 5.3.

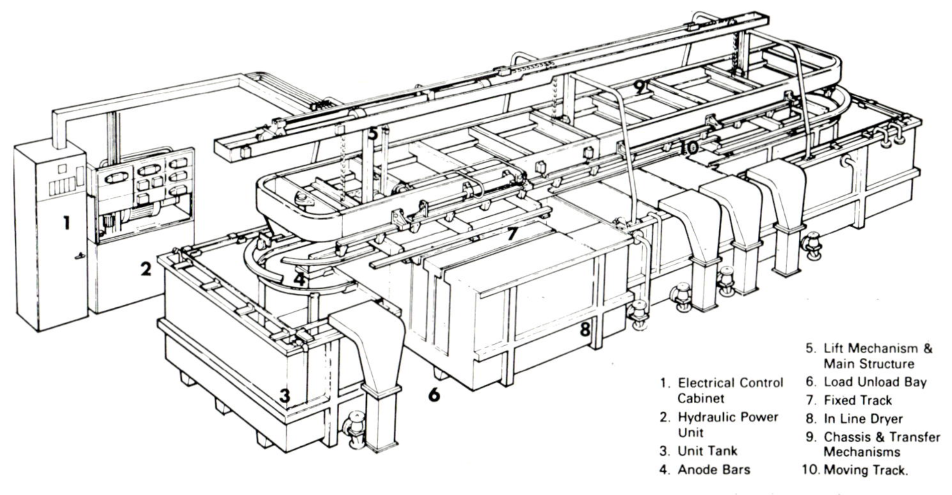

Fig. 5.3. Return-type automatic plating machine (the Udylite Processmaster).

The number of tanks in a plant of this type is generally about twenty; these are lined with materials such as plastic or rubber which are suitable for containing the solution with which each is filled. The hydraulic cylinders lifting and lowering the frame, which is balanced by means of counterweights, are located above tank level and hence are easily accessible.

Two designs are available which are essentially similar in principle, differing only in size and method of construction. In the first, the tanks are of multi-compartmented unit construction, the tank itself forming the base of the machine. In the second type, the tanks are independent of the base of the centre structure, and are not an integral part of it. The machines are designed to fail safe so that if any one movement is not completed the machine will stop. Since their original conception the return-type machines have been continuously developed and are now much more versatile and sophisticated than they were originally. In particular, they can now meet an extremely wide range of customer requirements. The following are some of the available modifications.

(1) *Extension Tanks and Multiple Indexing*

The basic machine can be fitted with an extension tank, usually for the main plating process, situated at the end of the machine beyond the main lifting section. This tank is fitted with a fixed chassis which carries the cathode track, pusher tees, etc., similar to the lifting chassis on the main machine. It has transverse movements only and these operate in unison with, and are synchronised with, those on the main machine.

For normal operation the extension tank traverse will take place only when the main machine pusher tees make their bottom traverse movement. However it is possible to double index in the extension tank, i.e., when the main machine makes its top traverse the extension traverse mechanism will operate thus giving two bottom traverse movements per cycle to one on the main machine. The effect is a 50% reduction in this particular process time and deposit thickness. If the overall plant cycle time allows, triple indexing is also possible; this produces a 66% reduction in deposit thickness where this is required. Multiple indexing can only be used by batching loads of work to the machine.

(2) *Automatic Load/Unload*

It is possible to fit automatic load/unload mechanisms to either single or two-row machines. Thus the rack size and weight need not be limited to that which a man can handle. Work flow to and from the machine can be automated, with the loader transferring racks between the plating machine and the shop conveyor to make a fully automatic unit integrated into a flow line mass production system. The hydraulically-powered loaders operate within the main machine cycle time and are synchronised with it.

(3) Delayed Set-Down

This mechanism is employed at any single stage process tank where the required immersion time is critical and is less than the other single stage process in the chemical sequence. It is hydraulically operated and its operations are integrated with those of the machine. Batch loading or individual rack selection is possible; where selection is random it is determined by inserting a selection pin in the heads of those racks of work requiring a delayed set-down. This mechanism can also be used to completely by-pass a single stage process tank. The required delay time is pre-set on a time clock mounted on the face of the electrical control panel.

(4) By-Pass Mechanism

The by-pass mechanism can be fitted on a machine where a multi-stage process tank or series of process tanks are to be by-passed by batches of work or individual racks. The mechanism consists of an additional fixed section of work track mounted at high level and parallel to the work track on the main machine; this track extends over the tank or tanks to be by-passed. Selected racks of work are transferred to and from the machine to the by-pass track by means of shuttle mechanisms mounted in the main plant-lifting chassis at the entry and exit to the by-pass. Selection is made, either for batched work or individual racks in a similar manner to delayed set-down selection. This principle can also be utilised on two parallel rows of tanks to give a choice of two. alternative processes.

(5) Dryers

Dryers operating on a recirculating hot air principle are usually incorporated into the tank layout as the final process prior to reaching the load/unload stage. Air is recirculated by an axial flow fan through a heat exchanger.

(6) Unload Extensions

The load/unload section, if at the end of the plant, can be extended to form a low level conveyor on which parts can be directly racked and unracked.

Fig. 5.5 shows the unload end of a similar plant for zinc plating brake components. An automatic load-unload unit is seen in Fig. 5.4.

5.3.2. AUTOMATIC RETURN-TYPE BARREL MACHINES

There are three sizes in this group of high output machines and although their barrel capacities vary, their principle of operation and method of construction are basically similar. The machines are of the return type, the load stage always being at the end with the unload stage adjacent to it.

The main structure of the plant consists of a series of centre columns mounted on a base frame. The process tanks, as in the case of the Senior machine, are separate

the interconnecting ducting to enable the humidity in the cabinet to be controlled.

As the barrels leave the drying sections they climb a further cam bringing the barrel to its unload position where the dry plated work falls under gravity into an unload chute which directs it into a pan. In order to ensure that the barrel empties completely it is rotated in its unload position by a separately-driven section of overhead worm.

Many features can be incorporated to suit process sequence requirements.

(1) Delayed Set-Down: This mechanism is applied to single stage tanks only and can be electrically or pneumatically operated. Batch or individual barrel selection is possible.

(2) By-passing: By-passing of multi-stage tanks or part multi-stage tanks is possible using fixed by-pass tracks at high level. Batch or individual barrel selection can be made.

(3) Automatic Loaders: These can be supplied to work in conjunction with this machine to make a fully automatic unit which can easily be integrated into a flow line system. Some of the above features are illustrated diagrammatically in Fig. 5.7.

Figs. 5.6 and 5.8 show an automatic unloading unit in operation on a barrel machine.

Fig. 5.6. Automatic barrel plating machine at International Harvester, Pullman Works, Chicago. The conveyor (foreground) transports the unplated parts to a hopper. A designated number of pieces are then automatically weighed and loaded into the barrel. Note that all loading, unloading, rectifier and machine controls are within easy reach of the operator.

[Courtesy Oxy Metal Industries Inc.]

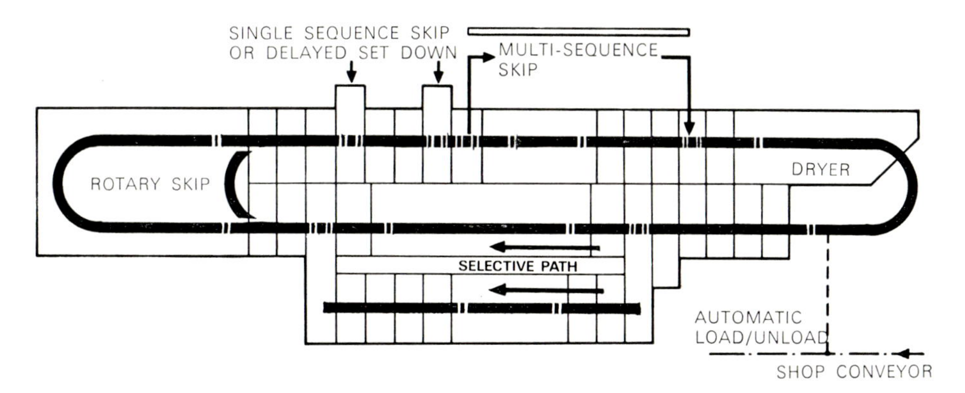

Fig. 5.7. Some of the typical features which may be found on an automatic plating machine. [Courtesy Oxy Metal Industries (GB) Ltd.]

Fig. 5.8. Another view of plant shown in fig. 5.6. The barrel at left is automatically positioned under the loading chute (1) while the other cylinder unloads itself onto a conveyor (2).

5.3.3. PROGRAMME-TYPE PLANTS

The programme machine is quite different in concept from the machines so far considered; it is perhaps the simplest machine in principle, but requires a high degree of sophistication of design if it is to give satisfactory results. The programme-type machines were introduced to meet the requirements for an even greater degree of versatility and flexibility, such as may arise from changing production requirements, than is possible with the return-type plants, including their modified versions as already described.

The programme-type machine consists essentially of a lifting and traversing carriage or carriages, operating over a series of tanks, designed to transfer work loads sequentially through the process cycle or cycles; the independent carriages operate on runways either overhead or alongside the tanks. A carriage has two reversing electric motors, each fitted with a disc brake, one driving a lift and lower mechanism through a fixed distance, the other operating the drive wheels to traverse either right or left as required. Accurate stopping of both mechanisms is the major design requirement.

Top and bottom limits of travel of the lift mechanism are controlled by magnetic switches. The traverse drive unit operates through a pair of polyurethane-tyred wheels using frictional traction on the main runway, the remaining pair of wheels being idlers. Interlocks are fitted throughout to ensure that the carriage can only travel if the hoist is fully up or fully down, and that it can only be raised or lowered when correctly positioned at a work station. All these movements can be manually controlled at the carriage itself if necessary, although normally all movements under production conditions are determined by the programmed control equipment.

The sequence of operations is for the carriage to lift a load from the load stand, traverse forward to the first process stage, lower the load into the tank, and then move on to carry out similar transfers elsewhere on the plant. At the end of the required treatment time in the first process tank the carriage returns, lifts the load out, travels to the next required tank (which may not be adjacent to it) and lowers the load. The repetitive series of events which are carried out by the carriages is termed a programme and it is from this concept that the machine received its name. In the majority of programme plants there is no horizontal movement of the work in the individual process tanks.

This type of plant is also ideal for work which requires cell plating with conforming anodes, large and heavy articles, and where variable outputs and finishes are a routine requirement. Where multi-process or mixed rack and barrel outputs are required because outputs are not sufficient to warrant separate machines, the programme machine is also especially suitable.

The essential requirement of all types of programme machine, irrespective of size and layout, is reliable control equipment. Unlike the return-type machines, neither the lift and drop mechanism nor the traverse mechanism operate over single fixed distances. All work movements are carried out singly and separately and the total number of movements in a cycle, before a repeat is achieved, may be as high as two hundred and variable from carriage to carriage and programme to programme.

There are two main control systems available. The first employs a photo-electric tape reader, the outputs of which control the carriage and ancillary equipment in a similar manner. The instructions are fed into the reader from a closed loop of perforated P.V.C. film, one row of perforations on the tape usually comprising one traverse and one lift/drop movement together with any interlock of ancillary equipment operation instructions. On completion of these movements the operation of proving switches on the plant will step the instruction on by one step. A change of programme is achieved by changing the tape loop. This is the simplest system and being relatively unsophisticated, it has for several years been widely accepted by the maintenance engineers of the industry, who frequently do not take readily to ' black box ' equipment in the form of solid state devices in which faults are not immediately apparent.

An alternative system which is suitable for users with greater experience of electronic equipment has recently been developed. This employs solid state techniques and incorporates, as the only electro-mechanical device, a standard Telex tape-reader. The rest of the switching is carried out by conventional logic systems based on integrated circuits. The inclusion of the Telex link enables programmes to be transmitted direct to the user of the plant. This system is a particularly reliable one but is, of course, more expensive.

Fig. 5.9 shows the load-unload station of an automatic plant for bright nickel and chromium plating at an electrical appliance manufacturer.

Fig. 5.9. 'Trojan' automatic plant for bright nickel and chromium plating at Belling & Co. Ltd. [Courtesy W. Canning Ltd.]

5.3.4. 'COMPACT' PLATING MACHINES

Over the years both the conventional return-type automatic plant and the programme machines have tended to become longer and hence more expensive and space-consuming. This is because the cleaning and rinsing cycles have become progressively more elaborate to ensure that even exceptionally contaminated articles are adequately cleaned. With programme-type machines a reduced length increase as compared with the return-type automatic plant is possible by making use of the greater flexibility of these machines which enables cleaning and rinsing cycles to be changed where this is possible by virtue of the nature of the work being processed. A further important contribution towards keeping the dimensions of programme machines in bounds whilst retaining adequate margins of safety is the Oxy Compact Plating Machine. The essential feature of the design is the provision of a single three-stage rinse unit, generally with counter-flow, in which all rinsing operations are carried out. The required water flow rate is reduced in this way to the cube root of that for a single stage rinse, or to 3/2 of the cube root of the requirements for a 2-stage counterflow rinse. Flight bars are transferred between the rinse stages independently of the trolley, which does not therefore have to spend an unduly large proportion of its time on this function.

A substantial proportion of rinse contamination occurs as a result of drips entering the rinse tank. In the compact system, this source of contamination is prevented by an automatic drip-catching tray attached to the trolley which is automatically positioned under the components as the jigs travel from the process to the rinse tanks. The liquids which collect in the trays are pumped out periodically. By the use of this system of operation, the space taken up by a programme machine can be reduced by up to 50 per cent as compared with a conventional machine, and fewer trolleys will be required; also, the rinse water requirements can be reduced by up to 80 per cent, with an appropriate saving in the cost of effluent treatment equipment. The drip-tray on the Compact machine is shown in Fig. 5.10. The mode of operation can be seen from Fig. 5.11, whilst Fig. 5.12 illustrates the saving in space which can be achieved by this system.

The possibility of passivation of the substrate in certain cases by the presence of hexavalent chromium in the unified rinse system of the Compact machine can be overcome by the interposition of a chromium reduction stage between the chromium processing solution and the rinse. This, incidentally, eliminates the need for a similar stage in the effluent plant. Stream separation which is required where continuous effluent treatment plants are employed can also be taken care of by a comparatively simple modification to the system used.

5.3.5. HORIZONTAL OSCILLATING BARREL

An interesting addition to the range of barrel machines is the Horizontal Oscillating Barrel. The oscillating barrel concept provides a means of processing work in a

Fig. 5.10. With the 'Compact' programme plating machine the drip-tray is used selectively to avoid contamination or to prevent equipment corrosion.

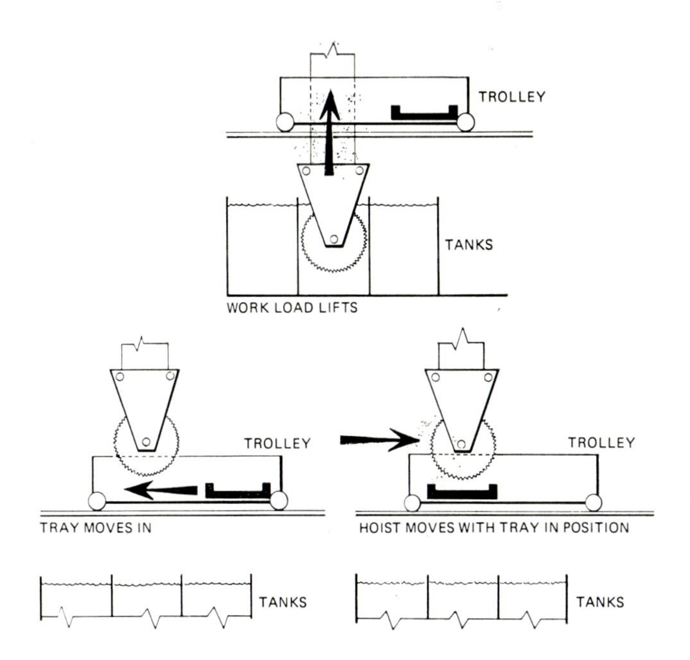

Fig. 5.11. Mode of operation of 'Compact' machine.

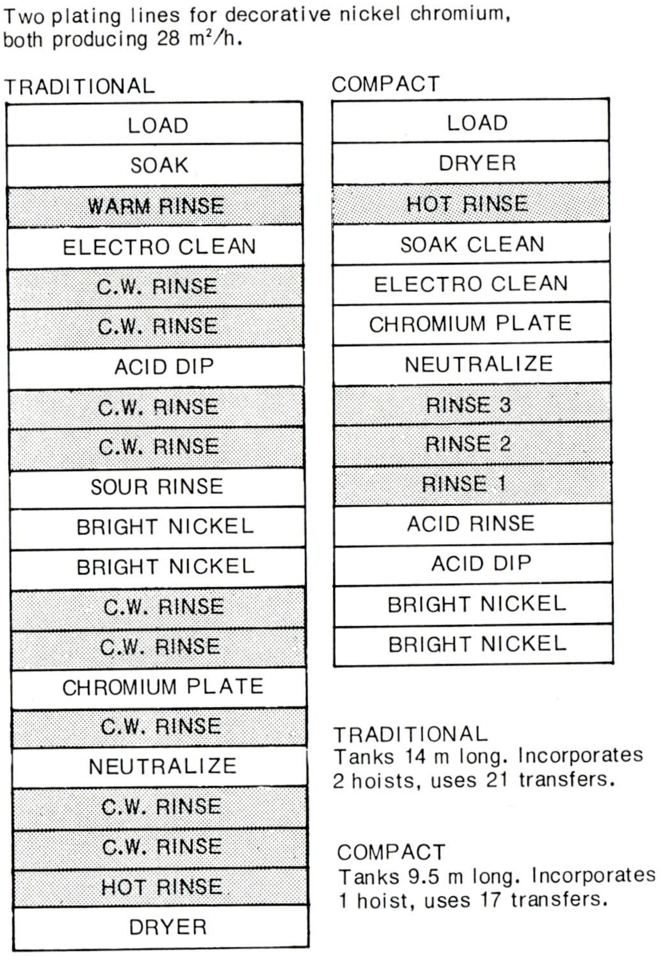

Two plating lines for decorative nickel chromium, both producing 28 m^2/h.

TRADITIONAL	COMPACT
LOAD	LOAD
SOAK	DRYER
WARM RINSE	HOT RINSE
ELECTRO CLEAN	SOAK CLEAN
C.W. RINSE	ELECTRO CLEAN
C.W. RINSE	CHROMIUM PLATE
ACID DIP	NEUTRALIZE
C.W. RINSE	RINSE 3
C.W. RINSE	RINSE 2
SOUR RINSE	RINSE 1
BRIGHT NICKEL	ACID RINSE
BRIGHT NICKEL	ACID DIP
C.W. RINSE	BRIGHT NICKEL
C.W. RINSE	BRIGHT NICKEL
CHROMIUM PLATE	
C.W. RINSE	
NEUTRALIZE	
C.W. RINSE	
C.W. RINSE	
HOT RINSE	
DRYER	

TRADITIONAL
Tanks 14 m long. Incorporates 2 hoists, uses 21 transfers.

COMPACT
Tanks 9.5 m long. Incorporates 1 hoist, uses 17 transfers.

Fig. 5.12. Comparison between number of plating stages in traditional and 'Compact' plating plants.

horizontal barrel without the need for a lid, making it possible to load and unload the barrels automatically and thus integrate the machine into a production flow line. It is also a relatively simple matter to make provision for drying the work in the barrel. This barrel is designed to operate with through- or return-type automatic machines, but it may also be used with manually-operated hoist lines. The oscillating motion of the barrel, which extends over 120°, is provided by a low voltage motor with a solid state current reversal mechanism.

The essential differences between the present standard hexagonal barrel and the H.O.B. are as follows:

(1) The shape of the barrel is circular instead of hexagonal, whilst it has a permanent opening across its full width. It is fitted with a tumbling strip diametrically opposite to the lid opening. This, combined with the oscillating movement, gives excellent mixing of work and an improved finish.
(2) The load capacity of each barrel is increased from 350 litres to 450 litres and the permissible work weight is increased to 150 kg.
(3) The total open area under the solution has been increased by almost 100% as a result of permanent opening; tests show that as much as 1500 amps. can be drawn in a standard zinc solution operating at 15V.

5.3.6. CELL-TYPE PLANT

The cell-type machine has been developed to meet the need for high production of large loads with long process cycles. Individual cell plating also facilitates variation of the process times and current and enables alternative processes with batch loaded or individually selected flight bars to be plated.

The plating racks are loaded into flight bars in the normal manner; each stage or cell in the machine has a flight bar, thus ensuring full machine utilisation. As in the case of the return-type machine, there is only one operating mechanism and all machine movements take place at the same time. All the process tanks on the machine are in the form of single cells, and as such it is possible to alter the anode configuration to suit particular applications.

5.3.7. PLANT FOR CONTINUOUS SHEET AND STRIP PLATING

Pre-coated metal strip is being used in increasing quantity because it offers many advantages to the manufacturer. Not only does it eliminate the need for specialised and expensive finishing facilities and the probable complication of effluent treatment equipment, but it also provides him with a material which is clean to handle, is less liable to corrode in storage, and is of a consistent quality. After fabrication, the coating thickness is also generally uniform over the surface of the article, which is frequently not the case when formed products are coated. By far the most important metal which is supplied in pre-finished form is steel strip, although brass, copper and aluminium are also pre-plated.

Tinplate

With a world production of many million tons a year, tinplate is by far the most important continuously coated steel product. Its main application is in the canning industry; it is increasingly used in the form of double-reduced strip in which the strip, after cold-reduction to 0.255mm, is annealed and further reduced to a thickness of 0.015mm. Almost all tinplate is electrotinned nowadays, since this enables thinner coatings to be applied than was possible by the earlier hot-dipping method. Moreover, coatings of different thickness on each side can be produced; this results in substantial cost savings, as tin is an expensive metal. The coils of steel weigh 12-15 tons each, and travel through the plant at speeds of 600m a minute or more.

The electroplating process most widely used in this country is the 'Ferostan' system using an acid tin sulphate electrolyte. The strip travels through the plant in a series of vertical loops in deep tanks. It is first cleaned in alkaline solutions and acid pickled before entering the tinning tanks. Tin anodes are placed between the parallel rows of steel strip. The deposit is about 0.0002—0.0025mm thick; as it emerges from the bath it is matt in appearance, the required bright finish being obtained by momentarily melting the tin by resistance or induction heating as it passes through the plant after rinsing. It is then quenched and oiled, either with cottonseed oil or with dibutyl sebacate. The 'reflowing' and oiling process also serves to reduce the porosity of the as-deposited tin.

Fig. 5.13 shows the control panel and the cleaning and plating sections of an electrotinning line.

Fig. 5.13. Cleaning and plating sections of an electrotinning line, showing control panel. [Courtesy Wean Engineering Co.]

Chromium-chromium oxide containers

A recent development is the introduction of a substitute material, in which a chromium-chromic oxide film is continuously deposited onto the steel strip instead of tin. It was originally developed in Japan where the interest arose from the fact that Japan has substantial indigenous chromium supplies, but no tin. Apart from being cheaper than tinplate, the chromium deposit has the advantage of providing an excellent bond for the application of lacquers with which the interiors of cans are generally coated nowadays.

The deposit is applied from a chromic acid solution containing a very low concentration of sulphates and fluorides. The resulting coating, which has good corrosion resistance, consists of a mixture of hydrated chromium oxides and metallic chromium.

The process has some limitations, in that it is difficult, if not impossible, to solder so that other methods of fabrication have to be used for cans made from it. The resistance to corrosion by carbonated beverages is also not as good as tinplate, whilst the temperatures to which it is subjected during print enamelling must be kept low.

Electrogalvanised steel

The amount of electrogalvanised steel sheet produced and the demand for it are increasing; there are several plants in the U.K., with others in France, Germany, Japan and the U.S.A. In one type of plant[5] the strip passes horizontally through alkaline cleaning, anodic pickling, zinc plating and phosphating or chromatic sections. The solution tanks are shallow, with slots sealed by rubber lips at each end through which the strip passes, any overflow being collected, filtered and returned. The plating solution used is of the acid sulphate type, with horizontal zinc anodes disposed on the bottom of the plating tank and on plastic grids a short distance above the strip. After plating the product is either chromate-rinsed or phosphated in the line.

The strip travels through the plant at 30m per minute or more, the zinc thickness being about 0.0025mm. Thicknesses of about half this amount are sometimes applied on one side. The strip is usually cut into sheets at the end of the line, but can be re-coiled if required. A continuous electrogalvanising line for steel strip is shown in Fig. 5.14. The sequence of operations is shown diagrammatically in Fig. 5.15.

Electrogalvanised steel sheet is almost invariably organically coated, as it has inadequate corrosion resistance otherwise. It is therefore phosphated or chromated in the line to provide a good bond for the subsequent painting operation. The great advantage of the zinc underlayer is that should the paint coating become damaged or the underlying steel exposed, rusting does not occur, so that progressive failure of the paint coating is prevented. The material thus finds application in the manufacture of kitchen equipment, partitioning, office furniture, and automobile components such as oil filters. It can be readily spot welded and fabricated. It has also been used for complete automobile bodies, since these can be stored or transported, if reasonably protected, for painting at a later date.

Fig. 5.14. Electrogalvanising line for steel sheets.
[Courtesy Oxy Metal Industries (GB) Ltd.]

5.4. Rinse Tanks

The quantity of solution taken out of plating baths into the subsequent rinse tanks is considerable. It varies with the shape of the articles and with the nature of the plating bath. In the case of complex shaped articles the volume is considerable and can amount to 20-80 ml/m^2. This corresponds to a loss of some 10% of the volume of the plating electrolyte per month by this means alone with two-shift working. In the case of decorative chromium plating it has been estimated that only eight to ten per cent of the chromic acid used finishes up in the form of electrodeposited chromium, the rest being lost mainly by drag-out, while some 20% is wasted in the form of spray through the exhaust system. To a certain extent these losses can, however, be reduced by the use of appropriate surface active materials in the chromium plating solution. It is thus seen that drag-out in the rinsing operation is an important factor in water conservation and effluent treatment. A good deal can be done to minimise the losses by such means as suitable racking of the work, adequate dwell times before rinsing, and attention to the design of work holders to ensure that solution retention is kept to a minimum[(6)].

The maintenance of an adequate degree of cleanliness of the water in rinse tanks is carried out on a somewhat arbitrary basis in most plants. In the older plating shops the conventional rinse tank, as often as not, consists of a vessel with a tap in one corner supplying the fresh water, together with a standpipe in the adjacent or opposite corner

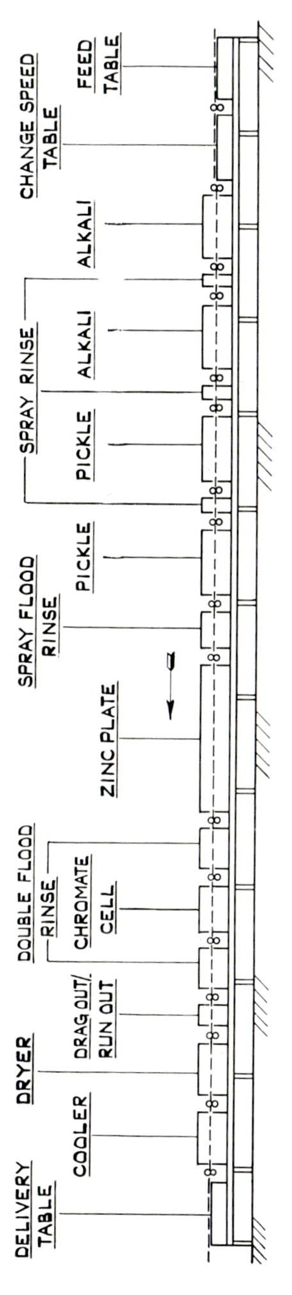

Fig. 5.15. Layout of electrogalvanising line for steel sheets.

into which the overflow runs. It is easy to see that a system of this kind is very unsatisfactory. In the first place, there is no control on the rate of flow of water (apart from the vagueness and uncertainty of looking into a tank to see that it is reasonably clean) which can result in excessive wastage on the one hand and insufficient rate of change of water on the other. Also, the efficiency of mixing of the clean water and the contaminated water is likely to be inadequate so that it is possible for dirty water water to be present in the bottom of the tank, the incoming fresh water flowing directly across the surface to the overflow pipe and performing no useful function.

Much better results can be obtained where the ingoing water is taken to the bottom of the tank by means of a pipe, but it is essential in such cases to have a break in the line to prevent the possibility of siphoning back into the main supply. Such a device has the possible disadvantage that it is not easily seen that the water is flowing; for this reason a tun dish is often fitted at the top of a standpipe to take the water to the bottom of the tank; such a device is satisfactory if it is of adequate size. Weir overflows are, however, the best method of keeping rinse tanks clean, and in these, the waste water overflows into a channel located along one side of the tank where it is run off. This system has the advantage of keeping the whole of the water surface clean and free from deposited particles, which can interfere with the production of smooth electrodeposits. This method is especially suitable for alkaline cleaning baths; here, however, there is no continuous flow of water, but when the level has to be made up periodically to allow for evaporation, the opportunity is taken to overflow the surface of the cleaner into the weir to remove grease and other foreign matter which may have accumulated and floated to the surface of the tank. Steam condensates and cooling water emanating from cooling systems used in rectifier installations, vapour degreasers, chromium plating and anodising plants, should also be returned for utilisation in rinse tanks.

Air agitation of all rinse tanks is particularly desirable since this prevents stratification of the water in the tank and hence local contamination; its use can lead to both improved rinsing and substantial water economies. Where it is necessary to change from an alkaline to an acid process, a minimum of two successive rinse tanks is essential to prevent carry-over of alkaline into acid solutions, or vice versa. Such carry-over must be avoided both on account of the possibility of the deposition of sparingly soluble salts on the metal being plated, and also to prevent contamination of the second plating bath itself.

5.4.1. CASCADE RINSES

In many automatic plating plants in particular, use is nowadays being increasingly made of counterflow or cascade rinsing for obtaining the maximum effective use of water, the rinses being coupled in series; Fig. 5.16 shows a cascade rinse system of this type in which the fresh water is fed into the bottom of the second of a pair of rinse tanks overflowing into a weir at the back and entering the previous tank in the line, afterwards leaving this through an overflow in a second weir at a slightly lower level.

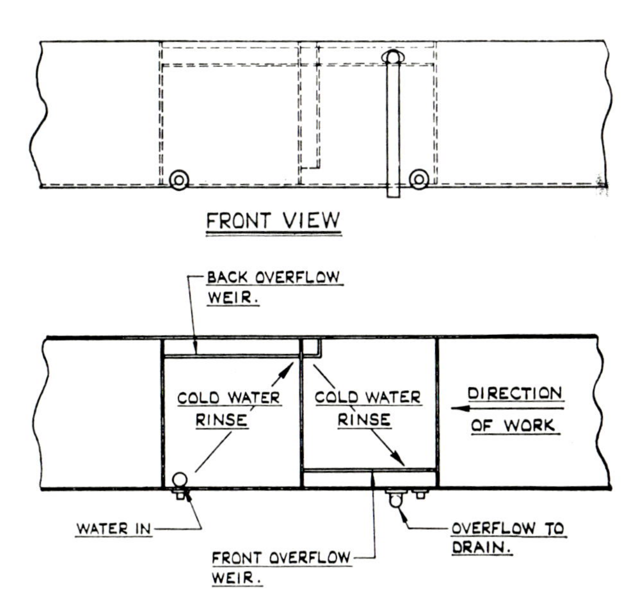

Fig. 5.16. Design of two-tank counter-flow rinse system.

This ensures that the second of the pair of rinses contains the cleanest water, which is clearly desirable; more than two rinses can be coupled in this way if desired. As an indication of the enormous saving which is possible by the use of the counterflow system Harris[7] gives the following example:

If it is assumed that the concentration of chemicals in the plating vat and final rinse under satisfactory conditions of rinsing are 2200 and 7.5 g/l respectively, and that drag-out is at the rate of 5 litres per hour, the rinse water requirements would be:

For a single running rinse	2250 l/hr.
For a double counterflow rinse	70 l/hr.
For a triple counterflow rinse	30 l/hr.

A rather more elaborate system for the multiple re-use of rinse water is shown diagrammatically in Fig. 5.17. This has been incorporated in an automatic plant for the cleaning and etching of large aluminium sheets, but the method is equally applicable to anodising processes. The process sequence is as follows:

1. Mild alkali cleaner (70°—85°C)
2. Cold water rinse
3. Weak acid neutralise (cold)
4. Cold water rinse
5. Acid etch (cold)
6. Cold water rinse

7. Cold water rinse
8. Cold water rinse
9. Re-circulated spray rinse

With the exception of the larger etch tank (No. 5) all the tanks are 0.5m wide, 2.5m long by 2m deep and have a capacity of about 2,000 l.

The spray pump serving the final rinse has a capacity of about 700 l/min at 2 kg/cm^2. Water for the previous running rinses is tapped from the pump system at the rate of 70 l/min, the loss being automatically made good by a ball valve in the base of the final spray tank. The water thus bled off is cascaded back through plastic pipes to the three previous immersion-type cold water rinse tanks (6, 7 and 8), and thence to the two further cold rinse tanks, which precede the acid etch (2 and 4). This is a particularly effective and economical system, since it makes use of the slightly acid rinse waters after acid etching as partial neutralisers following the preceding alkaline treatments. The total rinse water consumption for the entire plant is thus only 70 l/min.

Another method of conserving water is to use rinse tanks which are entirely static. Instead of using running water, they are periodically discarded and refilled when they have reached a given degree of contamination. Properly controlled, this method can give adequate rinsing with very substantial savings of water.

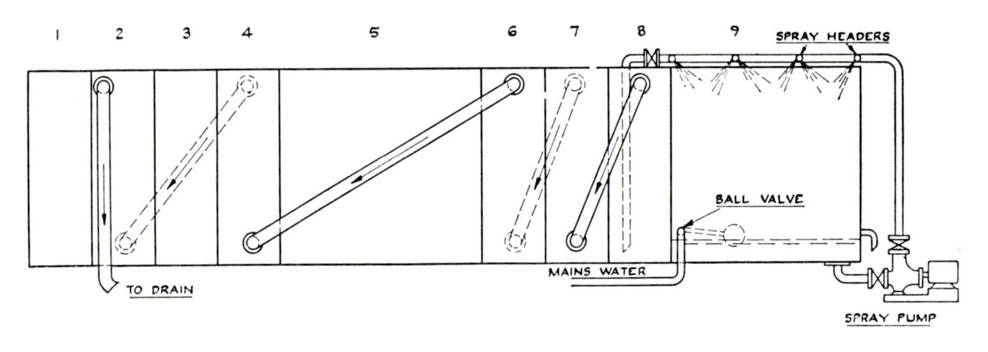

Fig. 5.17. Cascade rinse system used in automatic plant for cleaning and etching of aluminium.

5.4.2. RATE OF WATER CHANGE

The rate of change of the water in a rinse tank is a matter of trial and error and depends on a great many variables. The most important of these are: (1) the concentration of the plating solution, since clearly the salts introduced from a bath are greater in quantity if the solution used is a strong one, (2) the type of work being processed, as cup-shaped and recessed articles will cause a high degree of contamination of rinse water, and (3) the surface area of the work put through the plant. In general, three to four changes per hour are considered satisfactory for most purposes, but when rinse tanks are cascaded the total volume can be very much less.

Further economies can be effected by the introduction of one or more static drag-out recovery tanks immediately after the main processing tanks. From time to time the water from each of these tanks is used as make-up water for the plating tank, thus reducing overall metal loss and the amount of contaminated water to be treated in the effluent plant. Such drag-out tanks are invariably used in precious metal plating for obvious reasons, and also after chromium plating baths, because of the large amount of chromic acid normally carried on the surface of water leaving them. It is reasonable in these cases to have two such tanks in series. They are also now increasingly common after nickel plating, in which fairly concentrated and relatively expensive solutions are used. Where a single recovery tank is incorporated, the concentration of chemicals in it should not be allowed to exceed 20% of the bath concentration, and not more than 5% in a second tank. A single recovery tank can lead to a saving of around 60% of the chemicals which would otherwise be lost, while a pair of such tanks can give up to 80%[8]. Where the bath temperature is high so that considerable amounts of make-up water are required for the plating solution, the drag-out tank can be kept reasonably clean by this means, but this is not always the case and some of the static tank water must be discarded from time to time. Chromic acid drag-out can be removed and concentrated by evaporation before being fed back into the plating bath, and this is an entirely practical way of conserving both chromic acid and rinse water. Concentration of nickel solution collected in static drag-out tanks is also possible in this way, and the method has been employed to a very limited extent.

Drag-out tanks are best incorporated in new installations, as their addition to plants later on becomes very difficult because the space is, more often than not, unavailable. This applies especially to automatic plating machines.

5.4.3. SPRAY RINSING

Spray rinsing provides a very satisfactory way of making the maximum use of rinse water. The method consists of a small reserve tank served by a pump which supplies the water to sprays fitted to each side of the tank through which the work passes. A ball valve maintains the level of the water in the reserve tank in which the rate of change of water can be controlled as required. For a given quantity of water usage it is calculable that a lower degree of contamination of the rinse tank can be maintained than is the case when simple immersion rinsing is employed. The reserve tank can be quite small when adequately designed sprays and suitable pressures are used. this system can also be employed in conjunction with immersion rinse tanks, the sprays being located above them so as to impinge on the work before it is immersed. It is desirable to incorporate a filter in the pump circuit to prevent solid particles removed from the work from clogging the spray nozzles. Further savings of water and power can be effected in the case of automatic plants by incorporating limit switches at appropriate positions to start the spray pumps only when a rack of work is actually about to enter the rinse zone.

5.4.4. AUTOMATIC CONTROL

It is also possible to control the amount of water used for rinsing by employing automatic analytical control methods[9]. One of the most practicable systems is based on the conductivity of the water. On the assumption that salts, acids or alkalis introduced into rinse water must increase its conductivity, it is possible to arrange for a conductivity measuring device to release a valve when a predetermined contamination level is reached, thereby introducing fresh water into the tank until the conductivity of the water is again reduced to the desired figure.

Whether the maintenance and initial cost of such a system are worthwhile has to be decided in each case. It is not cheap, in any event, while the establishment of the appropriate safe conductivity level is also not always an easy matter to decide upon. A sufficiently wide operating range is, in any event, necessary to prevent " hunting " of the controller, while the conductivity measuring unit itself can also in certain circumstances be put out of action by the absorption of contaminants onto its surface. The system must also be temperature compensated, since quite small changes in temperature can alter the conductivity of the water considerably.

A further problem with the conductimetric control of rinse water is the fact that gross contamination with organic matter which can adversely affect plating in a radical manner may not change the conductivity of the water at all. This is a minor hazard at most stages in the plating sequence, but is likely to occur during rinsing after alkaline cleaning, since the solutions used for this operation are apt to accumulate grease on their surfaces, and this can then enter the rinse tank in small amounts. In such cases serious contamination of the rinse water can take place, without the controls functioning to introduce fresh water at the appropriate time. It is for these reasons that the system has only found limited application, although it has been installed in a few large automatic plants.

It is evident that very considerable economies can be achieved by re-using water in plating processes. Of the methods available, cascade rinsing and the full deployment of cooling water are the simplest and most easily incorporated, even in older plants. Ion exchange installations are costly initially and demand a fair amount of control; they can seldom be justified on financial considerations alone in this country, even when incorporated with effluent disposal plants, in the absence of special considerations. Other factors, such as the increasingly stringent requirements of local authorities, are rapidly becoming of greater significance; however, where de-ionised water is essential to the processes concerned, such installations can pay for themselves in a reasonable time.

5.5. Filtration and pumping of plating solutions

Filtration is nowadays regarded as an established part of the plating process, and is normally employed on nickel, copper and, to a lesser extent, on other solutions. It may be carried out continuously or intermittently. For bright nickel solutions continuous filtration is mandatory if a high and consistent quality of plating is to be maintained. A comprehensive review of commercially available filter equipment has recently been published(10).

The purpose of filtration is to remove from the solution solid particles which may find their way into it from the atmosphere, from imperfectly cleaned work, or from the anodes. There are other possible sources of contaminants, e.g. plating racks, pipe lines or even tank linings such as rubber, which may liberate foreign particles into the bath in some circumstances. The particles may be metallic or non-metallic, and range from the ultra-microscopic to the visible. In some cases they are semi-colloidal or gelatinous, and hence invisible, even when a strong beam of light is passed through the solution. In the case of nickel solutions in which ferrous metals are being plated, iron can accumulate in the bath due to the dissolution of the metal from recessed areas of the work and be liberated in the bath as gelatinous hydroxides, particularly at higher pH values.

The accumulation of insoluble material in a plating solution is undesirable in many ways. In the first instance, rough deposits are likely to be obtained, and this roughness will increase with thickness of plate because of preferential deposition onto the particles; these tend to settle, especially on horizontal surfaces. Secondly, the presence of such occlusions can result in porous deposits of relatively low corrosion resistance. These effects depend not only on the sizes of the particles of foreign matter concerned but on their composition. Certain types of foreign matter are more deleterious than others; thus, minute particles of metal oxide in a nickel plating bath are particulalrly prone to cause rough plate, whilst silicon particles of much larger size can be tolerated to a greater extent without trouble.

Filtration systems consist essentially of a pump, a filter, and associated pipework. A mixing tank is usually also provided to enable filter aids or activated carbon to be made into a slurry for the purpose of coating the filter itself when necessary.

The filters now generally used for plating solutions consist of a fabric (either natural or synthetic), a specially treated paper, or a wound element made from suitable fibres. Some types of filter use mesh or stainless steel discs to act as a support for a diatomaceous earth or similar filter medium.

5.5.1. CONTINUOUS FILTRATION

Filtration may be either continuous or intermittent. Continuous filtration is normally employed on nickel and copper plating solutions, whilst other baths are generally filtered at intervals. The frequency of periodic filtration can only be decided on the basis of the amount of contamination present. It has been estimated that complete filtration through a filter five times will remove 99% of all particles, whilst by filtering ten times 99.99% can be removed. In continuous filtration, the rate at which particles enter the bath is never equalled by the rate of removal in any practicable system, so that an arbitrary rate of filtration has to be decided upon.

Since the primary reason for carrying out filtration is to remove suspended particles quickly and hence avoid roughness, it is desirable to filter at the fastest practicable rate in the case of continuous systems. There is, of course, a limit inasmuch as the faster the rate of filtration which is required, the higher is the capital cost of the plant. Generally rates of flow aim at a complete turnover of the solution every two to three hours, but in cases where articles liable to introduce a high degree of contamination are being plated, tank turnovers of up to two to three times per hour have been used.

If the rate of continuous filtration is not adequate, particles will build up to such an extent that it will be necessary to filter the solution completely into a spare tank periodically. In this way it is possible to start with a completely clean bath. One useful scheme is to continue filtration during a period when the plant is not plating, since this enables the filter to deal with contaminants arising from anodes and chemical additions independently of particles introduced on the work. Certain types of component are particularly prone to contaminate plating baths, e.g., tubular work which cannot be adequately cleaned internally.

It is desirable that filters should have as large a surface area as possible so as to minimise the amount of cleaning and filter replacement needed. Rapid clogging of filters of small surface area cause a reduction in flow, and a build-up in internal pressures, which can result in excessive stresses on the pump seals. For this reason it is advantageous to provide a by-pass, but although this avoids damage, it does, of course, reduce the efficiency of filtration if the filter is not cleaned sufficiently frequently.

Filters should be installed as close as possible to the tank to reduce pressure loss, and also to avoid the possibility of contamination from piping. Rubber and plastic lined pipes are extensively employed for permanent installations, as are pipes made wholly of plastics, especially rigid PVC. Glassfibre reinforced pipework has also been used to a limited extent. Plastic piping necessitates the use of specially designed joints and fittings. As each section and bend has to be individually manufactured, however, the design of lined steel piping systems is difficult and the installation tends to be expensive. Moreover, changes are not easily made in the layouts if this should be required. Although flexible pipes are not subject to this limitation, they are much less durable and are not recommended. It is desirable to have the suction inlet of the filtration system close to the bottom of the tank where the sediment tends to accumulate, and the outlet at the top. A useful safety device is to include a small hole in the intake pipe just below the solution level so that, in the event of any leakage developing, the contents of the tank would not be inadvertently sucked out and lost. This is an important factor when filtration systems are allowed to work unattended.

In some instances use has been made of diaphragm tanks to prevent particles from the anodes entering the cathode compartment. The diaphragms are made of anode bag fabric, and supported on plastic or plastic-covered frames. This reduces the required rate of filtration but the arrangement can be troublesome to maintain. Diaphragms also have the disadvantage of taking up a fair amount of tank space.

5.5.2. PUMPS

All filtration systems depend on a pump to circulate the solution, and a wide choice of pump design and materials of construction is available. The types in general use include centrifugal, positive displacement, gear, screw, cam and diaphragm pumps, to mention only a few. They may be made of cast iron, steel, nickel-chromium alloys, bronzes or special alloys such as high-silicon iron. The latter material has high corrosion resistance but is unmachinable and can only be shaped by grinding. Pumps may also be made entirely of, or lined with, rubber, ebonite, carbon, glass, ceramics or plastics such as phenolic resins, polyesters or polyvinyl chloride. The range of sizes available is also considerable.

Pumps used in electroplating may be grouped into two main general classes, i.e. positive displacement and centrifugal pumps.

5.5.2.1. Positive displacement pumps

The positive displacement pumps are either of the reciprocating or rotary type. In the former a plunger or a diaphragm moves backwards and forwards, so that there is an alternate increase and decrease in the volume of the pump body. As the plunger is withdrawn liquid enters the pump by suction, and as it returns it displaces the liquid so that it is discharged. Backflow of liquid is prevented by suitable valve systems. Alternatively a diaphragm may be employed instead of the plunger, so that there is no direct contact between the driving mechanism and the fluid being pumped. The diaphragm, which may be of rubber or any suitable flexible material, is operated by hydraulic or air pressure or by mechanical means.

The rotary positive displacement pumps make use of devices such as engaging gears or screws of various types to seal off a series of chambers transferring the liquid being pumped from the inlet to the output side. Alternatively, a flexible tube in contact with a roller or cam is used in some types of pump of this kind. The Mono pump is a rotary positive displacement unit which has found application in electroplating. It makes use of a rotor of a special helical form, with a single scroll turning in a stator designed as a double internal helix. The stator is normally of rubber and the rotor of corrosion-resisting steel. The design is such that the rotor maintains a constant seal across the stator and the two elements engage with one another in such a way that the seal travels continuously across the stator producing a uniform, positive displacement. The line of contact results in a complete seal between the suction and discharge ends of the pump which is thus self-priming; the head developed is independent of the speed and the capacity is proportional to it. The flow is also uniform and non-pulsating, whilst the rubber stator enables foreign particles to enter the pump without causing damage.

The reciprocating or diaphragm pumps are simple, reliable and quiet in action, but have the disadvantages of being rather bulky in size whilst the materials of which they can be constructed is limited; they also tend to be relatively expensive. The flow is a pulsating one unless multiple cylinders are used, and they must be protected against excessive pressure build up. A further disadvantage is that they are not self-priming. Hence, the rotary types are preferable in that they are self-priming, cheaper and smaller in size. Many types require no valves. On the other hand they must be made to close limits so that only materials capable of maintaining fine clearance can be employed in their construction. In view of this they easily lose their efficiency through abrasion or corrosion. At low speeds their efficiency falls also, since the slip approaches the displacement. Pressure relief valves are necessary to avoid damage should the pressure become excessive.

5.5.2.2. Centrifugal pumps

The centrifugal pump is so-called because the fluid being pumped has velocity imparted to it by centrifugal force which is then partially converted to pressure. Liquid enters the centre of the rotating impeller and is thrown outwards either radially or axially. These pumps are essentially cheap and simple in that they have only one moving part, i.e., the impeller, and they are therefore nowadays the most widely used. The radial flow type generates the highest head of pressure, and the axial flow the lowest for a given impeller size and speed. There are also other types.

Centrifugal pumps require no close tolerance in manufacture and so can be made of rubber-lined steel, plastics and similar materials. They can be driven at high speeds, and are quiet. They require little maintenance, since they can tolerate a considerable amount of corrosion and erosion. They are not normally self-priming, but design modifications can be introduced to make them so. The materials used for the impellers should be as strong and light as possible. Also, care must be taken in sealing the gland through which the driving shaft passes, as this is often a source of trouble.

Centrifugal pumps suffer from the disadvantage that the axial flow power curve rises sharply at zero discharge; this means that overheating and breakdown can occur if the discharge is shut off or blocked for any reason.

5.5.2.3. The LaBour pump

The LaBour pump has established itself over a period of years as being very suitable for heavy duty in plating processes. It is a self-priming centrifugal pump with a set of discharge throats close to the impeller periphery. The casing is always kept partially full of liquid by means of a trap. During priming, the air in the casing and that drawn from the empty suction lines, is mixed with the liquid in the pump casing by the action of the impeller. This mixture is driven out of the primary throat and into a separating chamber where it re-enters the pump casing by way of a secondary throat. When priming is complete the flow in the secondary throat reverses and discharge continues through both throats. The pump does not make use of valves or sealing rings and the fact that the impeller is cast integrally with a length of shaft sufficient to extend into the bearing housing and has negligible end thrust, means that bearings are needed on one side only. Hence there is only one seal, which may be liquid sealed or water cooled.

The pump can be constructed from a wide range of materials, but for bright nickel solutions, including the high chloride type, a high nickel-chromium alloy has proved very satisfactory. It consists of nickel 55%, chromium 23%, copper 6%, molydenum 4%, tungsten 2%, silicon 4% and carbon 0.2%. Cyanide solutions can be handled satisfactorily by cast iron or steel pumps.

5.5.3. FILTERS

Three main types of filter are used today in the plating industry:

(1) Plate filter press
(2) Pressure leaf filter
(3) Cartridge filter

5.5.3.1. Filter press

The filter press is the oldest and most widely known type of filter; it has a large capacity, and is robust. It consists of a series of plates and frames, made of a corrosion-resisting material such as ebonite or plastic, with filter cloth or paper between them. Leakage is prevented by various means, such as the use of caulkage strips. They have the disadvantage that a good deal of labour is needed to clean and assemble the plates, but modern versions are available whereby back flushing and mechanical operation make the task less arduous.

5.5.3.2. Pressure leaf filters

These filters consist of uniformly spaced, horizontal or vertical leaves mounted in a pressure vessel. The plates, which may be in the form of a mesh or a ribbed plastic grid, form a support on which the filter medium (paper or fabric) is supported. The properties of some filter fabrics are given in Table 5.1.

TABLE 5.1
Characteristics of Various Filter Fabrics

Material	Sp. Gr.	Max. working temp. (°C.)	Melting point	Resistance to Acids	Resistance to Alkalies
Cotton	1.55	90°	150°	Poor	Good
Wool	1.32	100°	140°—150°	Good	Poor
Nylon	1.15	105°—110°	215°—250°	Poor	Good
Polyester	1.38	160°	257°	Very good	Medium
Acrylonitrile	1.16	140°	250°	Very good	Medium
Polypropylene	0.91	100°	170°—175°	Very good	Very good

The leaves are assembled on a central manifold through which the clean filtrate is discharged. The solution flow is upwards against a splash baffle before passing through channels in the leaves to the outlet manifold. The cylindrical pressure tank is rubber built. Fig. 5.18 shows a portable vertical leaf filter unit with slurry tank for coating the filter plates.

When the leaf is horizontal, the cake only forms on their upper surfaces, which reduces the filtration area although it is easier to keep the filter cake or filter aid in

Fig. 5.18. Portable plate-type filter press.

place. A horizontal plate filter unit in a Perspex housing is shown in Fig. 5.19. Liquid flows into the filter from a pump, fills a cylinder and passes through holes provided in a series of sludge rings. Half of the liquid from each ring is then passed through a filter disc beneath it and half through another disc on top. The sediment of sludge remains in the ring on the surface of the disc. The filtered liquid from each disc enters into a perforation screen adjacent to it, and then through channels in the screen to a common outlet in the centre. Thus, each disc acts as an individual filter. The pressure in the filter is uniform except where the discs touch the perforated screen. This type of filter provides a large amount of surface area, but has the disadvantage that all the plates and discs have to be dismantled for cleaning. Filter discs are available in a wide range of porosities, but they are made of cellulose and are frequently built up in such a way that the fibres are fairly loosely packed on one side, progressively becoming more densely packed through the thickness of the disc. Hence the outer surface catches the coarse particles whilst each additional layer stops finer particles as they penetrate further into the filter material. It is claimed that this type of filter can be operated for a long time, and does not need a filter aid. A special type of horizontal plate filter available in the U.S.A. employs plates mounted on a hollow shaft which rotates the leaves; this agitates the contents of the

Fig. 5.19. Horizontal plate filter unit in a Perspex housing.

filter tank, and discharges the filter cake either dry or re-slurried. Wash liquor trickles over the leaves.

Both the filter press and the leaf-type filter are suitable for use with filter aids or activated carbon. The latter lends itself more readily by sluicing or backwashing, and removal of the plates need only be carried out relatively infrequently.

5.5.3.3. Cartridge filters

Amongst the most extensively used filters nowadays is the cartridge type, which have the advantage of simplicity and ease of handling. Essentially they consist of a cartridge of pleated paper which is impregnated with a resin to impart strength. The resin is applied in such a way that the permeability of the paper is not greatly affected, whilst its water resistance is considerably improved. End caps of epoxy resin are cast onto the pleated paper cylinders to keep them in shape. A perforated centre tube made of corrosion resisting steel is inserted and provides an outlet for the solution. The cartridge is held in a rubber-lined cylindrical container. For higher capacities several cartridges can be combined in a single unit.

Cleaning of the cartridges is extremely simple since they need only to be lifted out and hosed down. With reasonable care the cartridges can be made to last for a considerable time. They usually fail as the result of a mechanical damage rather than through clogging up, particularly if a satisfactory filter aid is used.

An alternative type of filter cartridge consists of a wound fibre material which can be of polypropylene, cotton or Terylene applied on a stainless steel or plastic core. The winding is carried out in such a way that it is densest towards the centre of the tube and progressively more open towards the surface. Hence, larger particles are retained at the exterior, and the smaller ones held back within the wound filter. These cartridges are able to retain impurities within their structure rather than purely on the surface; hence a larger surface area is povided than with a simple cartridge of the same external superficial area. The particles retained within the filter are, however, rather more difficult to remove, although the cartridges can be backwashed; it is found that after backwashing the reduction in porosity of the cartridge is between 5 and 20%. The ratio between the external area of the filter and its available filtering surface is approximately 1:7. Such cartridges, which are relatively expensive, are available with surface areas ranging from 2.5 to 33m^2 and are capable of holding 30-220 grams of sludge respectively.

5.5.3.4. The Scheibler filter

The Scheibler filter, which is a plate filter, is becoming increasingly popular especially for large installations, because it can maintain a very high rate of filtration for long periods. In many respects it is of conventional design, but its originality lies in the fact that the supporting element for the filter bag is much smaller than the bag itself, the latter being three times its width. The extra length is accommodated by pleating the bag over the element frame; thus a plate of $1m^2$ nominal surface area will have an effective filtration area of $3m^2$. The result is that not only is the high rate of filtration maintained, but the frequency of changing the bags is also reduced. The supporting elements themselves consist of rectangular collapsible frames made of suitable materials provided with tensioning struts. The solution enters the frame through the bag and is caried away by the effluent duct, which is designed in such a way as to prevent the filter bag from being bunched up or pressed into the flow passage.

The joints between the individual filter elements and their respective outlet orifices are made by means of conical bayonet couplings or, in some cases, claw-lock couplings. The former requires no additional sealing joint, but with the latter a packing ring is needed. The filter bag itself has only a single opening at one corner. In operation the collapsed filter element is inserted through this aperture and is opened up in the interior of the bag. The sleeve is then tied to the effluent duct of the element by means of a cord, thus ensuring that the element and the filter bag need only sealing at the coupling position and the bag sleeve. A wide variety of fabrics, including synthetics, from which the bag can be made is available and the choice being determined by the nature of the solution being filtered. During operation, the liquid to be filtered is introduced into the filter tank under pressure, enters the exteriors of the bags, and then leaves the tank outlet orifices into the main filtrate-collecting channel via " spreader channel ' effluent ducts in the filter elements. The filter container itself may be rubber or plastic lined, in accordance with requirements, and is closed by means of a lid in the case of the vertical type, or by a door at the side for the horizontal type. Units are available with effective filtration areas of up to $150m^2$.

5.5.3.5. Automatic filters

With the rising cost of labour there is a growing interest in fully automatic filters which do not need the considerable amount of manual labour required for cleaning. Moreover, such a cleaning very often tends to be overlooked with the result that the quality of filtration suffers and production troubles occur. This is avoided by the use of automatic filters, of which one or two types are already available and others are under development. In the main, automatic filters work by having a continuous roll of filter material which is fed into the filter itself, a fresh filtering area coming into use continually. The feed of filter medium can be controlled by pressure, i.e. new filter surface is introduced whenever a predetermined pressure is reached, as a result of clogging of the filter medium or by time. In the latter event, the clean medium is introduced into the filter at regular time intervals. The filter medium itself is generally in the form of either a dry compressed paper type of material, or bonded fabric, which frequently contains added ingredients, such as filter aids and activated carbon. When the discarded filter medium is finished with, it is discharged from the machine, together with the filter cake.

The advantages of automatic filtration are clearly evident, but, of course, the equipment is necessarily more expensive, both in initial cost and in operation, than the conventional filter. The reasons are that the automatic filters have to incorporate a good deal of sophisticated control equipment, whilst the utilisation of non-reusable filter media must of necessity be relatively expensive.

Control of the filter movement is generally electrical, but one filter system has recently been introduced which employs a fluidic device. When the filter element in the working line is partially blocked by contamination, the increased back pressure diverts the filtrate to a second channel via a bistable gate or flip-flop on the wall attachment principle. The limit of persistence of the gate can be made as small as 0.15—0.25 kg/cm^2.

The advantages of fluidic over electrical controls are most marked where corrosive materials are being handled, or where the plant operates in an unfavourable environment.

5.5.4. FILTER AIDS

To maintain the permeability of filter material it is frequently the practice to use a filter aid. There are a number of types of these, some of which are in powder form, whilst others are fibrous. The most commonly used material is kieselguhr, which is a naturally occurring product consisting of practically pure silica formed from the residual skeletons of minute sea organisms, known as diatoms. (Fig. 5.20). There are an enormous number of diatomaceas, more than 10,000 having been recorded, each with its own size and shape characteristics. Hence it is important to ensure that the correct grade is used, as the product varies considerably. The structure of the kieselguhr is such that the finest grade available will remove slime or colloidal suspensions of rather less than 0.1 micron in diameter. The normal practice is to add about 3kg of kieselguhr to about 100 litres of nickel plating solution contained in an auxillary tank. This is then circulated through the filter unit until the solution runs clear. The aim is to build up a layer of powder about 5 mm thick over the whole of the filter cartridge. The low specific gravity of the material makes preparation of a slurry quite simple. It is also possible to add the filter aid directly to the plating bath when batch filtration is being carried out; a proportion of about 0.25% is generally adequate. In this way the rate of filtration is maintained for a longer period since clogging does not occur and the bed remains open during the whole of the filtration process. The kieselguhr and the filter cake when finished with can be readily removed by releasing the pressure or by backflushing at low pressure.

Since about 1950, the expanded perlite filter aids have become increasingly important. This material is of wholly mineral origin, being, in effect, a supercooled volcanic rock having no microcrystalline structure. It consists of a mass of small pellets about the size of lead shot with a dry bulk density of 30-100 g/l. Its great advantage lies in its low density, giving a 20-30% weight and cost saving. This may

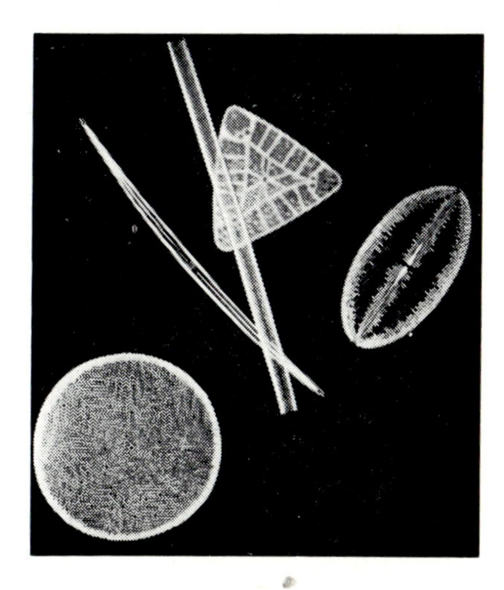

Fig. 5.20. Detailed structure of diatoms.
[Courtesy of Great Lakes Carbon Corp.]

not, however, be fully realised, owing to its greater tendency to " pack ". Perlite is not suitable for highly alkaline solutions, its use being restricted to around pH 4-9.

In preparing the pre-coat it is desirable to work with a relatively low pressure—about 0.75kg/cm^2 is a suitable figure with liquid flowing through at 5-10 litres per square metre per hour—to obtain a uniform coverage. The pre-coat forms in 20 minutes to half an hour. With a correct filter aid layer there should be no impregnation of the filter cartridge; if this is not the case, either the charge of kieselguhr is insufficient or the grade used is too porous for the nature of the contaminant.

One of the most useful fibrous filter aids is filter asbestos. The advantage of a fibrous filter aid is that it makes it possible to form a more tightly bound matrix over a relatively porous filter medium. Hence extremely fine filtration is obtained on a filter medium which is less likely to become clogged by fine slimes. However, the use of asbestos is now discouraged on health grounds.

Alpha-cellulose has also been employed as a filter aid either alone or in conjunction with asbestos. The material is available as a flour or as bleached cotton linters. Flour products have the advantage of being able to absorb materials like suspended oil, for example; it forms a good pre-coat very quickly. Up to a pressure drop of 1.5-2.0 kg/cm^2 the pre-coat holds its shape quite well but higher pressures are undesirable since the pores tend to close, owing to the softness of the material. Cotton linters are more " springy " and do not compress so tightly. Hence it is useful to combine the latter with the more absorbent and closely textured wood flour. Mixtures of fibrous-filter aids and diatomaceous materials have the great advantage of adding greater compressibility to the pre-coat.

To sum up, it could be said that fibrous filter aids are better than diatomaceous earths when rigid particles rather than slimes have to be removed. With rigid particles, coarser grades of diatomaceous earths can be employed and have the advantage of being cheaper than the finer materials. When slime is present, the addition of fibre may improve filtration rate and clarity.

5.5.4.1. Activated carbon

Activated carbon has been used in the chemical industry for many years, and is especially valuable in bright nickel and copper solutions because of its ability to remove breakdown products of brighteners. With certain types of brighteners it cannot be used at all, since the activated carbon may absorb them in their entirety. The material can also be used to remove traces of oil and grease which may enter the solution. Activated carbon is made from charcoal, various sources of which are used; coconut shells are known to give a particularly good charcoal of high density. The process used for activation of the charcoal consists in heating it out of contact with air in the presence of steam, hydrogen chloride or other agents. Exact details are not usually disclosed by the manufacturers. The degree of activation is variable, but can be tested by observing to what extent equal amounts of different grades will decolorise a dilute solution of methyl violet. Carbon treatment is most effective at low pH, and the carbon should be free from iron and sulphates; it should also not contain excessive amounts of ultra-fine material as this may prove difficult to remove. Activated carbons are either powdered or granular, and are rated according to the total surface area per gram. The larger areas indicate greater absorptive power, and typical grades range from 500 to 1200 m^2/g. Although powdered carbons are more absorbent, the granular types are easier to handle.

For the best results a substantial proportion of the powder should be 300-mesh sieve and the charcoal should also have a high bulk density.

Activated carbons will remove most organic contaminants, the procedure being to agitate the solution with it for some hours, preferably at a pH of 2 to 2.5. The pH is then raised and the carbon removed by filtering through a filter unit pre-coated with a diatomaceous earth; 200-300g charcoal per 100 litres are usually sufficient. For continuous filtration a layer of kieselguhr is formed on the filter, and over this a layer of activated carbon is built up by the normal slurrying method. About 5 kg of activated carbon is recommended for every m^2 of filtering area although the amount can be increased if the contamination is heavy. Removal of the contaminants is usually complete after one filtration and the charcoal is then discarded. For batch filtration, rather more carbon is required, i.e. 0.5-1 kg. per 100 litres of solution. The effectiveness of carbon treatment is best assessed by a Hull Cell test. Frequently a carbon treatment can be made more effective by the judicious use of hydrogen peroxide or a cationic surfactant; this technique makes the contaminants more readily adsorbable[9].

Recently it has become possible to obtain activated carbon in the form of porous sheet which can be made into filter candles or filter discs. The material is, however, rather expensive, and the advantages of using it for plating purposes have not so far proved convincing.

5.6. Heating and Drying

A variety of methods are employed for heating plating solutions. For a discussion of these and guidance concerning temperature calculations the reader is referred to the recent review by Alina[11].

In the treatment of ferrous and non-ferrous metals where wet methods are employed, it is generally necessary for drying to be carried out either as a final operation, or as an intermediate stage before further processing can take place. Frequently, the appearance of the product is not particularly important, but in electroplating, for example, steps must be taken to ensure that the finish is clean and stain-free after drying.

The main treatment processes where drying is required are the following:

(1) Pickling of metal components, sheet and strip for scale removal.
(2) Cleaning of articles with alkaline or other solutions to remove grease and drawing compounds.
(3) Cleaning of sheet, strip and wire before hot dip coating with zinc, tin or other metals.
(4) Electroplating, phosphating, anodising and related chemical finishes, such as passivation.

Moisture can be removed by evaporation, absorption or by physical means.

5.6.1. ATMOSPHERIC DRYING

One of the simplest, and probably the most widely used method of drying wet metal, is to allow the surface film of water to evaporate naturally after a final hot water rinse. It is the main method of drying used after pickling. To enable this to be done, the period of dwell in the final rinse tank must be sufficient to enable the article to take up enough heat to evaporate the water, which should be substantially free from dissolved salts so as not to leave deposits of the work where appearance is important. The method is therefore unsuitable for thin or lightweight components since, in such cases, the weight to surface ratio is too low. It is therefore used mostly for thick stampings or pressing, and particularly for castings of various types. The ratio of the weight to the surface area of parts to be dried by this method should not be less than 40kg per m^2. Also, of course, it is essential that the surrounding atmosphere should not be excessively saturated with water vapour to prevent drying. The following table shows the minimum thickness of sheet of various metals which is needed to enable water to be dried off on cooling from 85° to 70°C; the calculations are based on the evaporation of 4.5 litres of water per 100 m^2 of surface:

Steel	2.2 mm
Brass	2.7 mm
Zinc	2.2 mm
Aluminium	3.4 mm
Magnesium	4.9 mm

These figures are, of course, in the nature of a simplification since they relate to flat, vertical, drained surfaces. In practice they can be rated as conservative, since there are often recesses to contend with when finished components are being dried.

A cold air stream in itself does not assist drying unless it is powerful enough to actually blow the water off.

5.6.2. HOT AIR DRYING

An equally important method of drying is by the use of heated air in a suitably designed plant. The dryers may be of the batch or continuous tunnel type, and heated by gas, steam, high pressure hot water, or electricity.

5.6.2.1. Batch dryers

Batch dryers of the simplest design take the form of cabinets in which the work is suspended or placed on trays. They depend on convection to remove the moisture from the wet articles, but have the disadvantage that loading and unloading take a relatively long time. On the other hand, this type of drying unit is inexpensive. Vanes are located within the oven to ensure that the air current passes over the articles being dried.

Modern cabinet dryers are, however, generally equipped with a centrifugal fan to accelerate drying rates. A high and uniform rate of air movement of 150 to 300m per minute is essential for efficient operation.

In double-cased ovens, which are most widely used, the moist air leaving the inside of the cabinet passes through the space between the two casings, thus pre-heating the incoming air. A further refinement is the treble-cased oven, which is mainly used for paint stoving, where solvent vapour must not be allowed to enter the heating zone, and products of combustion must likewise not come into contact with the articles being treated. A rate of air change of 40 to 50 times a minute is generally employed. Thermal insulation is essential in all types of drying and stoving ovens to conserve heat and thus reduce operating costs.

Drying ovens of this type operate at 100°-200°C, with a high percentage of the air being re-circulated.

5.6.2.2. Tunnel dryers

Continuous dryers of the tunnel type are most frequently used nowadays. They employ fans and heaters as do batch ovens, but have the advantage of being simpler and quicker to load and unload. Cross-flow or mixed flow air currents can be employed, with a re-circulation rate of about 75%. The cross-flow system is to be preferred, since the air is maintained in contact with the articles for a shorter time than is the case with mixed flow, so that there is less of a drop in air temperature. The drying conditions are also more readily controlled. The air enters the tunnel through holes in the top or sides of the oven at 155-200 m/min, and is then drawn back with the fans over the heaters through the bottom.

Loss of hot air from the tunnel ends, especially in paint drying, where longer stoving times are required than for moisture control, is frequently minimised by using a "camel-back" design to retain the hot air in the central part of the unit.

A preliminary mechanical air blast is also often provided to remove trapped water where articles of complex shape have to be dried.

Sometimes articles are removed from the processing plant and put onto a separate monorail conveyor which passes through the drying oven, which may be located in any appropriate position in the system; it is even possible to suspend such an oven from the roof where it does not take up valuable floor space. This method has the advantage that a relatively long drying time can be provided at a lower capital cost than would be involved in incorporating a lengthy drying zone in the processing plant itself. It is also possible to combine both systems, with a relatively short drying unit in an automatic plating plant, for example, to remove much of the water, and an independent final drying tunnel on the main conveyor to complete the process. This can deal with work coming from several finishing operations in a shop. It may, in fact, be convenient to arrange for all wet articles to pass through a single tunnel to finish drying them after rinsing in demineralised hot water, which is not always adequate to ensure complete drying of light gauge articles. This procedure has been adopted with success in several plants.

5.6.2.3. Design factors

In the design of dryers the ratio of the humidity of the air used for drying to the humidity when it is saturated at the same temperature is the most important parameter, and corresponds closely to the relative humidity as determined by wet or dry bulb measurements. The total heat of an air-water vapour mixture is the sum of the sensible heats of the air and the water together with the latent heat of the water vapour at the wet bulb temperature of the mixture, and is independent of the dry bulb temperatures at constant wet bulb temperature. The relationships between the characteristics of moist air at various temperatures are given in Psychometric or Humidity charts, from which the necessary calculations are to be carried out.

The design of the ductwork of air dryers should be carefully carried out to minimise mass velocity losses as far as possible. Bends should be smoothed and special attention given to the entry and exit points, since the face area of the heater is often considerably greater than the cross section of the ducting. The transition can be smoothed out by means of diffuser vanes at the expansion zones; the face area of the heater can also be made smaller by using finned tubing. The rate of drying rises initially, with the rate of evaporation increasing as the work approaches the temperature in the dryer. This is followed by a constant rate period when the moisture assumes the temperature of the drying air. After this, the evaporation rate begins to fall as the work begins to dry and approaches the temperature of the drying air. Additional heat is then transferred to the drying surface at the dry areas.

Hot air dryers are very effective, and as the saturation vapour pressure of water in air rises rapidly with temperature, it is not necessary to use particularly high velocities or volumes. In the interests of economy, partial recirculation is arranged for to conserve heat. 10-12 air changes an hour are generally adequate for the drying of 50-100 kg of work an hour. Air losses through the entry and exit ends of the drying tunnel are sufficient to carry away the requisite amount of moisture. When high pressure hot water is used for heating the air a pressure of 20 kg/cm^2 is suitable, corresponding to a temperature of 95°-120°C. In the case of steam and high pressure hot water, finned tube heat exchangers are most satisfactory; they should be placed transversely in the recirculating air system so as to impede the air flow as little as possible, and have sufficient capacity to provide up to 10 or 15 times the heat theoretically needed to evaporate the residual water on the work completely.

An alternate method is to use pre-heated air; here, the incoming air is blown over an electric heating unit, or through a gas or steam-operated heat exchanger, with partial recirculation as before. It is not normally necessary for the air to be above 100°C to remove the water; in fact, for most purposes air at 70-80°C is sufficient. In exceptional cases, such as where there are seams or hollow sections, the only satisfactory method of ensuring complete water removal is to boil it off; this necessitates the use of high temperature air at 100°-120°C. In designing drying ovens it is usual to assume a 3 minute drying time, and to provide the equivalent of 2-4 lbs of steam per lb of water evaporated from the cold surface. If the articles enter directly from a hot water rinse, however, the amount of heat needed is correspondingly reduced. The following equation has been given by Pinkerton[12] as an approximation to estimate the amount of steam required to evaporate water from metal surfaces brought from hot rinses at 70°C into a dryer:

$$h = k \frac{0.02\,w - cr}{8}$$

where h = kg of steam required per sq m of total surface area.
w = litres of water/1000 sq m of total surface area.
c = specific heat of the metal.
r = ratio of the weight of the articles in kg to total surface in sq m.

The factor k is a function of the efficiency of the dryer, and varies from 2 to 4.

Drying ovens should be insulated to prevent heat losses, with tunnel openings being kept as small as possible. Where large openings are unavoidable, a heated 'air curtain' can be employed. The air is directed downwards across the opening into a slot at the bottom, whence it is returned to the blower via a heater. In some instances, e.g. where sensitive finishes such as chromate conversion coatings are applied on zinc and cadmium plated articles, the air must not be heated above 70°C, or the protective value of the conversion coating will be adversely affected. When wet metals at room temperature have to be dried by hot air blast, a modified equation must be used.

5.6.3. INFRA-RED DRYING

Infra-red heating, by means of either tungsten filament lamps of 250W or 500W output, or radiant elements, has also been used for drying. However, the method is slow when bright plated surfaces, or aluminium, are to be dried, since the high reflectivity in such cases makes the rate of heat transfer somewhat slow. If the radiant sources are enclosed in insulated ovens with air re-circulation, these become, in effect, hot air ovens and perform satisfactorily in many cases.

5.6.4. DRYING OF STRIP

The drying of electroplated strip presents a special problem, as it may have to be carried out at speeds of up to 200 m/min or even more in some cases, as in the electro-tinning of steel strip. Flash drying can be used satisfactorily for this operation. Hot water is applied to the strip immediately in front of a pair of squeegee rolls which spread the water over the surface as an even film. A cold air blast on the outgoing side of these rolls is then sufficient to dry the strip completely. The more conventional method of blowing heated air along a tunnel in a direction counter to that of the strip can also be employed, but is somewhat more costly and may result in a greater liability to staining of the strip.

5.6.5. DRYING ELECTROPLATED ARTICLES

In many respects the drying of electroplated components presents special problems, and can be expensive both in energy and plant. The reasons are several. First of all, when plated articles emerge after the final rinse, they may retain water in recesses, and it is desirable that this should be removed physically, if satisfactory drying is to be achieved in a reasonable time. Clearly, the less water which remains to be removed by evaporation, the more economical will the process be. In the case of hand-operated plant, there is no serious difficulty since, with a little manual dexterity, the operator can shake off excess water; but with automatic equipment, which is being increasingly used nowadays, other means must be adopted. Air jets strategically placed to blow off surface water, are frequently used, but these are not always entirely satisfactory because of the difficulty of locating the jets where they are needed. In certain circumstances, pivoted plating jigs have been used on automatic plants; these enable cupped articles, which may have to be plated with their open ends upwards to avoid the trapping of gases in their interiors, to be tilted by means of floats as they emerge from the rinse tank.

Removing as much water as possible before drying with hot air also helps to minimise the occurrence of staining.

Staining is a particularly important problem when dealing with articles which are bright, such as chromium plated components, since any form of water stain makes them unacceptable and has to be removed. As often as not, this may mean an additional buffing operation. To avoid residual mineral deposits on the work, demineralised water is commonly used in the final rinse. It is sometimes advantageous to add a 'rinse-aid' to the water to lower the surface tension, so that less is retained on the surface. This can be in the form of a surface active material which does not leave a solid residue, and hence does not result in staining. Additives of this kind are available commercially, but must be used with extreme caution.

5.6.6. SAWDUST DRYING

One of the classic methods of drying, which is especially useful in the case of highly-finished, relatively expensive articles such as silverware, for example, is in heated sawdust. This gives a clean, relatively stain-free finish, although it is somewhat slow and involves a fair amount of handling.

The simplest method of sawdust drying is to use a metal tray containing a layer of sawdust about 15 cm deep, heated by hot air, steam or a water jacket. The sawdust used is hardwood, which should not contain particles smaller than about 60 mesh, and be free from tannic acid. Maple, lime, poplar or beech sawdust are most satisfactory. The heat requirement is relatively small in comparison with oven drying. To prevent the sawdust from being saturated when in contact with the work, the work and the sawdust are frequently agitated in a box shaker tray. Oblique open-ended sawdust barrels are also available for the drying of small parts in batches. The barrels are fitted with a heating jacket.

Continuous sawdust dryers have also been designed. In these, a rotating horizontal cylinder contains a helical internal screw which moves the work and the sawdust along its length. The cylinder is only perforated at the discharge side, where the sawdust falls out and is conveyed to the input end of the drum after passing over heating elements. Hence, the barrel is continually replenished with hot dry sawdust. An important point in favour of sawdust drying is the fact that it can be used on small, light articles since it does not depend on the heat content of the articles to evaporate the water. Moreover, the sawdust leaves a light protective film on the dried work, which acts as a mild corrosion preventive.

5.6.7. CENTRIFUGAL DRYERS

Centrifugal dryers are very frequently used for drying small articles in bulk after cleaning, pickling or plating. The articles are placed in a perforated basket, and even if the centrifuge is unheated, the rotational force is sufficient to dry light articles after removal from the hot rinse, when the heat they retain is insufficient for atmospheric drying. These centrifuges are usually fitted with two-speed motors, and the direction of rotation is reversible. The drying of articles liable to nest together is facilitated by reversing the direction three or four times. Centrifuges are available in capacities of 30-150 kg. A centrifuge of about 60 kg capacity is a suitable size of unit for most normal applications. They generally operate at temperatures of 70°-180°C, with steam, gas or high pressure hot water heating, and mass air velocities of 150-600 $kg/cm^2/hour$.

Hot air is introduced into centrifugal dryers by means of a low pressure blower passing over a bank of heaters. Recirculation to the extent of up to about 65% can be incorporated. Whilst the heated centrifuge is satisfactory for drying most small components, its rate of rotation can result in articles with large flat surfaces, such as washers, or having hollow areas, to be clamped together in such a way that water cannot be flung out of the centrifuge basket completely, even by the reversal of the direction of rotation. In such cases, a drum dryer is more effective since it turns more slowly. This is a horizontal perforated barrel provided with internal vanes which constantly turn the work over, at the same time moving it forward through the drum so that it emerges dried at the opposite end from that at which it is loaded. Heated air is blown through the barrel at the same time.

The drum dryer does, however, have the disadvantage that it may cause damage to some articles such as those having threads. In such cases a shaker dryer through which hot air passes can be more satisfactory. As this type of dryer is of the batch type, it also has the advantage that successive loads can be treated separately so that there is no danger of mixing of work. In many modern automatic barrel plating plants, provision is made not only for automatic loading and unloading, but also for hot air drying in the plating barrel itself. In one typical plant, a solenoid-operated telescopic duct contacts the end of each oblique, open-ended polypropylene barrel as it passes the drying station containing 15-20 kg of work whilst it is rotating at about 1 r.p.m. Hot air is supplied at the rate of 4000 litres/minute by means of a fan. Drying of the work is complete in 7 minutes after the barrel emerges from the hot rinse.

5.6.8. SOLVENT DRYING

Another method which has been promoted in recent years for the stain-free drying of electroplated articles, involves immersing the wet article in a specially formulated boiling trichloroethylene-based mixture, so that the water is evaporated off as an azeotrope. The trichloroethylene contains about 1 per cent of a surface-active additive which causes the solvent to wet the metal preferentially, thus displacing water, together with any dissolved salts from the surface. The additive also increases the rate at which the water is removed from the metal surface. The additive is rinsed off in a clean solvent, which then evaporates. The method is suitable for all kinds of plated metal pressings and stampings, as well as for tubular and wire work. It can be used on finishes such as nickel, silver, gold and chromium plate, as well as on phosphated, anodised and electrolytically polished articles.

It is carried out in a more-or-less conventional stainless steel or otherwise protected two-compartment degreasing plant. A water separator and rim ventilation are incorporated. The trichloroethylene containing the additive is in one compartment of the plant whilst the second, rinse, compartment is filled with plain trichloroethylene. The water-wet articles are immersed in the first section for, say, 15-30 seconds, after which they are transferred to the rinsing section for a further 15 seconds or so. They are slowly withdrawn and after remaining in the vapour zone above the solvent for a short time, removed from the plant.

In the case of large articles, where a vapour rinse may be inconvenient, a three-compartment plant can be used; one tank is for drying, the other two are for rinsing. Since the azeotropic mixture of trichloroethylene and water, which contains 7% of water, boils at 73°C as against 87°C for the solvent alone, water in the drying compartment which has been removed from the work with the aid of the additive is rapidly volatilised. The solvent and the water are condensed on the coils of the plant simultaneously, the former being removed in the water separator whilst the latter is returned to the plant.

Perchloroethylene can be used with advantage for this purpose in place of trichloroethylene, since its azeotrope contains 16 per cent of water. It is also more stable, less toxic, and has a lower heat of vaporisation. There is little difference in the results obtained. Perchloroethylene has, however, a rather high boiling point (121°C), and is therefore to be used with caution on articles which may be affected by higher temperatures. Perchloroethylene is slightly more expensive than trichloroethylene, but the running costs are less when the former solvent is employed. This is because it only requires 1.5 kWh to evaporate 1 kg of water when perchloroethylene is used, as compared with 4.6 kWh with trichloroethylene. The proportion of solvent in the azeotropes also favours perchloroethylene. 19 kg of the former are required to evaporate 1 kg of water, whilst 108 kg of trichloroethylene are needed to evaporate the same amount of water. This makes perchloroethylene more than five times as effective for drying than trichloroethylene.

5.6.8.1. Plant for solvent drying

Solvent drying plants are best constructed from stainless steel; galvanised equipment is less satisfactory, as it tends to corrode as a result of solvent hydrolysis in the presence of water. Copper water-cooling coils are provided around the top of the plant to remove the heat from the boiling solvent and prevent the escape of vapour when it is not in use. Water should flow through the coils at a rate of about 60 cm per second; the inlet temperature of the water should be not more than 10°C, and the outlet temperature 40°C. Separators are also fitted to remove water which mostly enters the plant by condensation in the cooling zone. The water separator functions by condensing the water-solvent azeotrope in the cold zone of the plant. Here it separates into its individual constituents, the lighter water floating on top of the solvent, whence it is drained off through an overflow pipe. The solvent may then be either returned to the plant or to a distillation unit.

Solvent drying plants are generally heated by steam or high pressure hot water coils, the temperature of which must not exceed 118°C in the case of trichloroethylene and 148°C for perchloroethylene, these being the respective temperatures at which the solvents begin to decompose. Contact of the solvent vapour with gas flames must be avoided, as it can result in the formation of phosgene. A sump thermostat is set at these temperatures to prevent overheating, whilst another thermostat set at not more than 73°C (for trichlorethylene) or 110°C (for perchloroethylene) provides a safeguard in the event of failure of the cooling water supply around the top of the plant.

The principal uses of solvent drying include the removal of rinse water after electroplating, electrolytic and chemical polishing, bright dipping and anodising. The solvent pentrates completely into cracks and interstices, as the articles are completely immersed. It is therefore particularly suitable for the drying of intricate shaped articles which are threaded or have blind holes. Solvent drying has the advantages of being able to deal with intricately shaped components readily, whilst the labour and space requirements are low. It is also relatively rapid in operation. On the other hand, it involves the need for removing racked articles in the wet state, and transferring them to baskets for treatment, as the trichloroethylene attacks many types of rack coating . The cost of the solvent used is also not inconsiderable, whilst the process will only operate successfully if the articles are of sufficient mass to retain adequate heat to evaporate the residual solvent after rinsing. It is also unsuitable for chromated zinc or cadmium plated articles, as the solvent has the effect of dehydrating the conversion coating and reducing its effectiveness. Parts dried in this way are also exceptionally free from grease and oil, so that they may be more liable to develop corrosion, particularly in recessed areas where the plating is liable to be on the thin side.

5.6.9. CONCLUSIONS

A variety of methods for the complete drying of wet metal surfaces is available, but great care must be taken in the design of the plant where stain-free drying is to be obtained. The rectification of even slightly stained work is costly if only because it involves an additional handling operation. Fuel costs can be kept down by removing as much water as possible mechanically before drying by means of heat. Partial air recirculation is an important factor in obtaining economical drying.

5.7. Materials of construction for electroplating equipment

Electroplating equipment is subject to very considerable corrosion, and special attention must be given to materials of construction. This attack occurs from the nature of the solutions used and also acid fumes which are often generated. Many solutions are, moreover, seriously affected if they are contaminated by metallic impurities, so that should corrosion occur the effect on the plating operation itself may well show before any deterioration in the plant is evident. A further factor is the presence of electric currents which can result in the rapid attack even of normally resistant metals if they should become anodic. Stray currents must therefore be avoided by careful design to eliminate bipolar effects.

5.7.1. TANKS AND LININGS

For acid plating solutions, such as nickel, and cyanide solutions which include copper, zinc, and cadmium, and also for tanks for acid dips and rinses, rubber-lined steel is very widely used; the rubber is of the hard type and is vulcanised onto the steel. Such tanks will withstand temperatures of 70-80°C, but are unsuitable for oxidising acids such as nitric or chromic acid. The rubber must be carefully selected and be free from metallic impurities and some sulphur-containing accelerators which might be leached out by certain plating solutions (notably nickel) leading to dark and discoloured deposits. Linings should be tested electrically for continuity before being put into service.

Effective protection by suitable paints, such as those based on chlorinated rubber or epoxy resins, is necessary for tank exteriors, although the unit type of tank construction described above helps to avoid difficulties due to corrosion between tanks. Highly vulnerable small fittings such as brackets can be protected by plastic dipping techniques.

The other principal non-metallic tank lining which is of commercial importance is polyvinyl chloride, both rigid and plasticised. One method of lining tanks which gives an efficient bond consists in applying two or more coats of a suitable cement both to the tank wall and the plastic. This is allowed to dry for several hours, after which the tank is heated to a temperature of 140°C by applying local heat to its exterior by a welding torch. The PVC sheet is then pressed onto the tank wall, whereupon bonding takes place. Joints of all kinds can be made in the plastic by means of a hot-nitrogen welding gun using plastic filler rod; this technique is now well-established. An alternative method is also available which is more suited to lining partitioned tanks or those on site where some or all of the exterior surfaces cannot be heated; this consists in applying a cold setting adhesive both to the sheet and the tank. The PVC generally used for lining is a plasticised, filled grade, usually black, although lighter-coloured linings have some advantages in service. On the continent of Europe some use has been made of translucent plastic, it being claimed that the lining operation is more easily carried out with such materials as air pockets can be seen. However, dark grades of fillers are more widely used; these have some reinforcing effect and serve to strengthen the lining.

Vinyl linings have found an important application in the case of chromium plating tanks, which have in the past generally been lined with lead and heated by means of an external water-jacket containing heating coils. Lead linings have to be protected against short-circuiting by work coming into contact with them, reinforced glass sheets being mostly used for this purpose. Because of the large amount of space which a water-jacket takes up, this type of construction is extremely inconvenient in automatic machines, and PVC linings are therefore to be preferred, with heating coils immersed in the solution. Lead-covered steel coils have a limited life under such conditions, but titanium coils are entirely satisfactory with the normal type of chromic acid solution. Difficulty is, however, experienced with certain chromium solutions containing silico-fluorides, which tend to attack titanium. Tantalum coils can be used in such cases, but are relatively costly. PVC lined tanks cannot be satisfactorily heated by means of water-jackets owing to the low rate of heat transfer through the plastic, while the differential expansion at the liquid level line may cause blistering to occur.

Tanks can also be moulded in one piece or fabricated from phenolic resins reinforced with fabric, asbestos, or graphite. Graphite has good chemical resistance and high strength for acid solutions, and some grades also withstand oxidising acids in moderate concentrations.

Glass-fibre reinforced resins are finding increasing application in the manufacture of tanks and also of fume ducting and have the further advantage of not requiring maintenance on exterior surfaces; corrosion from the outside inwards can be serious, particularly in the case of tank bottoms which are not mounted clear of wet floors. Glass-fibre reinforced tanks can be made from flat laminated sheets, or better still built up on formers using glass-fibre mat or woven cloth. Jointing of the sheets can be done by means of the catalysed resin with or without a filler and is best carried out under pressure. They are extremely light in weight; thus a typical 60 × 100 × 100 cm tank only weighs 30 kg and a 15,000 litre tank weighs only 400 kg, and the advantage in erection and installation is readily seen. Such tanks, being translucent, have the advantage that the location of articles inside them can be seen. The tank walls must be of adequate thickness because the laminate has a relatively low rigidity modulus, although its tensile strength is high. A common way of strengthening tanks is to use a thickened rim at the top with reinforcing ribs around the sides; these may be of timber or steel embedded in the laminate.

Although glass-fibre tanks can be cold cured, they are now almost invariably given a final heat cure which results in a more consistent product than is obtainable by cold curing alone. Composite tanks, using epoxy-resin laminates on the interiors and polyester resins for the exterior surface, have been used to reduce cost where high chemical resistance is required. Such tanks are very useful not only for rinses but also for storage tanks for bright nickel plating solutions. Another useful combination consists of vinyl sheet bonded to glass-fibre reinforced resin.

Steel and plastic laminated sheet, consisting of steel to which a PVC film of about 0.050 cm has been bonded, has found some application for acid dip tanks. Such sheet can withstand a fair amount of forming and bending and it is therefore quite practicable to construct tanks by bending methods. Flanges can be assembled together by bolts using a suitable gasket material, or the seams can be welded by means of filler rods as with PVC fabrications. It can also be employed for fume exhaust ducting, the plastic face being the internal one.

5.7.2. PLATING BARRELS

The plating barrel used in automatic plating plants for handling small parts and also in manually-operated lines, presents a difficult problem in the choice of materials. The requirements are severe; in the first place the product used must be capable of being fabricated into polyhedral barrels, which may be quite large and perforated with numerous holes. The barrel must have high strength and impact resistance and be capable of withstanding the complete sequence of operation to which barrel-plated articles are subjected. These include caustic alkaline cleaners at elevated temperatures, acid pickling solutions, alkaline and acid plating baths, and hot rinses. A considerable degree of rigidity at operating temperatures is needed, since the barrels must carry a heavy load of work and even slight sagging can cause barrel doors to jam, whilst distortion may also interfere with the movement of the rotating and transfer mechanism.

One of the earliest types of barrel material was ebonite and this has given satisfactory service in simple applications; its impact resistance is, however, not good enough for the larger horizontal barrels. Barrels have also been made from fabric-reinforced phenolic resins; these have good acid resistance, but they do not withstand alkalis too well. They also have the disadvantage that when holes are drilled in them attack can take place along the exposed fibres of the reinforcement. This is also true of glass-fibre reinforced resins, which in other respects have adequate chemical resistance and mechanical strength, particularly when epoxy-type resins are employed; hence they are little used for this application.

Amongst the most widely used materials for barrels have been the methyl methacrylate resins, such as Perspex, which can be fabricated and cemented together to make barrels having good strength and chemical resistance. They have two main disadvantages, however: great care must be taken to see that they are not subjected to temperatures higher than 80°C, as above this they may soften and distort; secondly, certain types of organic brighteners employed have been known to attack the plastic and cause surface crazing. Nevertheless, with reasonable care such barrels have a long life, although their cost tends to be high.

The advent of polypropylene resins has provided very useful material for barrels, as it is unaffected by temperatures up to the boiling point of water, and is not attacked by most plating chemicals. Attempts have also been made to use it for tank lining, but bonding to steel presents difficulties. Other barrel construction materials used chiefly in the U.S.A. include some special rubber-resin combinations which, being of the thermosetting type, do not have any serious temperature limitations. Some use has been made of steel barrels coated with a vinyl plastisol by dipping, and quite large plants have been equipped with barrels of this description. They are mechanically strong and relatively inexpensive, although the coatings will evidently have a limited life; the barrels however, can be re-coated with plastisol when necessary.

5.7.3. EXHAUST DUCTING

There is an increasing tendency now for exhaust ducting and trunking to be made throughout of corrosion-resisting materials instead of the traditional steel coated with bituminous paints in order to reduce maintenance costs and avoid delays in production. PVC is currently widely used for this purpose because of the ease with which it can be fabricated. Initially, ducting was fabricated from flat sheet welded into box-shaped structures, but these leave a lot to be desired; curved fabrications are more satisfactory as they have greater strength and impede the air flow much less. Where more difficult conditions are encountered, laminates of PVC and glass-fibre are being used to some extent.

Glass-fibre exhaust ducting has now come into very widespread use as it is strong and does not soften with temperature. The strength and rigidity of installations made from these materials also makes them to all intents and purposes immune from the danger of mechanical damage to which many other plastics are subject. The fact that glass-fibre reinforced resin fabrications are not subject to limitations of shape makes them particularly attractive. On price they can be competitive with PVC and it is anticipated that they will be used still more widely as time goes on. Exhaust fans for use in extraction systems are generally of rubber-covered steel, plastic materials, or stainless steel.

5.7.4. PIPEWORK

A perennial problem in connection with many plating installations (particularly bright nickel) is the pipework system for connecting the plating tanks with pumps, filters, storage vessels, etc. These are commonly made of rubber-lined steel, and rubber-lined valves and fittings are used in conjunction with them. Pipes must be kept relatively short owing to the impracticability of lining excessive lengths, and the entire installation has to be individually designed. Ranges of vinyl resin pipework complete with moulded fittings are now available to the industry, and although they are subject to temperature limitations and mechanical damage they are being increasingly used on automatic installations.

Glass-fibre laminated pipes and fittings appear to have considerable advantages for this purpose, especially in bright nickel plating, as they are unaffected by the solutions, can be used at any normal temperature, and will withstand rough handling. Suitable piping has a wall thickness of about 1 cm and is heat cured. Epoxy resins are to be preferred to the cheaper polyesters because they are less subject to chemical attack.

Glass-fibre pipework can be joined by means of short lengths of oversized sleeving which are pushed over the straight tubes; bends and tees can also be fitted in this way, the joints being cemented together. Locating stops should be provided inside the sleeving to ensure correct overlapping. The gaps between the sleeving and the pipes cannot unfortunately be very accurately controlled, so that there is a certain amount of danger of leakage. For this reason specially manufactured screwed fittings are to be preferred.

5.7.5. RACKS AND RACK COATINGS

For the protection and insulation of plating racks PVC plastisols have now become the standard material, having the required properties of flexibility, ease of application, and chemical resistance. Application is by pre-heating the rack before immersion in the cold plastisol, and heat curing. Cements are often used to improve bonding of the plastisol to the brass or copper of which the racks are made. Coatings of polythene are also used to a lesser extent, application being best carried out by putting the heated racks into the plastic powder, which is fluidised by blowing air through it. The coating is then fused by heat. Though tough and chemically resistant, polythene does not have any adhesion to the basic metal so that solution is apt to enter where contacts are exposed.

Rack design is important in minimising plating defects[13], Mitchell has reviewed the design of racks for electroplating and anodising[14].

5.8. Electrical equipment

Electroplating processes require adequate supplies of low voltage current. For a long time this was provided by motor generator sets, but these are now all but obsolete, having been replaced by solid state rectifiers. A review of rectifier equipment available commercially has recently been published[15].

The earliest rectifiers were of the air-cooled copper oxide type, which were followed successively by selenium, germanium and silicon. The germanium vogue was shortlived, mainly because of its low operating temperature limit, and virtually all current for electroplating is provided by either selenium or silicon rectifiers.

The ideal rectifier should be compact in size, capable of operating for long periods with minimum maintenance under adverse atmospheric conditions, and have the best possible electrical efficiency. A plating current supply system consists of a transformer to convert the incoming high voltage supply to the desired low voltage, a rectifier to change the alternating to direct current, and a current control unit which can take the form of a simple tap switch, a variable transformer, or best of all, a thyristor or saturable reactor.

Selenium rectifiers are generally oil cooled, both the transformer and the selenium diodes being mounted in the same housing. The fact that the rectifier is totally enclosed in this way protects it from corrosion. Selenium rectifiers must be effectively cooled, since the operating temperature must not be allowed to exceed 55 °C. They are also subject to ageing, their output falling by 5-10 per cent after ten to twenty thousand hours of service. They can also stand considerable overloads of up to 50 per cent for limited periods. Fig. 5.21 shows an oil-coiled Selenium rectifier installation. The maintenance of rectifier equipment has recently been discussed by Mendel[16].

5.8.1. SILICON CONTROLLED RECTIFIER

Silicon rectifiers have a very high rating per unit area, and are thus more compact than the selenium rectifier. This carries the penalty of greater sensistivity to overloading, so that stringent precautions have to be taken to prevent this from occurring. This can be done very effectively by means of a thyristor, which also provides an ideal method of controlling output. Typically, the output voltage is compared with a reference voltage and the error signal is used to control the thyristor. The output voltage therefore remains stable regardless of load current. To vary the output it is only necessary to vary the reference voltage. This method enables voltage and current to be controlled to within 1 per cent.

Fig. 5.21. Oiled-cooled selenium rectifier installation.
[Courtesy Westinghouse Brake & Signal Co. Ltd.]

There are many advantages attached to the use of rectifiers of this type, which are likely to dominate the market in the future. Not the least of these is the automatic overload protection provided by the thyristor control, so that no fuses are even needed. A number of different circuit arrangements are possible, the choice being determined by criteria of optimum performance and minimum cost.

5.8.2. RIPPLE

One of the effects of rectification is ripple which can cause problems, especially in the case of chromium plating. Much more important than the RMS value is the instantaneous minimum voltage, or valley voltage, and momentary current interruption. Tests have shown that if this valley voltage is kept above 3 volts no adverse effect on deposition occurs. A ' whitewash ' effect can occur in chromium plating if this voltage falls below 3 volts; this can, however, be prevented by fitting a simple filter capacitor.

The use of water cooling is recommended for silicon rectifiers, since it enables the diodes to be sealed from the atmosphere, thus preventing corrosion and enabling the assemblies to be more compact and hence less expensive. The rectifiers can also be mounted close to the plating tanks in corrosive atmospheres without harm, so that there is a saving in the cost of low voltage busbars.

In one design an indirect cooling method is employed in which the surfaces to be cooled are exposed to a circulating coolant consisting of propylene glycol and distilled water. The intake water passes through a heat exchanger. When the heat sink attached to the rectifier plates reaches 54°C the water is turnd on, being shut off when the temperature falls to 51°C.

Fig. 5.22 shows a naturally-cooled 5000 amp silicon rectifier. Fig. 5.23 shows a water-cooled rectifier of the silicon type with constant voltage and current limit control.

5.8.3. RECTIFIERS FOR AUTOMATIC PLANT

The advent of the silicon-controlled rectifier has made the automation of electroplating plants considerably simpler. The most widely used systems incorporate voltage stabilisation and often automatic current density control. Automatic voltage control reduces rejects and enables maximum plating rates to be maintained; it is also readily adapted to automatic current density control, which prevents ' burning ' of deposits and over or under plating.

Silicon rectifiers should be conservatively rated. A 10 per cent higher voltage than is required should be specified, whilst the diode capacity should exceed the output current rating of the system by at least 20 per cent in view of the possibility of high ambient temperatures[17]. Under-rating the diodes will also extend their life and maintain their peak inverse voltage ratings.

Fig. 5.22. 5000 amp, naturally cooled, silicon rectifier. No controls.
[Courtesy Westinghouse Brake & Signal Co. Ltd.]

Fig. 5.23. A water-cooled silicon rectifier with constant voltage and current limit control. [Courtesy Oxy Metal Industries (GB) Ltd.]

Besides current overloads, the major cause of silicon diode failure is severe transient voltages. The diodes are usually installed with two or more in parallel, and since they have a very low forward electrical resistance, a serious current imbalance combined with a slight difference in diode characteristics can cause diode failure. The overload control must be capable of disconnecting the primary power in 6 cycles or one-tenth of a second to avoid damage.

The specification and choice of a rectifier demands careful consideration of all the relevant factors to obtain maximum efficiency, long life and minimum capital cost. Special attention should be paid to corrosion protection, especially of housings and transformers, and proper impregnation procedures are necessary to avoid deterioration in plating shop conditions.

REFERENCES

1. G. A. Cummins Jr., *Plating,* 1973, **60**, No. 9, 928-30.
2. M. Rubinstein, *Product Finishing,* 1973, **37**, No. 5, 46-53.
3. H. Silman, *Trans. Inst. Chem. Eng.,* 1961, **41**, 383; *Trans. Inst. Metal Finishing,* 1963, **41**, 11.
4. G. Lindner, *Metalloberfläche,* 1973, **27**, No. 4, 135-9.
5. P. W. Fruin, *Metal Finishing J.,* 1964, **10**, 392.
6. H. Silman, *Chemistry and Industry,* 1962, 2046.
7. E. P. Harris, 'A Survey of Nickel and Chromium Recovery in the Electroplating Industry'. DSIR, 1960.
8. R. Weiner, 'Effluent Treatment in the Metal Finishing Industry', Robt. Draper, 1963.
9. Robert W. MacKay, *Products Finishing,* 1968, **9**, 48.
10. *Metal Finishing Plant & Processes,* 1973, **9**, No. 3, 85-97.
11. W. Alina, *Products Finishing,* 1973, **37**, No. 6, 109-113.
12. H. L. Pinkerton, 'Electroplating Engineers Handbook' (Ed. A. K. Graham), 1968. 716.
13. L. Durney, *Products Finishing,* 1974, **39**, No. 2, 92, 94.
14. R. L. Mitchell, *Metal Finishing,* 1971, **69**, No. 10, 48-55.
15. *Metal Finishing Plant & Processes,* 1972, **8**, No. 5, 163-70.
16. T. Mendel, *Metal Finishing,* 1974, **72**, No. 4, 33-4.
17. J. A. Viola, *Plating,* 1970, **1**, 17.

BOOKS FOR GENERAL READING

1. 'Automation in the Metal Finishing Industry', Ed. R. R. Read, Portcullis Press, Redhill, 1976.

CHAPTER 6

Metal colouring and phosphating

6.1. Metal colouring

THE colouring of metals by chemical treatment in a large variety of solutions has been practised for centuries; in modern usage such colouring processes are applied largely for decorative purposes where it is desired to improve the appearance of a metal or to simulate a more costly metal, as in the bronzing of steel, for example. A degree of protection against atmospheric attack may also be obtained, but this is generally of limited value.

6.1.1. APPLICATIONS

While metals may be coloured by the application of paints or lacquers, the term ' metal colouring ' is usually applied to those processes in which the metal surface is itself treated so that it acquires a colour which differs from the original. Treatment, which is brought about by a chemical, thermal or electrochemical process, may result in the conversion of the metal surface into a coloured compound, e.g. an oxide or sulphide, or it may only occur to a very small depth so that intereference films are formed on the metal surface. These films are very thin (0.025-0.055 microns) and, although without colour themselves, give a coloured appearance to the metal surface due to the mutual extinction of certain wavelengths of light which are out of phase when reflection takes place from the metal and from the film surfaces. As the thickness of the film increases, the colours change from yellow to red, blue and bluish green, after which the colour of the film itself becomes evident. If the film is non-uniform in thickness, an iridescent or mottled coating will be formed.

6.1.2. GENERAL METHODS

The finishes to be described are produced generally by one or other of the following means:

(a) Treatment by immersion, rubbing or spraying with a chemical solution which converts the metal surface to one of its coloured compounds (e.g. oxide, sulphide, etc.).
(b) Heating in air or in a gaseous atmosphere to form a coloured oxide film.
(c) Immersion in a solution of the salt of a metal more noble than that being treated, whereby a chemical exchange reaction occurs with the deposition of a film of the second metal on the first. Such a film may or may not be metallic in appearance, depending on the form in which it is deposited.
(d) Conversion of the metal surface into one of its compounds by electrolysis; an example of this is the anodic oxidation of aluminium, whereby the metal surface is converted to a tenacious film of aluminium oxide.

The appearance of finishes so obtained is greatly dependent on the state of the initial surface of the metal being treated. The films are usually thin, so that they reproduce the surface condition of the underlying metal. Thus, entirely different effects can result from the same colouring treatment depending on whether the metal has been pre-treated by acid dipping, shot blasting, scratch brushing, etc. The coloured film may be altered by rubbing with wet pumice or by polishing to 'relieve' the highlights of the article.

Combined electroplating and colour processes may be employed. This is especially the case with steel objects, which are often electroplated with copper from a cyanide solution, the latter metal then being coloured. Copper is chosen because it is easily treated to give a variety of pleasing colours, as will be described later.

The process of deposition of one metal onto another by immersion is typefied by the copper coating produced on steel when the latter is immersed in a solution of copper sulphate. A simple chemical exchange takes place:

$$Fe + CuSO_4 = FeSO_4 + Cu$$

The deposits are relatively thin and have poor adherence.

In other cases, as when zinc is immersed in nickel sulphate solution, the nickel is precipitated in a finely divided form, appearing black to the eye so that a black finish results instead of a bright nickel deposit.

As a general rule, the coatings to be described in this chapter have relatively low durability and are unsuitable for exposure to severe atmospheric conditions for any length of time. They are, therefore, commonly coated with a transparent lacquer to improve their protective qualities. Alternatively, if they are of a porous type, they may be oiled, waxed or impregnated with a variety of different substances to obtain a satisfactory finish.

In some instances, colouring plays an important part in the finish, as in gun blackening. Here it is essential to darken the steel to avoid reflection of light which would render the weapon unduly conspicuous.

Some coloured finishes (for example, the oxide coatings on steel applied either by heat or by one or other of the chemical treatments to be described) are very useful on account of their hardness. These finishes are not easily scratched or damaged, despite the fact that they do not withstand weathering or corrosive conditions well. Thus they are useful on parts and fittings for indoor service which are subject to much handling, such as small tools. In this respect they are superior to organic finishes and to phosphate coatings.

Chemical colouring can be carried out by immersion treatments using a great variety of solutions. Some of these give good finishes, but many also give indifferent results. Many published solution formulations are either too critical so far as their operating conditions are concerned, or they demand a particular grade of metal or surface finish for satisfactory results to be obtained. Some of the more complex formulae which have been published from time to time appear to contain constituents which perform no useful function in the solution.

A detailed account of metal colouring processes is given by Fishlock(1).

6.1.3. COLOURING OF IRON AND STEEL

6.1.3.1. Immersion deposits of nickel

Nickel can be deposited on steel by simple immersion in a weak solution of nickel sulphate or nickel ammonium sulphate for 5 to 10 minutes. The solution contains 5 to 12 g/l of nickel sulphate and often a little boric acid to maintain the pH at around 3 to 4. The solution is used at a temperature of 70° to 80°C. The deposit obtained is very thin and somewhat variable. The process is used chiefly on very cheap articles as a finish, and as a pre-treatment before the ' direct-on ' vitreous enamelling process. For the latter purpose a nickel thickness of not more than 0.000005 cm to 0.000015 cm is required for maximum adhesion of the enamel.

Somewhat heavier well-adherent immersion deposits have been obtained(2) by immersion in a solution containing:

Nickel chloride	600 g/l
Boric acid	30 g/l

The pH is 3.5 to 4.5 and the temperature used is about 70°C. The work should be agitated. Cleaning is carried out cathodically in an alkaline solution. Acid pickling is not very satisfactory unless followed by an alkaline treatment as it seems to set up large anodic areas leading to uneven sporadic nickel coatings. The deposits are of low protective value.

6.1.3.2. Temper colours

The most important of the coloured finishes applied to ferrous metals are the blackening processes. For many applications a dark finish on steel is desirable for decorative purposes, whilst a durable black finish is essential for military equipment to avoid light reflection from the metal surface.

When ferrous metals are heated in air a range of colours is produced varying from yellow through orange to dark blue, depending on the temperature and duration of heating. These colorations are due to the formation of oxide films on the metal surface so that interference tints are developed. The higher the temperature of treatment, the thicker will be the oxide layer; the thicker coatings are more durable than the thinner ones. Thus, if steel be kept at the same elevated temperature for a prolonged period of time, the colour produced on the surface will change progressively as the film thickens. Table 6.1 shows the approximate nature of the colours formed on carbon steel when it is heated up gradually in air.

These ' temper colours ' are so called because they are normally formed on metals in the tempering process, but in order to produce thicker or more uniform oxide coatings, steel parts may be dipped in oil prior to heating. Tallow or linseed oil are traditionally employed in gun-browning, the grease being applied as a thin film to the parts, after which they are heated for about half an hour at a temperature of 200° to 400°C. The treatment may be repeated two or three times to obtain a good finish. By the introduction of sulphur into the oil, darker shades of colour can be produced. Another method consists in heating the metal for 1½ to 2 hours at 650°C in a carburising compound or in a smoky atmosphere. The process of barrelling in charred sawdust to obtain a black finish is also employed by gunsmiths. Very tenacious oxide coatings can be formed on iron and steel by subjecting the metal to heat in alternate oxidising and reducing atmospheres.

TABLE 6.1

Temper colours produced on steel by heating to various temperatures[5]

Temp. °C	Temper colour
225	Very pale to light yellow
235	Straw to deep straw
245	Dark yellow to yellow brown
258	Brown to reddish brown
270	Purple brown to light purple
280	Full to dark purple
295	Blue
315	Dark blue

6.1.3.3. Bower-Bariff process

In the Bower-Bariff process the steel is heated to about 800°C for twenty minutes, first in contact with air and then in superheated steam. A coating consisting of a mixture of Fe_2O_3 and Fe_3O_4 is formed, after which heating is continued in a producer-gas atmosphere which converts it entirely to Fe_3O_4. This coating is claimed to have high durability owing to the fact that both the underlying metal and the oxide have similar coefficients of expansion.

An adherent black oxide coating can be produced on steel or cast iron by exposure to steam superheated to between 500 and 700°C. At 700°C a film thickness of about 37 microns is obtained on cast iron in 1 hour[4].

A similar process[5] makes use of low-pressure steam partially dissociated by being passed through a hot iron tube. The metal is heated in this atmosphere at about 600°C for half an hour, after which it is heated in the presence of oil fumes for a further fifteen minutes.

The above methods are capable of giving finishes of good appearance and durability if carefully caried out, although they require a degree of skill and are not readily suited to mass production where speed and uniformity are required.

6.1.3.4. Black oxide finishes

For large-scale heat colouring the method of immersing the steel parts in molten salt mixtures is used. It enables close temperature control to be achieved and hence colours may be accurately and uniformly reproduced without any great skill being required on the part of the operator. Molten sodium nitrate at 300°C may be employed, but mixtures of sodium and potassium nitrates or sodium nitrate and sodium hydroxide enable lower treatment temperatures to be used.

However the most widely used black oxide finishes are those produced by immersion in aqueous caustic alkali solutions containing an oxidising agent. The finish produced by this method is now used for a number of applications including blackening car parts and engines, cameras, machine tools and aircraft turbines. There are a number of proprietary processes based essentially on this principle, some of which incorporate special additions of activating agents to the alkali-oxidising agent mixture which, it is claimed, give improved results and increase the life of the bath (e.g. peroxides[6], tannates, tartrates and cyanides).

Typical solutions can be made up by dissolving 120 to 150 g of caustic soda in about 100 cc of water and adding about 30 g of sodium nitrate. The concentrated solution thus produced is used at its boiling point, which is high. The temperatures are of the order of 150°C, depending on the concentration of salts. The steel parts are cleaned and immersed for a period of five to thirty minutes in the solution until a satisfactory colour is obtained. The coatings thus formed may be as much as 0.002 to 0.005 cm in thickness. A typical formulation is as follows:

Caustic soda	480 g
Potassium nitrate	300 g
Water	1 litre
Temperature	140°C
Time	10 minutes

As evaporation of water takes place, the concentration increases so that the boiling point rises; more water must therefore be added from time to time to maintain a constant boiling point, this latter providing a useful measure of the strength of the solution. Addition of water must be carried out with circumspection, as the solution is liable to splash when the water is introduced. Salts must on no account be added to the hot bath, since a great deal of heat is given out when caustic soda is dissolved in water, and an explosion may result. Heating of the solution, which is used in a cast-iron or steel tank, must be very uniform. When gas is employed, the heat should be directed on to the sides as well as onto the bottom of the vessel by suitable baffles to prevent stratification of the very dense solution and to maintain it gently at the boil. The solutions deteriorate with use, due to the absorption of carbon dioxide which converts the caustic soda into the less effective sodium carbonate, and also as a result of the accumulation of iron salts which retards their action and lengthens the processing time.

Many proprietary baths contain constituents which convert contaminants into insoluble compounds which precipitate to the bottom of the bath as a sludge, or which produce a flocculant compound which comes to the surface and can be skimmed off[7].

Other oxidising agents, such as potassium permanganate or sodium dichromate, have been used either together with or in place of sodium nitrate. In the ‘ Jetal ’ process[8], which is typical of black finishes of this description, a bath is made up of the following composition:

Sodium hydroxide	750 g
Sodium nitrate	250 g
Water	1 litre

This is operated within the temperature range of 120 to 150°C, an optimum temperature range of 135 to 145°C being preferred. The time of treatment may be up to thirty minutes, depending on the colour and thickness of coating required. After the bath has been in use for some time it will be found that it has become exhausted so that only a poor coloration or no colour at all is produced. It should be possible to treat about 0.6 square metres of surface in 1 litre of solution before this stage is reached. When the bath begins to lose its efficiency about 0.5 kg of sodium cyanide is added per 100 kg of bath and, if necessary, some caustic soda and more nitrate to replace drag-out losses. This regenerating treatment is repeated as needed, an addition of about 1 kg of sodium cyanide being required for each 65 m^2 of metal treated. The cyanide serves to remove excess iron ions from the solution by the formation of sodium ferrocyanide $Na_4Fe(CN)_6$, so that the iron goes into the anionic complex, the reactions which occur being represented thus:

$$Fe(OH)_2 + 2NaCN = 2NaOH + Fe(CN)_2$$
$$Fe(CN)_2 + 4NaCN = Na_4Fe(CN)_6.$$

The initiation of the reaction is sometimes slow, so that blackening does not readily occur; this is particularly the case with cast iron and sometimes with cold-rolled steel sheet. It has been recommended that the metal be immersed for a few minutes in a 5 to 10 per cent solution of hydrofluoric acid prior to treatment to expedite the reaction in such cases.

Krause[9] recommends the use of urea in alkali-nitrate solutions. With a mixture of 20 g/l of urea to a caustic soda solution containing 10 g/l of sodium nitrate, it is stated that iron was coloured a deep dull black and steel a glossy black in twenty to thirty minutes at 125°C. At 177°C the colours were produced in five minutes.

A completely automated modern plant for the chemical blackening (bronzing) of steel gun components has been installed in Belgium at Fabrique Nationale D'Annes de Guerre[C10]. The various stages through which the components pass are:

1. Ultrasonic degreasing in perchloroethylene.
2. Electrolytic degreasing. With minimum delay components are transported on racks to this tank and successively given cathodic, anodic and again cathodic cleaning treatments. After vigorous rinsing in cold water for 60 seconds the articles are carried through to the bronzing tank.
3. Bronzing is carried out by plunging the components into a sodium nitrate type of bath at a temperature of 140°C for approximately 2 minutes. The temperature and duration are at this stage controlled carefully within very narrow limits to ensure colour uniformity. After the bronzing treatment the components are passed through rinsing and drying stages and are then immersed in an oil bath to protect the finish by sealing its pores.

A notable feature of this plant is the emphasis which has been laid on safety features.

6.1.3.4.1. Characteristics of black oxide coatings

As has already been mentioned, chemical blackening or bronzing processes are more easily carried out than heat blueing or browning, whilst the methods are better adapted to large-scale production. For parts which are kept oiled and cleaned they are useful, but the corrosion resistance of the finishes under conditions of outdoor exposure is low and is largely ascribable to the oil film with which the oxide coating is generally impregnated. The oxide does not have a high absorptive capacity for oil in any case, so that this must be constantly renewed. The oxide coatings are, however, fairly hard and will resist a considerable amount of mechanical wear.

The alkali-oxidising agent method is unsuitable for the treatment of assembled parts containing folds, recesses, or soldered joints. The reason is that the highly hygroscopic salts accumulate within the joints and subsequently corrosion is likely to occur due to the seepage of the salts and consequent corrosive attack on the metal surface. After-treatment (e.g. immersion in weak acids) to remove retained alkali is of limited efficacy.

6.1.3.5. Blue finishes

Blue finishes on steel are highly popular and can be produced by chemical methods. One type of solution which is often employed consists of 120 g of sodium thiosulphate and 50 g of lead acetate dissolved in 1 litre of water. The solution is maintained at a temperature of about 80°C, treatment time being around 5 minutes.

An alternative process consists in immersing the parts in a boiling bath containing 5 g per litre each of potassium ferrocyanide and ferric chloride.

6.1.3.6. Oxidised finishes

The well-known ‘oxidised’ copper and silver finishes which are applied to steel and also to non-ferrous metals are decorative and very suitable for indoor conditions where they are capable of giving good service.

‘Oxidised’ copper is produced on steel by first plating the latter in a cyanide copper solution to give a thin copper deposit, which does not usually exceed 0.0003 cm in thickness. The article is then brushed with, or immersed in, a dilute solution of ammonium polysulphide. The duration of immersion and the concentration of the solution determine the depth of colour produced, but much depends on the skill of the operator. It is desirable to use dilute sulphide solutions since these enable greater control to be exercised over the darkening process, whilst the sulphide films obtained are also more adherent.

After rinsing and drying, the parts may be improved in appearance by a ‘relieving’ operation, which consists in removing the sulphide deposit from the high-lights by hand-rubbing or wet scratch brushing using pumice powder.

The ‘oxidised’ silver finish is likewise produced by plating with a thin coat of silver and then immersing in a sulphide solution in the same way. Barium sulphide solutions have also been successfully employed in conjunction with silver deposits. Silver can also be ‘oxidised’ in a dilute solution of platinum chloride which enables attractive tones to be produced by the deposition of finely divided platinum on the silver surface: $4Ag + PtCl_4 = Pt + 4AgCl$. This process is, however, more costly than those previously described, and is relatively little used.

The ease with which a whole range of browns and greys, and also black, can be produced by sulphide treatment makes this method of finishing a popular one. The sulphides lack stability, however, and are prone to undergo re-crystallisation changes, so that local discoloration of the coating, sometimes called ' sulphur spotting ', is liable to take place.

6.1.3.7. Miscellaneous methods of colouring iron and steels

Other processes for the colouring of iron and steel are less commonly employed than those described above. Like the ' oxidised ' finishes, they do not usually produce good results by simple immersion. Scouring, rubbing or brushing whilst the metal is in contact with the solution are essential to the development of a satisfactory colour.

The selenious acid method finds application when a dark grey or black colour is required. The solution employed consists of:

Selenious acid	10 g
Copper sulphate (cryst.)	20 g
Nitric acid (sp.gr.1.2)	8 ml
Water	1 litre

Another method makes use of the interference colours produced by a film of lead oxide precipitated onto the steel surface. A blue colour is obtained by immersion in a solution consisting of:

Lead acetate	30 g
Sodium thiosulphate	100 g
Water	1 litre

The solution is employed at a temperature of 70°-80°C, the articles being pre-heated in hot water and immersed in the thiosulphate bath for 3-5 minutes. By controlling the time of immersion, a series of colours, ranging from yellow through red to purple and blue-black, can be obtained.

For colouring steels containing chromium and manganese, an aqueous solution of 100 g/l oxalic acid, 100 g/l citric acid and 25 g/l sodium chloride may be used[11]. A blue colour results on immersion of the parts in the solution.

A steel bronze finish can be produced on ferrous metals by treatment in a warm solution of arsenic, which is deposited on the metal surface slowly. A solution of this type contains:

Arsenious oxide	120 g
Copper sulphate	60 g
Ammonium chloride	12 g
Hydrochloric acid	1 litre

The arsenious oxide is dissolved in the hydrochloric acid, whereby a solution of arsenic trichloride is produced, after which the other constituents are added. The deposit can then be wet scratch brushed or lightly buffed and lacquered. This solution will also produce a similar finish on tin, zinc, and on cadmium deposits, but should be diluted ten or fifteen times for use on these metals.

A number of attractive colours and effects can be obtained on copper (cf. Section 6.1.5), and these can be produced similarly on iron and steel if the metal is first copper plated in a cyanide solution. Where better protection is required, a greater thickness of metal is applied from an acid copper sulphate bath after the initial deposit of cyanide copper.

6.1.4. STAINLESS STEEL

Stainless steel has for many years proved very difficult to colour because of the protective oxide which is normally present on the steel and gives it its corrosion-resisting properties. However, a recently developed process[12] appears to have overcome the main problems and it is claimed that stainless steel can be coloured red, gold, green or blue without in any way diminishing the good appearance or protective nature of the oxide film.

The process involves the immersion of stainless steel in a hot concentrated solution containing 250 g/l of chromic acid and 490 g/l of sulphuric acid at 70°C, the desired colour being produced by controlling the duration of the treatment. The film is hardened by means of a second electrolytic treatment in a solution similar to but more dilute than that used for colouring. Although the process can be successfully operated over wide limits of solution concentration and temperature, these variables, together with the duration of treatment, must be closely controlled in order to achieve good colour matching. The finish has good wear resistance, excellent colour retention, and is said to resist boiling water for prolonged periods.

Stainless steel can be blackened by thermal and chemical treatments. It is possible to produce a durable black coating by immersion in molten sodium dichromate at a temperature of 385 to 400°C. Treatment time is 20 to 30 minutes, after which the work is allowed to cool and rinsed in hot water. Stainless steel may also be blackened by immersion in a solution containing:[13]

180 parts sulphuric acid
50 parts potassium dichromate
200 parts water.

The work is immersed for 20 to 30 minutes at 90 to 100°C and then thoroughly washed in clean water.

In another process[14] a paste is prepared which contains sodium dichromate and sodium alginate. A thin coating of the paste is applied to the parts which are then heated to between 320 and 400°C. Finally, the surface is burnished with an oil-impregnated mop to produce a uniformly black finish. This process has been successfully employed in blackening a.c. data transmission motors.

6.1.5. COPPER AND ITS ALLOYS (OTHER THAN BRASS)

Copper and copper alloys are usually coloured various shades of brown, as this is the type of colour that is most readily obtained and is attractive in appearance. With copper alloys, the development of the colour depends almost entirely on the copper constituent, since the other metals usually present do not form coloured compounds during the treatment.

When copper is immersed in solutions of sulphides, combination takes place:

$$4Cu + 2H_2S + O_2 = 2Cu_2S + 2H_2O$$

Air and moisture are essential for the reaction. Solutions of sodium and potassium sulphides and ammonium polysulphides are all effective at room temperature. As the temperature increases, the reaction is accelerated; thus, a 10°C rise of temperature approximately doubles the rate of reaction. Increase in the concentration of the solution has a similar effect.

A typical solution consists of about 12 g/l potassium sulphide, the bath being maintained at a temperature of 70°C. The parts are immersed and brushed until the desired shade of colour is obtained after which they are rinsed, dried and 'relieved' if required. The exact shade of colour produced can only be gauged by experience, since it is determined not only by the temperature and time of treatment but also by the composition and physical structure of the alloy. The colours obtained on castings generally differ from those on wrought metal, and it is not usually practicable to match the two shades. In such cases it may be desirable to copper plate all the parts concerned if a true match is desired.

The colouring treatment is usually carried out in the form of several 'cycles' until the required depth of shade is obtained, viz. cleaning in an alkaline solution followed by rinsing, immersing in the sulphide solution, rinsing again, and then immersing for a few seconds in a 2 per cent sulphuric acid solution to neutralise residual alkali. The articles are finally washed once more and dried, after which they are scratch-brushed or 'relieved' using wire or tampico brushes, wet pumice or sometimes buffing compounds on calico buffs. As the durability of these finishes is poor unless they are protected, a good quality transparent lacquer is finally applied.

Hogaboom[(15)] describes a method for producing a range of sulphide colours on bronzes or high copper alloys generally. A sand-blasted surface is 'oxidised' in a sulphide solution and scoured with sand. A clear lacquer, high in gum, is then applied and dried in an oven at a temperature which turns the lacquer a light brown. While still warm, the article is brushed with a goat's hair wheel to which is applied a wax containing a pigment. If turmeric is mixed with equal parts of bees' and carnauba wax and applied in this way, an attractive shaded golden bronze can be produced.

6.1.5.1. Thioantimonate solutions

Solutions of sodium thioantimonate react with copper and brass with the deposition of bronze-coloured deposits having an attractive uniform and glossy appearance. The thioantimonate can be applied either by dipping or brushing in strengths of 10 to 20 g/l together with 10 g/l of potassium sulphate. Acidification of the solution accelerates the reaction by liberating hydrogen sulphide, but this procedure removes one of its advantages, i.e. the fact that it does not evolve the unpleasant and toxic gas in the absence of acid. To accelerate the treatment, it is better to dip the article alternately into the thioantimonate solution and then into a very weak solution of sulphuric acid.

6.1.5.2. Fused salts

Although not commonly carried out, copper and its alloys can be coloured by treatment in molten salts. Thus, copper acquires a red colour by immersion in sodium nitrate for five minutes at 350° to 450°C. Potassium nitrite can also be used, as can mixtures of the two salts. The parts must be dry and free from grease as ignition of such foreign material on the metal surface may take place with splashing of the fused mixtures. The deposit consists of oxide and it may be lacquered or, better, dipped in molten paraffin wax, buffed, and then lacquered.

The use of fused salt demands fairly elaborate equipment and careful handling of the parts being treated. The method is generally limited to the treatment of small components.

The oxide finishes tend to be more durable than the sulphide coatings owing to the instability of the latter compounds and their liability to internal re-crystallisation, as has already been pointed out.

6.1.5.3. Black colours

Copper and its alloys can be coloured blue-black or a deep matt black by immersion in a solution containing:

Caustic soda	50 g
Ammonium persulphate	10 g
Water	1 litre

The bath is used at boiling point, a period of about five minutes being required for the development of the colour. A violent evolution of oxygen takes place from the solution during the process. The parts are rubbed after treatment, when a black matt finish is obtained. The solution decomposes gradually, so that additional persulphate must be added from time to time.

Potassium permanganate in alkaline solution can likewise be employed as the oxidising agent, but the colours lack the dense black appearance of those obtained from the persulphate bath. They are, however, useful where brownish shades are not objectionable. Solutions of ammonium sulphate or nitrate produce slate-grey colours on copper and its alloys.

Copper and its alloys may also be blackened by anodic oxidation[16]. The process consists of anodically treating copper at a current density of 1 A/dm^2 in a solution containing 150 g/l sodium hydroxide and 1 g/l ammonium molybdate. The colour produced is dependent on the solution temperature, varying from grey at 20°C to velvety black at 85°C.

6.1.5.4. Green patinas

The green patina which develops on old copper after prolonged ageing in the atmosphere consists almost entirely of the basic sulphate of the metal, and approaches the composition $CuSO_4.3Cu(OH)_2$.[17] When formed under natural conditions the coating is tenacious and durable and protects the underlying metal against further attack very well. This accounts for the fact that copper roofs hundreds of years old are still giving service in various parts of the world. After a prolonged period the patina appears to show little increase in thickness, reaching a stable condition. Artificially patinated copper sheets are being produced on a commercial scale.

The appearance of the patina can be approximately imitated by various treatments, although such coatings do not have the weather resistance of the deposit naturally developed under suitable conditions. One method recommended by Hitchin[18] makes use of a solution consisting of:

Sodium chloride	180 g
Ammonium chloride	140 g
Ammonium acetate	140 g
Water	1 litre

This is applied by brush, being allowed to dry on to the metal, the treatment being repeated as desired.

A method for producing an artificial coating approaching the natural patina in composition consists in treating the article twice daily for a week with a 10 per cent solution of ammonium sulphate, and then for a further two days with a 10 per cent solution of copper sulphate containing about 1 per cent of sodium hydroxide. No further protection is then required. The method is said to be particularly successful with arsenical copper.

Vernon[19] also describes a superior method for producing the patina. This consists in treating the metal anodically in a solution of:

Magnesium sulphate	10 per cent
Magnesium hydroxide	2 per cent
Potassium bromate	2 per cent

The solution is kept at about 95°C, treatment being continued at a current density of about 0.04 A/dm^2 for fifteen minutes. Initially, the deposit has the composition $CuSO_4.Cu(OH)_2$, but gradually increases in basicity on exposure until the composition of the patina is reached.

6.1.5.5. Other colouring methods

Other methods of colouring copper and its alloys make use of various solutions containing certain salts of metals such as lead, tin, antimony, arsenic, etc., from which deposits of the sulphides of these metals are precipitated. The salts employed include tartrates and thiosulphates, and colours ranging from light brown to grey and black can be obtained.

A typical solution is made by mixing equal parts of solutions containing 120 g/l sodium thiosulphate and 96 g/l lead acetate respectively. By immersion in this solution at temperatures of 40° to 60°C, colours ranging from yellow to blue can be produced depending on the temperature and time of immersion. By increasing the thiosulphate content the rate of colouring is increased, and this is also the case when a small excess of tartaric or acetic acid is present. The parts are well rinsed and dried after treatment.

6.1.6. BRASS

Colours can be produced on brass by processes similar to those employed in the case of copper. Owing to the fact that, with a few exceptions, zinc compounds are not coloured, practically the entire effect is due to the copper constituent of the alloy. Special difficulties may arise with certain types of brasses. Thus, the free-machining lead-bearing brasses may cause trouble during sulphide treatment because of the development of black spots at local inclusions of lead. Likewise, the pre-cleaning treatment before colouring may bring about changes and non-uniformity in the initial colour of the brass which will result in variations in the final appearance of the finish. Cathodic alkaline cleaning may result in dezincification of the metal and re-precipitation of metallic zinc on the surface, with a lightening of the colour of the film. Acid dipping, if it is to be carried out, may, on the other hand, redden the colour of the brass. This is not caused directly by preferential dissolution of the zinc but is a secondary effect resulting from the re-precipitation of copper which has dissolved in the cleaning acid. Preliminary cleaning processes therefore demand careful attention.

Besides the treatments already described for copper, brass can be given a pleasing golden brown coloration by a short immersion in a solution of neutral copper acetate maintained at a temperature of 70°C. A black colour can be obtained by immersion in an ammoniacal solution of copper carbonate; the black usually has a bluish cast.

A typical solution has the composition:

Copper carbonate	200 g
Ammonia solution (sp. gr. 0.88)	1 litre
Water	1 litre

6.1.7. CADMIUM

Both cadmium and zinc deposits are frequently chromated (or 'passivated') by immersion in chromate solutions. These conversion coatings range in colour from very pale yellow to brown and olive green, and are applied to prevent the formation of white corrosion products on exposure to the atmosphere rather than for appearance. Cadmium can be blackened by immersion in a hot lead acetate-sodium thiosulphate solution, similar to that used in the treatment of copper alloys, e.g.:

Lead acetate	1.5 g
Sodium thiosulphate	72 g
Water	1 litre

The deposits are, however, inclined to be powdery and to have poor adhesion. A copper sulphate-potassium chlorate-sodium chloride bath gives a denser black finish which is more adherent[20]. The solution used consists of:

Copper sulphate	15 g
Potassium chlorate	18 g
Sodium chloride	18 g
Water	1 litre

The temperature is of the order of 80° to 95°C and the solution is used in a rubber-lined or stoneware container. If there is any tendency for basic salts to be precipitated, resulting in a 'bloom' on the work, a few drops of sulphuric acid should be added. The time of immersion need only be a few seconds. A three-stage reaction takes place. Copper is first precipitated by interaction with the metallic cadmium, and this is oxidised by the chlorate, first to the reddish cuprous oxide, and then to black cupric oxide. Rapid movement of the articles whilst in the solution results in the deposition of the copper in a compact form so that the film obtained is more adherent. After treatment, the parts may be lightly brushed with steel wool or pumice, and lacquered. This produces an attractive imitation pewter finish.

A brown colour can be produced on cadmium by treatment in a solution of:

Nitric acid (sp. gr. 1.2)	6 g
Water	3 g
Potassium dichromate	1 litre

The articles are immersed in the bath which is maintained at a temperature of 60° to 70°C. Additions of dichromate and nitric acid have to be made to the bath from time to time, but excessive amounts of nitric acid are to be avoided or the film will be poor. A pale yellow colour is developed at first, but on continued immersion for two to four minutes, a rich mahogany colour appears. The film has good adhesion and durability. Longer immersion times produce still darker colours. The appearance can be improved by polishing the highlights on a buff using a tripoli composition. The finish applied on cadmium plate has been used for imitating old silver; however, cadmium plate must not be used on utensils coming into contact with foodstuffs, owing to the toxicity of the metal.

An antique brown shade on cadmium can also be produced by immersion in a hot solution containing:

Copper sulphate	1.5 g
Ferric chloride	12 g
Hydrochloric acid	60 ml
Water	1 litre

The solution is used as hot as possible, the articles being immersed and allowed to dry without rinsing. The treatment is repeated until the desired shade of colour is obtained.

Permanganate solutions for producing durable colours, ranging from yellow to deep brown, are worthy of mention[21]. These solutions, in addition to the permanganate, contain cadmium nitrate, potassium chlorate and sometimes also ferric chloride or copper nitrate. A deep reddish brown is produced from a solution containing:

Potassium permanganate	320 g
Ferric chloride	20 g
Cadmium nitrate	100 g
Water	1 litre

The bath is used hot (about 80°C). If the solution becomes turbid, a few drops of nitric acid should be added. For best results it is desirable that the colouring operation be carried out directly after plating and before there has been an opportunity for an oxide film to form on the surface.

A highly durable black finish on cadmium, which is equally applicable to zinc, can be obtained by immersion for a few seconds in a bath containing a soluble copper salt and a chlorate followed by hydrogen peroxide solution, which serves to increase both the adhesion and the blackness of the finish[22]. An intermediate permanganate treatment can also be used. The three steps are repeated about 10 times, each dip lasting for about 10 seconds.

The molybdenum colours are useful both on cadmium and on zinc plate. Thus immersion in a solution of:

Ammonium molybdate	12 g
Ammonium chloride	25 g
Potassium nitrate	6 g
Boric acid	6 g
Water	1 litre

for a few seconds results in a fine, adherent black deposit of molybdenum sesquioxide. The solution is used hot at a temperature of 60° to 70°C.

A black finish can be produced electrolytically on cadmium by treatment in a solution consisting of:

Nickel sulphate	75 g/l
Nickel ammonium sulphate	45 g/l
Zinc sulphate	37.5 g/l
Sodium thiosulphate	15 g/l

The bath is operated at 50° to 55°C with current densities of 0.5 to 2.5 A/dm^2. The pH is kept within the limits of 5.6 to 5.9 to avoid precipitation of zinc. The process gives a range of colours between light green and black, depending on the conditions of treatment.

6.1.8. ZINC

The colouring of zinc, unlike cadmium, presents difficulty owing to the fact that the salts of the metal are practically all colourless. A method of producing a black finish makes use of immersion deposits of nickel obtained by dipping the articles in a solution of nickel sulphate or nickel ammonium sulphate. Additions of sodium thiocyanate, iron or ammonium salts are sometimes made. The colours thus obtained are black or grey, depending on the exact composition of the solution and time of immersion.

Buffered solutions of molybdenum salts may also be employed; these are kept slightly on the alkaline side, at a pH of 8.5 to 9.5. They may be used hot or cold, the time of immersion varying from one to five minutes. The time is important as excessively long immersion times cause flaky coatings to form, whilst poor or irregular coloration results from too short an immersion period. The coatings, when correctly applied, are quite tenacious and adherent. A slightly acid solution of ammonium molybdate will colour zinc black, addition of chlorides or fluorides accelerating the reaction.

Zinc deposits, like those of cadmium, can also be coloured black in copper sulphate-potassium chlorate solutions, and in the permanganate types of bath.

Zinc-base die-castings are sometimes copper plated in a cyanide solution and the latter metal is then coloured. This is not a really satisfactory procedure, however, owing to the fact that the copper tends to be gradually absorbed by the zinc, so that the colour disappears.

Perhaps one of the best methods of colouring zinc black is to apply a 'black nickel' electrodeposit. This is done by plating nickel on to the zinc at low current density (about 0.2 A/dm^2) from a solution of a nickel salt containing zinc sulphate and thiocyanate. A representative formula for such a solution is:

Nickel sulphate	100 g
Zinc sulphate	25 g
Sodium thiocyanate	12 g
Water	1 litre

The solution may be used cold, the anodes being of nickel or stainless steel. The deposit consists largely of nickel sulphide together with some occluded organic substances.
A method of depositing a green nickel sulphide has been described[23]. The solution used consists of nickel ammonium sulphate 100 g, sodium thiosulphate 10 g, sodium citrate 15 g, water 1 litre; pH 6.4. The current density ranges from 0.01 to 1.5 A/dm^2, depending on the colour and structure of the deposit required.

6.1.9. CONVERSION COATINGS ON CADMIUM AND ZINC*

Chemical conversion coatings are frequently applied on zinc-base die-castings or on zinc or cadmium plate to reduce the tendency of these metals to form white corrosion products. The processes can be conveniently subdivided into three categories:
(a) The single dip, non-polishing solutions producing a yellow, bronze, olive or black finish.
(b) The single dip polishing solutions producing a colourless or lightly coloured finish.
(c) Anodic processes.
One of the first commercially successful processes of this type was the 'Cronak' process introduced in 1936. It enables an iridescent golden film to be applied on zinc, or zinc- cadmium-plated components, which has considerable protective value. It does not however, greatly delay the onset of rust under outdoor conditions if the zinc plate is of inadequate thickness[24]. Essentially the method consists in immersing the zinc, after cleaning in the usual way, in a solution consisting typically of:

Sodium dichromate	200 g/l
Sulphuric acid	6 to 9 ml/l

The time of immersion is about ten to twenty seconds; the temperature of the solution should be kept at about 25 °C for best results. After treatment, the articles are rinsed thoroughly and dried, preferably by means of a warm air blast at not too high a

* See also Chapter 12.

temperature (below about 50°C). Hot-water drying is not recommended as this may result in the formation of films of low protective value.

The film consists of basic chromium chromate, $Cr_2O_3.CrO_3.xH_2O$, or $Cr(OH)_3.Cr(OH).CrO_4$. A typical film weighs about $1.5 \times 10^{-4} g/cm^2$ and is 1 micron thick. The protective value is said to depend on the slow release of the hexavalent chromium from the film to the water which comes into contact with it, and this has an inhibiting effect on the development of corrosion.

Table 6.2 gives some test results under experimental conditions of exposure. The production and dyeing of zinc chromate films is dealt with fully in Chapter 12, cf p. 433.

6.1.10 LEAD

When exposed to weather lead acquires a grey patina which is normally very thin, dense and highly durable. Artificial patinas, on the other hand, are much softer and less durable, although they may be useful in some cases, especially if lacquered. Before any chemical treatment is applied, lead must be cleaned in an alkaline solution of a mild type such as trisodium phosphate with or without the addition of sodium carbonate.

Black or dark colours are produced by dipping in a solution of yellow ammonium sulphide or in a solution containing 10 per cent of ferric chloride in a 50 per cent solution of hydrochoric acid. A jet-black finish can be obtained by treating the metal with hydrochloric acid and then heating gently with a blow lamp.

An antique green patina can be produced by dipping the lead into a warm solution containing:

Copper nitrate	60 g/l
Ammonium chloride	30 g/l
Acetic acid	30 g/l
Chromic acid	6 g/l

Alternatively the solution may be brushed on to the lead surface after which the article is rinsed, dried and lacquered.

TABLE 6.2

Exposure tests on chromate-treated and untreated zinc

Exposure to	Time to first white salts (days)	
	Untreated	Cronak treated
95°C steam	1	12
40°C steam	1	182
Salt spray	8 hours	91-104 hours
Distilled water	1	38

6.1.11. NICKEL

Nickel is a difficult metal to colour satisfactorily. The persulphate method is one of the best for obtaining a black film, a suggested solution consisting of:

Ammonium persulphate	200 g
Sodium sulphate	100 g
Ferric sulphate	10 g
Ammonium thiocyanate	6 g
Water	1 litre

The solution is adjusted to a pH of 1 to 2 and is used at room temperature.

6.1.12. ALUMINIUM

This metal is best coloured by anodising (see Chapter 13), followed by dyeing or impregnating the film with a suitable pigment. The dyed films must be sealed by after-treatment, e.g. in hot water at a pH of 6, in a solution of sodium silicate or in a hot solution of a heavy metal acetate (e.g. cobalt or nickel acetate). Unlike most other metal-colouring processes, the films so produced are hard and durable and require no further lacquering or other protective treatment.

There are also some non-anodic methods of colouring aluminium. Thus an antique silver finish can be produced by treatment in a boiling alkaline arsenate solution containing 60 g of arsenious oxide and an equal amount of sodium carbonate per litre of water.

Aluminium may be coloured blue by immersion in a solution containing:

Ferric chloride	450 g
Potassium ferricyanide	450 g
Water	1 litre

This is used at a temperature of 70°C. The colour results from the precipitation of blue iron salts in the pores of the thin oxide layer normally present on aluminium.

A black colour, which is not very dense, however, can be produced by immersion in an alcoholic solution of a cobalt salt and burning-off the alcohol. The cobaltammines are useful compounds for this purpose; they are formed by adding ammonia to a solution of a cobalt salt until the precipitate which first forms just dissolves. The molybdate, permanganate and copper sulphate solutions which have been referred to for the production of black finishes on zinc and cadmium (p. 245ff.) have been successfully employed with aluminium.

6.1.13. GILDING

The deposition of gold onto metals, either electrolytically or by immersion, is a common method of obtaining a pleasing colour. Coloured gold deposits can be produced by deposition from solutions to which various substances have been added. Thus, small amounts of silver salt will enable ' green gold ' to be deposited, the depth of green depending on the amount of silver present. Rose-gold is produced by the addition of sodium hydroxide or sodium carbonate to the plating solution, whilst copper salts darken the coloration obtained. This is the basis of ' Russian ' gold finishes. Attractive effects can also be produced in these processes by rubbing or relieving the high-lights of the articles.

6.1.14. ELECTROLYTICALLY-PRODUCED INTERFERENCE COLOURS

Processes in which colours are produced by interference in very thin chemically- or electrolytically-formed oxides have been developed. Different colorations are obtained according to the conditions of formation or deposition. Some very attractive finishes can be obtained in this way but, again, their durability is low and they must be lacquered if useful service is to be obtained. Stareck[25] describes the ' Electrocolor ' process in which a lactate solution is employed. Plating is carried out in the normal manner, using a voltage of about 0.5 V and a cathode current density of 0.03 to 0.3 A/dm^2. The deposit consists essentially of cuprous oxide, Cu_2O.

$$2Cu^+ + 2OH^- = Cu_2O + H_2O$$

The colours result from interference tints in the thin oxide film and the actual colour obtained depends on the current density used, the composition and temperature of the solution, and especially on the duration of plating. As the film thickens, its colour changes through the spectral range, after which the cycle recommences. The cycle can be repeated up to fifteen times; with each succeeding cycle the colours become less brilliant, although the film thickens and hence becomes more durable. The solution is claimed to have an excellent throwing power, and the cathode efficiency is said to be 100 per cent.

It is rather difficult to maintain a standard colour from the bath, and this militates against the use of the method for mass production. It is therefore best suited to individual specialised articles where an attractive finish is required. The ' Electrocolor ' process may be applied to any metal which can be normally electroplated, but aluminium and zinc require first to be plated with copper.

6.2. Phosphating

Since their original development by Coslett in 1906[26] phosphate coatings have been increasingly used in the protection of ferrous metals where a very high degree of rust resistance is not required. A number of proprietary phosphate coating processes are now available as a result of the considerable amount of research which has gone into their development and application.

6.2.1. PHOSPHATING SOLUTIONS

Phosphate coatings are applied to metals by treating them in dilute aqueous solutions containing phosphoric acid and one or more of the primary phosphates of iron, manganese or zinc with or without the addition of materials acting as accelerators. Treatment is carried out by dipping or spraying after careful cleaning of the metal surface, the time of treatment depending on the solution used, the coating thickness required and the operating temperature. After treatment the metal is rinsed and may be given a passivating dip in a chromic acid or chromate solution, dried and impregnated with an oil or other protective medium. The phosphating process lends itself to automatic operation and many fully automatic plants are in use.

The phosphating solution reaches a state of equilibrium which depends on its temperature and the relative concentrations of the constituents present. Primary iron, manganese or zinc phosphates dissociate to form secondary or tertiary phosphates. Zinc phosphate, for example, dissociates as follows:

$$3Zn(H_2PO_4)_2 = 3ZnHPO_4 + 3H_3PO_4$$
$$3ZnHPO_4 = Zn_3(PO_4)_2 + H_3PO_4$$
$$3Zn(H_2PO_4)_2 = Zn_3(PO_4)_2 + 4H_3PO_4$$

With the exception of the secondary zinc phosphate the salts produced are only slightly soluble, and in the absence of free phosphoric acid are precipitated as a sludge by any factor that causes increased dissociation, e.g. rise in temperature or increased dilution. Free phosphoric acid must therefore be present to drive the reaction to the left. When a ferrous metal is immersed in the bath, the iron dissolves and hydrogen ions are discharged. The reduction in hydrogen ion concentration in the solution layer adjacent to the metal then causes the equilibrium to shift to the right with the result that the solubility products of the secondary and tertiary phosphates are exceeded locally. These salts precipitate from the solution at the metal surface and the phosphate coating forms. Restoration of the original state of equilibrium takes place, partially at any rate, by the action of the phosphoric acid which is formed.

It is recommended[27] that components for phosphating should be of such construction that liquids cannot be trapped in blind holes, cavities and unsealed tubes. Drainage holes should be of adequate size to permit complete draining of the solution in the time available. A single large hole is more effective for this purpose than several smaller ones.

6.2.2. ACCELERATORS

The acid zinc solutions are capable of retaining in solution the higher ferrous phosphates (which are produced by interaction with the iron). The bath therefore becomes progressively loaded with soluble phosphates, so that it begins to lose efficiency.

By the introduction of oxidising agents, the basic ferrous phosphates are gradually oxidised to the ferric state, and as these salts are practically insoluble, they are precipitated and the life of the solution is correspondingly increased. The coatings consist primarily of tertiary zinc phosphates with small amounts of secondary and tertiary iron phosphate.

The use of modified zinc processes of this type is universal, and many proprietary processes are available. The addition agents used in these solutions are termed 'accelerators' since they accelerate the phosphating process. The accelerators which are employed have mild oxidising properties and probably function to some extent, although not exclusively, by oxidising the hydrogen evolved at the metal surface. Strong oxidising agents are unsuitable as constituents of a phosphating bath since they convert any ferrous phosphate to the ferric stage, immediately disturbing the balance of the solution and thus impairing the functioning of the bath. Suitable oxidising accelerators are alkali metal nitrates (e.g. sodium nitrate), nitrites, copper salts, and chlorates. The concentration of the accelerator in relation to the concentration of iron salts and free acid in the bath plays an important part in the type of coating obtained and its thickness.

By the use of accelerators the time of treatment can be reduced from an hour or more to ten minutes or even less. Increasing the concentration of oxidising agents progressively reduces the amount of gas evolved, and hence the time for the formation of the coating. There is, however, an optimum concentration, and if too much accelerator is added the phosphate layer may become coarse and streaky. By increasing the free phosphoric acid, however, good coatings can again be obtained. This effect has been noted by Machu[(28)], who points out that excessive concentrations of oxidising agents cause the surface to become passive and that this can be counteracted to some extent by increasing the concentration of free phosphoric acid.

Nitrates are the most widely used accelerators; they are used either alone or in combination with nitrite or chlorate in zinc phosphating solutions prior to painting. The nitrate ion acts as a depolariser, being itself reduced in the process, and it is likely that these reduction products play a very important part in the cycle. It is the only oxidising agent compatible with ferrous iron, the other accelerators commonly used, such as nitrite or chlorate, oxidising the ferrous iron to the ferric state, forming insoluble ferric phosphate, so that the bath is substantially iron-free. The concentration of nitrate is usually 1 to 3 per cent. Nitrates are also very valuable in retaining in solution the ferric phosphate which is the primary constituent of the sludge formed in phosphate coating baths[(29)]. The nitrate ion is reduced during the phosphating operation and the reduction products have a pronounced depolarising action on the metal. The concentration of nitrate ions used ranges from 1 to 3 per cent, and at this level the nitrate ion is reduced by the iron to nitric oxide. Further reduction takes place at the interface, with the evolution of ammonia.

In addition to functioning as accelerators, nitrates greatly increase the solubility of ferric phosphate in the bath, which is an advantage since the latter product is the primary cause of sludge formation. In this way, utilisation of the sludge is ensured since the ferric iron enters the coating.

Nitrites are not used as constituents of concentrates employed in the preparation of phosphate baths, although they may be added from time to time, since excessively rapid decomposition would occur if they were introduced in bulk. An initial addition of about 2 grams per litre is generally made when the bath is prepared, and further additions are made when tests show the presence of ferrous ions. Sodium nitrite solution is often added in conjunction with caustic soda which serves to neutralise the free excess acid liberated when the ferrous phosphate is oxidised to the ferric form by the nitrite. A slight excess of nitrite is maintained, based on tests to determine the absence of ferrous ion. In addition to its depolarising action, the nitrite increases the solubility of the ferric ion, and hence reduces the formation of sludge, which consists largely of ferric phosphate. The excess nitrite forms an oxide film on the metal surface, and this reacts with the phosphoric acid as follows:

$$FeO + 2H_3PO_4 \rightarrow Fe(H_2PO_4)_2 + H_2O$$

Little or no gas is liberated, and as no free acid is liberated within the film the reaction terminates in a few minutes, resulting in the formation of the phosphating film. The accelerating effect appears to depend on the amount of nitrous acid available in the solution to form the initial oxide coating. Baths containing nitrites are operated at a rather lower temperature than those with nitrates alone, to reduce nitrous acid loss. They are therefore more suitable for spray operation in the pre-treatment of metals before painting, as it is expensive to maintain sprayed solutions at high temperatures.

Chlorates have been used as accelerators in phosphate solutions for many years. It was found, however, that gelatinous suspensions of ferric phosphate formed in chlorate-accelerated solutions, so that the bath was difficult to operate. This objection was overcome by Darsey[30] who found that a combination of nitrate and chlorate eliminated the tendency for precipitates to form, since nitrates increase the solubility of ferric phosphate. About 1 per cent of nitrate and 1 per cent of chlorate are generally present in the bath, the solution being prepared by adding about 10 per cent of zinc or potassium chlorate to the concentrate used in its preparation. Chlorate-type solutions have the disadvantage of being corrosive, the combination of chlorate and nitrate ions in acid solution being a particularly aggressive one towards most metals.

Copper salts act as accelerators by increasing the rate of dissolution of iron. The copper is deposited in small particles on the steel surface so that numerous minute electrochemical cells are formed which increase the rate of formation of the coating. The presence of copper in the finished surface has, however, an adverse effect on its protective value, so that the use of this type of acclerator has now been largely discontinued.

Another process employs the so-called 'nitro-accelerators'[31], of which typical examples are the nitro-derivatives of aliphatic compounds, e.g. nitro-urea, nitroguanidine and nitro-urethane, and nitro-derivatives of aromatic compounds, e.g. nitro-benzene, the nitro-benzoic acids and picric acid. Organic and inorganic cyanides, sodium nitroprusside, and hydroxylamine have also been proposed as accelerators. Nitroguanidine is claimed to be not only an accelerator of the reaction but also a stabiliser of the zinc acid phosphate. It must be intimately mixed with the phosphate to the extent of 5 to 15 per cent.

Improved results are obtained by increasing the concentration of accelerator actually at the metal surface. This is done by applying the accelerator before the coating solution[32]. In practice, the surface to be treated is dipped into, or sprayed with, a solution of the accelerator, after which the coating solution is applied by a dipping or spraying operation. The accelerators which have been found especially suitable for this purpose are sodium or potassium nitrate. The metal surface is first cleaned thoroughly and then immersed in, or sprayed with, the accelerator, which consists of a hot solution of up to about 0.5 to 1 per cent concentration of the activating material. The metal is then immersed directly into the phosphate bath. In this way a small amount of accelerator is introduced into the main phosphating solution by drag-out from the previous treatment tank. This has a further advantage, because if accelerators, such as nitrites for example, are put directly into the phosphate solution in any concentration, they tend to be decomposed before actually coming into contact with the metal being treated. By the method described the accelerator concentration in the phosphate bath is maintained at about 0.002 per cent.

6.2.3. SLUDGE

All phosphating baths tend to form sludge continuously during their operation; this sludge may be deposited on the work, causing uneven coatings which interfere with subsequent finishing operations, especially painting. The sludge in the commonly used zinc solution consists almost entirely of tertiary zinc and iron phosphates. The use of oxidising accelerators in theory increases sludge formation by virtue of the fact that they oxidise the soluble ferrous ions to the insoluble ferric salt. Their use is, however, justified by the fact that the coating forms so much more rapidly than in the absence of an accelerator.

6.2.4. COMMERCIAL PROCESSES

Phosphating solutions are supplied under a wide variety of commercial names and are offered as concentrates containing up to 50 per cent of phosphate, with nitrates or other additives supplied separately. Sometimes additional phosphoric acid is introduced where it is desired to remove rust, but the coating formation rate is impaired by this and the quality suffers, whilst the cost of treatment is also higher. A well-balanced zinc-phosphate-type concentrate should contain 35 to 40 per cent of phosphate and 15 to 20 per cent of nitrate or other additive. A recent commercial development[33] is the direct production of jet black zinc phosphate coatings by immersing the components in a proprietary solution between the final rinse of the cleaning operation and the actual zinc phosphating treatment.

The dihydrogen orthophosphates of the metals are prepared commercially by dissolving the metals in a hot concentrated solution of orthophosphoric acid (60 to 75 per cent strength) to form a soluble phosphate, cooling the solution, and then crystallising out the salt, which is removed from the mother liquor by centrifuging[34]. Ferrous dihydrogen phosphate, $Fe(H_2PO_4)_2$, is made in this way, whilst the manganese and zinc salts, $Mn(H_2PO_4)_2$ and $Zn(H_2PO_4)_2$ respectively, may be produced similarly. The zinc salt melts at 60°C so that care must be taken to avoid overheating the crystals in the course of manufacture.

Iron dissolves with difficulty in phosphoric acid, so that the phosphate solution is not easily prepared. Ferrous sulphide dissolves much more easily; advantage is taken of this fact in the Walterisation process[35] in order to prepare iron phosphate solutions more readily. The iron and manganese salts are produced together from ferro-manganese, which is a readily-available raw material containing some 80 per cent of manganese and 6 to 10 per cent of iron together with impurities of various kinds. By dissolving the ferro-manganese in a 65 per cent solution of phosphoric acid, the iron and manganese salts are formed; by drying the resulting crystals after centrifuging, the ferrous salt is oxidised to the insoluble ferric phosphate while the manganese salt does not readily oxidise and can be taken up into solution again. By filtering and re-crystallisation, the manganese salt is obtained in a practically pure form[36]. During the phosphating process, iron dissolving in the manganese phosphate bath results in the formation of acid iron phosphate in the solution; the coating consists of a mixture of both the iron and manganese phosphates. The best type of coating should contain at least half as much manganese again as iron. To achieve this result the bath needs to contain at least one-third as much again manganese dihydrogen phosphate as ferrous dihydrogen phosphate.

The well-known ' Parkerizing ' treatment for ferrous metals makes use of a solution of acid manganese phosphate. This solution will produce a satisfactory phosphate film on steel in a considerably shorter time than Coslett's original iron phosphate solution. For the treatment of steel prior to painting or electro-painting, however, the zinc phosphate processes are preferred because of the fine and smooth coatings produced by them. Manganese phosphates have one advantage in that they have rather less tendency to oxidise and form insoluble sludges than the zinc phosphates, so that the primary acid phosphate concentration of the bath is maintained. The solutions are used at or near the boiling-point, the time of treatment being about an hour, during which time a coating of some 0.0007 cm to 0.0012 cm in thickness is produced. Under these conditions both iron and manganese solutions precipitate coatings of secondary and tertiary phosphates on the metal surface.

The manganese phosphates are superior from the protective point of view to the secondary and tertiary ferrous phosphates as the latter are gradually oxidised to the ferric state in air. The change in the crystalline form resulting from this leads to lack of adhesion of the coating and increased crystal size, leading in turn to a greater permeability to air and moisture and reduced protective value.

As the phosphate is removed from the solution by the reaction it must be replaced by the addition of more primary acid phosphate. Sludge is removed periodically from the bottom of the tank. The commercially-available salts manufactured for phosphate treatment generally consist of complexes containing more phosphoric acid than is required by the formula of the primary acid phosphate so that the acidity of the bath is maintained without the specific addition of free phosphoric acid.

For satisfactory phosphating the concentrations of the metal, hydrogen and acid phosphate ions must be such as to ensure that deposition of insoluble phosphates will be confined as far as possible to the liquid-metal interface. This can only be achieved if the solubility products of the secondary and tertiary phosphates are high, but are not exceeded.

When an iron surface is immersed in such a solution the hydrogen ion concentration at the interface is reduced by the liberation of hydrogen, so that secondary and tertiary phosphates of the metal are precipitated locally. More hydrogen ions are again liberated to restore equilibrium in the bath and the process is then repeated.

Jernstedt(37) investigated a series of solutions with the object of obtaining a pre-dip treatment which would increase the corrosion-resistance of a phosphate coating. He found that immersion of zinc-coated or plain steel in a 1 per cent solution of disodium phosphate containing about 0.001 per cent of a titanium salt gave a considerable increase in the corrosion resistance of the phosphate film, even when the latter was unprotected by an organic coating. In preparing the titanium dip it is apparently necessary to add the titanium salt to the original disodium phosphate solution at the time it is made up and to evaporate to dryness; the simple addition of a soluble titanium salt to a 1 per cent solution of disodium phosphate results in a bath which is relatively inactive. As previously indicated, addition of a titanium-phosphate complex to cleaners used before phosphating also activates the metal surface and refines the crystalline formation(38).

Thick phosphate coatings can be built up more rapidly by the use of electric current in conjunction with the phosphate solutions. In the Electro-Granodine process the parts are suspended in the hot acid phosphate solution and an alternating current at a pressure of about 20 volts is applied. A resistant coating approximately 0.0012 cm thick can be built up in about 3 minutes; after rinsing and drying, the parts may be stained and oiled or painted. Electrolytically produced phosphate coatings appear to have rather better durability than those formed by immersion methods, but the plant required is rather more complicated and the process has limited application.

6.2.5. OPERATING CONDITIONS

Modern phosphating processes are simple to operate and may be carried out completely automatically. The solutions are prepared by dissolving the appropriate chemicals in water, boiling for about an hour and commencing treatment immediately. The iron and manganese phosphates are supplied as the crystalline orthophosphates, but owing to their hygroscopic nature, zinc acid phosphate mixtures are generally marketed in the form of a concentrated aqueous solution, and this is diluted and employed for treatment as required.

In applying phosphate coatings to steel the pre-cleaning of the metal is important. A common procedure is to degrease the parts in trichloroethylene, whilst, if the metal be at all rusted or scaled, barrel cleaning may be successfully employed. Rust must be carefully removed, since zinc phosphating solutions give very unsatisfactory coatings where rust is present and the coatings on rusted metal are apt to have lower durability than those produced on clean steel. Pickling in strong acids or cleaning in very strong alkalis, or indeed any processes in which the metal is subjected to the action of powerful electrolytes, are to be avoided. Such materials appear to activate the grain boundaries of the metal in such a way that the phosphates are formed in an irregular manner, so that rough and granular coatings of large grain size, which have poor protective qualities are produced. A grain-refining addition containing titanium phosphate may be incorporated in a mild alkali cleaner, but this is not always satisfactory. It is better to incorporate the grain refinement material in a pre-dip treatment.

Mild alkali cleaners are generally used at 20° to 50°C principally in conjunction with cold spray phosphating solutions which operate at low concentrations and assist in the formation of uniform and fine-grained zinc phosphate coatings of 10 to 25 mg/dm^2 coating weight. Two rinses are generally included after alkaline cleaning and before phosphating to reduce carry-over of alkali.

Single- or di-phase emulsion cleaning is a satisfactory method of cleaning prior to phosphating and many solutions used nowadays are effective and do not have any adverse influence on the quality of the coating. It is useful to filter the cleaning solutions continuously in the case of spray plant to reduce the danger of clogging of the spray nozzles.

To reduce the number of operations in the phosphating sequence, combined cleaning and phosphating solutions using sodium or ammonium acid phosphate solutions may be used. Alternatively, organic degreasing media may be emulsified with the phosphating solution, or they may float on the surface of it. Another compromise is to use a type of cleaner which can be safely carried over into the bath without affecting it adversely, thus eliminating one operation.

The coarsening effect of pickling on the grain size of the phosphate coating can be seen in Fig. 6.1. The left-hand photograph shows the coating obtained on a steel surface deagreased by solvent only, the right-hand illustration shows the large-grained coating produced when the same steel surface was acid pickled. These photographs were produced by a replica technique. The phosphated surface is

Fig. 6.1. Coarsening effect of pickling on the grain size of the phosphate coating. Steel surface on left has been degreased by solvent wiping, surface on right has been pickled prior to phosphating.

moistened with a suitable solvent after which cellulose acetate tape is pressed on to the area to be examined. The tape is allowed to dry and removed, when a replica of the crystal structure is obtained which can be examined by transmitted light or photographed. This enables a much clearer picture of the structure of the coating to be obtained than when the examination is carried out by the reflected light method.

6.2.5.1. Effect of bath concentration

The concentration of the phosphating bath is an important factor in the type of coating obtained. For optimum results a zinc phosphate solution should contain:

Primary zinc phosphate	1.5 to 3 g/l
Phosphoric acid	3 to 4 g/l
Accelerator	3 to 4 g/l

The smoothness of the coating improves as the concentration decreases within the above range.

The concentration of the oxidising accelerator is also important, since if it is inadequate in amount the hydrogen liberated will cause poor coatings, whilst too much accelerator will give rise to streaked and uneven coatings.

6.2.5.2. Effect of pH

Gilbert[39] has demonstrated by the use of pH titration curves that the successful formation of a phosphate coating depends on (1) the pH for the point of incipient precipitation of the metal phosphate or phosphates in the solution and (2) the rate at which the reaction between the metal and free acid in the solution raises the initial pH to the pH for the point of incipient precipitation. Any measures which reduce the amount of pickling before this point is reached will accelerate coating formation. Metal saturated solutions appear to have an operational advantage over solutions containing free acid in excess of that required to maintain the metal content. Nitrates do not appreciably influence the pH for the point of incipient precipitation.

6.2.5.3. Effect of temperature

The temperature at which the phosphating bath is operated is important since it is desirable to produce a phosphate of small crystal size. Coatings of large grain size, such as are formed when the bath temperature is lowered, have poor corrosion resistance. Thus Macchia[40] has determined the resistance to rusting in a 3 per cent sodium chloride solution of steel treated in a 3 per cent solution of manganese dihydrogen phosphate; the results are shown in Fig. 6.2. It is seen that as the temperature falls from 98°C to about 80°C the corrosion resistance is reduced by half. Machu also found that the treatment time increased correspondingly. Thus, at 98°C the time taken was ten minutes; at 88°C the time was increased to forty minutes; whilst at a temperature of 70°C the reaction was incomplete even at the end of three hours.

The rate of formation of the phosphate coating can be increased by bringing the solution into contact with the metal surface by pressure, such as can be applied by

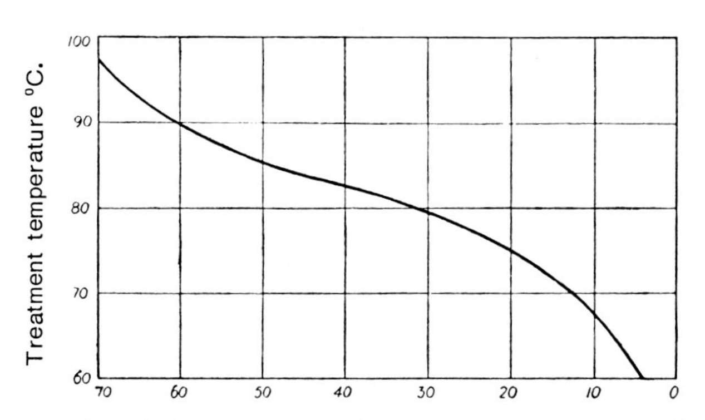

Fig. 6.2. Influence of treatment temperature on corrosion resistance of steel phosphate-coated in 3 per cent manganese dihydrogen phosphate solution.

rubber rollers. This method is particularly applicable to the treatment of sheet, and the process time can be reduced to three to ten seconds. Brushes can also be employed for this purpose. Besides accelerating the rate of formation of the coating, the use of brushes or rollers tends to make the phosphate layer more fibrous in structure, and thus enables it to withstand deformation. It has, in fact, been claimed that sheet phosphated in this way and painted can then be fabricated without injury to the coating[41].

6.2.6. BATH CONTROL

The ratio of combined to free phosphoric acid in the solution must be carefully controlled if satisfactory results are to be obtained. It is essential to maintain as high a proportion of the insoluble tertiary phosphate in the coating as possible, and to keep the treatment time to a minimum. A low proportion of free phosphoric acid is desirable, although too low a ratio may lead to excessive hydrolysis with the result that the insoluble phosphates are precipitated in the body of the solution. This is wasteful of chemicals and undesirable in other respects. A ratio of about 7 or 8 to 1 of combined to free phosphoric acid gives a good coating in a reasonable processing time.

In the case of a zinc phosphate bath containing an accelerator, a decrease in the acid ratio (excessive free acid) causes a progressive darkening of the coating and the formation of large, sparkling crystals. It is found that the latter contain increasing quantities of iron phosphates which are formed as a result of the etching action of the more concentrated solutions. Simultaneously, the protective value of the coating decreases as it becomes darker or more crystalline. The change in weight of the metal being treated is determined by the relative amounts of iron dissolved and phosphates precipitated. As the acid ratio increases, so does the loss in weight. At low concentrations precipitation of secondary and tertiary phosphates takes place not only at the metal surface but throughout the bath, causing sludge formation and waste of materials. Conversely, excessive total concentrations result in etching, and may cause the equilibrium changes to be delayed. The acid produced when precipitation takes place tends to re-establish equilibrium, and it may also dissolve the coating. In baths of high or low concentrations the treatment time is prolonged and sludge formation is high, whilst the corrosion resistance of the coatings is also adversely affected.

The proportion of manganese in an iron-manganese phosphate coating is generally higher than that present in the treatment solution. Some of the iron dissolved in the bath forms insoluble phosphates which are deposited as sludge and part of it remains as ferrous dihydrogen phosphates dissolved in the bath. As the bath is used it becomes less rich in manganese and richer in iron (dissolved in the form of dihydrogen phosphate). The appearance of typical phosphate coatings can be seen from the photomicrograph in Fig. 6.3 and a selection of typical Parkerised components is shown in Fig. 6.4.

Fig. 6.3. Heavy zinc phosphated mild steel surface ($\times 1000$).

6.2.7. IMPREGNATION AND AFTER-TREATMENTS

Phosphate coatings tend to be powdery on the surface, and they have relatively little corrosion resistance when used alone. To be of maximum value the film must be impregnated, one such treatment being to stain it black in a boiling solution of a dye to improve its appearance, and subsequently to immerse it in hot oil, such as linseed or a mineral oil. On wiping or centrifuging off excess oil, a pleasing finish is obtained, which is extensively employed on small parts such as nuts, washers, bolts and springs, and on components operating in the presence of oil or where conditions of exposure are not unduly severe. It has been found that oil films applied by immersion of the articles in hot concentrated aqueous oil emulsions followed by drying are capable of giving exceptionally good protection. Phosphate coatings of this description can give very satisfactory service, although they do not withstand much abrasion. Phosphate finishes are very useful for springs because of the considerably lower tendency for hydrogen embrittlement than in the case of plated finishes. These oil treatments are usually applied to the manganese phosphate types of coating, which are thicker and rather more absorbent than that produced by the zinc processes. The coatings can also be impregnated with solutions of waxes or with dilute lacquer solutions (e.g. plasticised shellac lacquers), and in this way satisfactory finishes can be obtained which are suitable for conditions of service where the presence of an oil film may be a disadvantage.

It has been established that the addition of 0.1 to 0.2 per cent of chromic acid in the rinse water after phosphating improves the protective value of the coating. Concentrations in excess of 1 per cent will dissolve the coating, so that the chromic acid content should be kept at below 0.5 per cent. Radiometric investigation on the mechanism of adsorption of the chromic acid indicates that some kind of chemical combination takes place since the adsorbed acid is not removed to any appreciable extent by hot or cold water[42]. Adsorption is also helped by the presence of phosphoric acid in admixture with the chromic acid. Thus, for the optimum corrosion inhibiting effect, a solution consisting of 1.75 g of CrO_3 or a mixture of 1 g of CrO_3 and 1 g of H_3PO_4 per litre should be used. This type of solution is used in preference to the chromic acid rinse, and is often followed by a demineralised water mist spray, this water being returned to the chromic acid-phosphoric acid stage.

6.2.8. COLD FORMING

An important application for phosphate coatings is in the cold forming and drawing of metals, where they are used on a large scale. The coatings themselves act as lubricants and also as carriers for oil and soap films used as lubricants. The zinc phosphate coatings are most widely used for this purpose, partly because of their ability to react with some of the lubricant soaps. The crystalline phosphate and soap combination is stable under high pressures, and up to four successive draws can be made without the need for applying more lubricant.

Fig. 6.4 Selection of typical Parkerized components.

6.2.9. PRE-TREATMENT FOR PAINTING

The zinc phosphate processes are preferred to the manganese phosphate coatings when organic finishes are to be applied subsequently. Besides being formed much more quickly, they are thinner and less granular in nature, so that a smooth paint film is more readily obtained. The adhesion of paint to steel pre-treated in this way is much better than to bare steel, while the lateral spreading of rust areas under any breaks that may occur in the paint film is slower on pre-treated surfaces (Fig. 6.5). The effectiveness of phosphate coatings as undercoats for paint is based on their large specific area and high porosity which cause better wetting and adsorption of the paint to take place. Improved mechanical adhesion also results from the crystalline nature of the coating, even when it is very thin. Sherlock and Shreir[43] have studied the adhesion of polymers to steel surfaces as a function of thickness and type of phosphate film. It was found that the thinner the underlying phosphate film the greater the adhesion of organic coatings.

In a recent paper[44] Steinbrecher has discussed the pre-treatment of metals prior to electroplating, dealing with methods of evaluating pre-treatment processes, optimum methods of cleaning, rinsing, phosphating and passivation, the effect of drying at different temperatures prior to electropainting, and rinsing after painting. A number of panels were treated by different techniques and these were then tested for corrosion resistance by salt spray and humidity, and for adhesion. The results indicated that the degree of dissolution of zinc phosphate coatings during the electrodeposition process varied with the pH of the paint and the voltage applied. Paints in the lower pH range 7.4-7.8 gave the highest coating weight losses (25-50%) whilst pH systems above 9.0 gave the lowest weight losses (0.3%). The zinc phosphate dissolved appears to become an integral part of the paint coating itself; analysis of the used paint baths in the laboratory showed the complete absence of zinc phosphate in the paint bath proper, and this is confirmed by work done in large-scale plants in the American automobile industry. It appears that zinc phosphate does not accumulate in the paint.

Fig. 6.5. Benefits of conversion coatings. [Courtesy Pyrene Chemical Services, Ltd.]

The panels right were prepared under strict laboratory control as follows:

All panels were degreased and right hand side masked off, left hand side processed as described.

Series A – steel – zinc phosphate (crystalline)
Series B – steel – iron phosphate (amorphous)
Series C – hot dip galvanized – zinc phosphate (crystalline)
Series D – aluminium – chromate (amorphous)

Prior to powder coating the masked sides were again degreased. Dry film thickness of powder coating ≡ 2.5 mils. All panels were cross-scored down to base metal and subjected to 1000 hours ASTM B117 salt spray.

The panels were then taped across the full length of the score line and the tape was then removed in one rapid movement.

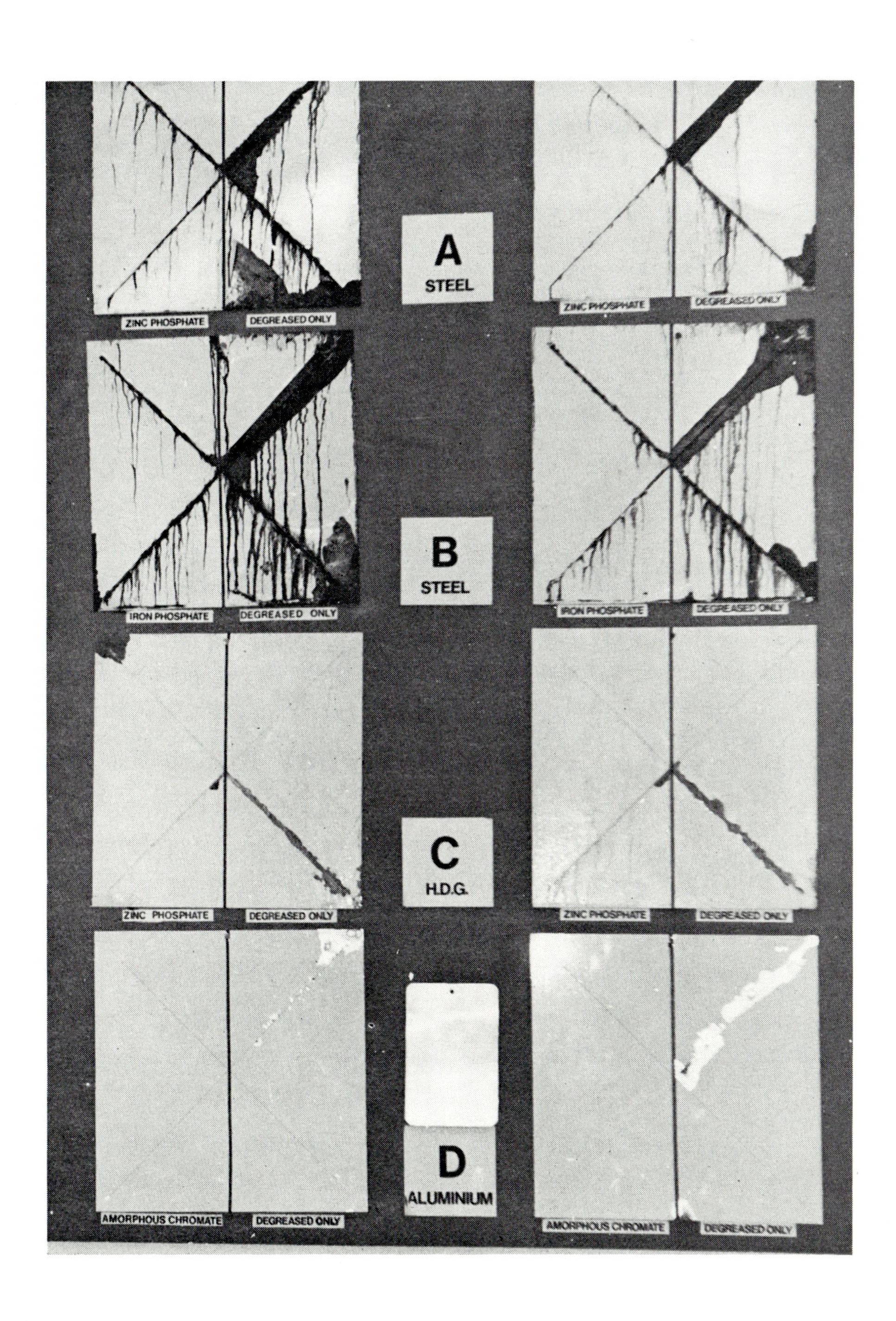
A
STEEL
ZINC PHOSPHATE
DEGREASED ONLY
ZINC PHOSPHATE
DEGREASED ONLY
B
STEEL
IRON PHOSPHATE
DEGREASED ONLY
IRON PHOSPHATE
DEGREASED ONLY
C
H.D.G.
ZINC PHOSPHATE
DEGREASED ONLY
ZINC PHOSPHATE
DEGREASED ONLY
D
ALUMINIUM
AMORPHOUS CHROMATE
DEGREASED ONLY
AMORPHOUS CHROMATE
DEGREASED ONLY

Zinc phosphate coatings are preferred to iron phosphate coatings, although the latter are acceptable where the lower corrosion resistance is adequate and where a single-coat paint system is used because of gloss considerations. The zinc phosphate coatings should have a weight of about 15 to 17.5 mg/dm^2. The use of chromate rinses is more effective in improving the corrosion resistance of the final finish if the work enters the paint bath after low temperature drying, rather than after conventional stoving.

The use of a deionised (demineralised) water rinse is always recommended as a final step in the pretreatment process, since the presence of any concentration of chloride or chromate in the paint will reduce the quality of the finish drastically. Deionised water application after the passivation chromate rinse will, in general, not affect the salt spray resistance. The time between the application of the chromate rinse and the deionised rinse does not appear to be significant over the range of 10-300 seconds. Oven-drying the zinc phosphate coating before painting is carried out in the general range of 150 to 200°C for 5-10 minutes. This is because of changes in the degree of hydration of the zinc phosphate crystals.

6.2.9.1. Etch primers

Phosphoric acid is the basis of water-based etch paints which form an excellent primer on steel and non-ferrous metals. These contain vinyl resins dissolved in alcohol solvents; phosphoric acid is added as the etching medium while basic zinc chromate is frequently included as the corrosion-inhibiting pigment. Etch primers can be applied to steel or aluminium structures which cannot be conveniently phosphate-treated, and combine a pre-treatment and a primer coating in one operation. They have proved very successful in many fields.

6.2.10. LOW TEMPERATURE PHOSPHATING

In recent years low temperature processes have become increasingly employed, both on account of the high cost of fuel and the need to conserve energy. This may mean temperatures of up to 50°C. and since articles are often warm when put into the solution, and a certain amount of heat is evolved as a result of the chemical reaction, the temperature can often be maintained without additional heat input.

The lower temperature, especially in the spray-type of plant, is a great advantage. This is because scale tends to deposit rapidly on steam coils, so that they become covered with an insulating layer which is very difficult to remove and slows down the rate of heat transfer very considerably. At lower temperatures, such sludge as is produced in the reaction is lighter in texture and non-adherent. It does not, therefore, accumulate in the nozzles and pipes of spray plant, and thus the need for chipping and hammering to remove scale accumulations is elimininated. The sludge can be very largely washed away when necessary by flushing with water.

One low temperature process uses a solution of acid zinc phosphate containing hydrogen peroxide (or a compound capable of forming hydrogen peroxide in solution, such as sodium percarbonate or perborate), to which periodic additions of an alkali such as sodium hydroxide or sodium carbonate are made. The object of the alkali is to neutralise the free phosphoric acid produced by the hydrolysis of the zinc phosphate in solution, and also as a result of the oxidation of any ferrous phosphate and its subsequent precipitation as ferric phosphate. The alkali is also used to keep the solution in a slightly supersaturated condition with respect to zinc phosphate; this can be achieved by maintaining a pH of more than 0.1 above the equilibrium pH. The zinc content of the solution is kept at between 1 and 5 g/l with a hydrogen peroxide concentration of 0.04 to 0.24 g/l.

Another low temperature process[45] makes use of a zinc solution containing sodium nitrite and fluoride. The pH of the solution must be kept within defined limits for statisfactory results. As has already been pointed out, the decreased rate of coating formation at low temperatures cannot be counteracted by increasing the free acid concentration; it has, in fact, been shown that the amount of free acid required to keep tertiary zinc phosphate in solution at low temperatures is considerably less than that required at the boiling point. Fluoride compounds are useful in maintaining the pH constancy of this type of bath, and a low free acid concentration is necessary.

6.2.11. PHOSPHATED GALVANISED COATINGS

Phosphated (as well as chromate-treated) electrogalvanised coatings on steel are widely used where greater resistance to corrosion is required than is obtainable by direct phosphate treatment of unplated steel. The ferrous metal part is plated with a very thin zinc coating, of the order of 0.00003 to 0.00007 cm thick, but in any case not exceeding 0.0002 cm, which is then phosphate-treated, so that part of the coating is converted to insoluble phosphate, by immersion for about 15 seconds in an acid zinc phosphate solution at a temperature of about 60°C. About half of the zinc coating is converted to a phosphate coating of small crystal size which provides a good undercoat for paint and does not interfere with spot welding. In this way finishes of high durability can be obtained, and large plants for the continuous treatment of steel sheet and strip in this way are in operation. The phosphate film can be further improved by immersion in 2.25 g/l of chromic acid and an equal amount of phosphoric acid at about 80°C. Rinsing is not carried out after this treatment; the chromic acid probably acts as a passivating agent in the pores of the coating and inhibits the corrosive effect of any accelerators which may be left in the phosphate coating.

Machu[46] has discussed the special problems encountered when attempts are made to produce phosphate coatings on composite metal articles such as car bodies, etc. (e.g., steel and aluminium or galvanised steel and aluminium). Aluminium dissolves into the phosphating bath and acts as a negative catalyst. When present in concentrations as low as 0.3 g/l it can completely prevent the formation of a phosphate coating on steel. If only small amounts of aluminium are involved, for example, after treating galvanised zinc coatings containing aluminium, the addition of small quantities of fluorides and sometimes nickel salts to the phosphating bath will eliminate the adverse effect of the aluminium. The function of the fluoride ion is to complex the simple aluminium cations into the form AlF_6^{3-}, so preventing them from interfering with the phosphating treatment. For the treatment of articles in which the proportion of aluminium surface area is greater than about 5-10%, a phosphating solution containing a relatively large proportion of fluoride as well as other additions has been developed. It is claimed that fine-grained, dense, homogenous and strongly corrosion-resistant coatings are produced on work pieces comprising any proportions of steel, zinc or aluminium.

6.2.12. TESTING PHOSPHATE COATING

The testing of a phosphate coating is not easy to carry out, whilst the evaluation of the results is to quite a considerable extent subjective. A common method of assessing the completeness of the process is to draw the thumbnail across the surface, when a uniform and unbroken streak should be visible. An alternative method is to boil the test piece in aerated water, using a glass diffusing disc, for at least 24 hours. The specimen is immersed in the water and heated at such a rate that the water begins to boil not less than 15 minutes and not more than 20 minutes after the initial application of heat. The degree of rusting which has taken place during this test is subsequently assessed.

A British Standard Specification (B.S. 3189) exists for the phosphate treatment of iron and steel for protection against corrosion and covers 5 classes of coating, ranging from heavyweight coatings for wear resistance down to extra lightweight coatings for indoor applications such as office furniture, toys, etc.

6.2.13. PHOSPHATING PLANT

Plant for carrying out phosphate coating has been developed to a tremendous extent. Very large conveyorised installations have been built both for immersion and for spray treatment on steel pressings for motor bodies, wings, headlamps, etc. The parts are attached to special racks or frames and travel through the plant continuously. Modern phosphating plants for treatment prior to painting carry out cleaning, rinsing, pre-dipping, phosphating, chromate rinsing and final water rinsing operations. An additional high-pressure cold water spray may be incorporated before cleaning to remove deposits from the work to be treated. The water is recirculated, and the operation serves to keep the rest of the plant clear and to reduce spray nozzle blockages. The installation and operation of several fully-automatic plants for phosphating have been described in the literature[47].

In smaller plants, the treatment is carried out in steel tanks into which the articles are lowered on racks, wires, or in baskets. Rotating perforated baskets are also

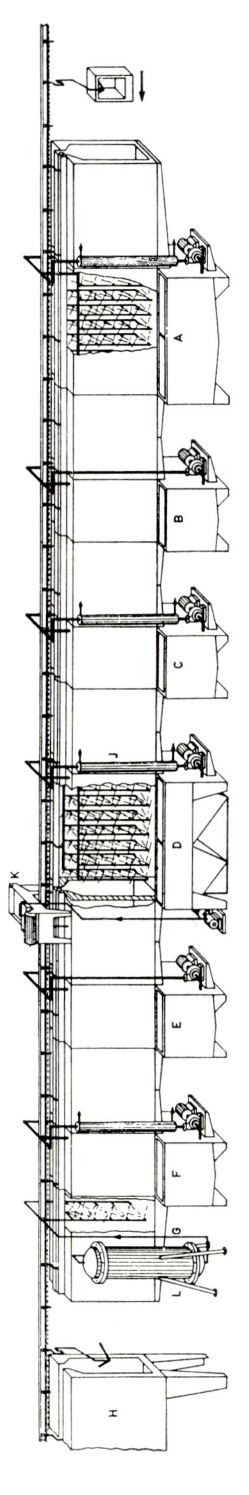

Fig. 6.6. Scheme of 6-stage installation for spray phosphating.

A. degreasing
B,C. water rinses
D. phosphating
E. water rinse
F. passivating post-treatment
G. spray rinse (demineralised water)
H. drying oven
J. heat exchanger for heating phosphating bath
K. filter for phosphating solution
L. water demineraliser

Fig. 6.7. Immersion phosphating installation. [Courtesy Sietam.]

Fig. 6.8. Immersion phosphating installation. [Courtesy Soc. continentale Parker.]

Fig. 6.9. Articles emerging from a spray phosphating plant.
[Courtesy Pyrene Chemical Services, Ltd.]

employed to provide more intimate contact of the solution with very small components; however, the perforations tend in time to clog up with crystallised salts, reducing their efficiency. An ingenious type of automatic plant for dealing with small articles is the tipping basket machine in which the articles are transferred successively through a series of baskets, and thus through the whole treatment cycle.

After prolonged use steel tanks become coated on the inside with a layer of insoluble phosphates, but such tanks will last for several years before failure eventually occurs due to perforation of the steel by the solution. Stainless steel is not attacked by phosphating solutions but is, of course, relatively expensive. The best type of stainless steel for resistance to phosphate solutions is the molybdenum-containing austenitic nickel-chromium steel since this is less subject to pitting corrosion than the normal 18:8 nickel-chromium stainless steels. An added disadvantage is that welds may be less resistant due to galvanic action with the stainless steel surface. Nevertheless, large plants have been constructed using stainless steels.

Perhaps the most satisfactory material of construction, especially for spray-plants, is hard rubber-lined steel. This has, however, the disadvantage of mechanical weakness and may be damaged if jigs fall on to it or when the plant has to be entered for cleaning purposes. To reduce this risk, rubber mats may be laid on the tank floor, or the bottom may be lined with acid-resisting tiles.

Fig. 6.10. Spray phosphating installation. [Courtesy Triton S.A.]

Fig. 6.11. Exit end of a spray phosphating plant.
[Courtesy Pyrene Chemical Services, Ltd.]

The heating of phosphating plant presents considerable difficulties. Usually, copper steam coils or plate heaters are employed, but the sludge precipitated in the solutions during working forms hard, coherent deposits which can only be removed with great difficulty, usually by chipping them off. The sludge as initially formed is light, but rapidly becomes extremely hard and reduces the heat transfer efficiency of the coils. There is some advantage in circulatinng the solution by means of a pump through a relatively coarse filter to remove sludge, and this procedure has been adopted with success in some installations, although filters tend to become clogged rather quickly. The new low-temperature processes produce much less sludge and deposits in coils.

Fig. 6.6 shows diagrammatically the sequence of operations in a typical spray-phosphating plant. Figs. 6.7 to 6.11 are views of different types of phosphating installations.

REFERENCES

1. D. Fishlock, 'Metal Colouring' Robert Draper, Teddington, 1962.
2. W. A. Wesley and H. R. Copson, *J. Electrochem. Soc.* 1948, **94**, 20.
3. A. D. Merriman, 'A Dictionary of Metallurgy', p.352. McDonald and Evans, London, 1958.
4. P. Szeki, *Engrs. Digest* 1959, **20**, 11.
5. H. Silman, *J. Electrodepos. Tech. Soc.* 1945, **20**, 77.
6. K. C. Scrivastava, *Electroplating (India)* 1959, **1**, No. 2.
7. A. J. Mitchell, *Metal Finishing* 1973, **71**, No. 3, 41-44, 49.
8. U.S.Pat. 2,077,450.
9. H. Krause, 'Metal Colouring and Finishing', p.157, 1938.
10. R. Thonnard, *Industrial Finishing* 1971, **23**, No. 282, 10.
11. *Z. Metallk.* 1961, **52**, No. 2, 141.
12. T. E. Evans, A. C. Hunt, H. Jones and V. A. Smith, *Trans. Inst. Metal Finishing* **1972**, 2, 77-79.
13. 'Surface Finishing Stainless Steel', p.63. pub.S.A.D.A., London, 1971.
14. Brit. Pat. 834,834 (1960).
15. G. B. Hogaboom, Proc. 26th Ann. Conv. Am. Electroplaters' Soc., 1938, 180.
16. S. G. Clarke and J. F. Andrew, *Trans. Inst. Met. Finishing.*
17. W. H. Vernon and L. Whitby, *J. Inst. Met.* 1929, **42**, 181.
18. J. N. Hitchin, *Met. Finishing J.* 1973, **19**, No. 214, 19-21.
19. W. H. Vernon, *J. Inst. Met.* 1932, **49**, 153.
20. W. J. Erskine, *Met. Industry (New York)* March, 1939, p.123.
21. H. Krause, 'Metalfärbung', Carl Hanser Verlag, Munich, 1951.
22. P. H. Masquities, *Proc. Am. Electroplaters' Soc.* 1937, **71**, 225.
23. W. J. Young and H. Kerstan, *Trans. Electrochem. Soc.* 1937, 71, 225.
24. E. A. Anderson, *Proc. Amer. Electroplaters' Soc.* 1943, p.6.
25. J. E. Stareck, *Proc. Amer. Electroplaters' Soc.* 1941, p.48.
26. Brit. Pat. 8,667.
27. C. M. Postins, *Met. Finishing J.* 1969, **15**, No. 180, 426.
28. W. Machu, *Korr u. Metallschutz* 1941, **17**, 157.
29. L. O. Gilbert, *Proc. American Electroplaters' Soc.* 1956, p.195.

30. V. N. Darsey, U.S. Pat., 2,293,716.
31. Brit. Pat. 510,684.
32. Brit. Pat. 519,823.
33. *Industrial Finishing* 1971, **23**, No. 280, 50.
34. Brit. Pat. 270,820.
35. Brit. Pat. 419,487 and 447,176.
36. Brit. Pat. 270,680.
37. G. Jernstedt, *Trans. Electrochem. Soc.* 1943, **83**, 361.
38. H. A. Holden, *Corrosion Technology* 1961, **2**, 40.
39. L. O. Gilbert, *Proc. Amer. Electroplaters' Soc.* 1956, **43**, 195.
40. O. Macchia, *Korr. u. Metallschutz* 1936, **12**, 211.
41. V. A. Darsay and W. R. Cavanagh, *Trans. Electrochem. Soc.* 1947.
42. S. C. Eister, *Industrial Finish* 1957, **9**, 818.
43. J. C. Sherlock and L. L. Shreir, *Brit. Polymer J.* 1969, **1**, 34.
44. L. Steinbrecher, 'Aspects of Pre-Treatment of Metals Prior to Electropainting', Fourth International Congress on Metallic Corrosion, Amsterdam 7-14th Sept., 1969.
45. A. Foldes, *Korr. u. Metallschutz* 1943, **19**, 281.
46. W. Machu, *Trans. Inst. Met. Fin.* 1971, **49**, 214.
47. *Product Finishing* 1968, **21**, No. 10, 36; *ibid* 1968, **21**, No. 4, 58 1968; *ibid* 1968, **21**, No. 5, 36; *Metal Finishing J.* 1964, 10, 467.

BOOKS FOR GENERAL READING

1. 'Phosphating of Metals', G. Lorin, Finishing Publications Ltd., Teddington, 1974.

CHAPTER 7

Deposition of the noble metals

NOBLE metals are used as electrodeposits for both decorative and industrial applications because of their resistance to atmospheric and chemical attack. It has been estimated that in 1973 the electronic industries in the United States alone consumed precious metals worth $50M.[1] Electrodeposits of the principal noble metals, viz. gold, silver, platinum, palladium and rhodium, are also used on jewellery, scientific apparatus, and in numerous other applications where their special characteristics are often irreplaceable. As the noble metals are normally cathodic towards the underlying basis metal, it is essential that the deposits should be free from imperfections in the form of pores, pits and cracks.

The performance of data-handling, computer and automated equipment is dependent to quite a large extent upon the satisfactory functioning of noble-metal-plated electrical contacts. Noble metal deposits have made it possible to produce contacts which can operate with very low voltages and contact pressures. Almost 75% of all electrodeposited gold, for example, is used in the manufacture of such contacts, where the prime requirements are low contact resistance and the ability to withstand heat and corrosive attack[2].

Of the metals of the platinum group, those of main interest so far as commercial electroplating is concerned are rhodium, palladium and platinum, of which the first is the most important. A general review of the electrodeposition of platinum group metals has recently been given by Benninghoff[3].

7.1. Rhodium plating

Rhodium plating has been found of value when a hard, non-tarnishing deposit of pleasing colour is required on silver or on silver-plate, gold, white gold alloys or platinum. The metal can also be plated directly on to nickel. In the case of copper- or zinc-base alloys, a nickel deposit should first be applied and should be of adequate thickness to protect the underlying metal from the action of the highly acid rhodium-plating solution.

Rhodium deposits are used for electrical contacts in corrosive atmospheres. Rhodium in contact with gold wipers provides one of the best wiper-contact combinations available.

For telecommunications equipment, where high frequencies are encountered, silver is the preferred undercoat owing to the need for high surface conductivity for this application. The surface skin effect is pronounced at very high frequencies. The thickness of rhodium may also have to be restricted in such cases since its conductivity is only about one-third that of silver. Silver is also a good undercoating for rhodium where deposits of over 2.5 μm thickness are required, because it does not add to the inherently high stress levels of rhodium coatings. Gold undercoatings are used to obtain maximum protection of the basis metal against corrosion. Low internal stresses in thick deposits are favoured by high metal and acid concentrations in the solution and high operating temperatures, whilst efficient cleaning and the absence of contaminants in the bath are also important factors. Reid[4] established that the tensile stress of rhodium deposited from a sulphate electrolyte was over 150 kg/mm^2. He also found that additions of aluminium or manganese prevent crack formation in coatings of moderate thickness but do not reduce stress. However, nearly stress-free deposits can be obtained from a sulphamate bath[5].

Rhodium has an exceptionally high resistance to corrosion. It has also an attractive appearance and thin deposits are used to protect silver and silver-plated articles against tarnishing, and on jewellery.

The reflectivity of the metal is high (about 78 per cent), and this, coupled with its resistance to abrasion and to oxidation at high temperatures, makes it useful for the manufacture of reflectors where conditions are severe, e.g. cinematograph reflectors. Its hardness also makes rhodium-plated articles highly wear-resistant.

7.1.1. PLATING SOLUTIONS

The principal plating solutions currently employed are: (a) solutions of rhodium sulphate, $Rh_2(SO_4)_3 \cdot 12H_2O$, containing 2 to 20 g/l of metallic rhodium together with some free sulphuric acid; (b) phosphate baths with a similar concentration together with a suitable concentration of free phosphoric acid; and (c) phosphate-sulphate baths containing rhodium phosphate and free sulphuric acid. Of these baths the sulphate bath is the most important and is widely used. The phosphate bath finds application where thin decorative deposits of rhodium are required. Both these types of solution are relatively easy to operate and have good throwing power.

Thorough cleaning is essential before rhodium plating in these solutions, particularly where thicker deposits are specified. Normal alkaline cleaning procedures are employed, but anodic etching in cyanide prior to silver plating or immersion in a dilute hydrochloric acid solution before nickel plating are helpful. Bright nickel is used as an undercoat for decorative rhodium as it improves the brightness of the deposit and prevents attack on the basis metal. In the case of nickel, activation of the undercoat by cathodic treatment in sulphuric acid (10% V/V) at 2.25 A/dm^2 for half a minute helps to promote adhesion. Articles so treated should be transferred to the sulphate bath without rinsing; adhesion on nickel, even with these treatments, is inferior to that obtained on silver[6]. Stopping-off materials which may be used include chlorinated rubber paints stoved at 72°C or cellulose lacquers, as well as many waxes and plastisols formulations. Great care must be taken in the choice of screening material as the presence of even very small amounts of organic matter can produce highly-stressed deposits[7].

7.1.1.1. Sulphate bath

The rhodium sulphate bath is most conveniently prepared from the commercial concentrate. The concentrate should always be added to a sulphuric acid solution in order to prevent the re-precipitation of some of the rhodium by hydrolysis, and this may be difficult to re-dissolve.

The bath consists essentially of an acid solution of rhodium sulphate which is a well-defined salt having the composition $Rh_2(SO_4)_3\ 12H_2O$. According to Reid[8] the metal is probably present in solution partly as a simple hydrated cation $Rh(H_2O)_6^{3+}$ and partly as an anion $Rh(SO_4)^{3-}$ with possibilities of intermediate forms such as $Rh_3(OH)^{++}$ or $Rh(OH)_2^{+}$.

The actual concentration of rhodium in the plating solution is kept low so as to reduce the capital cost of the bath and keep drag-out losses to a minimum. However, it is evident from Fig. 7.1 that the cathode current efficiency is greatly reduced at low concentrations, particularly where high current densities are employed. Apart from wastage of electrical power, low cathode efficiencies are undesirable because of the attendant reduction of hydrogen which increases stress in the deposit. Since an increase in acid concentration in the bath favours the discharge of hydrogen ions it is not surprising to find that it causes a decrease in efficiency. Agitation promotes high cathode current efficiency, by maintaining the rhodium concentration in the vicinity of the cathode.

For the production of heavy deposits of rhodium (20–50 μm), where high cathode efficiency is important, solutions containing 10 to 20 g/l of rhodium are recommended[9]. Baths containing 10 g/l of rhodium should have a sulphuric acid concentration of about 50 ml/l[10] which is sufficient to prevent hydrolysis at 50°C and with current densities of between 1-2 A/dm^2 gives satisfactory results. Solutions having lower concentrations are also commonly used.

Rhodium sulphate solutions are operated with insoluble platinum or platinised titanium anodes and the metal content is replenished by the addition of the rhodium sulphate concentrate. Regular replenishment is important to avoid reduction in the deposition rate which occurs if the solution becomes too weak. The solution must be operated in an acid-resisting tank. Glass is very satisfactory for small tanks, whilst for larger baths hard-rubber- or PVC-lined steel gives good service. Heating can be carried out by electric immersion heaters with fused silica casings or, alternatively, a water jacket can be employed. The latter procedure has the advantage that the valuable solution will not be lost if the container should fracture or develop a leak.

Regular jogging of the work rod during plating, either mechanically or by hand, minimises the possibility of pin holes in the plate due to adherence of gas bubbles.

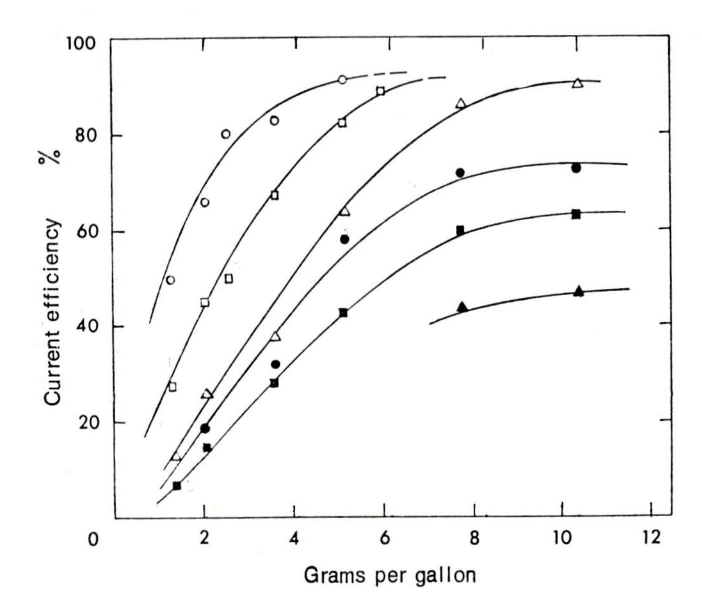

Fig. 7.1. Effect of rhodium concentration on the current efficiency of agitated sulphate solutions. 49°C 26.4 ml/l sulphuric acid.

○ = 1.08 A/dm²   ● = 6.47 A/dm²

□ = 2.16 A/dm²   ■ = 8.62 A/dm²

△ = 4.31 A/dm²   ▲ = 12.92 A/dm²

The concentrated acid spray given off during deposition makes it desirable that provision is made for the removal of spray by extraction.

7.1.1.2. Phosphate bath

Phosphate plating solutions produce very white deposits and are most suitable for applying thin coatings. A suitable bath may contain 2 g/l of rhodium metal and 40-60 ml/l of phosphoric acid (85 per cent), but the acid concentration is not critical. The phosphate solution does not consist of a single compound but contains a series of complex phosphates of ill-defined composition. This may possibly account for the fact that the colour of rhodium as plated from a phosphate bath appears to vary with the concentration of metal in the solution. Generally, however, the colour lacks the faint bluish tinge of rhodium as plated from the sulphate solution. The bath is usually operated at about 40° to 50°C.

The phosphate solution has a lower conductivity than the sulphate bath so that higher working voltages are needed, but it is operated in much the same way as the sulphate bath.

7.1.1.3. Phosphate-sulphate bath

The phosphate-sulphate solution contains about 2 g/l of rhodium as phosphate and 50-100 ml/l of sulphuric acid. The solution has a lower cathode efficiency and is rather more difficult to control than the sulphate solution.

Other solutions occasionally used are the fluoborate solution and the sulphamate bath.

7.1.1.4. Low-stress rhodium plating baths

Rhodium deposits are normally so highly stressed that fine cracks appear in coatings thicker than about 30 μm. These stresses may be so high as to cause copper to lift off printed circuit boards being plated with rhodium.

A number of low-stressed rhodium plating electrolytes have been proposed, notably sulphate and phosphate baths containing selenium (in the form of selenic acid or alkali metal salts) [(11–13)]. Concentrations of between 0.5 and 1.0 g/l selenium are normally employed in solutions containing 10 g/l rhodium. Sound low stress deposits can be obtained from such solutions operating at about 50°C with a current density of about 10 A/dm^2.

Rhodium solutions are sensitive to organic contaminants, which can be introduced into the solution from printed circuit boards or stop-off lacquers; contamination should be removed by treatment with activated carbon, otherwise cracking of the deposit may occur. Iron, copper, silver, tin, lead and zinc are also harmful in concentrations of more than 5 p.p.m.

Rhodium baths may also become partly oxidised during operation, their trivalent rhodium being converted to the quadrivalent or hexavalent state at the anode. This process is essential for satisfactory deposits to be produced, but if the higher oxidation state rhodium ion concentration becomes excessive the bath changes in colour from reddish-brown to greenish-brown, and then almost black. The bath is then unfit for use, but can be regenerated with hydrogen peroxide, which is subsequently removed by heating.

Some recommended deposit thicknesses of rhodium are as follows:

Anti-tarnish applications	0.2-25 μm
Contacts (light load)	0.3–0.5 μm
Contacts (heavy load)	2.5–3.0 μm
Contacts (heavy wear)	0.5–2.0 μm
Corrosion protection	0.25–1.0 μm

Rhodium deposits can also be applied locally to such items as the commutators of large motors by the tampon technique, and also by chemical displacement from hydrochloric acid solutions.

7.1.2. APPLICATIONS

Rhodium deposits have high hardness and wear resistance and are therefore used on slip rings and switches in telecommunications equipment and in high-speed computer switching. They are specially suitable for sliding contacts which have to remain inoperative for long periods.

7.2. Platinum plating

Platinum deposits are used to a limited extent for decorative applications, but apart from high temperature contacts and valve grids platinum is tending to be replaced by rhodium (which has a much lower specific gravity) and by gold or palladium. A general review of the electrodeposition of platinum has been given(14) and Weser(15) has described a process which it is claimed will give ductile platinum coatings.

The early baths used for the electrodeposition of platinum were based on chloroplatinic acid, a typical bath composition being:

Chloroplatinic acid	$H_2PtCl_66H_2O$	4 g/l
Disodium phosphate	$Na_2HPO_412H_20$	100 g/l
Diammonium phosphate	$(NH_4)_2HPO_4$	2 g/l

Later, platinum diammino nitrite $Pt(NH_3)_2.(NO_2)_2$ baths were used (16), but produced thin, dull coatings.

The complex ammino-phosphate baths(17) are based on $(NH_4)_2PtCl_6$ or Na_2PtCl_6, Na_2HPO_4 and $(NH_4)_2HPO_4$ and give better deposits, whilst the solutions have a longer life. The bath contains, typically: disodium phosphate, 100 g/l; diammonium phosphate, 20 g/l; platinum chloride, 7.5 g/l; ammonium chloride, 25 g/l.

The diammino nitrite bath has also been used in an acid version, sulphuric or phosphoric acid being added, when low-stress deposits are produced.

7.2.1. HYDROXY-PLATINATE SOLUTION

Platinum can be plated at almost 100% efficiency from a sodium hexahydroxy platinate bath which contains:

Sodium platinate	20 g/l
Sodium hydroxide	5.5 g/l
Sodium oxalate	5.5 g/l
Sodium sulphate	33 g/l

The bath is operated at temperatures between 65 and 85°C with current densities of 0.5–1.5 A/dm^2. Although the bath has several disadvantages which include liability to absorb carbon dioxide from the air, sensitivity to impurities such as cyanides and silicates and the need for regular analysis of the solution, it can provide very high quality deposits at thicknesses up to 25 µm which are dense, very bright and almost indistinguishable from those of rhodium.

7.2.2. SULPHATO-DINITRO-PLATINITE SOLUTION

Platinum can be successfully electrodeposited on to copper, brass, silver, nickel, lead and titanium from a solution containing sulphatodinitroplatinous acid, $H_2Pt(NO_2)_2SO_4$, which is available commercially under the name of D.N.S.[18]. The bath is stable and does not deteriorate on standing. However, it cannot be used for plating directly on to iron, tin, zinc or cadmium owing to its high acidity. Good deposits are obtained with platinum concentrates of 5 g/l at 30–70°C and at current densities of 0.5–1.0 A/dm². The coatings are bright and lustrous at all thicknesses and have a hardness of 400–450 D.P.H. They appear to be free from porosity at thicknesses up to 7.5 μm, but may crack when more than 25 μm is deposited. The pH of the bath should be below 2. The deposition rate is 25 μm in 2 hours at 0.5 A/dm², but this time can be reduced to 30 minutes by using a 15 g/l bath at 2 A/dm².

7.2.3. APPLICATIONS

One of the most important uses of platinum plating is in the manufacture of platinised titanium anodes which are used for the cathodic protection of ships, marine structures and pipelines. They are also useful as anodes in electrodeposition, especially of the noble metals.

The use of platinum plating in the electrical and electronics fields has declined since the introduction of new rhodium plating solutions.

7.3. Palladium plating

Palladium is relatively low in cost, and low-stress deposits are readily obtained from a number of solutions. However, the metal cannot be readily soldered; also, its rather high resistivity restricts its use for VHF applications. It has low surface contact resistance and is relatively hard (about 300 DPH), so that it is suitable for sliding contacts which operate under high mechanical pressures. Palladium does not diffuse readily into certain baser metals which makes it suitable, for example, as a barrier between gold and copper alloys.

The original palladium plating bath was of the phosphate type, similar to the phosphate platinum bath in composition, consisting of:

Palladium chloride	$PdCl_2.2H_2O$	3.7 g/l
Disodium phosphate	$Na_2HPO_4.12H_2O$	100 g/l
Diammonium phosphate	$(NH_4)_2\ HPO_4$	20 g/l
Benzoic acid	$C_6H_5.COOH$	2.5 g/l

The solution is boiled after making up until the dark red colour changes to bright yellow, when the palladium complex is formed.

The solution is operated at 50°C at a current density of 0.2 to 0.25 A/dm² and a cell voltage of 1 to 2 volts. Bright deposits are first obtained, which tend to become dull as they thicken. Palladium, platinum or platinised titanium anodes are used, but the solution must be replenished by additions of palladium chloride followed by boiling.

7.3.1. DIAMMINO-DINITRITE-PALLADIUM BATHS

One of the most important palladium plating electrolytes is based on diammino-dinitrite-palladium ($Pd(NH_3)_2(NO_2)_2$[8, 19]. The bath is operated at 50°C and contains:

Diammino-dinitrite-palladium		8 g/l
Ammonium nitrate		100 g/l
Sodium nitrite		10 g/l
Ammonia	to	pH 9–10

Palladium exists in the solution in the form of the tetrammino-palladous complex ion $Pd(NH_3)_4^{++}$. With a current density of 1.0 A/dm² deposition efficiencies of about 70% are obtained. The palladium or platinum anodes used are insoluble, so that the metal content of the bath must be replenished at intervals by adding more diammino-dinitrite-palladium.

Palladium can also be deposited from an alkaline bath based on diammino-dinitrite-palladium and ammonium sulphamate which is used for producing 5–7.5 μm thick deposits. It is also claimed[20] that non-porous ductile palladium is deposited from a bath containing 16.32 g/l $Pd(NH_3)_2Cl$, 65–250 g/l NH_4Cl and NH_4OH to give a pH of at least 8.8. Palladium-nickel alloys containing 20% palladium can also be deposited.

Palladium plating baths are particularly sensitive to copper, which will produce dark deposits in concentrations greater than about 0.03 g/l.

Palladium is characterised by its great capacity for absorbing hydrogen and if much hydrogen liberation occurs at the cathode, non-adherent dark-coloured deposits result.

7.3.2. CHEMICAL DEPOSITION

Palladium can also be deposited without application of electric current by chemical reduction onto catalytic surfaces. Thick deposits can be built up, as the deposited palladium is also catalytic. They are ductile and can be applied uniformly on complex shapes[21].

A suitable solution for chemical deposition onto nickel consists of:

Tetrammino palladium chloride $Pd(HN_3)_4Cl_2$	5.4 g/l
Ammonium hydroxide	350 g/l
Disodium salt of ethylenediamine tetracetic acid	33.6 g/l
Hydrazine	0.3 g/l

The bath is used at 80°C; the rate of deposition is about 20 μm per hour.

Deposition can also be carried out satisfactorily onto aluminium, chromium, steel, silver and other metals. The solution is kept in a plastic container onto which the metal will not deposit.

7.3.3. APPLICATIONS

Palladium is used in the electronics industry in end connector applications, such as the plating of printed circuit patterns, and in telephone equipment. In some circumstances it can, however, form insulating films on rubbing contacts, which is a disadvantage.

The resistivity of palladium, although low, is considered too high for high frequency applications.

7.4. Ruthenium and osmium plating

Ruthenium has only recently become available in the form of practical plating solutions. The deposits are useful under high arcing conditions, and have found some application in jewellery and the watch-case industry. Both ruthenium and osmium have also been employed in the tool industry being used to coat the tips of drills and cutting tools[22].

Ruthenium is hard and its low cost in comparison with rhodium has engendered interest in it as a substitute for rhodium. Although it forms a superficial oxide at elevated temperatures, the oxide film has a conductivity not dissimilar from that of the plated metal, so that it can be used for contacts where higher temperatures than normal prevail, and as coatings on electrodes in chlor-alkali cells.

A nitroso salt solution is one of those commercially available. It contains:

Ruthenium (as ruthenium nitroso chloride)	8 g/l
Sulphuric acid	80 ml/l

The bath is operated at 2–3 A/dm^2 at 45°–75°C.

A ruthenium sulphamate bath containing 20 g/l of the salt and 20 g/l of sulphamic acid gives good deposits at 25°C–40°C and 1–2.5 A/dm^2. The cathode current efficiency is, however, low (below 20 per cent).

Deposits of 2.5 μm of ruthenium are hard, corrosion resistant, and have considerable possibilities for contacts.

Notley[23] has described the deposition of osmium from hexachloro osmate solutions containing also KCl and operated at low pH. A current efficiency of approximately 25% is claimed and it is said that bright deposits up to a thickness of 10μm can be obtained. Deposits which are highly adherent and bright up to a thickness of 50μm have been obtained from osmic acid solutions[24].

7.5. Iridium plating

Iridium coatings are highly resistant to corrosion by concentrated acids. Good deposits up to 10 μm thick can be obtained from iridium bromide solutions at a cathode current efficiency of 65 per cent. Chloride solutions have also been used, but the baths appear to lack stability.

7.6. Deposition of platinum group metals on to active and refractory metal substrates

Palladium and platinum are chemically stable in all but the most severe environments. Thus, electroplated coatings of these metals (at suitable thicknesses to avoid porosity) are extremely useful for protecting refractory metals against corrosion and oxidation at high temperatures. Without such protection, metals such as molybdenum, tantalum, niobium and vanadium would undergo catastrophic oxidation at temperatures well below those at which they are satisfactory from the point of view of structural and mechanical soundness.

Cramer and Schlain[25] have studied the deposition of platinum and palladium onto the refractory metals and titanium in detail. The electrolytes recommended are given below, together with details of the operating conditions:

	Constituents	*Concentration*	*Temp. °C*	*Cathode current density*	*Anode*
Electrolyte for deposition of Pd[26]	Pd (as $PdCl_2$)	50 g/l	50	1.5 A/dm^2	Palladium
	NH_4Cl	20 g/l			
	HCl(conc)	150 ml/l			
Electrolyte for deposition of Pt[27]	Pt(as $Pt(NH_3)_2$	40 g/l	75	1.5 A/dm^2	Platinum
	$(NO_2)_2H_2NSO_2OH$	80 g/l			

If adherent coatings of palladium are to be obtained it is essential to prepare refractory metal substrates by means of a fused salt pretreatment. In this pretreatment the substrate is made the cathode in a fused cyanide electrolyte containing salts of a platinum group metal. According to Cramer and Schlain[25] adherent and coherent deposits of palladium and platinum up to about 60 μm in thickness can be produced on molybdenum, niobium and tungsten following the fused salt treatment.

For the satisfactory deposition of platinum and palladium onto titanium the substrate should either be given an electrolytic pretreatment in a glacial acetic acid-hydrofluoric acid solution or be immersed in an etchant prior to plating. A recommended formulation contains:[27]

H_2PO_4	(85%)	300 ml/l
HF	(48%)	190 ml/l
NH_4OH	(28%)	64 ml/l

The bath is operated at a temperature of 25°C with immersion times of 2–4 minutes.

7.7. Silver plating

7.7.1. CYANIDE PLATING SOLUTIONS

It was as long ago as 1840 that the Elkington brothers first patented silver plating from cyanide solutions, this being the earliest application of electrodeposited metal coatings for decorative purposes. The deposition of silver, although to a very large extent carried out on tableware, musical instruments and the like, has also found important applications in the manufacture of electrical and electronic equipment, contacts and reflectors. Thin silver deposits are also applied to steel and other metals and then treated with sulphide solutions and 'relieved' to produce 'oxidised silver' finishes.

For cutlery and tableware, the basis metal most commonly used is nickel silver (an alloy of copper, nickel and zinc) because of its white colour and its suitability for direct silver plating. Silver is also plated onto brass, an undercoat of nickel being first applied; this improves the colour and wear properties of the final surface, and also the adhesion of the silver to the basis metal. Anodised aluminium can also be coated with silver providing the anodic film thickness is controlled[28].

Satisfactory silver deposits cannot be obtained from solutions of simple silver salts such as silver nitrate, and all commercial silver plating is carried out from a complex cyanide solution. A typical formulation contains:

Silver cyanide	40 g/l
Potassium cyanide	55 g/l
Potassium carbonate	35 g/l

This gives a solution with a free potassium cyanide content of about 35 g/l. Silver cyanide is insoluble in water and must be dissolved in a solution of potassium cyanide to give the double salt $KAg(CN)_2$:

$$AgCN + KCN = KAg(CN)_2$$

The bath can be made up more readily with potassium silver cyanide, which is available as a high purity salt containing 54 per cent of silver.

Occasionally silver chloride is used in the preparation of the bath:

$$AgCl + 2KCN = KAg(CN)_2 + KCl$$

The silver is present in these baths mainly in the form of an anion $Ag(CN)_2^-$.

7.7.1.1. Strike solutions

On many metals silver has a tendency to deposit by immersion from a cyanide solution, which results in poor adhesion. This is particularly the case with brazed or soldered articles, which are commonly silver plated. For this reason, a strike solution is often used prior to plating in the normal plating bath. This enables quick covering of the basis metal to be obtained, after which plating is carried out in the usual bath. Such striking solutions have a relatively low silver ion content and are operated at high current densities. They may contain 3 to 6 g of silver cyanide and 120 to 150 g of sodium cyanide per litre of solution; the precise composition to be used depends on the type of work being plated, and must be determined by practical tests. The strike deposit must be sufficiently heavy to protect the underlying metal during subsequent electro plating or else unplated and thinly covered areas will result, whilst adhesion generally may also be poor. Very often, if a good nickel deposit is applied before silver plating, the need for a strike deposit can be eliminated. A preliminary copper strike from a cyanide copper plating solution has also been found very useful and has received a fair amount of application.

At one time a momentary pre-dip in a weak solution of a mercury salt was used to improve the adhesion of silver deposits on brass and copper, for example. The solution is prepared by dissolving 3 g of mercuric oxide in 1 litre of water to which 18 g of sodium cyanide have been added. This mercury 'quick', as it is called, has the effect of protecting the basis metal from oxidation before immersion in the silver bath; it also prevents the formation of an immersion deposit of silver on the basis metal, as mercury is electro-negative to silver. There is no simple analytical method of controlling the 'quicking' solution because of the interference of the other heavy metals which dissolve in it as it is used; most platers, therefore, rely on the appearance of the mercury deposit as a guide to the working of the 'quick'. Owing to mercury pollution problems and toxic hazards, however, and the propensity of the mercury to produce stress cracking in brass, the use of the mercury dip is tending to be discarded.

7.7.1.2. Free cyanide

The presence of free cyanide (i.e. cyanide in excess of that required to form the $KAg(CN)_2$ complex) and carbonate, within limits, improves the conductivity of the electrolyte and promotes regular anode dissolution. Potassium cyanide is preferred to the sodium salt in silver solutions, despite the lower cost of the latter, as there is a reduced tendency for 'burned' deposits to occur than with the sodium salt. If sodium cyanide is used, the carbonate concentration in particular must be kept low. Again, sodium carbonate, having a relatively small solubility at low temperatures, tends to crystallise out of the solution, especially in cold weather. This causes trouble in plating as the small crystals attach themselves to the cathodes, so that rough or pitted deposits are produced. Excessive carbonate concentrations can also result in a lowering of anode and cathode efficiency. Carbonate can be removed by freezing or by treatment with calcium sulphate or calcium carbide.

7.7.1.3. Addition agents

The effects of various constituents and additions to the silver cyanide solution have been studied. Potassium salts, including the sulphate, carbonate and nitrate, have been found to have a favourable influence on deposition, the corresponding sodium salts being in general much less satisfactory. Chlorides and sulphates markedly increase the hardness of the deposit while other additions produce a brighter deposit (bright plating solutions). A bright solution which produces hard silver deposits contains a brightening agent made by boiling together antimony trioxide, glycerin and sodium hydroxide.

Carbon disulphide is still one of the most widely used brightening agents in silver plating; it has been known since 1874. It can be added in the form of a dispersion made by shaking a small amount of the disulphide with a little of the plating solution and adding it to the rest of the bath. Alternatively, as suggested by Hutchinson(29), it may be added by mixing 10 ml of carbon disulphide with 15 ml of diethyl ether and adding this mixture to the plating solution at the rate of approximately 1 ml/500 litres/day. Such additions of disulphide result in the formation of semi-bright deposits which are easily buffed to full brightness. The main drawbacks to its use are its comparatively short bath life and its unsuitability for plating on to soldered articles owing to the inability of the treated bath to cover the soldered areas.

Egeberg and Promisel (30) expressed the opinion that the brightening effect of carbon disulphide was not due to the presence of this compound itself but to some reaction product of the disulphide with the solution. The effects of a range of possible reaction products, such as urea, thiourea, guanidine and more complex compounds, were therefore examined. It was found that thiourea in the proportion of 35 to 40 g/l gave a brightening effect superior to that obtained with disulphide, the optimum brightening action being obtained within the current density range of 0.5 to 0.8 A/dm^2. The thiourea remained stable in the solution for a considerable period. Active brightening effects were also obtained with urea, potassium thiocyanate and thiosemicarbazide. Particularly good results were obtained with certain substituted dithiocarbamates, quantities of the order of 0.1 ml per litre having a marked brightening action. One such compound recommended has the composition

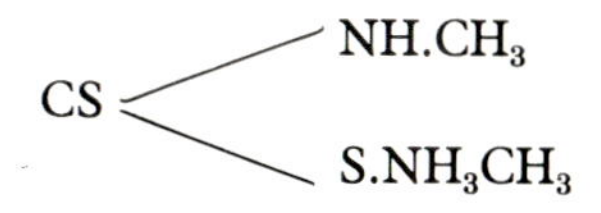

The addition of sodium thiosulphate to silver cyanide solutions has been advocated by Pan (31) as giving better brightening than carbon disulphide. However, only minute concentrations may be employed so that regular small additions must be made. Another difficulty encountered in the use of thiosulphate as a brightening agent is the tendency for the silver anodes to become passive.

Selenium appears to present some advantages as a brightening agent as a replacement for sulphur and may be added in a variety of forms, e.g. as selenites (32) together with condensation products of albumen and fatty acids, and a wide range of divalent selenium compounds(33) which are used in conjunction with small additions of lead and antimony compounds.

7.7.1.4. Plating conditions

Silver cyanide solutions are usually operated at room temperature, although it is desirable not to let the temperature drop too low in cold weather to avoid crystallisation. Excessively high temperatures, on the other hand, cause cyanide decomposition and result in matt deposits although, within reasonable limits, higher temperatures enable higher current densities to be employed; this, in turn, improves the throwing power of the solution. The optimum plating temperature appears to be in the region of 25–30°C.

Current densities are generally low, deposition being usually carried out at an average of 0.5 to 1.5 A/dm^2, although considerably higher current densities are employed in strike solutions. Current densities of more than 1.5 A/dm^2 are seldom used commercially owing to the danger of producing burnt deposits. Moreover, the deposits applied are not usually very thick and this, coupled with the high electrochemical equivalent of silver, means that the time of deposition need not be excessively long even at low current densities.

The silver metal concentration plays an important part in the feasibility of plating at higher current densities, although it is more expensive to work with solutions having greater metal contents because of their increased initial cost and higher drag-out losses. Nicol [34] recommends a concentration of 20.5 g/l of metal, and cites the effects of increased silver content on the maximum current density (Table 7.1). Higher metal and cyanide concentrations and higher operating temperatures and cathode rod agitation permit the use of higher current densities.

TABLE 7.1

Effect of silver content on maximum current density in cyanide silver plating solutions

Silver content (g/l)	approx. maximum current density A/dm^2
20.5	0.6
27.3	0.7
34.2	0.8
41.0	1.0
47.8	1.0

Usually the anode and cathode efficiencies of silver solutions approach 100 per cent, although in strike solutions the cathode efficiency is appreciably lower. The cyanide solution has good throwing power, but covering may be poor on some metals. Apart from the effects of cathode polarisation, the conductivity of the solution has an important influence on the throwing power. Increased potassium carbonate concentration results in improved conductivity of the solution.

Silver plating solutions do not demand a great deal of filtration, but periodic filtering helps to avoid rough or pitted deposits. Metallic impurities such as copper, nickel and zinc can produce 'treeing' of the deposit, whilst high iron concentrations may produce a yellowish colour. Pure silver anodes must be used since even small quantities of contaminants in the metal may result in a black film forming on them; this disintegrates and migrates to the cathode so that rough or pitted deposits may be produced. A high-temperature annealing treatment of silver anodes followed by quenching has been found helpful in improving anode dissolution and in preventing the formation of anode films in some cases. Homogeneity of the structure of the anodes appears to be more important than the actual grain size.

Sometimes, silver anodes are inclined to 'shed' or produce microscopically small particles of silver which may enter the solution and cause rough deposits. High purity annealed silver anodes, freedom from organic contamination of the solution and anode current densities above 5 A/dm^2 reduce the tendency for sludge formation to occur. Careful analytical control of the cyanide and carbonate concentrations must also be carried out.

The anode and cathode areas should be approximately equal in silver plating.

Silver plating is mainly carried out in plastic- or rubber-lined steel tanks, but care must be taken to use a type of lining which is highly alkali resisting and will not cause contamination of the solution. Cathode bars are usually connected to a reciprocating mechanism which imparts an oscillating movement to the articles being plated; this enables higher current densities to be employed, and produces more uniform deposits.

Barrel plating of silver on to transistors has recently been carried out in Japan[35] using a NaCN–KCN ratio of 3.5–7.0:1.

7.7.2. OTHER SILVER PLATING SOLUTIONS

The possibilities of a nitrate-containing bath have been pointed out by Promisel and Wood[36]. It is claimed that with the nitrate bath some of the anode corrosion is due to the nitrate ion, so that a lower range of free cyanide concentration can be used, even in the presence of carbonate, despite the known adverse effect of carbonate on anode corrosion. No carbonate is added to the solution to begin with, but this builds up as the bath is used. The nitrate solution contains:

Silver (as metal)	16 to 18 g/l
Free sodium cyanide	15 to 22 g/l
Sodium carbonate	22 g/l (max.)
Potassium nitrate	113 to 150 g/l

The presence of potassium nitrate is claimed to have a favourable effect on the appearance of the deposit; higher current densities can also be employed with a solution of equal silver concentration than in the absence of potassium nitrate. Sodium nitrate, on the other hand, is very undesirable in silver plating solutions, and if sodium cyanide is used to supply the low concentrations of cyanide required, the potassium nitrate concentration must be kept high relative to the total sodium ion concentration if the maximum advantage of the presence of the nitrate is to be obtained. However, the solution has received little practical application.

Silver solutions free from cyanides have been recommended from time to time, including sulphate, sulphite, ferrocyanide, nitrate, thiosulphate and fluoborate baths, but have not been employed on a large scale. Andreev *et al*[37] have investigated pyrophosphate, cyanide-sodium carbonate, and thiocyanate-cyanide electrolytes and found that the last solution, which contained Ag 100 g/l, KCNS 100 g/l, K_2CO_3 35 g/l operated at 20–25°C and a current density of 1.6–1.8 A/dm^2, gave high-quality smooth deposits with a surface finish of 0.1 μm.

7.7.3. PROPERTIES OF SILVER ELECTRODEPOSITS

Silver has a high reflectivity (up to 99%) for visible light. The reflectivity falls with reducing wave length. The specific electrical conductivity of hard and bright deposits is less than that of pure silver, but the conductivity can be improved by heat treatment at a temperature of 100°–150°C. The contact resistance of deposits obtained from sulphide-containing electrolytes is not very different from that of pure silver. Sulphur causes tarnishing with the formation of very thin sulphide films; the presence of moisture is necessary for this to occur, the process being accelerated by the presence of ammonia. The contact resistance depends on the moisture content of the sulphide film[38]. Certain sulphur compounds are strongly adsorbed on silver.

Annealed silver has a hardness of 27 kp/mm^2, but deposits obtained from additive-free solutions have a hardness of 80–90 kp/mm^2, whilst bright deposits reach 100–130 kp/mm^2. Figures of up to 100 kp/mm^2 can be obtained from special electrolytes. The hardness falls on prolonged storage[39].

7.7.3.1. Electrodepolishing silver by reverse current treatment

Silver can be electropolished by reverse current treatment in a cyanide plating solution. If an alternating current is used with an anode current density of 3 to 20 A/dm^2 (the current must be sufficiently low to prevent gassing), little or no metal is removed. A recommended solution suitable for both plating and polishing contains 32 g/l silver, 40 g/l free potassium cyanide, 50 g/l potassium carbonate and 10 g/l potassium hydroxide. The pH is 11.5[40].

7.7.3.2. Tarnishing of silver

Silver plate tarnishes when exposed to the atmosphere due to the formation of a black sulphide film. The deposition of a hard film of beryllium oxide on to the silver has been proposed as a method for preventing this tarnishing. The beryllium is deposited from a solution of 3.4 g/l of beryllium sulphate, $BeSO_4$, $5H_2O$, adjusted by the addition of ammonia to a pH of 5.83, corresponding with the formation of the basic sulphate $BeSO_4$,$Be(OH)_2$. A current density of 50 μA/cm^2 is used, the time of treatment being fifteen minutes. This treatment, although a very good one, has fallen into disfavour as a result of the hazards of working with beryllium salts.

An invisible tarnish-resistant film may be produced on silver by cathodic treatment in a solution of alkali chromates and carbonates. High current densities are employed and treatment time is of the order of five minutes[41].

An immersion treatment in a solution of 0.5–5 g/l of chromic acid has also been used to produce a tarnish-preventing film on silver. Good chromate passivation of silver for electrical applications without interference with solderability can be obtained by the use of complexing agents in the chromic acid solution[42].

7.7.4. APPLICATIONS OF SILVER PLATING

Increasing use of silver deposits on steel is being made for industrial purposes, especially in the production of electrical contacts. Deposits about 0.7 μm thick are used on plug and socket contacts where a low contact resistance is required. Thicker deposits may be used for contacts on heavy duty switchgear. For cutlery, silver deposits are quoted in terms of the number of grams per dozen articles, and the qualities produced range from 18 g/doz. for cheap cutlery to as high as 30 g/doz. for high-grade hotel plate.

In the case of good silver plate an extra deposit is applied to points of maximum wear, such as the bowls of spoons. This is usually carried out as a separate operation by means of auxiliary anodes.

7.8. Gold plating

Gold plating has been used for very many years because of the attractive appearance and complete resistance of the metal to tarnish and oxidation, even at high temperatures. The high price has also favoured the idea of using it in the form of a thin electrodeposit. Such gold-plated articles as jewellery, 'silver' ware, cosmetic containers, watches and pen nibs are widely used.

Jewellery is thinly plated with pure (24 carat) gold, but special colours can be produced by the deposition of gold alloys containing small amounts of copper, silver, cadmium, etc.

A great impetus to interest in gold plating has arisen since about 1960 because of the requirements of the aerospace and electronics industries which began to demand deposits of greater thickness which would have high corrosion resistance and hardness and the ability to withstand abrasion and wear. Gold has recently been deposited onto satellites[43] and also been used for solar energy purposes[44]. The result of all this interest has been the advent of a large number of new processes, mostly in the form of proprietary baths developed through the research and development work carried out by commercial firms. For a comprehensive account of gold plating technology the reader is referred to the recent work by Reid and Goldie[45] and also to the publications of Page[46–47].

7.8.1. PRETREATMENTS

The treatment of metals prior to gold plating is little different from that employed in the deposition of other metals. Adequate cleaning is essential to obtain adherent deposits. This is especially important in the case of industrial gold plating, as when the deposit is applied to electrical contacts, poor preparation of the basis metal can lead to rapid failure.

Gold can be plated directly on to metals such as brass and copper, but this is to be avoided because of the low protection afforded to such metals by the thin gold deposit, and because there is some tendency for the gold to diffuse into the basis metal and become absorbed; a nickel undercoat is greatly to be preferred.

Clarke and Chakrabarty[48] have demonstrated the importance of the undercoat in relation to the porosity of the gold coating. Variations in both the thickness of the undercoating and the bath conditions influenced the porosity of subsequently applied gold coatings. In many cases, such as in the plating of semi-conductor assemblies, other metals such as solder, brazing alloys, etc. may be present, so that special pretreatment procedures have to be adopted.

An initial gold 'strike' is always advisable before gold plating; this is applied from an alkaline or an acid gold bath (see Section 7.8.5).

7.8.2. GOLD PLATING SOLUTIONS

The principal solutions in use for gold plating may be broadly classified into (a) alkaline, (b) sulphite, (c) neutral and (d) acid electrolytes.

7.8.2.1. Alkaline gold plating solutions

The alkaline gold plating solutions are based on sodium or potassium gold cyanide. The typical alkaline plating solution is prepared by dissolving commercial gold potassium cyanide $KAu(CN)_2$ in water. If this salt is not available in sufficiently pure form, $68\% \pm 0.1\%$ Au, the bath may be prepared by dissolving gold, in the form of the chloride or cyanide, in a solution of potassium cyanide. The potassium salt is preferable to the sodium salt as it is more soluble. Gold chloride can be used for making up the bath; a precipitate of aurous cyanide is first formed on adding the cyanide to the chloride solution, and this dissolves in an excess of potassium cyanide to produce the double aurous cyanide:

$$AuCl_3 + 3KCN = AuCN + C_2N_2 + 3KCl$$
$$AuCN + KCN = KAu(CN)_2$$

The solution produced in this way also contains potassium chloride and the method is little used nowadays.

Some typical cyanide gold solution compositions are shown in Table 7.2.

Solutions containing higher gold concentrations are used for the deposition of heavier coatings, as they can be used at higher current densities. Stainless steel anodes are usually employed rather than gold, so that the bath has to be replenished periodically by the addition of potassium gold cyanide. Carbonate builds up in the solution as a result of the reaction of the cyanide with atmospheric carbon dioxide. A small amount is desirable in new baths to prevent excessive fluctuation of the pH. It also serves to

TABLE 7.2

Typical cyanide gold plating solutions

		A(g/l)	B(g/l)	C(g/l)
Potassium gold cyanide	$KAu(CN)_2$	2.0	6	-10
Potassium cyanide	KCN	15.0	30	12
Potassium carbonate	K_2CO_3		20	—
Sodium phosphate	$Na_2HPO_4,12H_2O$		30	—

improve throwing power and conductivity. The addition of sodium phosphate is sometimes practised as this acts as a brightener, and also helps to stabilise the pH.

The free cyanide content of the bath need not be high but a minimum quantity is necessary to keep the gold anodes clean and bright, to promote adequate anode corrosion, to increase throwing power, and to improve conductivity 2 to 5 g/l is usually sufficient and must be controlled by analysis.

The metal content of the bath is kept as low as is practicable in commercial practice so as to reduce drag-out losses. On the other hand, within limits, increased concentration of metal results in improved cathode efficiency at higher plating rates; the efficiency is generally of the order of 90–100%. The effect of gold concentration, bath agitation and current density on the cathode efficiency is shown in Fig. 7.2[49]. The gold concentrations most commonly used in practice are normally within the range of 1–5 g/l at pH values of 9.0–13.0. Cyanide solutions retard the co-deposition of base metals, which is useful in the jewellery industry, for which 16 and 18 carat processes have been developed.

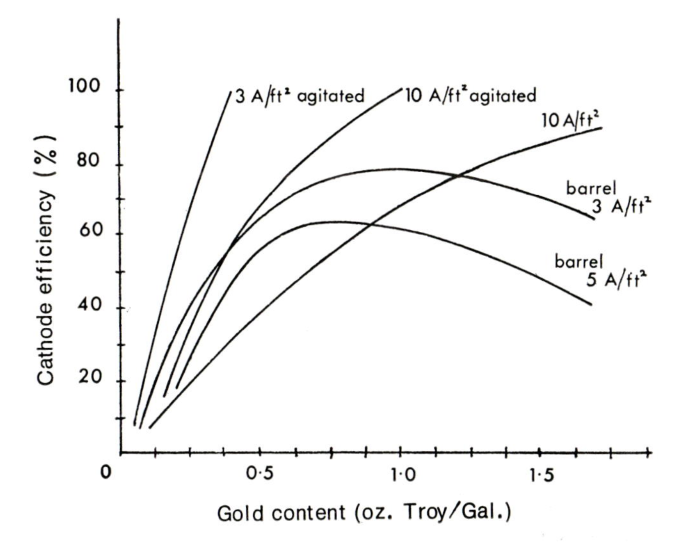

Fig. 7.2. Effect of gold concentration, bath agitation and current density on the cathode efficiency in cyanide gold plating solutions.[49]

Cyanide solutions are operated at fairly high temperatures (50 to 70°C), the current densities employed being 0.1 to 0.5 A/dm^2 with voltages of 2–5V. Higher temperatures enable higher current densities to be employed, but this is of no great consequence unless exceptionally heavy deposits are being applied for individual applications.

The alkaline cyanide baths have excellent throwing power but have disadvantages in the electronics industry since the high alkalinity can result in attack on plastic and ceramic substrates, and especially on printed circuit board laminates.

Brighteners are used in many alkaline gold plating solutions, and may be metallic or organic. Metallic brighteners are codeposited with the gold, and increase the hardness. Antimony has been used as a brightener for gold deposits [50] and also silver. A commercial process has been described in which potassium silver cyanide is employed together with nickel cyanide. Hard, bright gold deposits containing 1% silver are obtained from this bath.

Deposits obtained from alkaline solutions free from brightening additions normally require scratch brushing in order to brighten them. The wet scratch brushing treatment is applied to the gold deposit after only a short time of plating. Deposition is then continued with more intermediate scratch brushing treatments until a bright, adherent deposit of the desired thickness is obtained.

Primary brighteners are codeposited with the gold, thereby reducing its purity and increasing its hardness. Secondary brighteners are not codeposited but alter the crystal structure of the gold, enabling pure gold to be deposited uniformly and smoothly without roughness or nodulation.

7.8.2.2. Sulphite solutions

Gold is an example of a plating process where toxic cyanide solutions have been replaced on a considerable scale with advantage, although the effluent problem is less serious than with other processes owing to the high cost of the solutions and the comparatively small volumes used.

The alkali metal gold sulphite baths, which may also contain phosphates, citrates and chelating agents, have become increasingly widely used in recent years, either to produce pure gold deposits or gold alloys such as those with copper, cadmium, nickel or cobalt. These systems generally have better throwing power and distribution than the cyanide types, as well as better tolerance to impurities; the deposits also possess greatly improved ductility, enabling heavy deposits to be built up without cracking. They are particularly suitable for the deposition of alloys.

The deposition of gold-copper alloys from a sulphite electrolyte was studied by Losi *et al*[50] using a solution consisting of:

Gold (as gold sulphite complex)	0.051 m/l
Copper (as EDTA.$Na^2Cu.xH_2O$)	0.039 m/l
Ethylene diamine	0.051 m/l
EDTA.$Na_2H_2.2H_2O$ (free)	0.055 m/l
Alkaline or ammoniacal sulphite (total)	0.25 m/l

Solutions of this type are operated at temperatures of 30°–40°C and a pH of 3.2–4.0.

The effects of the various constituents were investigated mainly by current density/ electrode potential curves. The composition of the alloys obtained and their main characteristics were only slightly affected by changes in the operating parameters, and the system was found to be readily adaptable to practical operation. Fig. 7.3 shows the variation in the cathode efficiency of the system and the composition of the alloy obtained as a function of the gold concentration in the solution.

Other types of sulphite baths which have appeared on the market include one which is operated at pH 8.5–10.0, which permits alloy formation by the codeposition of base metals, enabling a range of coloured golds to be produced, and another type working at pH 6.0–8.0 capable of depositing low carat gold alloys of high corrosion resistance.

7.8.2.3. Neutral and acid baths

The other important gold solutions are the neutral and the acid baths, both of which contain cyanide complexes. The neutral gold solution also contain phosphates, chelates and brighteners but no free cyanide, as this would not be stable in the operating pH range of 6.0–8.0[51].

A typical neutral bath contains:

Potassium gold cyanide	6 g/l
Monosodium phosphate	15 g/l
Dipotassium phosphate	20 g/l
Potassium nickel cyanide	1.0 g/l

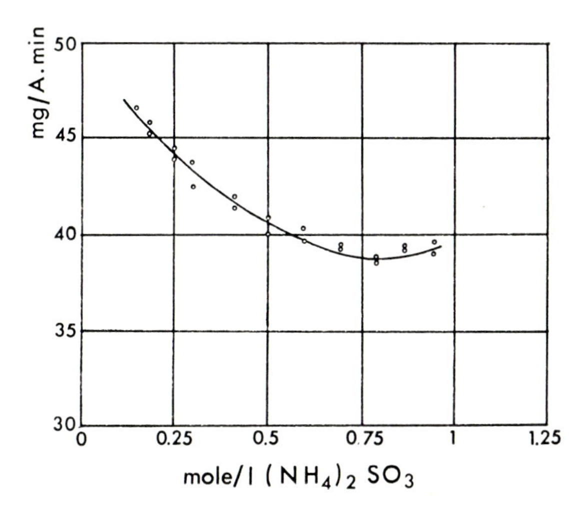

Fig. 7.3. Effect of gold concentration in solution on cathode efficiency and alloy composition in electrodepositing gold alloys from a sulphite solution[50].

The bath is operated at 65°C–75°C and a pH of 6.3–7.5 with a current density of about 0.5 A/dm^2.

The acid baths contain gold potassium cyanide, which is not decomposed within the operating pH range of 3.0–6.0, weak organic acids, phosphates and chelates, the latter serving to increase conductivity, to improve throwing power and to act as inhibitors to certain base metals which may be present. When contaminated, the efficiency of the acid solutions may fall off, but the fact that they are easy to control and that bright deposits of low porosity can be obtained from them makes them very useful in the electronics industry.

An acid bath containing citrates which is suitable for producing smooth, fine-grained deposits on printed circuits contains:

Potassium gold cyanide	25–45 g/l
Dibasic ammonium citrate	25–50 g/l

The solution is used at a pH of 3–6 at current densities of up to 1 A/dm^2.

A phosphoric acid-based solution developed by Atwater and Julich contains:

Potassium gold cyanide	10 g/l
Phosphoric acid (85%)	20 ml/l
Potassium hydroxide	to pH 1.8

The solution is operated at about 35°C.

The operating characteristics of cyanide and sulphite solutions have been compared by Mason and Blair[52].

7.8.3. ANODES

Gold anodes tend to dissolve at too high a rate in cyanide solutions if they are used excessively. They cannot be used satisfactorily in bright solutions since the dissolution of the gold disturbs the balance of the brighteners. Gold anodes also exhibit a tendency to become passive under certain conditions. This is promoted by the presence of metallic impurities such as lead, silver, bismuth or arsenic either in the anodes or in the solutions, or by sodium salts where these are used.

Nowadays insoluble anodes of carbon, platinum, platinised titanium or stainless steel are preferred for convenience and security. Stainless steel anodes are used in bright alkaline and in neutral solutions; carbon is used only in bright acid processes. Platinised titanium anodes are most favoured since their cost is low as compared with platinum, whilst they can be used in cyanide, neutral and acid baths.

7.8.4. IMPURITIES; COLOURED DEPOSITS

Impurities, especially of the metallic type, must be kept out of gold plating baths as they may adversely affect the colour of the deposit. Thus, at a sufficiently high concentration, silver and cadmium turn the deposit green, whilst nickel and zinc whiten it to some extent. The contaminants can enter the solution from the preceding plating stages or as a result of attack on the basis metals. Some metallic materials are complexed by the bath components and are not codeposited. However, silver, cadmium, zinc, lead and copper are codeposited with the gold from both alkaline, cyanide and neutral baths. This makes it possible to produce coloured gold deposits. Such deposits contain only a low percentage of the added elements, but their effect on the colour of the deposit is very marked. For green gold, about 1 g/l silver cyanide may be added, whilst a small proportion of copper cyanide results in the deposition of red gold. White gold is deposited from a solution containing about 4 to 5 g/l of nickel (as cyanide). The presence of lead is deleterious as it results in black, non-adherent deposits. (See also Section 7.8.7. Gold Alloy Deposition.)

Organic contaminants can enter the bath by drag-in from previous baths, from rack coatings and stop-off lacquers, and by degradation of cyanides or organic additives. Their presence results in low cathode current efficiency and in a poor appearance of the deposit, especially in areas of high current density. Contamination by organic substances can usually be dealt with by treatment with activated carbon, and in some cases with hydrogen peroxide.

7.8.5. GOLD STRIKES

Two types of gold strike solution are employed, i.e. the alkaline cyanide and the acid type, the latter being preferred. The strike serves to promote adhesion of the main gold deposit by eliminating the possible formation of immersion films of less noble metals. They also reduce contamination of the gold bath by the drag-in of solutions from previous processes.

In the printed circuit industry the acid strike, operated at pH 3.0–7.0 and at temperatures of 30°–60°C, is generally preferred on plastic-containing basis materials. The metal content of the bath is only about 0.5–4.0 g/l, so that the cathode current efficiency is low and deposition is accompanied by copious gas evolution. The voltage applied is 4–12V; platinised titanium anodes are most suitable.

The alkaline strikes are effected in baths maintained at pH 8.0–13.0 and at tempera- of 40°–60°C, with voltages of 3–8V. Stainless steel anodes are generally used.

7.8.6. IMMERSION GILDING

For the gilding of cheap articles and novelties, an extremely thin coating of gold can be deposited by simple immersion in a gold solution. Brass and copper are the most suitable basis metals for such treatment, but if gilding of other metals is required these may be copper-plated first.

In the old salt-water process of contact gilding the gold solution is contained in a porous pot which is in turn suspended in a copper vessel containing sodium and ammonium chloride solutions heated to about 70°C, in which a zinc plate is immersed. On suspending the articles to be plated in the gold solution and connecting them to the zinc plate with a copper wire, a primary cell is formed and thin, but uniform gold films are deposited without an external source of current. The process is subject to the disadvantage that the colour of the plate tends to become progressively darker in shade as the solution is used, so that it has to be discarded after a time, and the gold recovered by refining. This effect is a result of the build-up of copper in the solution, since 63 grams of copper must dissolve for each 193 gram of gold deposited.

Modern gilding solutions, which work more rapidly and give superior deposits, are now available. A proprietary amine bath is based on the reduction of a gold amine complex, and can deposit gold on brass at the rate of about 10 mg/dm^2/min at 60°C.

Hypophosphite solutions are widely used for the chemical deposition of nickel, and are also suitable for gold.

A typical bath contains:

Potassium gold cyanide	2.5 g/l
Ammonium chloride	60 g/l
Sodium nitrate	50 g/l
Sodium hypophosphite	10 g/l

The solution is maintained at a pH of 7.0–7.5 by means of ammonia additives; the rate of deposition depends on the temperature.

7.8.7. GOLD ALLOY DEPOSITION

Gold alloys are deposited to reduce cost, to produce deposits of improved hardness or wear resistance, and particularly to obtain gold finishes of different colours. The main alloys used in practice are those of gold and silver, palladium, copper, cadmium and zinc, either singly or in combination. Multi-layer deposits can be used, with the final layer containing most gold.

Gold-silver alloys are amongst the most widely employed. It is possible to produce deposits as low in gold as 10–12 carat from cyanide solutions, but 16–18 carat is more common. The deposits are yellowish-green in colour.

Gold-palladium alloys are produced from the sulphite system. White gold coatings are readily obtained, the palladium being deposited preferentially. The stress increases with coating thickness. Gold-palladium-copper deposits are plated on spectacle frames.

Gold-copper alloys have been very fully investigated, the colours obtainable varying from pink to red depending on the gold content, which may correspond to 14 carat or more. The properties of foils deposited from various gold-copper electrolytes have been studied[(53)]. The tensile strength was proportional to the hardness; higher current densities gave deposits of finer grain and greater hardness.

A 16 carat pink gold-copper alloy deposit is produced by a commercial sulphite-type bath operating at pH 6.5–7.0 at 60°C. The current density is about 1.0 A/dm^2, and the deposit is fully bright and highly corrosion resistant.

A process producing ductile bright pink ternary gold-copper-cadmium 17.5–18.0 carat deposits having a hardness of 250 DPN is also available.

Gold-nickel alloys are deposited from acid cyanide baths. The deposits are hard and white and can contain up to 35 per cent nickel. As little as 7 per cent nickel is sufficient to produce a white gold. The deposit is highly stressed, and the cathode current efficiency of the bath is low. The nickel content of the deposit increases with increase in current density and decreases with rise in pH and temperature.

Gold-cadmium alloys which are white to greenish-yellow in colour are readily deposited from sulphite solutions at cathode current densities of about 1.0 A/dm^2 at temperatures of 50–60°C and pH values of 9.0–10.0. The deposits are harder than the gold-silver alloy deposits (about 240 DPN).

Gold-zinc alloys are not as white as gold-cadmium and tend to have inferior resistance to corrosion.

7.8.7.1. Process control

Alloy plating requires close control of plating conditions to maintain the colour and avoid excess costs. Analytical control of the solution is essential whilst the cathode current efficiency of the bath and the deposit thickness must also be checked periodically.

Corrosion resistance can be determined by a simple nitric acid test, or by a salt-spray or perspiration test. The latter is carried out in a 5 per cent lactic acid and sodium chloride solution on wrist-watch cases[54].

Wear resistance on small articles such as watch cases and bands can be determined quite simply by rumbling the articles in a container of sand and measuring the weight loss after a definite period of time.

Many types of automatic gold plating machines are used in the electronics industry to control the deposit thickness, and to plate certain areas selectively to reduce gold consumption.

REFERENCES

1. *Metal Progress*, 1975, **108**, No. 2, 45, 47.
2. M. Antler, *Plating* 1967, **54**, 915.
3. H. Benninghoff, *Oberflache Surface*, 1972, **13**, No. 6, 119-125.
4. F. H. Reid, *Trans. Inst. Metal Finishing* 1956, **33**, 105-140; **36**, 74-81. 1959,
5. A. Korbelak, *Trans. Inst. Metal Finishing* 1964, **42**, 153.
6. R. R. Bertram, *Platinum Metals Review* 1961, **5**, No. 1, 13.
7. J. Hill, *Plating* 1965, **52**, 417-9.
8. F. H. Reid, *Metallurgical Reviews* 1963, **8**, No. 3, 167.
9. E. A. Parker, *Plating* 1955, **42**, 882.
10. D.T.D. Specification 931 (1961).
11. Brit. Pat. 808, 958.
12. French Pat. 1, 174, 957.
13. U.S. Pat. 2,866,740.
14. *Galvano Tecnica*, 1973, **24**, No. 4, 72-77.
15. A. Weser, *Electroplating and Metal Finishing*, 1976, **29**, No. 2, 6-8.
16. W. Keitel and H. E. Zschiegner, *Trans. Electrochem Soc.*, 1953, **59**, 273.
17. '*Galvanotechnik*', 1949, p. 940.

18. N. Hopkin and L. F. Wilson, *Platinum Metals Review* 1960, **4**, No. 2, 56, 57-58.
19. F. H. Reid, *Metal Finishing Journal* 1969, **15**, 124.
20. U.S. Pat. 3,920,526 (1974).
21. R. H. Rhoda, *Trans. Inst. Metal Finishing* 1959, 36,82.
22. Brit. Pat. 393,115 (1971).
23. J. M. Notley, *Trans. Inst. Met. Finishing*, 1972, **50**, 58-62.
24. L. Greenopan, Plating in the Electronics Industry, Third Symposium of Amer. Electroplaters Society, Inc., 1971, Feb. 3-4th.
25. S. D. Cramer and D. Schlain, *Plating* 1969, 56, 516.
26. E. M. Wise and R. F. Vines, U.S. Pat., 2,457, 021.
27. R. Duva and E. C. Rinker, U.S. Pat., 2,984,603.
28. L. Damnikov, *Metal Finishing*, 1972, **70**, No. 11, 44-47.
29. P. Hutchinson, *J. Electrodepositors, Tech. Soc.*, 1950, **25**, 189.
30. B. Egeberg and N. A. Promisel, *Trans. Electrochem. Soc.*, 1938, **74**, 211.
31. L. C. Pan, *Trans. Electrochem. Soc.*, 1931, **59**, 329.
32. Weiner, *Metallwirtschaft* 1943, **22**, 472.
33. B. D. Ostrow, U.S. Pat., 2,77,810 (1957).
34. A. E. Nicol, *J. Electrodep. Tech. Soc.*, 1940, **16**, 15.
35. Jap. Pat. 4,722,341 (1971).
36. N. E. Promisel and D. Wood, *Trans. Electrochem. Soc.*, Sp. Vol. p. 297 (1942).
37. R. P. Andreev, V. I. Molodkin and A. N. Zinkin, *R. Zh. Masch.* 1972, 128, 493.
38. Askiel, A. v. Krusenstjern and G. Veil, *Metall.* 1966, **20**, 592.
39. G. Heilmann, *Metall.* 1957, **11**, 515.
40. U.S. Pat. 2,416,294.
41. H. W. Dettner, *Plating* 1961, **48**, 285.
42. T. Baeyens and J. E. Melse, *Electroplating and Metal Finishing* 1959, **12**, No. 8, 303.
43. J. Ross, *Ind. Finishing*, 1972, **24**, No. 291, 32-33.
44. B. J. Brinkworth, *Gold Bull.*, 1974, **7**, No. 2, 35-38.
45. F. H. Reid and W. Goldie (editors) *Gold Plating Technology*, Electrochemical Publications, Ayr, Scotland, 1973.
46. R. J. Page, *Metal Finishing Journal*, 1973, **19**, No. 225, 274-279, No. 226, 304-308.
47. R. J. Page, *Metal Finishing Journal*, 1974, **20**, No. 228, 4-9, No. 229, 33-46, No. 239, 62-65, No. 231, 87-90, No. 232, 122-125.
48. M. Clarke and A. M. Chakrabarty, *Trans. Inst. Met. Finishing*, 1972, **50**, Pt. 1, 11-15.
49. W. T. Lee, *Corrosion Technology* 1963, April, 59.
50. S. A. Losi, F. L. Zuntini and A. R. Meyer, *Electrodeposition and Surface Treatment* 1972, 1, No. 1, 3-20.
51. D. Gardner Foulke, *Plating* 1963, **50**, No. 1, 39.
52. D. R. Mason and A. Blair, *Trans. Inst. Met. Finishing*, 1977, **55**, 141.
53. J. R. Cady and P. G. Willcox, *Plating* 1971, 60, 3, 139-145.
54. J. Underwood, K. Carvallho and A. McKinlay, *Trans. Inst. Metal Finishing* 1971, **49**, 123.

CHAPTER 8

Copper and brass plating

8.1. Copper plating

Copper is deposited for five main purposes:

(a) As an undercoat for nickel in the plating of metals, such as zinc-base alloy die-castings, to prevent attack on the basis metal by the nickel solution. Copper undercoats are sometimes employed prior to the nickel plating of steel, usually with the object of reducing polishing costs since the copper deposit is more easily smoothed and polished than the harder steel surface.
(b) In the production of bronze coloured finishes on steel and other metals, the parts are first copper plated, after which the copper is coloured by some such treatment as immersion in sulphide solutions.
(c) For stopping-off specific areas on steel components to be subjected to carburising treatments. The copper deposit prevents diffusion of carbon atoms into underlying areas.
(d) In electroforming and in the production of printing plates by the electrotype process.
(e) In the production of printed circuit boards.

Two types of solution have been in use for many years, viz. the alkaline cyanide and the acid solutions. The first type consists essentially of a solution of cuprous and alkali metal cyanides with or without various addition agents. Cyanide solutions have excellent throwing power; they are, however, unsuited in general to the building-up of deposits of substantial thickness. Their main advantage is that they are directly applicable to ferrous metals. The acid solution contains copper sulphate and sulphuric acid, and is used mainly for plating onto those metals which are not attacked by the solution chemically, and especially where substantial thicknesses are required, as in electroforming. It is not possible to plate directly onto steel or onto zinc-base alloys from the acid solution(1).

The copper pyrophosphate bath, which offers a number of advantages, has found considerable use in the production of through-hole printed circuit boards for the electronics industry, whilst the fluoborate bath (see p. 315) is used for electroforming.

Jelinek[2] has recently reviewed the history of copper plating and discussed later developments.

8.1.1. ALKALINE CYANIDE COPPER PLATING SOLUTIONS

The alkaline cyanide copper bath is employed when steel or zinc are the basis metals since the copper is in the form of a complex anion and deposition by immersion does not take place. This type of solution is employed almost exclusively where copper is required as an undercoat prior to nickel plating.

The chief components of the solution are copper cyanide (CuCN) and sodium or potassium cyanide (NaCN or KCN). Together these form several complex salts which dissociate to give rise to complex ionic species. The formation of the complex salts from copper and sodium cyanides can be represented by:

$$CuCN + NaCN \rightleftharpoons NaCu(CN)_2$$
$$NaCu(CN)_2 + NaCN \rightleftharpoons Na_2Cu(CN)_3$$

Although $Na_2Cu(CN)_3$ is the most important complex salt formed from sodium and copper cyanides, further complexing may occur to produce small concentrations of $Na_3Cu(CN)_4$ according to the equation:

$$Na_2Cu(CN)_3 + NaCN \rightleftharpoons Na_3Cu(CN)_4$$

Any cyanide which is not bound up in a complex salt is referred to as 'free' cyanide. Since most of the copper is present in the form of the complex salt $Na_2Cu(CN)_3$, the free cyanide content is calculated by subtracting from the total cyanide content the quantity of cyanide required to complex copper in this form. In solution, the complex salts dissociate to give rise to complex copper cyanide ions such as $[Cu(CN)_3]^{2-}$, $[Cu(CN)_2]^-$, $[Cu(CN)_4]^{3-}$, $[Cu_2(CN)_3]^-$ and $[Cu(CN)_5]^{2-}$ [3, 4]. The relative proportions in which they are formed depend mainly upon the free cyanide concentration. In most cyanide copper plating solutions, $[Cu(CN)_3]^{2-}$ is the most important complex[3, 5]. It is produced by the dissociation of the predominant complex salt, thus:

$$Na_2[Cu(CN)_3] \rightleftharpoons 2Na^+ + [Cu(CN)_3]^{2-}$$

A very important feature of this complex is that its degree of complete dissociation is extremely small:

$$[Cu(CN)_3]^{2-} \rightarrow Cu^+ + 3CN^-$$

$$K_{diss} = \frac{[Cu^+]\,[CN^-]^3}{[Cu(CN)_3{}^{2-}]} = 5.6 \times 10^{-28}$$

Hence, the effective concentration of free cuprous ions in the solution will be very low. For this reason the equilibrium potential of copper in cyanide baths will be much more negative (by up to 1 volt) than in acid sulphate solutions, and copper will not be displaced from cyanide plating baths by iron or zinc. Hence the suitability of such electrolytes for the direct deposition of copper on steel and zinc-base die-castings.

Whilst the deposition mechanism of copper from copper cyanide solutions is not fully understood it is extremely unlikely that deposition occurs directly from simple cuprous ions. This can be readily appreciated if it is realised that the concentration of these ions (at unit molar concentration of CN^- and $Cu(CN)_3{}^{2-}$) is of the order of 10^{-26} g/l. This could not possibly provide sufficient copper ions at the electrode surface to enable copper to be deposited at rates as high as 10 A/dm^2. It is more likely [6] that copper is deposited by adsorption and breakdown of the complex $[Cu(CN)_2]^-$ which forms on partial dissociation of the complex $[Cu(CN)_3]^{2-}$, itself formed by dissociation of the predominant complex $Na_2[Cu(CN)_3]$

$$[Cu(CN)_3]^{2-} \rightarrow [Cu(CN)_2]^- + CN^-$$

It is evident that the free cyanide content of the bath will tend to suppress the decomposition of the complex $[Cu(CN)_2]^-$ by shifting the equilibrium to the left. The decrease in the concentration of free copper ions increases cathode polarisation. The much higher degree of cathode polarisation associated with deposition from complex cyanide solutions than from simple sulphate electrolytes is shown in Fig. 8.1. The

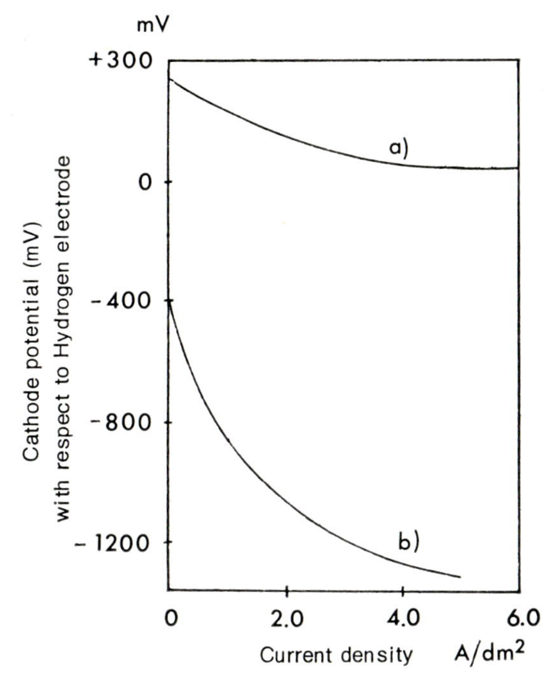

Fig. 8.1. Cathode polarisation curves for copper deposition from (a) acid copper sulphate solution and (b) alkaline copper cyanide electrolyte[5].

high cathode polarisation in conjunction with good electrical conductivity and decreased cathode efficiency with increased current density, impart good throwing power to cyanide solutions. The high cathode polarisation is also responsible for the hard, extremely fine-grained deposits that are obtained from these solutions.

In order to ensure satisfactory working of the bath, an optimum concentration of free cyanide must be maintained. Free cyanide not only increases the electrical conductivity of the bath, but also ensures dissolution of CuCN which forms at the anode and is insoluble in water. In the absence of an adequate concentration of free cyanide in the bath, a film of CuCN would soon form and cover the entire anode surface, thus leading to complete anode passivation. On the other hand, there is a practical upper limit to the free cyanide concentration. If it is too high the deposition complex concentration will be reduced to such an extent that the corresponding reduction in cathode efficiency becomes unacceptable. Under conditions of low cathode efficiency, the rate of hydrogen evolution at the cathode surface will be high and this might lead to blistering of the deposit as a result of hydrogen penetration.

There are three types of copper cyanide solutions in use: the low efficiency bath, the high efficiency cyanide bath and the Rochelle salt bath.

8.1.1.1. Low efficency copper cyanide baths

The low efficiency baths are nowadays principally used for strike plating or where a relatively thin undercoating of copper is required. A typical commercial solution consists of:

Copper cyanide CuCN	25 g/l
Sodium cyanide NaCN	40 g/l
Sodium carbonate Na_2CO_3	10 g/l

Since the formation of the complex salt $Na_2[Cu(CN)_3]$ requires 1.093 g of NaCN for every 1 g of CuCN, this allows for approximately 12.5 g/l free cyanide.

Rolled, extruded, or cast copper anodes are employed. Cast anodes are the least satisfactory owing to the possible presence of occluded oxide particles, which may lead to sludge formation and consequent rough deposits.

The solution is normally operated at about 30–50°C, with cathode current densities of up to 3 A/dm^2. Under these conditions the cathode efficiency will be of the order of 30–50% and the plating speed relatively low.

Though not essential, it is usual to add sodium carbonate to low efficiency copper cyanide baths to increase the electrical conductivity and to buffer the solution. In any case some carbonate will form in the bath due to hydrolysis and oxidation of cyanide, and also by the action of atmospheric carbon dioxide. If the carbonate concentration increases to excessive levels it can be reduced either by discarding a portion of the solution or, better, by cooling, when some of the carbonate will crystallise out.

An alternative way of removing excess carbonate has been recommended by Ross[7]. In this method 0.6 g/l of calcium carbide is added for each 1 g/l of carbonate. The carbide reacts with the water to form calcium hydroxide, which in turn precipitates the sodium carbonate as calcium carbonate. The acetylene which is liberated is removed by the plant exhaust:

$$CaC_2+2H_2O=Ca(OH)_2+C_2H_2$$
$$Ca(OH)_2+Na_2CO_3=2NaOH+CaCO_3.$$

Some free carbon and sulphides are also precipitated from impurities in the carbide, but can be removed by decantation or filtration. The calcium carbide should be added in small amounts sufficient to give the desired free caustic soda content; it is best carried out on part of the solution in a separate tank from which the sludge can be removed by decantation, and washed.

Calcium sulphate in a finely divided form can also be employed for carbonate removal. The amount required is 1.6 times the weight of sodium carbonate to be removed.

8.1.1.2. High efficency copper cyanide baths

The use of more concentrated copper cyanide solutions was advocated by Pan[8] in 1935 and since this time copper baths containing 75 or more g/l of copper cyanide have been increasingly used. Such baths, when operated at temperatures of around 70°C, allow very high current densities to be employed (3–10 A/dm^2). Due to the increased concentration of the deposition complex the cathode efficiency is raised to almost 100%. This has enabled the time for copper plating motor car parts to be reduced. The bath is also used for producing heavy coatings of copper on steel wire for electrical transmission purposes. The throwing power of such baths is lower than that of the low efficiency cyanide baths.

8.1.1.3. Rochelle salt solution

Apart from its beneficial effect of promoting anode dissolution via the formation of tartrate complexes, the addition of Rochelle salt to low cyanide plating solutions also results in considerably increased cathode efficiency. According to Wagner and Beckworth[9] the optimum concentration of Rochelle salt is 21 g/l, the efficiency falling at higher concentration. It is, however, usual to work with larger amounts to make up for loss through drag-out in the course of operation. A typical Rochelle salt cupro-cyanide bath contains:

Copper cyanide	30 g/l
Sodium cyanide	35 g/l
Sodium carbonate	35 g/l
Rochelle salt	45 g/l

The bath is operated at about 65°C, and current densities of up to 6 A/dm^2 may be employed. Sodium carbonate is added to reduce anode polarisation and to stabilise the pH value. A pH value of 12.2 to 12.8 is generally considered to give the best plating conditions; the lower value is difficult to maintain as the solution is poorly buffered, so that it is usual to work within a pH range of 12.5 to 12.8. At too high a pH value, the anode efficiency is considerably reduced. A further advantage of lower pH values is that in the difficult operation of copper plating onto zinc-base die-castings, the tendency to blistering is progressively reduced as the pH decreases[1, 10].

The throwing power of Rochelle salt baths is considerably better than that of the high efficiency solutions. This is largely due to the decrease in cathode efficiency with increase in current density that occurs in the Rochelle bath.

The carbonate content of the Rochelle bath builds up in due course, especially at high working temperatures; when the concentration reaches 100 g/l the solution may either have to be partly discarded or the carbonate crystallised out by cooling, as with normal cyanide baths.

High temperatures favour high efficiencies, but temperatures in excess of 70°C are not practicable owing to the high rate of decomposition of the cyanide; at temperatures above 60°C exhaust equipment is necessary in view of the fumes which are evolved. Under these conditions, current densities as high as 6 A/dm^2 can be successfully employed with cathode efficiencies approaching 70 per cent.

The anode current density must also be carefully controlled, and it is desirable to have an anode area equal to at least twice the cathode area so as to avoid an excessively high anode current density. The free cyanide content must be controlled by regular analysis and kept as low as possible consistent with the maintenance of film-free anodes.

Periodic filtration is helpful in maintaining a smooth and non-porous deposit but, as in the case of practically all cyanide solutions, a degree of roughness accompanied by a tendency towards blister formation develops if an attempt be made to build up thicknesses much above 0.002 cm.

8.1.1.4. Chromium Contaminants

There is a danger of contamination of cyanide copper solutions with hexavalent chromium compounds, particularly where automatic plants are used, the chromium being introduced via the plating racks. The effect of these compounds is to cause a reduction in efficiency with the formation of streaked deposits, as little as 0.3 parts per million in the bath producing noticeable effects. Increased chromium leads to blistering, or even complete suppression of copper plating.

Removal of chromium compounds from cyanide solutions is not easy. One method described by Love[11] which appears to be highly effective, consists in the precipitation of the chromium in the form of a chromium lake by means of an addition of caustic soda-alizarin mixture also containing stannous sulphate to reduce the chromic acid to the trivalent state; this is followed by removal of the lake by filtration through activated charcoal.

Treatment with sodium hydrosulphite, $Na_2S_2O_4.2H_2O$ is another effective method of purification, the chromium being reduced to the trivalent state and precipitated[12].

8.1.1.5. Addition agents to copper cyanide baths

Small concentrations of sodium thiosulphate, sodium bisulphite or sodium sulphite are sometimes added to cyanide copper solutions to improve the brightness of the deposit. These addition agents are said to reduce cathodic polarisation in copper cyanide electrolytes [13, 14] (Fig. 8.2).

It has also been reported[15] that addition of sodium thiosulphate or sodium sulphite to Rochelle salt baths reduces the grain size of the copper deposits. Nowadays, this type of addition agent is not so widely used since decomposition of the additive to sulphide may occur, resulting in the formation of dark and brittle deposits. However, many other brightening agents have been suggested and patented; these include tri-phenylmethane dye[16], cetyl-α-betaine[17], selenium bisdithiocarbamate[18] and other selenium compounds[19], as well as potassium thiocyanate[20, 21], particularly for high concentration cyanide baths.

For a general review of levelling in acid and cyanide plating baths the reader is referred to the short article by Pushpavanam and Shenoi [22]

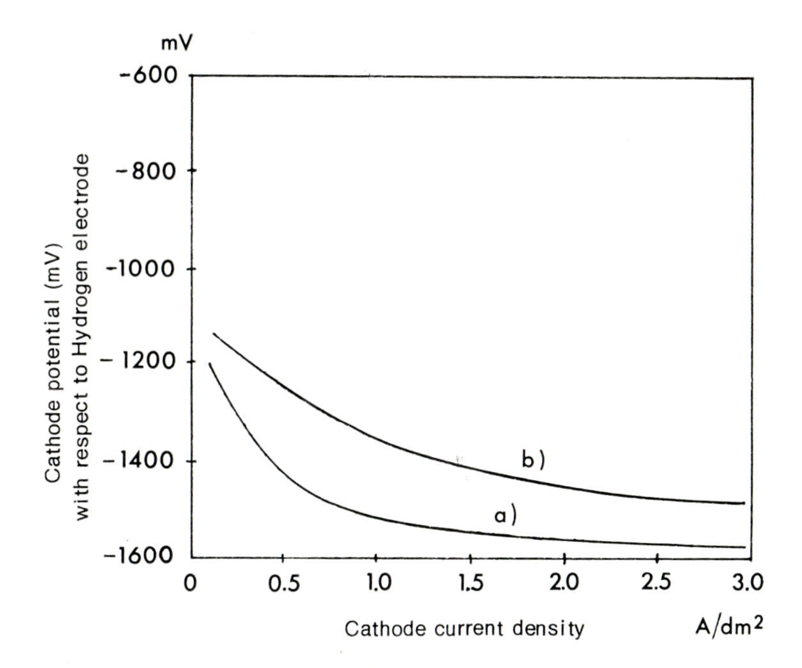

Fig. 8.2. Effect of sodium thiosulphate on the cathode potential/current density curves in cyanide copper solutions[14]

(a) 8g/1 Cu; 8g/1 KCN

(b) 8g/1 Cu; 8g/1 KCN

2g/1 $Na_2S_2O_3$

8.1.1.6. Ultrasonic vibration

The use of ultrasonic vibrations in a copper cyanide bath decreases the cathode overpotential[23], and thereby increases the cathode efficiency for metal deposition[23, 24] (Fig. 8.3) because, under these conditions, less hydrogen gas is evolved. Essentially, the effect of the ultrasonic vibration is to reduce concentration polarisation at both the anode and the cathode, and in this respect it is more effective than conventional agitation. The technique is used only on a limited scale on account of its adverse effect on the throwing power, and the relatively high capital cost of the equipment involved.

8.1.1.7. Periodic reverse current plating

The use of various forms of alternating current in plating processes has been practised for very many years; in particular, alternating current superimposed on direct current has been proposed as a means of increasing the rate of deposition, though this method is still used to only a limited extent. It was first proposed by Jernstedt[25] in 1948. The current is reversed at regular time intervals so that metal is plated on to the work during a portion of the cycle and then partially dissolved anodically during the remainder of the cycle. The overall plating speed is, of course, reduced owing to the anodic dissolution of the electrodeposited metal, but higher cathodic current densities can be employed. The electrodeposits obtained are much smoother and finer grained than those produced by normal de-plating. Although the timing cycle can be varied over a wide range it is usual to use plating cycles of the order of 5 seconds cathodic and 2 seconds anodic. Smoothing or levelling is improved by increasing the length of the anodic period in relation to that of the cathodic period. Fig. 8.4 shows how the smoothness of the deposit varies with the varying times of the

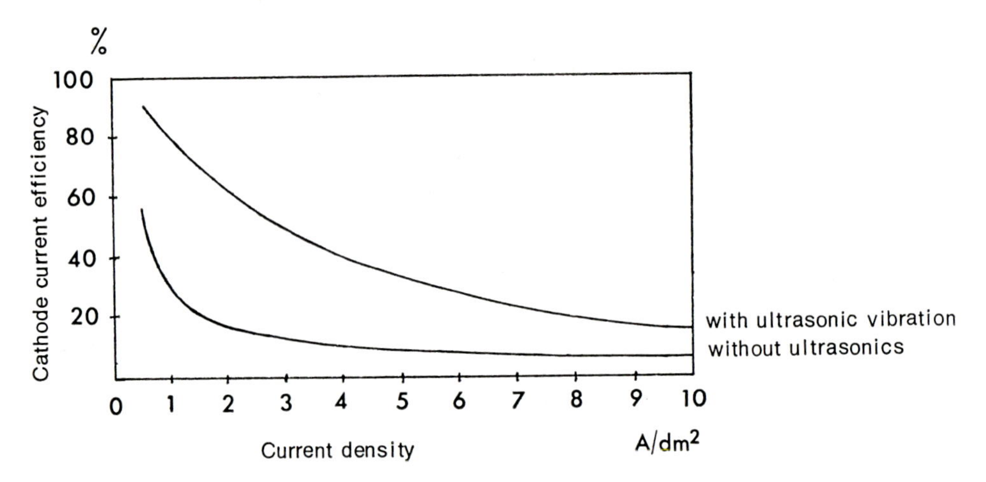

Fig. 8.3. Effect of ultrasonic vibrations on the cathode efficiency in a cyanide copper plating solution at 35°C. Ultrasonic frequency 38 kc/s, acoustical intensity 0.25 W/cm^2.[23].

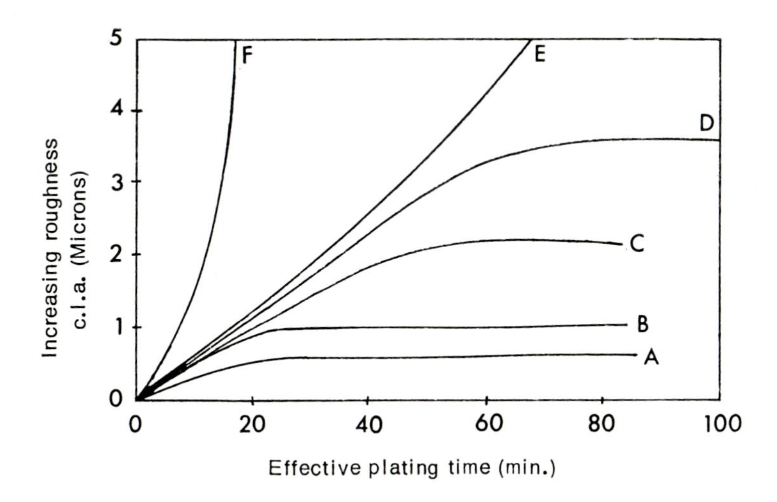

Fig. 8.4. Progressive roughening of copper plated by direct current and a various periodic reverse cycles in cyanide copper solution.[26].

A. Periodic reverse 8.7 s cathodic 6.3 s anodic
B. Periodic reverse 9.4 s cathodic 5.6 s anodic
C. Periodic reverse 10.0 s cathodic 5.0 s anodic
D. Periodic reverse 11.2 s cathodic 3.8 s anodic
E. Periodic reverse 12.5 s cathodic 2.5 s anodic
F. d.c. plating.

anodic and cathodic part cycles. Deposit roughness is indicated by reference to the centre line average (c.l.a) in microns.

In a recent study Pinkerton and Smith[27] showed that periodic reverse of current enabled alkaline cyanide plating solution to be used for printed circuit board applications. The epoxy glass and plastic boards were first coated with electroless copper.

8.1.1.8. Advantages of cyanide electrolytes for copper plating

The cyanide bath is still the most widely used solution for the deposition of copper. Its main advantages are:

(1) The bath is simple to operate and relatively cheap.
(2) Both ferrous and non-ferrous metals may be copper plated in the bath.
(3) Higher throwing power is obtained than with other copper plating solutions; hence cyanide electrolytes are more suitable for plating components of complex shape.
(4) The weight of copper deposited per ampere hour is often high (depending on the cathode efficiency) since deposition occurs by discharge of monovalent cuprous ions.

8.1.2. ACID COPPER PLATING SOLUTIONS

Acid copper plating is carried out from a solution consisting essentially of copper sulphate and sulphuric acid. Copper deposition in acid sulphate baths can be effected at high current densities so that the electrolyte is suitable for producing thick deposits. The main disadvantages of these baths are their low throwing power and their inability to deposit copper directly on to iron and steel. On immersing articles made of steel, iron displaces copper from solution according to the equation:

$$CuSO_4 + Fe \rightarrow FeSO_4 + Cu$$

The precipitated copper does not adhere to the substrate. For this reason, a thin layer of copper must first be applied from a cyanide bath. The article can then be transferred to the acid bath where a coating of any required thickness can be satisfactorily deposited. Pretreatment processes have been described to enable copper to be plated directly onto steel from acid solutions. In one method, an immersion copper deposit is first applied in the presence of a strong inhibitor, acetyl thiourea[(28)]; it is claimed that the copper deposit is as adherent as one obtained from a cyanide solution, whilst the use of toxic cyanides is avoided.

Much more careful cleaning of the basis metal is required before acid copper plating than is the case with cyanide solutions since, unlike the latter, the sulphate bath does not possess detergent properties. The usual range of composition of this type of solution is:

Copper sulphate ($CuSO_4 5H_2O$)	160–220 g/l
Sulphuric acid	50– 75 g/l

A high concentration of metal is desirable. The presence of sulphuric acid in the solution lowers the solubility of the copper sulphate, however, so that the concentration must not be so high that crystallisation takes place. The effect of sulphuric acid on the solubility of copper sulphate in water at 25°C is shown in Fig. 8.5.

The sulphuric acid fulfils the following functions:

(1) It prevents hydrolysis of the copper sulphate during electrolysis which would subsequently result in the formation of basic copper salts;

(2) It increases the conductivity of the bath, thereby (a) reducing the cost of the electric power supplied, and (b) reducing the tendency for rough and treed deposit formation to occur at high current densities.

(3) It may prevent the formation of cuprous oxide on the cathode, which reduces cathode efficiency[(29)].

Nominally the cathode reaction consists of the direct reduction of simple cupric ions:

$$Cu^{2+} + 2e \rightarrow Cu$$

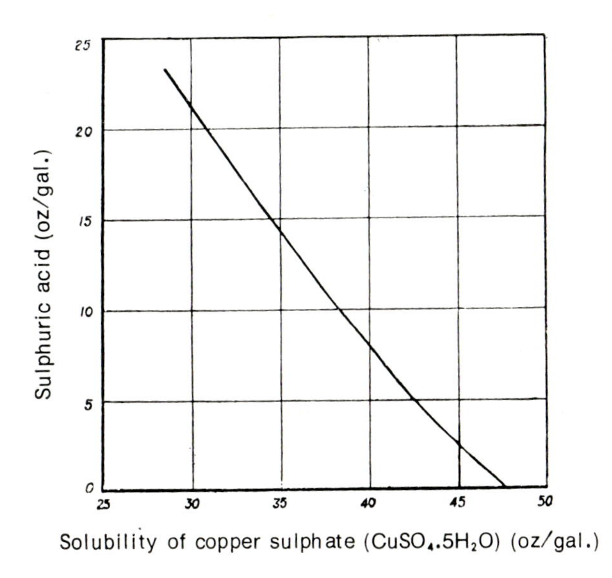

Fig. 8.5. Effect of sulphuric acid on solubility of copper sulphate at 25°C.

but kinetic studies[30] have shown that cupric ions are first reduced to intermediate cuprous ions:

$$Cu^{2+} + e \xrightarrow{slow} Cu^{+}$$

These cuprous ions are transient and are reduced as soon as they form by the faster reaction:

$$Cu^{+} + e \xrightarrow{fast} Cu$$

The reverse overall process occurs at the anode, where copper ionises thus:

$$Cu - 2e \rightarrow Cu^{2+}$$

Both cathodic and anodic current efficiencies are close to 100% under normal operating conditions, so that as much copper goes into solution at the anode as is plated out at the cathode, leaving the copper sulphate and sulphuric acid concentrations constant. In practice, however, the copper sulphate content increases and the sulphuric acid decreases. This is due to the fact that copper ions can exist in these solutions in both oxidation states (Cu^{+} and Cu^{2+}), and although the bulk of copper ions formed at the anode is divalent, a small proportion of monovalent copper ions is also formed:

$$Cu - e \rightarrow Cu^{+}$$

The monovalent ions, of course, require only one half of the electrical energy to deposit as compared with divalent Cu^{2+}. If the Cu^{+} ions were to reach the cathode and be plated out as such, then there would be no change in the copper sulphate content of the bath. However, the Cu^{+} ions are oxidised in the solution to Cu^{2+} ions:

$$Cu_2SO_4+H_2SO_4+\tfrac{1}{2}O_2 \rightarrow 2CuSO_4+H_2O$$
$$\text{or } 2Cu^{+}+2H^{+}+\tfrac{1}{2}O_2 \rightarrow 2Cu^{2+}+H_2O$$

In addition, cupric ions are also produced by the disproportionation reaction:

$$Cu_2SO_4 \rightarrow CuSO_4+Cu$$
$$\text{or } 2Cu^{+} \rightarrow Cu^{2+}+Cu$$

The finely divided metallic copper thus formed settles at the bottom of the vat and becomes part of the anode sludge.

The overall result of the oxidation of Cu^{+} ions is an increase in the copper concentration of the bath, since electrically the removal of one divalent ion is equivalent to the removal of two monovalent ions. The excess copper which builds up in the bath is removed either (a) by rejecting a proportion of the bath and replacing it with dilute sulphuric acid solution; or (b) by replacing a proportion of the copper anodes by lead anodes. The anode reaction on lead anodes results in the formation of sulphuric acid and oxygen:

$$SO_4^{2-}+H_2O-2e \rightarrow H_2SO_4+\tfrac{1}{2}O_2$$

so that a reduced proportion of the anode current is available for the dissolution of copper.

Because cathode polarisation with increase in current density, associated with copper deposition from acid sulphate baths is low (Fig. 8.1) the throwing power is low—much lower, in fact, than the throwing power of cyanide solutions. The throwing power of the copper sulphate bath is improved by increasing the acid concentration and reduced by increasing the copper sulphate content or raising the temperature.

Normally, acid copper baths have a high tolerance to impurities, but, if present in large enough quantities, arsenic, antimony and iron can cause rough deposits.

8.1.2.1. Operating conditions

Acid copper solutions are generally operated in hard rubber- or plastic-lined tanks. Current densities of 1 to 6 A/dm^2 are commonly employed, although under special conditions higher current densities are practicable. Suitable conditions for high current density plating can be arranged in automatic plants, where an adequate degree of movement of the cathode relative to the electrolyte can be obtained.

Acid copper solutions are usually operated at room temperature, although sometimes temperatures as high as 45°C are employed. Elevating the temperature results in a softer copper deposit, as does also an increase in the metal content or a decrease in the sulphuric acid concentration of the solution. Air agitation increases the current density which can be successfully employed but, as indicated earlier, introduction of oxygen into the solution causes a gradual increase of copper sulphate content.

8.1.2.2. Anodes

A common problem in acid copper baths is the formation during electrolysis of an anode sludge consisting largely of minute particles of metallic copper. These settle on the cathode when stirred up, and can cause rough or nodular deposits. These particles are formed in the anode area through disproportionation of cuprous ions, as already explained, and by anode disintegration.

Although high purity copper anodes have been recommended in the past, anodes which contain 0.02–0.03% of phosphorus are now often used. This concentration of phosphorus is in excess of that required to deoxidise the copper [31, 32]. Under normal polarisation conditions, a black adherent anode film is formed which encourages uniform anode dissolution and reduces particle formation due to anodic disintegration. Because the formation of anode particles is avoided, the deposit produced from phosphorised anodes is relatively free from nodule and tree growths and has a finer grain size.

8.1.2.3. Addition agents

Since dull deposits are obtained from the acid copper bath, much effort has been devoted to investigating and developing brighteners. Bacquias [33] has recently reviewed bright acid copper plating solutions and the influence of additives on levelling. Gelatin has been used[34] since 1915 as an addition agent for acid copper baths. When added in quantities up to 0.2 g/l it produces smooth finer grained deposits. Thiourea[35] is also used for this purpose, but above a certain concentration the deposits produced become brittle so that it is usual to include a further addition agent to reduce the tendency to embrittlement. It has been shown that thiourea forms sparingly-soluble cuprous complexes with copper, which have a strongly inhibiting action at the cathode, as a result of periodic supersaturation. The structure of the copper deposit is also independent of the orientation of the surface crystallites at the substrate[36]. The secondary agents used with thiourea include glycerol[37] which produces a relatively soft copper deposit, the sodium salt of 2,6-naphthalene disulphonic acid[38] and molasses[39].

Derivatives of thiourea may also be employed; acetyl thiourea and allyl thiourea are both used, together with secondary agents such as glycol, ethylene glycol, pyridine and glycerol. More comprehensive tests of addition agents for acid sulphate baths are given by Pinner[40] and Brimi and Luck[41].

Few studies of a fundamental nature have been directed towards obtaining an understanding of the mechanism of levelling and brightening in the acid copper sulphate bath. There is no single simple explanation that can account for the action of addition agents, and each individual agent requires separate study and consideration. In most cases adsorption of the addition agent or the formation of complex ions with the addition agent are involved.

Shreir and Prall[42] have investigated the action of benzotriazole in acid copper sulphate solutions. The results obtained from this study suggest that benzotriazole reacts at the metal/solution interface with the transient cuprous ion (formed as an intermediate during the discharge of cupric ions) to produce the insoluble complex Cu(1)BTA. The adsorption and co-deposition of this complex with copper results in a brightening of the deposit and also a reduction in its rate of corrosion in acid solution. In effect, benzotriazole incorporated in the deposit acts as a built-in corrosion inhibitor.

The conditions under which dendritic or powdery growths appear in acid copper solutions and the influence of thiourea and benzotriazole additives in agitated, concentrated solutions have been studied[43]. It was found that though they affect the crystal structure, they have little effect on the onset of dendritic structures.

8.1.2.4. Advantages of the acid copper sulphate bath

The main advantages of the acid sulphate bath are:

(1) Sound, heavy deposits can be produced at high plating rates; this makes the bath suitable for electroforming applications.
(2) The solutions are more stable than those based on cyanide; they are also easier to operate and their composition is less critical.
(3) The deposits are levelled more easily than those from other copper plating solutions.
(4) Effluent problems and health hazards are less serious.
(5) Compared to the pyrophophate solution the sulphate bath is more consistent, has higher plating speeds, is more stable and gives rise to less pollution[44].

The main limitations on the acid sulphate bath are its poor throwing power and inability to plate directly on to iron and steel.

8.1.3. PYROPHOSPHATE COPPER PLATING BATHS

Solutions based on pyrophosphates are used for building-up deposits on zinc-base die-castings prior to bright nickel and chromium plating. As with the copper sulphate bath, it is first necessary to deposit a thin strike of copper from a cyanide bath. The pyrophosphate bath is very suitable for electroforming, and other applications include the production of paint spray masks and moulds for making PVC toys. The bath has also found much use for plating printed circuit boards[45]. The good throwing power of the bath results in high 'hole-to-land' ratios with plated-through holes[46]. Mohler[47] has reviewed the use of copper pyrophosphate baths for printed circuit through hole plating.

Basically pyrophosphate solutions contain copper pyrophosphate, $Cu_2P_2O_7$, and alkali metal pyrophosphate, $K_4P_2O_7$. In solution the copper is present predominantly as the 1–2 anion complex $Cu(P_2O_7)_2^{6-}$ which has a pyrophosphate : copper ratio of 5.48:1. As with cyanide solutions, the complexant (pyrophosphate) must be present in excess. This promotes efficient anode dissolution, prevents insolubles from forming and ensures that complexing is complete. Potassium pyrophosphate is used in preference to the corresponding sodium salt owing to its higher solubility and better electrical conductivity. The weight ratio of pyrophosphate to copper is kept in the range 7–7.5:1. Within this range, sufficient pyrophosphate is present to form the complex and to provide the necessary excess. A lower ratio than 7:1 will render anode dissolution much less efficient so that the bath will become depleted in copper ions, but a higher phosphate-copper ratio is not detrimental.

In addition to the two basic constituents, one or more other compounds are normally added to the bath in order to improve the properties of the deposit and to increase the rate of deposition. These include ammonia, nitrates, aliphatic acids and organic addition agents. The presence of ammonia and some organic compounds such as citrates and oxalates improves the brightness of the deposit; citrates and oxalates also buffer the solution to some extent.

A typical plating bath consists of

Copper pyrophosphate $Cu_2P_2O_7 \cdot 3H_2O$	110 g/l
Potassium pyrophosphate $K_4P_2O_7 \cdot 10\ H_2O$	400 g/l
Ammonia (0.880)	3 g/l
Citric acid	10 g/l

Operating conditions:

pH	8.6–8.9
temperature	50°C
Cathode current density	up to 5 A/dm^2

Pyrophosphate solutions need to be vigorously agitated, otherwise a brownish non-adherent film forms which decreases the operating current density. Mechanical agitation may be used, but air is generally preferred owing to the greater simplicity of operation. Pyrophosphates, unlike cyanides, undergo no chemical decomposition and carbonates are not formed. The cathode efficiency is normally close to 100%.

Fine-grained, strong and reasonably ductile deposits are obtained from pyrophosphate solutions. The ductility of the deposit is, however, adversely affected if the ammonia concentration and pH are allowed to increase excessively[(48)]. The effect of operating conditions on the ductility is shown schematically in Fig. 8.6.

The pyrophosphate solutions possess levelling properties and have a throwing power approaching that of the cyanide bath. The effect of bath variables on the throwing power and cathode current efficiency is summarised in Table 8.1[(49)].

8.1.4. FLUOBORATE COPPER PLATING BATH

Copper fluoborate is highly soluble; solutions can be prepared which contain almost twice the amount of copper possible in acid sulphate solutions. Owing to the extremely high current densities which can be employed and the high cathode efficiencies (nor-

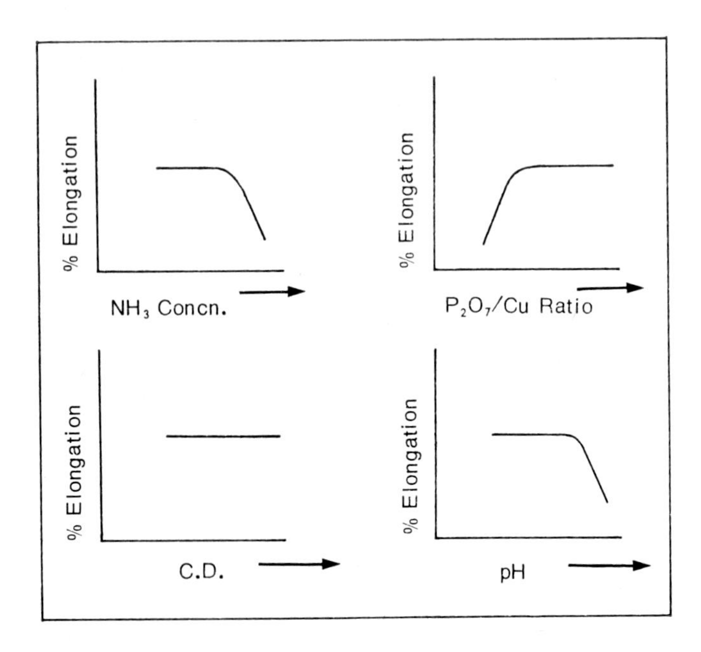

Fig. 8.6. Schematic representation of the effect of plating variables on the ductility of pyrophosphate copper deposits[48].

mally close to 100%) the bath is used for electroforming and in the printing industry, where rapid deposition of thick deposits is required.

A typical 'high' concentration fluoborate bath in use in the U.K. contains:

Copper fluoborate $Cu(BF_4)_2$	430 g/l
Free fluoboric acid HBF_4	30 g/l
Boric acid H_2BO_3(buffer)	30 g/l
pH in the range	0.3–0.6

TABLE 8.1.

Effect of bath variables on the cathode current efficiency and throwing power of pyrophosphate copper solutions[49]

Variable	Effect of increasing variable on	
	Throwing power	Efficiency
Pyrophosphate/copper ratio	Increase	Slight increase
Copper metal	Increase	No change
Ammonia	Decrease	No change
pH	Decrease	Slight decrease
Current density	Decrease	Decrease
Temperature	Decrease	Slight increase
Agitation	Increase	No change

The permissible cathode current density is determined by the temperature of the bath and the degree of agitation. Temperatures normally used range from 25 to 50°C. With vigorous agitation current densities of up to 45 A/dm^2 can be used at the higher temperatures.

The fluoborate bath produces fine-grained, smooth deposits, but the presence of lead in the solution, even in small concentrations, will cause brittleness. If lead is inadvertently introduced into the bath it can be removed by addition of sulphuric acid to precipitate it as sulphate.

8.1.5. ELECTROTYPING AND ELECTROFORMING

Copper deposition is widely used in the preparation of electroforms for various purposes, especially for printing. Rotary printing machines use electrotypes on a large scale. A mould is first produced from PVC by pressing the printing block into the plastic. The mould is then rendered conducting by spray silvering, which involves spraying with two solutions, one containing a silver salt and the other the reducing agent, from a twin spray gun. The silvered mould is first copper plated in an acid copper sulphate bath. A low current density is necessary as the thin silver film has limited current-carrying capacity. When a sufficiently thick film has been produced, the work is transferred to a second 'building-up' bath. Instead of acid copper sulphate solutions, fluoborate electrolytes can be employed for rapidly increasing the thickness of the deposit.

Electroforming is one of the main applications of copper plating; it is useful for the production of parts with complex shapes not easily fabricated by other methods. The techniques used in electroforming involve the preparation of a mandrel made of stainless steel, perspex or other plastic. Mandrels are polished to a high degree and very carefully degreased. In the case of plastic mandrels a conducting film is first deposited before electrodepositing the copper onto them.

Thin copper sheet is produced on a commercial scale[50] by a continuous process. The process is, however, rather difficult to operate, particularly from the point of view of maintaining a uniformly thick sheet.

8.2. Brass plating

The electro-deposition of brass is today an important industrial process. Brass is probably the earliest commercially-plated alloy; according to Faust[51] electroplating of brass was accomplished as early as 1841. The commonest brass deposit is the alloy with a 70:30 copper:zinc ratio; this brass has a distinctive yellow colour which is a useful criterion for judging whether the correct composition is being maintained. Brass plating is applied for decorative purposes, its attractive appearance making it useful for coating steel fittings and similar articles for service under indoor conditions. The protection afforded by a brass deposit against corrosion is not high, so that it is unsuitable for outdoor exposure.

An important application of brass plating is in the bonding of rubber to steel; an extremely strong bond is obtained when rubber is vulcanised on to the brass-plated metal. The bond is stronger than that obtainable by any other method and is, in fact, greater than the strength of the rubber itself. Bonded rubber of this type finds extensive use in engineering for flexible and anti-vibration mountings, couplings, etc. and for bonding the wire in steel-reinforced automobile tyres. Some motor car bumpers have, since 1957, contained electro deposited white brass between copper and nickel layers. It is claimed[(52)] that this gives increased durability.

8.2.1. CYANIDE BRASS PLATING SOLUTIONS

The standard type of brass plating bath contains copper and zinc in the form of cyanides, the relative concentrations of the two metals being approximately the same in the solution as in the alloy which it is required to deposit. A typical brass plating bath of this kind is as follows[(53)]:

Copper cyanide CuCN	30 g/l
Zinc cyanide $Zn(CN)_2$	12 g/l
Total cyanide	45.0 g/l

This bath is operated at a current density of up to about 1 A/dm^2. Current densities in excess of this cause a marked reduction in cathode efficiency. The pH used lies between 10.3 and 11.0. In general, increasing the pH seems to increase the percentage of zinc in the deposit, but lower pH values improve the current efficiency.

The pH may be controlled by means of caustic soda and sodium bicarbonate. The addition of hydroxide raises the pH, whilst the bicarbonate lowers it. Sodium carbonate is also often added to buffer the solution and maintain its pH.

Brass anodes containing about 70 per cent of copper and 30 per cent of zinc are employed. The anode and cathode efficiencies are relatively low, viz. about 75 per cent. Anode corrosion is good, whilst the throwing power and conductivity of the solution are also reasonably satisfactory. As in the case of cyanide copper baths, the free cyanide is the excess of cyanide over that bound up with the metals in the complexes $Na_2Cu(CN)_3$ and $Na_2Zn(CN)_4$. A certain amount of free cyanide (7 to 10 g/l) is essential in the solution to promote anode corrosion, but this should be kept low, since otherwise the cathode efficiency will be adversely affected. On the other hand, if the cyanide is too low, poor adhesion and blistered deposits may result. Plating defects in brass plating may also arise from many other causes e.g. traces of electrolyte left in pores in cast parts, and dips and passivation etc.[(54)]

8.2.1.1. Temperature

The temperatures used generally in brass-plating practice are of the order of 25 to 35°C; raising the temperature tends to increase the cathode efficiency and permits higher current densities to be employed. Excessively high temperatures result in cyanide decomposition and rapid build-up of carbonate in the solution.

8.2.1.2. Addition agents

Ammonia is commonly added to brass solutions to improve the colour of the deposit, although ammonia and ammonium salts are normally liberated in the bath during decomposition of the cyanide. It is, however, difficult to maintain an adequate concentration in the usual type of bath without periodic additions of ammonia, especially at high temperatures. The ammonia prevents any change in the proportion of zinc in the deposit, which tends to decrease when the current density and temperature are raised in the absence of sufficient ammonia.

Brighteners that have been used in brass plating baths include [41] up to 0.01 g/l arsenic trioxide, 0.1 g/l gelatin, 2 g/l sodium nickel cyanide and 1 g/l polyvinyl alcohol.

8.2.1.3. Rapid plating baths

Attempts to improve the speed of brass plating baths by raising efficiency and operating current density have been directed chiefly at increasing bath concentration, operating at high pH values and keeping the concentration of free cyanide low. The effect of free cyanide content on the cathode current efficiency in a concentrated brass plating solution[55] is shown in Fig. 8.7. Increasing the free cyanide content reduces the cathode current efficiency and hence plating speed.

The technique of rapid brass plating was studied during the Second World War for the purpose of applying 70:30 brass deposits to steel shell cases[56]. The solution ultimately developed for this purpose consisted of:

Zinc cyanide $Zn(CN)$	54 g/l
Copper cyanide $CuCN_2$	21 g/l
Sodium carbonate	20 g/l
Free cyanide	4–8 g/l
pH (Tropaeolin O)	12.6–12.8

This solution contains 30 g/l of zinc and 15 g/l of copper, which is approximately in inverse ratio to the composition of the deposit. It is thus seen that in this respect it differs fundamentally from the kind of solution hitherto used for the deposition of 70:30 brass. As has been indicated, the increased cathode efficiency at high current densities results in an increased copper content of the deposit, and this can be countered by the use of addition agents, especially ammonia. In the above solution, however, the same result is obtained by reducing the concentration of copper ions relative to that of zinc ions.

This bath has been used on a large scale and has given highly satisfactory results. The cathode efficiency is 55 to 60 per cent at a current density of 2 A/dm^2 and a temperature of 55 to 60°C. Under these conditions a deposit of 0.0008 μm can be applied in about twenty minutes. At higher temperatures the copper content increases and the deposit darkens in colour. No ammonia is added to the bath since the ammonia resulting from the decomposition of the cyanide is adequate for maintaining the conditions for good deposition.

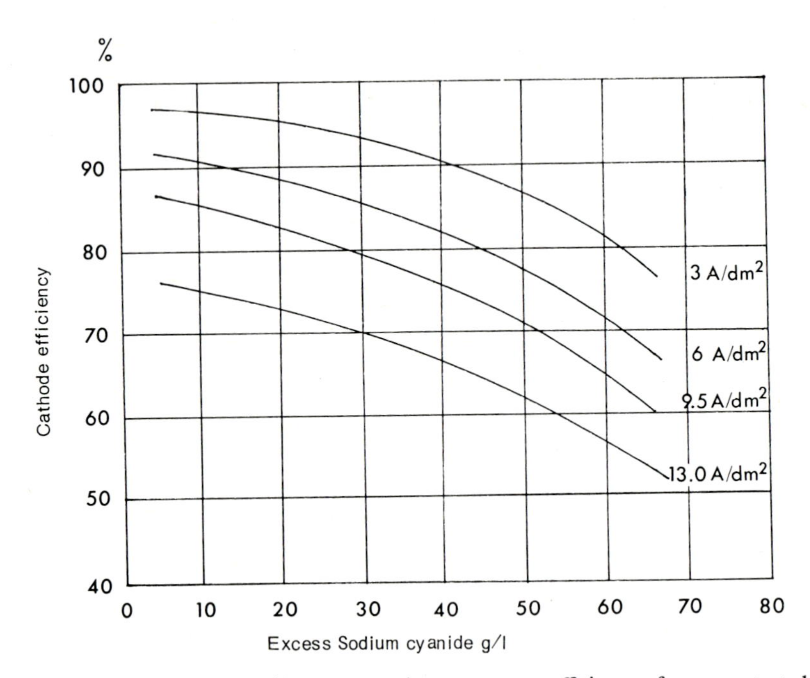

Fig. 8.7. Effect of free cyanide concentration on current efficiency of a concentrated (rapid plating) brass plating solution containing copper cyanide 90 g/l and sodium hydroxide 60 g/l[55].

Brass anodes tend to polarise at relatively low current densities (around 0.5 A/dm^2). Below this value they operate at about 100% efficiency, and since the efficiency of deposition is less than 100%, metal concentration of the bath builds up with a corresponding depletion of free cyanide and the formation of sodium hydroxide. This makes the balance of the solution difficult to maintain, since polarisation of the anodes reduces their efficiency very greatly.

Another method of facilitating brass deposition at high current densities, which also avoids the polarisation of the brass anodes which is liable to occur and which tends to throw the bath out of balance, is to use a solution employing high concentrations of sodium and copper cyanides together with sodium hydroxide in appreciable amounts to give a high solution conductivity, and a very low zinc ion concentration[55]. Typical operating conditions are:

NaCN (total)	100–140 g/l
Copper cyanide	95–105 g/l
Sodium hydroxide	50– 75 g/l
Zinc oxide	30– 95 g–l
Free NaCN	35–190 g/l
Cu:Zn ratio	10 to 30:1
Temperature	75–95°C
Cathode c.d.	2.5–15 A/dm^2

Anodes of the same composition as the deposit (75% copper and 25% zinc) are used and the deposit contains 75 to 83 per cent copper. This bath is said to give good service for the plating of steel strip with brass at current densities of up to 12 A/dm². The high copper content is essential for high efficiency, the colour and composition being maintained by means of the zinc content of the bath.

8.2.1.4. Dichromate dip

A dip in a 5 per cent solution of sodium dichromate after plating is useful in preventing staining of the deposit and especially in preventing the formation of dark-coloured spots due to traces of residual cyanide. This dip is not desirable if rubber bonding is to be carried out onto the brass. In this case the brass deposit should be washed and dried and the rubber bonding carried out with the least possible delay.

8.2.2. PYROPHOSPHATE BRASS PLATING SOLUTIONS

According to Shree and Rama Char [57, 58] brass can be deposited from a pyrophosphate bath having the following composition:

Zinc	120 g/l
Copper	10 g/l
Pyrophosphate	80 g/l

Cathode current densities of up to 4 A/dm² can be employed at a temperature of 60°C and the deposits obtained contain 50–90% copper. The advantages over cyanide brass solutions include higher cathode efficiencies, simplicity of control, fewer effluent problems and deposits of finer grain size. The bath is, however, relatively expensive and comparatively little used.

REFERENCES

1. P. Badet, *Product Finishing* 1969, **22**, No. 8, 33.
2. M. Jelinek, *Surfaces*, 1972, **11**, No. 70, 55-60.
3. H. P. Rothbaum, *J. Electrochem. Soc.* 1957, **104**, 682.
4. R. Penneman and L. H. Jones, *J. Chem. Phys.* 1956, **24**, 293.
5. G. Gabrielson, *Metal Finishing* 1954, 52, 60.
6. E. Raub and K. Müller, 'Fundamentals of Metal Deposition', Elsevier Pub. Co., Amsterdam, London, New York, 1967.
7. H. F. Ross, *Metal Finishing* 1949, **47**, 47.
8. L. C. Parn, *Trans. Electrochem. Soc.*, 1935, **68**, 471.
9. R. M. Wagner and M. M. Beckwith, Am. Electroplaters Soc. Ann. Conv., 1938, p. 141.
10. M. F. Maher, Jr., *Proc. Am. Electroplaters Soc.* 1941, **29**, 28.
11. E. Love., *Proc. Am. Electroplaters Soc.* 1959, **46**, 188.
12. M. M. Beckwith, *Proc. Am. Electroplaters Soc.*, 1941, **29**, 80.
13. G. E. Gardam, *Proc. 3rd Intern. Electrodep. Conf.* 1947, p. 203.
14. E. Raub, *Z. Electrochem.* 1951, **55**, 146.
15. F. J. LaManna, *Metal Finishing*, 1960, **58**, 66.

16. U.S. Pat. 2,805,194.
17. D. Holt, U.S. Pat. 2,255,057.
18. U.S. Pat. 2,873,234.
19. U.S. Pat. 2,701,234.
20. H. L. Benner and C. J. Wernbund, *Trans. Electrochem. Soc.* Sp. Vol. 0. 173, 1942.
21. Brit. Pat. 528,762.
22. M. Pushpavanam and B. A. Shenoi, *Electroplating and Metal Finishing*, 1976, **29,** No. 10, 11-12.
23. C. B. Kenahan and D. Sehkain, *Plating* 1961, **48**, 37.
24. S. R. Rich, *Plating* 1955, **42**, 1407.
25. G. W. Jernstedt, *Proc. Am. Electroplaters Soc.* 1949, **36**, 63.
26. A. Hickling and H. P. Rothbaum, *Trans. Inst. Metal Finishing* 1957, **34**, 199.
27. H. L. Pinkerton & J. W. Smith, *Plating*, 1972, **59,** No. 7.
28. B. Pantscher and Ch. Kosarev, *Metalloberflache* 1970, **24**, 109, 383.
29. E. Mel Yamanko, *Zhur. Fiz. Khim* 1958, **32**, 2479.
30. E. Matteson and J. O'M. Bockris, *Trans. Far. Soc.* 1959, **55**, 1586.
31. K. E. Bacon, J. J. Hockstra, B. C. Sison and D. Trevich, *J. Electrochem. Soc.* 1959, **106**, 382.
32. T. Zak, *Metal Finishing* 1963, **61**, 65.
33. G. Bacquias, *Galvano*, 1973, **42,** No. 430, 119-22.
34. A. Sieverts and W. Wippelmann, *Z. Anorg. Chem.* 1915, **91**, 1.
35. S. C. Barnes, *Metal Finishing* 1961, **59**, 60.
36. E. Raub and T. Schiffner, *Metalloberflache* 1971, **25**, No. 4, 114.
37. U.S. Pat. 2742412.
38. L. Domnikov, *Metal Finishing* 1959, **57**, 66.
39. L. Serota, *Metal Finishing* 1959, **57**, 72.
40. R. Pinner, 'Copper and Copper Alloy Plating', Copper Development Assocn., 1962.
41. M. A. Brimi and J. R. Luck, 'Electrofinishing', p. 97, Elsevier Pub. Co. Inc. N.Y., 1965.
42. L. L. Shreir and J. K. Prall, *Trans. Inst. Metal Finishing* 1964, **41**, 29.
43. Trans. Inst. *Metal Finishing* 1971, **49**, 17.
44. Products Finishing, 1975, **39,** No. 8, 64-6.
45. W. I. Flack, *Automation Progress* 1958, **2**, 236.
46. J. W. Dini, *Plating* 1964, **51**, 119.
47. J. B. Mohler, *Metal Finishing*, 1972, **70,** No. 9, 49-50.
48. C. J. Owen, H. Jackson and E. R. York, *Plating* 1967, **54**, 821.
49. J. W. Dini, H. R. Johnson and J. R. Helens, *Plating* 1967, **54**, 1337.
50. B. H. Strom, *Eng. and Mining J.* 1933, **134**, 281.
51. C. L. Faust, *Trans. Electrochem Soc.* 1940, **78**, 383.
52. W. A. Donakowski and W. S. Springer, *Plating*, 1971, **58,** No. 11, 1094-8.
53. H. Strow, *Prod. Finishing* 1957, **21**, 24, March.
54. *Galvanotechnik*, 1976, **67,** No. 5, 367-80.
55. E. J. Roehl, E. Michel and L. R. Westbrook, *Proc. Am. Electroplaters Soc.* 1955, **42**, 3.
56. J. Kronsbein and A. Smart, *J. Electrodepositors Tech. Soc.* 1944, **19**, 107.
57. T. J. Rama Char, *Electroplating and Metal Finishing* 1957, **10**, No. 12, 391, 408.
58. V. Shree and T. L. Rama Char., *ibid* 1959, **12**, No. 10, 385.

CHAPTER 9

Chemical and electrolytic deposition of nickel

NICKEL is one of the most important of the electrodeposited metals, and is unrivalled for its general usefulness as a coating for steel, brass, and zinc-base die-castings where decorative as well as protective properties are required. It is almost exclusively employed as an undercoating for chromium deposits on these basis metals.

Although nowadays nickel can be satisfactorily plated from a variety of baths, the original sulphate bath proposed by Watts in 1916 is still the basis of the most widely-used solution. It is easy to operate and maintain and is considerably less expensive than other nickel plating baths. By incorporating special additives in the solutions, bright coatings can be obtained which need little or no buffing, and practically all decorative nickel plating used in conjunction with chromium is today carried out from solutions of this type.

Other nickel plating baths include the high chloride bath and baths based on nickel sulphamate or fluoborate. The last two find their main application in heavy nickel plating for electroforming[(1)].

9.1. Cleaning prior to plating

Metal surfaces must be thoroughly cleaned prior to nickel plating in order to obtain satisfactory adhesion. Good adhesion is most important particularly where the nickel deposit is followed by chromium because the stressed nature of the latter deposit may lift the underlying nickel if there is any deficiency in this respect. In the case of steel articles, cleaning in a hot alkaline solution followed by anodic etching in 10–70% sulphuric acid solution for a minute or so prior to plating is often resorted to. Copper and brass are cleaned in alkaline solutions and may then be dipped in a weak cyanide solution followed by a dilute hydrochloric or sulphuric acid dip to remove tarnish. Thin copper deposits applied from a cyanide solution are useful on brass before nickel plating, particularly when the brass is of the leaded free-machining type. In the case of bronzes or where any appreciable amount of tin is present in the alloy, the cyanide dip is best omitted as it may lead to poor adhesion. Thin copper undercoatings (7.5 μm thick) have been used on steel to minimise porosity and to increase the corrosion resistance of the deposit, but the present view is that little improvement results from such thin undercoats which may tend, moreover, to reduce the adhesion of the nickel.

The degree of finish of the basis metal and the absence of local inclusions and defects has an important bearing on the durability of the nickel plate. For obtaining satisfactory nickel deposits on nickel suitable surface activation procedures must be employed[2].

9.2. The Watts bath

The Watts bath[3] has undergone a number of changes since the original composition was patented in 1869, but fundamentally it still consists of a solution of nickel sulphate containing some chloride and boric acid. Boric acid was added by Weston[4] in 1878, although its buffering action was not fully appreciated at the time. The addition of chloride to promote anode corrosion did not come until considerably later[5].

A typical bath consists of:

Nickel sulphate	$NiSO_4,7H_2O$	250 g/l
Nickel chloride	$NiCl_2,6H_2O$	40 g/l
Boric acid	H_3BO_3	25 g/l

Operating conditions:

Temperature	30–40°C
pH	5.2–5.8
Current density	2.0–4.0 A/dm^2

It is desirable to have a relatively high nickel sulphate content together with agitation of the solution or movement of the work being plated. This ensures a constant supply of nickel ions to the cathode, thus enabling an adequately rapid rate of plating to be obtained. Where the shape of the articles results in heavy drag-out losses it may be necessary to effect a compromise by working with a solution of lower concentration on economic grounds. Some chloride is necessary in the solution to promote anode corrosion; nickel anodes tend to become passive at high anodic current densities, so that evolution of oxygen occurs rather than the dissolution of nickel[6, 7]. Nickel chloride is preferred to sodium chloride, as there is evidence that alkali metal ions are not a desirable constituent of nickel plating solutions; moreover, the nickel ions contributed by the former salt perform a useful function in the solution. Chlorides also appear to help cathode efficiency and the production of smooth deposits. These advantages have, indeed led to some use of solutions based entirely on nickel chloride (see Section 9.6 below).

Boric acid acts chiefly as a buffering agent, aiding the maintenance of the pH of the solution. The dissociation constant for boric acid is very small, i.e.

$$\frac{\text{activity of } H_2BO_3^- \times \text{activity of } H^+}{\text{activity of undissociated } H_3BO_3} = 5 \times 10^{-10}$$

so that any hydrogen ions removed from the plating solution by gas evolution will be replaced by further dissociation of the acid: $H_3BO_3 \rightarrow H_2BO_3^- + H^+$. Since there are many more undissociated boric acid molecules than dissociated ones, the lost hydrogen ions are in fact replaced without significantly affecting the activity of the undissociated boric acid.

In addition to its buffering action, boric acid forms complexes with $Ni(OH)_2$ which tend to accumulate close to the surface of the cathode. The formation of complexes of the type $Ni(OH)_2.2H_3BO_3$ may considerably reduce the amount of colloidal nickel hydroxide present. Recently it has been claimed[8] that nickel acetate is more effective than boric acid for buffering Watts nickel solutions.

9.2.1. EFFECT OF pH

Watts nickel solutions may be operated at pH values ranging from less than 3 up to 6. A pH range of 5.2–5.8 is commonly used for dull baths, but for bright nickel deposits lower figures (2.4–4.8) are more usual. Too high pH value solutions should not be employed, since this promotes the formation of solid nickel hydroxides. In solutions where there is a tendency for iron to accumulate, such as in the plating of steel parts, basic iron compounds form and fail to redissolve at higher pH values. On the other hand, at low pH values more gas evolution occurs at the cathode which results in lower plating efficiency and may lead to pitting of the deposit.

The hardness of the deposited nickel is influenced by the pH of the solution, a high pH tending to favour the production of harder deposits owing to the co-deposition of hydroxides[9]. The pH may be measured electrometrically (using the quinhydrone or the glass electrode) or colorimetrically. Permanent colour standards can be employed, and special disc comparators are available for nickel-plating solutions; the method of use is to fill the two 10 ml tubes provided with the plating solution and to add to one of them 0.2 ml of the indicator solution (e.g. chlorophenol red, bromocresol purple, etc., depending on the pH range of the solution). The dial is then turned until one of the discs, when in position before the tube of solution not containing the indicator, produces a colour match with the other tube when both are viewed by transmitted light. The pH can then be read off from a scale on the instrument. It is to be noted that, as generally determined, the colorimetric pH value is usually some 0.5 pH higher than the true electrometric figure. In plating practice it is necessary to determine the pH value of the bath at frequent intervals and to add sulphuric acid from time to time, since the pH usually tends to rise during operation.

9.2.2. ANODES

Under modern plating conditions using hot solutions and air agitation, suitable nickel anodes must be employed to obtain a uniform rate of dissolution and to prevent their disintegration. The use of auxiliary nickel anodes assists in obtaining uniform coatings on difficult shapes[10].

The chief types of anodes used for nickel plating are:

(a) Depolarised rolled anodes

(b) Carbon-bearing anodes

(c) Electrolytic nickel squares or other shapes which are contained in anode baskets.

A review of nickel anode properties has recently been given by Parkinson[11]

9.2.2.1. Rolled anodes

With the advent of solutions operating at higher current densities, hot-rolled anodes have replaced the earlier cast type as the latter tended to disintegrate at higher current densities. The solubility of rolled anodes in a given solution can be controlled by the degree of rolling to which they are subjected in manufacture. The denser the structure, the lower is the solubility and the greater the tendency for passivity to occur.

9.2.2.2. Depolarised anodes

At high current densities and pH values, satisfactory anode corrosion is difficult to obtain even in solutions of high chloride content; for operation under these conditions the depolarised type of anode has, therefore, been developed and gives a uniform rate of corrosion even at very high current densities. Depolarised anodes are rolled, and contain a carefully regulated proportion of nickel oxide and sulphide which prevent the formation of a passive film and aids corrosion at relatively low anode potentials.

Even in chloride-free solutions, reactive nickel of this type will dissolve at a low potential until either the anode current density or the pH is raised above a critical value[12]; when, however, chloride is present as is the case with most nickel plating solutions, the reactive nickel dissolves at a low potential over a wide range of pH and current density.

The oxide content ranges between 0.25 to 1.10, and the amount of sulphur present in the nickel must also be veiy carefully controlled if uniform anode dissolution is to be obtained. Sulphur is normally present to the extent of about 0.003 per cent; the exact concentration is critical, since excessive amounts may cause the anode to disintegrate owing to the segregation of nickel sulphide at the grain boundaries, whilst an insufficient quantity causes ineffective depolarisation.

The effective depolariser appears to be sulphur, most other elements being without effect. It has been suggested[13] that the mechanism of the effect of sulphur is a result of the formation of microgalvanic cells between nickel and Ni_2S_2. In practice, it is necessary for the sulphur content to be high enough for depolarisation to occur, and low enough to ensure workability. The amount should be less than 0.01 per cent, and is generally controlled at 0.002 to 0.01 per cent. Addition of traces of copper to these anodes to promote uniform corrosion has also been claimed[14]. The metal is subsequently homogenised by rolling, hammering and forging, whereby the cross-section is reduced to about two-thirds. A disadvantage of these anodes is that as dissolution proceeds, the oxide is liberated as a fine brownish slime which is insoluble in the plating solution. Watson[15] has discussed the practical aspects of using 'S' nickel anodes.

9.2.2.3. Carbon-bearing anodes

The advent of bright nickel plating processes has raised problems in relation to anodes. Thus, certain organic additive agents increase the activity locally, leading to corrosion in depolarised anodes and releasing particles of metallic nickel which result in rough deposits. The pH values at which these solutions operate are also lower than those at which such anodes give the best results.

It has been found that the addition of carbon in the form of a uniform dispersion throughout the nickel enables anodes to be produced which are particularly suitable for operation with bright nickel solutions; these are available both in the cast and extruded forms, the latter being preferred as they have a smaller grain structure and are more uniform. The carbon content may be as much as 0.4 per cent, but 0.2 to 0.3 per cent is more usual. As the anode corrodes, an adherent film of carbon and silica (which is also present to the extent of about 0.3 per cent) forms a continuous sheath on its surface. This film is highly tenacious and serves to prevent nickel particles becoming detached and finding their way into the plating solution. Such anodes have been used without bags where air agitation is not employed, but it is advisable to use bags as a safeguard in the event of portions of the anode film breaking away. The carbon sheath thus formed persists for the life of the anode and due to anodic oxidation reactions within it, the nickel goes into solution more readily. It also prevents the adsorption of brightening compounds on to the metallic surface of the nickel and hence eliminates the tendency for the anode to pit and release metal particles into the solution, which can lead to rough deposits; the amount of metal loss is generally less than 0.1 per cent.

A typical extruded carbon-bearing anode contains:—

Carbon	0.26 to 0.32 per cent	
Silicon	0.28 to 0.35 per cent	
Copper	0.20	per cent (max.)
Iron	0.05	per cent (max.)

The action of the carbon is modified by the presence of the silicon; the sludge, which is rather voluminous, has no appreciable resistance to the passage of current, and tends to adhere to the anode surface. The anode should not be cleaned during its life and it is desirable to use a slightly longer anode bag than usual to accommodate the sludge which falls to the bottom as the anode wears. If it should collect around the tip, the sludge could lead to 'necking' of the anode above the solution level. Such anodes operate best at around 3 A/dm^2 and at pH values of up to 4.5 to 5.0.

9.2.2.4. Anode bags

Nickel anodes are generally bagged to prevent particles finding their way into the plating solution. The bags are usually made of cotton twill or calico, but heavier, closely woven fabrics of the sailcloth-type are also favoured, these being more retentive of fine particles. Anode bags of nylon or Terylene are also very satisfactory. The fabric must be free from dressing to avoid contamination of the solution by organic matter, and it is a wise precaution to boil anode bags in water before use. A practical improvement has been the introduction of bags in which the bottom few inches have been impregnated with a synthetic resin to make them impervious. This makes the bags less liable to break and more retentive of the minute particles which accumulate in the bottom. Where such bags are used they should, as already mentioned, be rather longer than the anodes.

Anode bags of woven glass fibre have been used on a small scale, these materials having the obvious advantage of not introducing organic material into the solution. They are somewhat expensive, however. Glass fabric anode bags should also be thoroughly degreased before use to remove the lubricants used in the spinning of the glass fibres.

9.2.2.5. Nickel anodes in baskets

The use of anodes in the form of small pieces of nickel retained in wire cages offers a number of advantages[16], namely:

(a) Specially shaped anode metal is not required, and the price of the primary nickel is lower;
(b) The anode cage can be placed in position in the bath and filled *in situ*, thus obviating the need to move it;
(c) The length of the anode does not vary;
(d) As the anode material (squares, pellets, etc.) wears down it packs naturally so that only minimal changes in surface area occur;
(e) Little maintenance is required;
(f) No time is lost in lifting out anodes for inspection as the level of nickel in the basket is readily visible.

As against this, however, the total amount of nickel required is generally higher, which involves an increased initial investment.

Up to several years ago it was not a really viable proposition to use nickel anodes in this form on account of the lack of a suitable basket material. However, with commercial availability of titanium, it was found that this metal becomes passive in nickel plating solutions and does not dissolve anodically (even at quite high anode potentials). Its suitability for making anode baskets thus became obvious and over the last decade or so it has been increasingly used for this application (Fig. 9.1). The baskets are filled with nickel pellets or squares and are enclosed in anode bags. Nickel squares for use in anode

Fig. 9.1. Titanium anode basket.
[Courtesy Oxy Metal Industries (G.B.) Ltd.]

baskets consist either of pure electrolytic metal or of nickel containing an activator such as sulphur.

9.2.3. PLATING CONDITIONS

In the operation of the Watts type of bath the conditions selected should be such as will favour the maintenance of an adequate concentration of nickel ions in the vicinity of the cathode. If this condition is not obtained, either pitted or burned deposits will result. The maintenance of the requisite concentration of nickel ions is achieved by:

(i) A high nickel content in the solution
(ii) Relatively high operating temperatures
(iii) Movement of the solution to prevent stratification.

The nickel content should be as high as possible consistent with the need to avoid excessive costs resulting from the loss of nickel by drag-out, this being naturally much higher in the case of hollow and recessed articles than for those having relatively plain surfaces. The hardness of the deposit is affected by the plating temperature and a compromise must be made between a tendency for burning to occur if the temperature is too low at a particular current density, and for the deposit to be too soft and hence difficult to finish if the temperature is too high.

The normal Watts solution is usually operated at a pH range of 5.2 to 5.8, as has already been stated. Within this range less hydrogen is liberated and there is, therefore, less tendency towards pitting from this cause. On the other hand, operating a high pH solution may result in the precipitation of basic compounds, which may also cause pitting.

The use of wetting agents to lower the surface tension of the solution is helpful in reducing the tendency for hydrogen bubbles to adhere to the deposit. Current densities of 2 to 4 A/dm^2 ($20\text{-}40 A/ft^2$) are usual with the Watts solution; Fig. 9.2 shows the rate of deposition of nickel at various current densities assuming a cathode current efficiency of 95 per cent.

In the absence of brightening agents, nickel deposits become progressively duller as their thickness increases, and they subsequently become distinctly rough, even when most carefully filtered. This is due to the micro-crystalline structure of the deposit which may be columnar or even treed in form. It is the crystal aggregates rather than

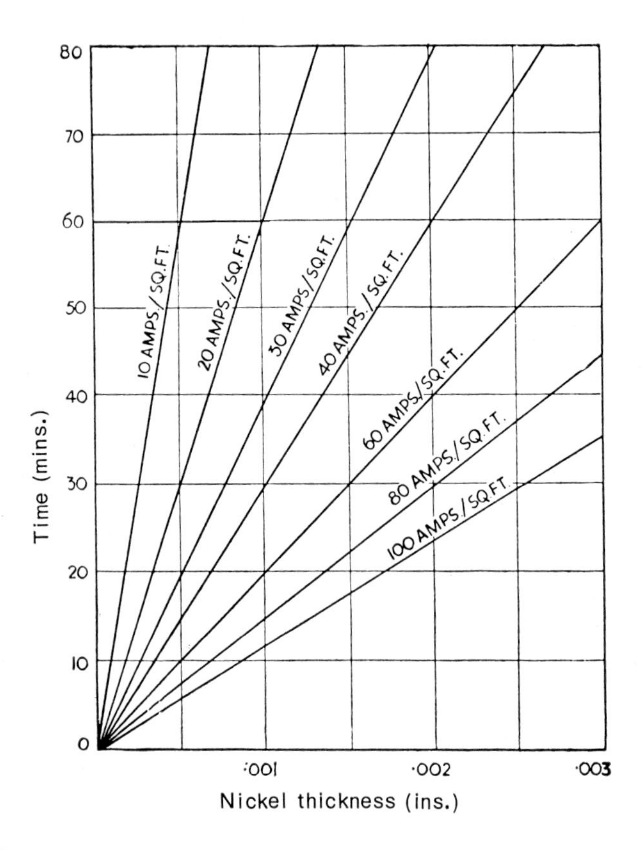

Fig. 9.2. Rate of deposition of nickel at various current densities and a current efficiency of 95%.

the sizes of the individual crystals which cause this progressive increase in mattness of nickel deposits. However, with modern bright nickel solutions, smooth deposits free from burning can be obtained even at very high current densities.

The influence of agitation and filtration on the quality of nickel deposits produced has recently been discussed by Meyer and Kreisel[17]

9.2.4. THROWING POWER

The throwing power of a nickel solution is usually moderately good, but constant efforts have to be made to improve this characteristic by suitable arrangements of anode-cathode distances and the selection of plating conditions. The protective value of a nickel deposit is greatly dependent on its thickness, so that those factors which make for uniformity in the deposit are of great importance.

Watson[18] has measured the throwing power of several nickel plating baths and has found that the throwing power of the Watts solution is slightly inferior to that of others. The results obtained (Table 9.1) were expressed on a scale based on perfect throwing power being expressed as +100% and the worst possible throwing power as —100%. All three solutions showed the same tendency of falling throwing power with increasing current density. Complexing nickel with amines and flourides did not result in any improvement in throwing power.

The presence of organic addition agents has been found in general to reduce the throwing power considerably.

9.3. Black nickel plating

A process worthy of mention is black nickel plating which is carried out on specialised products such as optical instruments where a protective, non-reflecting coating is needed. The deposits are usually thin and contain a considerable amount of sulphur, present in the form of Ni_3S_2 and an unidentified phase[19].

Samartsev and Andreiyeva[20] consider that the black colour of the coatings is not due to the presence of any actually black compounds but is related to the structure of the deposit.

TABLE 9.1

Throwing power of nickel plating solutions[18]

Solution	Average current density (A/dm²)	% Throwing Power at primary current distribution ratios of		
		5:1	12:1	25:1
Watts	4.32	8	7	14
Sulphamate	4.32	11	13	19
All-chloride	4.32	18	18	27

A typical solution for black nickel plating consists of:

Nickel sulphate	$NiSO_4,7H_2O$	75 g/l
Nickel ammonium sulphate	$Ni(NH_4)_2(SO_4)_2,6H_2O$	45 g/l
Zinc sulphate	$ZnSO_4,7H_2O$	37 g/l
Sodium thiocyanate	NaCNS	15 g/l

The solution is operated at a pH of 5.6 to 5.9, a temperature of 50°C to 55°C and a current density of 1.0 to 1.5 A/dm^2. The process is rather difficult to control and the deposits tend to be brittle.

An improved process which is claimed to be easier to operate and gives a coating of improved ductility and adhesion has been developed[21]. The solution used consists of:

Nickel chloride	$NiCl_2 6H_2O$	75 g/l
Ammonium chloride	NH_4Cl	30 g/l
Sodium thiocyanate	NaCNS	15 g/l
Zinc chloride	$ZnCl_2$	30 g/l

The bath is maintained at a pH of 5.0 and operated at room temperature; the current density is not critical but the nickel chloride content is more so, high concentrations requiring higher current densities for satisfactory results.

All black nickel coatings have poor corrosion resistance and should be lacquered where they are subjected to adverse atmospheric conditions or to outdoor exposure.

9.4. Bright nickel plating and related processes

Bright nickel plating is now to all intents and purposes the standard method of depositing the metal when it is applied as an undercoat for chromium. The advantages of being able to plate nickel bright without the need for polishing are many, the chief being:

(1) The elimination of the considerable cost of nickel finishing.
(2) Avoidance of the waste of nickel in the buffing operation, during which as much as 30 per cent of the metal deposit may be removed with a corresponding loss in the protective value of the deposit.
(3) Nickel plated articles can be directly chromium plated without the need for drying, unracking, re-wiring, etc.

In order to produce a bright deposit the growing cathode surface must be inhibited by the addition to the electrolyte of specific agents, generally organic surface active substances. The adsorption of these substances on to the cathode surface inhibits the deposition process and increases the overpotential at any given current density[22, 23]. This can be clearly seen from the polarisation curves given in Fig. 9.3, which show the effect of adding various brightening agents to a Watts nickel solution. Some brightening agents are adsorbed preferentially at active sites, inhibiting local growth and forcing

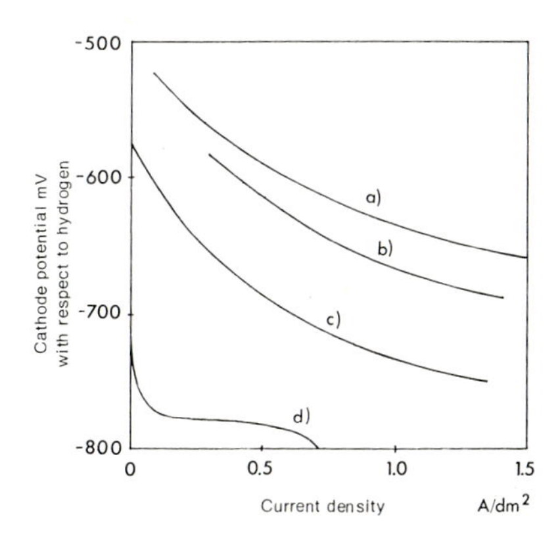

Fig. 9.3. Effect of adding brightening agents on nickel polarisation curves.

(a) Nickel bath without additives.
(b) 0.04g/l α-naphthol.
(c) 2g/l naphthylamine – sulphonic acid, stirred.
(d) 2g/l quinoline – 8 – sulphonic acid, stirred.

fresh nucleation to occur at adjacent sites. This process results in the formation of an extremely fine-grained, randomly orientated deposit. Other types of additives are adsorbed over the entire cathode surface. Reduction of metal ions then occurs at any instant at a relatively few sites which constantly change owing to the very labile nature of the adsorption.

Attempts to produce a lustrous finish on nickel plate by the addition of various materials to the plating solution have been described in early publications; cadmium salts have long been known to act as brighteners in nickel baths, as have colloidal materials such as glue and gelatine. These additions did not find extended use, however, owing to their uncertain effects and the difficulty of controlling the plating solutions containing them.

Since then a large number of solutions have been developed and many have been operated commercially on a very large scale. Virtually all the bright nickel baths employed in industry are proprietary, and are based fundamentally on the Watts nickel sulphate solution with various additions. A list of some of the more important patents is given in Table 9.2, with an approximate indication of the nature of the brighteners employed.

Normally additions are made in the form of a liquid concentrate, but an alternative method has been introduced in which automatic dosing pumps for adding the correct amount of brightener as determined by an ampere-hour meter are available. In one process all of the necessary additions for a bright nickel bath (other than chloride or boric acid, since their consumption is normally small) are contained within a single tablet which is added to the bath at suitable intervals[24].

TABLE 9.2

Bright Nickel Plating Patents

Solution			Brighteners used
Schlotter	U.S. Pat.	1,972,693 (1934)	Nickel benzene disulphonate; napthalene trisulphonate.
Weisberg and Stoddard	U.S. Pat.	2,026,718 (1936)	Cobalt sulphate, sodium formate and formaldehyde.
Harshaw	U.S. Pat.	2,029,386 (1936)	Sulphonated oleo resins
	U.S. Pat.	2,029,387 (1936)	Sulphonated terpenes.
Hinrichsen	Brit. Pat.	461,126 (1937)	Cobalt sulphate and sodium formate.
Hull	U.S. Pat.	2,085,754 (1937)	Organic ketones.
Harshaw	U.S. Pat.	2,125,229 (1938)	Selenious acid and naphthalene disulphonate.
McGean	U.S. Pat.	2,122,818 (1938)	Zinc or cadmium and naphthalene sulphonic acids.
	U.S. Pat.	2,114,006 (1938)	Zinc or cadmium, and sulphonated aminotoluenes.
Harshaw	U.S. Pat.	2,198,267	Naphthalene sulphonates, amino-polyaryl methanes (e.g. fuchsin) and sodium lauryl sulphate. Sulphonamides and sulphonimides.
Udylite	Brit. Pat.	529,825 (1940)	Ketones and aldehydes, e.g. formaldehyde and chloral hydrate.
Udylite	U.S. Pat.	2,191,813; 2,240,801; 2,231,182 (1940-46)	Zinc or cadmium and sulphonamides (e.g. *p*-toluene sulphonamide); halogenated aldehydes, toluene sulphonic acids.
	Brit. Pat.	525,847 (1940) 529,825 (1941)	
Harshaw	U.S. Pat.	2,238,861 (1941)	Azines, oxazines, indamines and indophenyl dies; fuchsin.
Udylite	U.S. Pat.	2,402,801 (1946)	Thallium sulphate and *o*-benzoyl sulphonimide and *p*-toluene sulphonamide.
Udylite	U.S. Pat.	2,466,677; 2,467,580 (1949)	Chlor- and bromo-substituted ethylene sulphonamides; aryl sulphone sulphonates.
Harshaw	Brit. Pat.	741,096 (1953)	Coumarin compounds; aromatic sulphur compound and methane derivatives.
Udylite	Brit. Pat.	736,556 (1953)	Quinolium chloride plus sulphonic acid compound
Udylite	Brit. Pat.	758,162 (1953)	Coumarin sulphate, benzene sulphonamide and saccharin.
Harshaw	U.S. Pat.	2,782,152; 2,784,152 (1954)	Coumarin and amino polyarylmethane.
Udylite	U.S. Pat.	3,041,256 (1962)	Sodium 1,4-dihydroxy-2-butene-2 sulphonate.
W. Canning	Brit. Pat.	973,423 (1964)	Coumarin, 4-hydroxy coumarin.
Dehydag	French Pat.	1,393,139 (1964)	o-toluene sulphonamide, alkyl pyridine compounds, sodium dodecyl sulphate.

TABLE 9.2 (continued)

Solution		Brighteners used
M & T Chemicals	Brit. Pat. 971,165 (1964)	Saccharine, o-sulphobenzaldehyde, sulphonated, dibenzothiophene-dioxide.
Dehydag	DAS 1217170 (1966)	Toluene sulphonimide, guanidine, formaldehyde and dihydroxy decylsulphate.
Udylite	U.S. Pat. 3,661,596 (1972)	Selenium compounds.
Udylite	U.S. Pat. 3,898,138 (1975)	Alkylene diols, butene, alcohol, metasulphobenzoic acid.

9.4.1. BRIGHT NICKEL PLATING PROCESSES

The bright nickel processes are based essentially on the Watts type of bath, but a solution with a high nickel chloride content has been found very useful where there is much tendency towards contamination by zinc.

An examination of the compositions of bright nickel plating solutions shows that the brightening agent usually consists of two or more constituents which may be organic or inorganic. It can be said, broadly speaking, that the classes of materials employed fall into two groups. The first group consists of substances which act as very effective brighteners, but which must be used in carefully regulated quantity. Typical of these are materials such as cadmium and zinc salts, aldehydes, selenium, tellurium, glue, gelatine and dextrin. The range of concentrations within which these brighteners can be employed is so narrow that they cannot be used satisfactorily under commercial conditions. Excess quantities soon reduce the cathode efficiency, particularly in low current density areas, so that there may be no deposit at all in recessed parts of articles being plated. Moreover, very slight changes in the operating range of the bath or in brightener concentration may lead to dark and brittle deposits. For these reasons, secondary addition agents are used which, although they may or may not be brighteners in themselves, have as their chief characteristic the ability to make the solution tolerant of a much higher concentration of the primary brighteners already mentioned. Among the substances of this group are cobalt salts, naphthalene sulphonates, polysulphonic acids and aryl sulphonamides. One of the most important addition agents of this type is saccharin, which is also valuable in reducing stress in bright nickel deposits.

A combination of suitable materials from these two groups is used in appropriate concentrations to give a solution which will have the maximum plating range as regards temperature and current density, and will be reasonably insensitive to the presence of impurities in the solution. The addition agents should obviously have adequate solubility and should not precipitate out of the solution.

Many addition agents which have good brightening properties are unsatisfactory as they may, for example, have too high a vapour pressure, resulting in excessive consumption rates, or give adsorbed films on the deposit which can adversely affect its adhesion to the basis metal. To obtain addition agents which will have all the required properties is not a simple matter, and very large numbers have been evaluated and rejected for one reason or another.

Many of the bright nickel solutions using organic addition agents are not air agitated, because attempts to do so may result in excessive foaming. The work may, however, be subjected to rod agitation to reduce the tendency to pitting of the deposits. In spite of this, pitting may occur unless a small concentration of a suitable wetting agent is present. In such solutions hydrogen peroxide cannot be added as an anti-pitting agent owing to the likelihood that it will oxidise the organic brighteners present. Amongst the wetting agents employed in commercial solutions are sulphonated alcohols and alkyl-substituted aryl sulphonates. Such wetting agents reduce the surface tension of the plating solution, so that any colloidal particles or hydrogen bubbles which form do not adhere to the cathode surface. The presence of wetting agents may, however, cause difficulty by emulsifying oils or other contaminants, necessitating filtration or treatment with activated carbon. No practical chemical method for controlling the wetting agents in solution is available since they are only present in very small concentrations; control is, therefore, best carried out by surface tension measurements, using a stalagometer or a similar apparatus[25].

9.4.2. ORGANIC BRIGHTENER SOLUTIONS

The solutions containing organic addition agents are, as has already been stated, almost all proprietary and give a high degree of brightness. Amongst the earliest of the commercially exploited organic brighteners were the alkyl naphthalene sulphonic acids, but these were soon succeeded by the unsubstituted polysulphonates with 2 or 3 sulphonic acid radicals, notably the naphthalene disulphonic acids developed by Schlotter[26]. These compounds were added to a Watts type of solution, a typical bath consisting of:

Nickel sulphate	175 g/l
Nickel chloride	85 g/l
Boric acid	20 g/l
Sodium naphthalene trisulphonate	35 g/l

Operating conditions are:

pH	2.5 to 4.5
Temperature	50° to 55°C

Another important group of brighteners was the aryl sulphonamides with or without zinc or cadmium additions. The aryl sulphonic acids were introduced in 1946 and the aryl sulphone sulphonates in 1949; these have the property of providing good tolerance to metallic contamination and a high degree of brightness. Unsaturated compounds are the key to many brightening agents, chiefly those containing C=O, C=C, C=N, N=N and N=O groups. The unsaturated sulphonic acids give good brightness at low current densities and provide a high tolerance for metallic and non-metallic impurities.

C=N group compounds are useful in conjunction with primary brighteners although if used alone they may result in streaky, brittle deposits. The N=N group appears in the azo-dye additive range, whilst the N=O group is used in the sulphonated form to confer solubility.

Virtually all bright nickel solutions contain compounds of the C–SO_2–type as primary brighteners. Cathodic reduction of such compounds results in the incorporation of sulphur in the nickel deposit largely in the form of Ni_3S_2[27]. According to Wernick[28] as much as 6% of sulphur may be present, but in modern proprietary solutions the figure is normally very much lower than this. A useful discussion of the chemistry of bright nickel plating solutions has been given by Saubestre[29].

Modern organic brighteners are relatively easy to control, but there is a tendency for the additives to break down, so that periodic treatment with activated carbon to remove degradation products is necessary. Where large-scale plating is being carried out the economic advantages of the organic baths are very great. The baths operate over a very wide range of conditions, and show little tendency to burning even at very high current densities. Control of the addition agents is usually carried out by empirical methods and by plating tests with a Hull Cell (see Section 9.4.13 below).

9.4.3. SEMI-BRIGHT SOLUTIONS

In addition to the bright baths, there is a considerable field of application for deposits which are very easily buffed and have the advantage of being relatively ductile. For certain applications where plated articles are liable to be stressed during attachment or in service, the ductility of semi-bright nickel plate is a useful asset. Motor-car bumpers and windscreen surrounds are typical applications for this type of deposit. Organic addition agents are generally employed, whilst a fair amount of chloride is also usually present in the bath. One semi-bright solution[30] makes use of 0.15 g/l coumarin together with 0.1 g/l of sodium lauryl sulphate as the wetting agent; a pH of 3.5 is used, the temperature being 60°C; the operating current density is 5 A/dm^2. Melitotic acid is the principle reduction product from coumarin containing solutions and must be removed periodically by charcoal treatment. Recently a method for removing it continuously and automatically has been described[31].

9.4.4. CORROSION RESISTANCE

The results of the intensive effort which has been put into research on improving the corrosion resistance of nickel-chromium deposits in recent years have now become manifest. The finish became the subject of much criticism with regard to its behaviour on exposure to corrosive atmospheres, especially when used on automobile components.

The new process developments have mostly originated in the laboratories of the supply industries and, to a lesser extent, in those of the motor manufacturers, especially in the United States. The problem of the inadequate durability of commercial electroplating was accentuated by the introduction of bright nickel, which is known to be more corrodible under conditions of atmospheric exposure than the earlier dull nickel coatings. The difference was masked, however, by the use of improved plating standards which led to an increase in the thickness of nickel which was generally applied. Failures were especially marked in industrial areas where acid-containing moisture condensation occurs, and especially in cities where salt is used for snow and ice dispersal in winter. Corrosion troubles were most serious in the U.S.A. where vast quantities of salt have always been used on the roads in winter; in this country the use of salt has also increased dramatically. An important factor in the failure of bright nickel is the presence of sulphur in the deposits, originating from sulphur-containing brighteners.

9.4.4.1. Causes of failure

As a result of investigations it is generally accepted that the inherent corrodibility of the nickel deposit, accentuated by electro-chemical effects resulting from the superposition of the chromium layer, is the dominant factor in the corrosion of the deposits. Under modern conditions of plating the influence of porosity of the nickel (upon which undue emphasis has almost certainly been placed in the past) is of secondary importance. As the improvement which can be effected by increasing the total plating thickness has already been approaching the limit of practicability for some time, most of the recent work has been concentrated on finding ways of increasing the life of the finish by changing the characteristics of the nickel and the chromium layers themselves.

The most promising methods of achieving this have involved the use of successive deposits of different types of nickel; these may, if desired, be employed in conjunction with double chromium deposits plated respectively under different conditions. The practical application of some of these systems is also relatively straightforward, as the plating sequences are only marginally longer than those hitherto employed and they can be operated in automatic installations. The problems of adhesion between the separate deposits of nickel, and likewise of chromium, have also been more readily overcome than was anticipated during the early stages of the introduction of these multilayer systems. (See also Section 9.11.)

9.4.5. DUAL NICKEL DEPOSITS

The duplex or dual nickel system owes its origin to the Ternstedt Division of General Motors Corporation which produces a vast range of plated car components, many of which are zinc-base alloy die-castings which are particularly susceptible to atmosperic attack. In view of the fact that it is known that the presence of sulphur in bright nickel deposits makes them more susceptible to corrosion, it was thought that an improvement could be effected by first applying a nickel deposit from a semi-bright bath (which may contain levelling agents but is free from sulphur) before the fully-bright nickel. The semi-bright deposits are smooth and, having good levelling properties, do not demand a high degree of polish on the underlying metal. The work is then transferred directly to a fully-bright bath from which the final nickel is plated. The latter layer need only constitute a small proportion of the total deposit, which must still be kept to full thickness specification if satisfactory results are to be obtained.

This system of plating is now being operated fairly extensively on such articles as bumpers and zinc-base alloy die-castings. The improvement obtained by this technique is more marked in certain atmospheres than in others; it shows up particularly well in accelerated corrosion tests such as the CASS and Corrodkote tests. It is found on sectioning corrosion pits which have developed in dual nickel deposits of this kind that the rate of penetration is delayed, and that when it does occur the corrosion tends to spread laterally at the interface between the two nickel layers. Hence attack through to the basis metal is considerably reduced.

Experiment has shown that with dual nickel coatings a ratio of about 90% semi-bright sulphur-free nickel to 10% bright nickel overlay gives the best corrosion protection results when about 0.25 μm to 0.125 μm of chromium is used as the final finish. It is, however, difficult to maintain this ratio in practice without resorting to buffing of the final nickel, which in most cases is wholly impracticable. As a result, compromises have to be made: on steel the tendency has been to use an approximately 70-30 ratio, and for zinc die-castings, which are usually of a more intricate shape than steel articles, even a 60–40 ratio may have to be employed; it is difficult to obtain adequate brightening in recessed areas unless at least 30% of the nickel plate is bright nickel. A 50–50 ratio gives markedly inferior results. Further substantial improvements can be obtained by applying a copper deposit of reasonable thickness under the nickel.

Bright nickel has a lower resistance to corrosion than dull nickel, particularly when chromium plated. This has been ascribed to its sulphur content, which is introduced by reduction and cleavage of the sulphur compounds in the addition agents[32]. The sulphur-free nickel produced from either a Watts bath or a semi-bright nickel gives better protection than a bright solution. Deposits produced by plating the fully bright nickel over a semi-bright deposit give better protection than a wholly bright nickel deposit. The fully bright part of the combined deposit is usually about 25 per cent.

Corrosion pitting in nickel under pores and other discontinuities in chromium is not eliminated when 0.1 μm of chromium is plated over a dual nickel deposit[33]. It does, however, appear to be retarded at the interface between the semi-bright and the bright layers [34, 35]. The pits which penetrate the nickel grow preferentially in a lateral direction which delays basis metal corrosion.

9.4.6. TRIPLE NICKEL

By interposing a thin nickel strike having a relatively higher sulphide content than bright nickel between the semi-bright sulphur-free nickel and the final bright nickel it is possible to obtain a further improvement in corrosion protection. This intermediate layer should contain from 0.1 to 0.2% sulphur as nickel sulphide, a minimum thickness of 2.5 μm being preferred although as little as 0.0025 μm is beneficial. Normally a time of one to two minutes at 4-5 A/dm^2 is required to apply the strike, but this may be increased somewhat with advantage if the parts to be plated are of complex shape. The method gives the best protection with 50–50 and 60–40 ratios of semi-bright sulphur-free nickel to bright nickel. The higher sulphide-containing interposed thin nickel deposit results in a sharp increase in anodic activity between the two principal nickel layers, and this tends to confine any corrosion attack on the nickel preferentially to the sulphur-containing nickel. The result is that the first sulphur-free nickel layer is less vulnerable to attack, with any corrosion pits developing laterally in the sulphur-containing nickel once the attack has penetrated to the semi-bright region. They therefore proceed downwards into the basis metal only very slowly, the improvement being very apparent for both industrial and marine exposure.

It has been observed that, as with dual nickel, the triple deposit will withstand more prolonged accelerated tests, and that corrosion pits may become quite broad before any attack on the underlying semi-bright deposit shows itself. On exposure to industrial atmospheres the surface pits are very small and closer together than in the accelerated tests, and cannot be observed with the naked eye.

The high sulphur-containing layer becomes more anodic to the first nickel layer than the bright nickel, and may even be undercut in the course of corrosion; this, however, leaves a projecting bright nickel deposit over the pore as a sacrificial protection for the exposed sulphur-free layer. It is believed that this accounts for the relatively slight attack which occurs on the sulphur-free nickel even when the pits are quite broad. The triple nickel system is best used over buffed copper on steel articles such as bumpers and on copper-plated zinc-base alloy die-castings. It has shown up particularly well in accelerated tests of the CASS and Corrodkote types.

The method is also better than dual nickel from the point of view of practical operation. For example, the bright nickel component of a dual nickel system must not contain more than about 0.08 per cent of sulphur if the brightness of the plate is to be maintained without impairing its ductility and adhesion, which makes the process somewhat critical to operate. On the other hand, the intermediate strike method enables a high percentage of sulphur to be introduced, whilst maintaining excellent adhesion and a high degree of tolerance towards organic and inorganic impurities. The process is dependent on the presence of suitable addition agents in the strike solution to maintain a uniform distribution of sulphur in the thin intermediate deposit over a wide range of current densities. The bath, however, cannot be used for the whole of the final nickel, any attempts to do so leading to cloudy nickel deposits with poor levelling properties.

9.4.7. SATIN NICKEL PLATING

A satin finish can be obtained by mechanically scratch brushing the nickel deposit, but this method has the disadvantage of possibly leaving the thickness of nickel deposit undesirably low and reducing its corrosion resistance. Furthermore, mechanical finishing is relatively costly since it involves several additional operations such as racking and cleaning. Satin finishes can, however, be directly electroplated[36] from semi-bright or bright nickel baths containing dispersions of fine bath-insoluble particles such as oxides, sulphates or carbides. Complete uniformity of finish and day to day reproducibility are obtainable by this method.

The proprietory 'Satylite' process, enables satin nickel to be produced directly from the plating bath uniformly and consistently over a wide range of current densities. Satin finishes have been used for a long time on expensive articles, such as cameras, but the process enables such finishes to be applied to a much wider range of lower cost products.

The nature of the powder and its particle size are critical in the Satylite process since a proper distribution of the particles, particularly on vertical and near-vertical surfaces, is required in order to avoid non-uniformity. The inert particles are occluded on the surface of the deposit and it is their presence which is responsible for the matt finish. Subsequently the nickel is chromium-plated in the normal manner but, if required, attractive two-tone effects can be produced by buffing the satin nickel at local areas prior to applying the chromium.

Another method for the direct plating of satin nickel finishes has been described by Baker and Christie[37]. The solution used is a Watts nickel bath containing two addition agents, one a brightener and the other a compound to produce the satin finish. This latter additive is fully soluble at low temperatures, but when a critical temperature is reached (normally 22°C) it precipitates in the form of fine droplets which influence the nickel deposition in such a way as to produce a matt velvet effect. Fig. 9.4 shows the surface of such a nickel deposit and the numerous characteristic indented craters producing a broken surface, and hence reduced reflectivity.

Fig. 9.4. Surface of satin nickel deposit. × 300.

9.4.8. THE 'NICKEL-SEAL' PROCESS

A direct result of the work on satin nickel has been the development of the Nickel-Seal process which enables a particularly high degree of corrosion protection to be obtained. It was found early on during accelerated corrosion tests on satin nickel plus chromium deposits that an exceptionally high corrosion resistance was being obtained. On the face of it, this should not have been the case, because the presence of foreign occlusions in nickel deposits generally affects their performance adversely. Subsequent investigations have established that the improvement was due to the effect of the type of fine particles present in the nickel surface on the characteristics of the subsequent chromium layer, whereby the electrochemical relationship between the latter and the underlying nickel was favourably altered.

The next step following on these findings regarding the influence of occluded particles on corrosion resistance, was to determine whether it was possible to produce a fully bright nickel-chromium coating with the same advantages. After extensive tests with a very large number of powders it was found possible to achieve this, and to induce the required fine porosity pattern in the usual 0.25 μm thickness of chromium. The bright nickel deposit obtained in this way has something of the order of 1 million sub-microscopic particles per cm^2 of surface, although the presence of these particles in no way detracts from the brightness of the nickel. The thin chromium when subsequently applied does not cover the particles embedded or partially embedded in the bright nickel surface completely, with the result that the required porosity is developed in an indirect manner uniformly over the chromium surface, including recessed areas. Auxiliary anodes are not required and only one to three minutes of chromium plating are needed for excellent results to be obtained.

The fully-bright nickel which is deposited with the sub-microscopic inclusions in this way remains very bright for up to about three to five minutes; only some five minutes of plating time at 4 to 5 A/dm^2 are required for the maximum degree of corrosion protection when bright nickel of this type is applied over a single or dual nickel deposit, and is followed by a final chromium coating of about 0.25 μm thickness. The effect on the deposit is to produce a larger exposed anodic area of nickel. It can be applied on steel, which may with advantage be previously copper plated, and on zinc-base alloy die-castings. Maximum protection for a given thickness of nickel is produced by applying the Nickel-Seal process over a triple nickel deposit.

9.4.9. MECHANISM OF THE BRIGHTENING EFFECT

The causes of the brightening action of addition agents in electroplating solutions have been the subject of a good deal of speculation and several mechanistic theories have been proposed. The first characteristic of bright deposits is that brightness is maintained through indefinite thicknesses. The deposits appear to be virtually amorphous in that the crystallite size is smaller than the smallest wavelength of visible light. Brighteners therefore act by interfering with crystal growth. It is generally considered that this crystal growth modification is preceded by the adsorption of addition agents (or their breakdown products) on the cathode surface[38, 39]. This adsorption is controlled by diffusion across the cathode layer. Additives such as saccharin are in true dynamic adsorption equilibrium with the changing deposit surface whereas compounds such as thiourea react irreversibly with the deposit[40].

It was suggested by Hendricks[39] that bright plating in the presence of organic addition agents is brought about by a mixture of organic bases and inorganic colloids in the cathode film, which transforms normal deposition into a periodic phenomenon, reducing the grain size of the depositing metal, and restraining perpendicular grain growth by their adsorption. The banded structure in the case of bright nickel deposits based on sulphonic acids or similar compounds would then be the result of a reduction cycle, whereby these compounds are reduced to mercaptans at points of high current density. The polar sulphur group is then adsorbed to form an insulating inhibitor film. The non-polar portion of the molecule is next electrolytically decomposed to sulphur, which is left in the deposit; the preferential attack on this constituent by the etching acid gives the characteristic banded structure when a sectioned bright nickel deposit is etched. Hoar [41] has suggested that either the molecules rapidly adsorb and desorb whilst maintaining a nearly complete equilibrium monolayer at the cathode, or else in the case of certain molecules, they adsorb preferentially where selective deposition may occur.

The adsorption theory is supported by the fact that addition agents are lost at a greater rate than can be accounted for by drag-out and are found in deposits of bright nickel by analysis. Dissolution of bright nickel deposits containing sulphonates leave a white residue.

There appears to be no overall relation between preferred orientation in a bright nickel deposit and the brightness of the plate[42, 43]. This applies to a very large number of brightening agents which have been tested. X-ray diffraction patterns have shown that it is possible to have as high a degree of fibering in a very dull deposit as in a bright one, or to have any degree of fibering in both bright and dull deposits.

According to Evans[44] the predominant orientations in matt, semi-bright and bright deposits were found to be [100] or [110] or a mixture of both, and for any given electrolytes the [110] orientation was shown to be associated with the higher values of internal stress and hardness; the use of higher deposition temperatures, corresponding to a decrease in internal stress and hardness resulted in the [100] texture.

Brightness could not be correlated with either a high degree of orientation or a grain size smaller than the wave length of light. Neither was there any connection between brightness and orientation, the grain sizes of matt, semi-bright and bright nickel deposits being similar. Brightness appears to be the result of a smoothing action during the formation of the deposits, the free surface of the grains being flat and parallel to the general surface direction. The fact that the deposits also showed a high degree of preferred orientation is incidental to, and not the cause of, the brightness.

9.4.10. LEVELLING

An important development in bright nickel plating was the introduction of levelling agents. These are usually organic materials which reduce the tendency for the deposit to reproduce scratches and imperfections in the basis metal. Hence, as the deposit builds up, smooth bright nickel is obtained even when the degree of polish of the basis metal is relatively poor. There is evidence that in levelling, growth of the coating is weaker in the lower than in the middle and upper portions of the profile, with the result that voids may be formed and subsequently covered by deposited metal[45].

Watson and Edwards[46] investigated the levelling action of coumarin, thiourea and saccharin and came to the conclusion that levelling is due to variations in cathode potential between peaks and recesses of a rough surface brought about by differences in the rate of diffusion of addition agents to these points once a diffusion layer has been established. Like brighteners, levelling agents greatly increase cathode polarisation when they are present within a certain concentration range. The concentration of levelling agent is very small and appears to be critical if a simple additive is used; thus the maximum levelling power of butyne diol is achieved at a concentration of 8 millimoles/litre [47]

Rogers, Ware and Fellows[48] used a radio-active tracer technique to show that more thiourea is adsorbed on surface peaks than in recesses. Due to the higher concentration of levelling agent on the peaks more of it is deposited with nickel at these points, thus forcing nickel to deposit preferentially in the recess. Deposition from levelling solutions generally gives rise to deposits with coarser laminations at recessed points. Fig. 9.5 shows the structure of this type of nickel deposit, but in some cases the columnar structure of unimpaired growth may be found.

Coumarin is a very useful addition agent producing semi-bright levelled deposits

Fig. 9.5. Deposit from a bright nickel plating solution.

TABLE 9.3

Effect of some addition agents on stress in nickel plating[50]

Addition agent	*Formula*	*Concentration in g/l to give* 0.001 M	— *Group*
Saccharin (sodium salt)	$C_6H_4SO_2$ N Na . CO . $2H_2O$ (ring closure bracket)	0.24	2
p-Toluene sulphonic acid	CH_3 . C_6H_4 . SO_2OH	0.17	2
Coumarin	C_6H_4 . O . CO . CH:CH (ring closure bracket)	0.15	2
Thiourea	NH_2 CS NH_2	0.076	3
Thioacetamide	CH_3 CS . NH_2	0.075	3
Thiosemicarbazide	NH_2 . NH CS NH_2	0.091	3
Urea	NH_2 CO . NH_2	0.060	2
Acetamide	CH_3 CO NH_2	0.059	2
Semicarbazide	NH_2 . NH CO NH_2	0.075	2
Succinonitrile	CN . CH_2 . CH_2 CN	0.080	1
Succinimide	$(CH_2\ CO)_2$ NH	0.099	2
Glycine	NH_2 CH_2 COOH	0.075	2
Adipic acid	COOH $(CH)_{24}$ COOH	0.15	2
Sucrose	$C_{12}H_{22}O11$	0.34	2
Acetone	CH_2 CO CH_2	0.058	2
Chloral hydrate	CCl_3 $CH(OH)_2$	0.17	3
Hexamine	$(CH_2)_6N_4$	0.14	1

with good mechanical properties. Unfortunately coumarin and many of its derivatives break down during plating to form decomposition products which adversely affect the properties of the coating and reduce the levelling action of the solution[49]. Melilotic acid molecules are the main decomposition product and these are incorporated in the nickel deposit[50].

9.4.11. EFFECT OF ADDITION AGENTS ON STRESS IN NICKEL DEPOSITS

With respect to their effect on internal stress, addition agents can be divided into three groups[51] depending on whether they:

(1) Continuously increase stress with increasing concentration
(2) Reduce stress to a fixed value which is then unaffected by a further increase in concentration
(3) First lower then raise stress as concentration is increased.

The effect of a large number of addition agents on the internal stress in nickel deposits has been reported in the literature[51–52]. The results obtained by Watson[51] who studied the effect of some 22 addition agents are summarised in Table 9.3, and the behaviour of three of these substances (one from each group above) is shown graphically in Fig. 9.6.

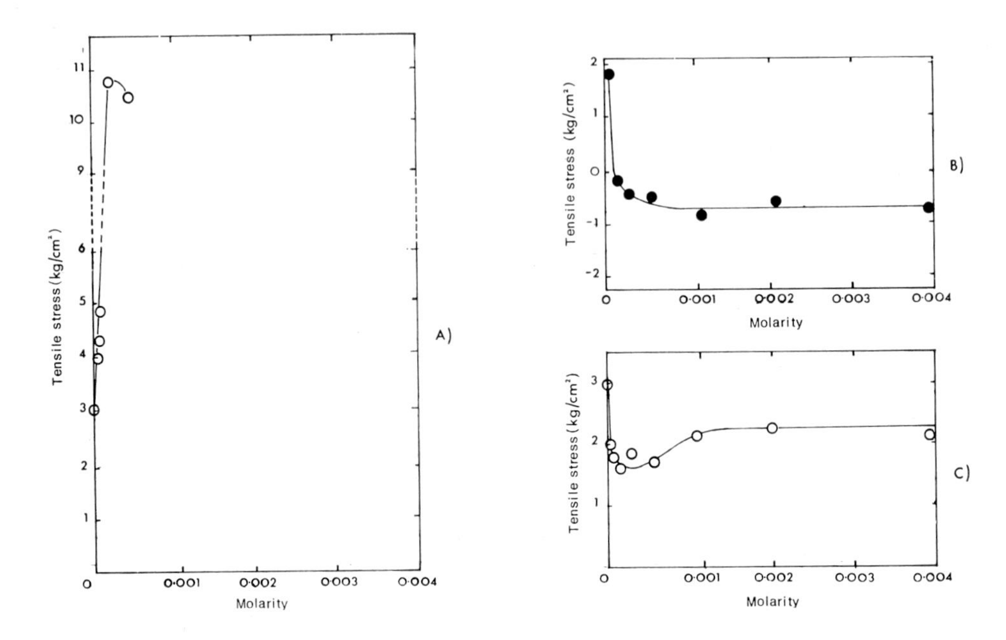

Fig. 9.6. The effect of addition agents on the internal stress of nickel deposits (A) Succinonitrile (group 1). (B) Saccharin (group 2). (C) Chloral hydrate (group 3).

Since the stress due to the pressure of one particular substance can be cancelled out by addition of a second agent which affects the stress in an opposite sense, it is normal commercial practice to add a further substance besides brighteners and levelling agents to counteract stresses in the deposit.

9.4.12. EFFECTS OF IMPURITIES AND THE PURIFICATION OF NICKEL PLATING SOLUTIONS

All bright nickel solutions are susceptible to the effects of impurities. Roughness results from the presence of suspended matter, whilst dissolved substances may also affect the character and brightness of the deposit. Nickel solutions are probably more sensitive to impurities than almost any other types of solution in commercial use. The chief sources of contamination are: (a) anode sludges; (b) metallic impurities, such as copper, iron and lead, introduced from metals in contact with the solution or by articles being plated; (c) dust and grease from the atmosphere, from compressed air, or from substances entering the solution fortuitously on articles being plated; (d) insoluble constituents arising from the use of hard water for making up the solution or present in plating chemicals; (e) organic contamination, resulting from the breakdown of organic addition agents.

Anode sludges can be largely prevented from entering the solution by the use of suitable anode bags, but a certain proportion of fine sludge seeps through the material of which the bags are made, whilst accidental tearing of the bags may cause trouble. All forms of insoluble materials of this kind tend to settle on the cathodes, particularly on horizontal or inclined surfaces, despite agitation of the solution, resulting in rough deposits.

The effect of metallic contamination (Cu^{2+}, Zn^{2+}, Cr^{6+} and Fe^{3+}) on the visual appearance of nickel deposits has been assessed by Dennis and Fuggle[54]. A scanning electron microscope technique was used to examine Hull Cell panels (see Section 9.4.13) and the following conclusions were drawn:

(a) Copper has the most drastic effect on nickel solutions and results ultimately in the production of black deposits of dendrite form
(b) Zinc brightens dull deposits under some conditions and does not lead to dendrite growth
(c) Chromium, even in low concentrations, inhibits the deposition of nickel
(d) Iron affects the appearance of the deposit much less than copper or zinc.

Iron is a common constituent of nickel solutions, particularly where ferrous metals are being plated. Normally, the presence of a little iron is not deleterious, but in high-pH solutions basic ferric hydroxide particles may cause roughness. The trouble is especially prone to occur in air-agitated baths used for plating on steel, as the air oxidises the ferrous salts to the insoluble ferric state. Small amounts of sodium fluoborate can be added to such baths with advantage to sequestrate the iron. Copper and zinc adversely affect throwing power by acting as cathode depolarisers when they are present in excessive amounts. They are particularly undesirable in bright nickel solutions. As little as 0.03 g/l of zinc manifests itself by causing some brightening and embrittlement of the nickel deposits, especially at low current densities. More than 0.3 g/l of zinc in a warm nickel solution is sufficient to brighten the deposit; cracking and poor corrosion resistance develop and reach serious proportion at about 0.6 g/l of zinc. Streaky deposits appear at zinc contents below this figure, but the streakiness is less evident at higher bath temperatures. Zinc can be removed most economically by electrolysis at current densities of 0.2 to 0.4 A/dm^2 down to a concentration of 1 mg/l.
Chemical methods are available for the purification of nickel solutions which have become contaminated. Precipitation of iron, copper, zinc and trivalent chromium can be effected by adding nickel carbonate until a pH of about 6.2 is reached, followed by filtration of the precipitated compounds. A pH of 6.3 should not be exceeded since nickel hydroxide commences to precipitate at this stage. This method will, however, not remove zinc when concentration is less than 250 mg/l. Copper in a nickel plating solution progressively reduces the salt spray corrosion resistance of the deposit; 10 mg/l of copper in a Watts' solution in one series of experiments reduced the corrosion resistance by 30 per cent, whilst at 40 mg/l of copper the corrosion resistance of a nickel deposit 25 μm thick was halved[55].

It is desirable wherever possible to arrange for transfer from the bright nickel to the chromium tank without unracking, since it is in this way that the greatest economies are obtained and this procedure is, in fact, universally adopted in automatic nickel and chromium plating machines.

Care must, however, be taken to ensure that the racks do not trap solutions. If this occurs, chromic acid may eventually be introduced into the nickel solution where it will reduce the efficiency of deposition very considerably, and if sufficient enters the bath its effect may be to render the bath unusable. Rack insulation kept in good condition minimises this danger, but it is a useful precaution to dip racks in an oxalic acid or sodium hydrosulphite solution after passage through the chromium tanks in order to reduce any chromic acid that may have been retained in them.

According to Macnaughton and Hammond(56), the presence of 0.025 to 0.20 g/l of chromic acid in a nickel solution caused a decrease in cathode efficiency, and at 0.220 g/l deposition ceased whilst the evolution of hydrogen increased; 0.01 to 0.025 g/l resulted in an increase in the stress of the deposit with some exfoliation. Chromium sulphate, when added to give a chromium concentration of 0.0149 g/l caused the cathode efficiency to fall to 81 per cent. The deposit increased in brightness with a tendency to peel, but the anode efficiency was not affected. Hothersall and Hammond(57) found that chromic acid increased the evolution of hydrogen at the cathode and did not stop pitting. The lower the pH of the nickel bath, the more hydrogen was evolved, and the higher the stress in the deposit.

Chromium contamination is difficult to remove. Electrolysis at a pH of 2, using very high current densities in conjunction with lead anodes, enables the chromium to be removed as lead chromate(58). Chromium can also be removed by precipitation as lead chromate by the addition of lead carbonate(59) but the introduction of lead salts is best avoided, certainly in the case of bright nickel plating solutions.

In current practice, filtration of the solution is an essential method of maintaining the required degree of freedom from contamination for satisfactory operation of the process, and this may be carried out either continuously or intermittently. Diatomaceous earth is commonly employed in filters to remove suspended matter, but where contamination with organic material has occurred, the use of activated carbon is recommended. This will remove substances such as emulsified oil, organic degradation products of addition agents, and size which may have been introduced on anode bags. The usual procedure is to add about 0.5 to 1 per cent of activated carbon by weight to the bath, heat to about 60°C, agitate for a period, and then filter, when the contaminant will be removed together with the carbon.

The sensitivity of nickel plating solutions to the presence of foreign metals and the difficulty of finding metals to withstand the solution, especially in the presence of possible stray electric currents, has led to a search for suitable non-metallic materials from which pumps, filters, etc., can be constructed; rubber, glass, stoneware and plastic materials have all been employed to some extent. However, titanium has proved a useful metal for the manufacture of equipment (such as heat exchangers and anode baskets) for use in nickel plating solutions. Titanium is not attacked by nickel compounds when it is anodic, cathodic or electrically neutral.

9.4.13. CONTROL BY PLATING CELLS

A useful method of controlling impurities in bright plating solutions is by means of specially designed test plating cells. One of the best-known types is the Hull cell (Fig. 9.7). The function of this cell is to provide by geometric configuration a cathode deposit on a flat surface which reproduceably records the character of the deposit formed at all current densities within the operating range. The cathode plate is inclined at an angle to the anode, so that by measuring the distance of the bright plating range from the end of the plate the limiting current density can be derived from a curve. Polished cathodes about 6 cm by 10 cm are satisfactory and the current used should be about 3 amps at 12 volts. A certain amount of experience is required in interpreting the results of thes tests.

A simple alternative to the Hull Cell is to plate a brass strip bent to an angle directly in the bath.

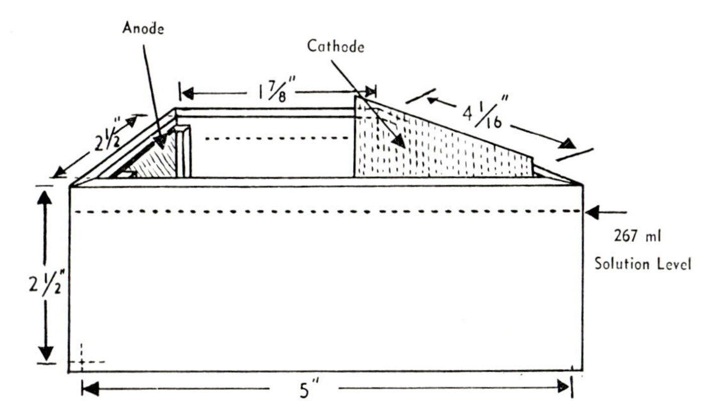

Fig. 9.7. 267 ml capacity Hull cell. All measurements given are internal and in inches.

9.4.14. PLATING OF ZINC-ALLOY DIE-CASTINGS

Zinc-alloy die-castings are now very widely used for many components in the automobile and other industries. The cheapness and rapidity with which such fittings as motor-car horns, handles, radiator grilles, and innumerable components can be produced make this technique a very attractive one.

Die-cast parts are employed for mechanical components inside assemblies, and in such cases little or no protection of the metal is needed. A chromate conversion treatment is sometimes used, however, to inhibit the formation of white corrosion products on the metal surface, especially where close fits on moving parts are concerned.

For external use, some finishing is essential, and although organic finishes are employed, chromium plating is probably the most generally useful and serviceable finish to apply. For plating, die-castings should be free from surface imperfections and have a good dense skin. Normally this skin is only about 300 μm thick, so that if the surface of the casting is not smooth it may be cut through in polishing, thus exposing the porous interior of the casting. Within limits thin walls are helpful inasmuch as they promote rapid cooling of the casting and hence, in a sense, a smooth surface. Flat, plain surfaces should be avoided in design as they emphasise slight surface imperfections. The introduction of curves or fillets is helpful in obtaining a good appearance after plating. Likewise sharp corners should be avoided, whilst deep recesses are undesirable as they make the castings difficult to clean and liable to trap polishing compound, which is not easily removed before plating. The introduction of zinc into the plating solution may also occur as a result of chemical attack on such areas. If such recesses are required to have a bright finish they may also prove difficult to reach with the polishing wheel.

9.4.14.1. Polishing

The polishing of zinc-alloy die-castings is fairly simple and is best carried out with a tripoli composition on calico wheels. The most important point is to avoid excessive pressure as, if the the surface skin be penetrated, the porous interior of the casting may be exposed with increased liability for plating difficulties, such as blistering and pitting, to occur; the durability of the plate will also be poor. More recently, abrasive belt polishing has become increasingly popular. After polishing down to about 220 mesh abrasive, the castings are buffed on soft muslin wheels to produce a high surface finish. Small parts may be barrel-polished prior to plating.

9.4.14.2. Cleaning

In cleaning prior to plating it is desirable to degrease the polished castings. Vapour degreasing may be adequate in certain cases, but when intricately shaped articles are involved a liquor-vapour plant is essential to facilitate the removal of polishing composition residues. Ultrasonic vibration is useful in conjunction with solvent cleaning, particularly where any length of time elapses between polishing and cleaning. The degreasing operation should be followed by electro-cleaning in a suitable hot alkaline solution to disperse any non-greasy material left after solvent degreasing, together with soap or other films on the metal surface. Cathodic cleaning is most commonly used, since this gives the maximum gas liberation and hence the greatest dispersing action; sometimes anodic cleaning is employed. In some plants the work may with advantage be made cathodic followed by anodic cleaning for a few moments to remove any residual substances, which may be electrolytically deposited during the cathodic phase.

The use of anodic cleaning, as opposed to cathodic, has been advocated by Lewis[60] to reduce blistering on the ground that cathodic cleaning is likely to lead to hydrogen absorption to which some types of blistering may be ascribed.

Strong alkalis attack zinc alloys with the formation of discoloured films on the metal surface. Hence only mild alkaline cleaners having a pH not in excess of about 10.2 should be used. A satisfactory cleaner may be made by dissolving 10 to 20 g/l of trisodium phosphate in water and operating the bath at a temperature of 80°C to 95°C.

The surface tension of such solutions is relatively high, but soaps cannot be successfully used as a cleaner ingredient to reduce this undesirable feature of the cleaning solution, since they are decomposed at pH values of less than 10.2. Some of the synthetic wetting agents which are stable in the low pH range are useful in such cleaners, although those excessively prone to foaming are best avoided for practical reasons. A number of proprietary cleaning compositions are available which give excellent results; metasilicate-containing materials are to be viewed with suspicion, however, as they may cause undesirable film formation on the metal surface.

The time of cleaning is important, since blistering and poor adhesion of the deposit sometimes occurs on surfaces which have been either over-cleaned or insufficiently cleaned. Zinc is a reactive metal and all pre-plating treatments should be as short as possible.

After cleaning, the articles are well washed in running cold water and then dipped momentarily in a dilute solution (about 5 per cent) of sulphuric or hydrochloric acid, until gas evolution is observed. A solution of hydrofluoric acid is favoured in some quarters for this purpose. The acid dip removes the slight oxide film usually left on the metal surface, since such films are highly subject to hydrogen embrittlement. Such a brittle interlayer between the plate and the basis metal would lead to poor adhesion. The articles are then again thoroughly washed, and are now ready for plating.

9.4.14.3. Plating

Originally direct nickel plating from a high-sulphate solution was used for zinc-plating die-castings, but the process was rendered obsolete by the introduction of bright nickel. Prior to bright nickel plating a copper undercoat is essential and this is applied from a cyanide solution of the usual type, but special solutions giving fine-grain deposits have been developed for this purpose. The free cyanide is controlled at about 6 g/l and the temperature at 40°C. Current densities of 1.5 to 2.0 A/dm^2 are usually employed.

The Rochelle salt copper plating bath is also useful; with this solution, current densities of up to 6 A/dm^2 can be used, smooth deposits being obtained. The thickness of copper required depends on the shape of the articles and to some extent on the pH of the nickel solution employed. Bright copper solutions can be employed under nickel, but smoothness is more important than brightness for this application. With bright nickel solutions operating at pH 4.3 to 4.6, it is unsafe to have less than about 7 μm of copper undercoating. It is found that copper deposits diffuse into zinc quite rapidly, so that an adequate thickness of this metal must be present not only to prevent attack on the zinc by the nickel solution but to ensure that the copper undercoat is not absorbed during the lifetime of the article. If this occurs, failure of the nickel due to lack of adhesion may subsequently ensue. After coppering, the die-casting is rinsed successively in hot and cold water, dipped in 5 per cent sulphuric acid to neutralise alkali, rinsed again, and transferred to the nickel plating bath.

A thin copper undercoat adds little or nothing to the protective value of the plating and the amount of nickel required to meet particular service requirements is not reduced.

9.5. Sulphamate nickel plating

Nickel was first deposited from a solution of nickel sulphamate by Piontelli and Cambi in 1938[61]. The bath, although more expensive to prepare than the Watts bath, is nowadays used on a substantial scale particularly for electroforming components in the engineering industry. The good throwing power of the solutions and low internal stress in the deposits produced make it very suitable for this purpose. Electroforming applications of the sulphamate nickel bath include the manufacture of printing plates and electrotype[62] and the preparation of gramophone record matrices[63]. The bath also finds use in the production of decorative nickel finishes[64]. Nickel sulphamate ($Ni(NH_2SO_3)_2$) is a salt of the strong monobasic sulphamic acid NH_2SO_3H. The acid is similar in chemical structure to sulphuric acid:

Sulphamic acid	Sulphuric acid
O ‖ OH—S—NH$_2$ ‖ O	O ‖ OH—S—OH ‖ O

One hydroxyl group in the sulphuric acid molecule is replaced with an amino group in the sulphamic acid. In aqueous solutions, the sulphamate ion $NH_2SO_3^-$ hydrolyses to give ammonium and sulphate ions:

$$NH_2SO_3^- + H_2O = NH_4^+ + SO_4^=$$

which can increase the internal stress of the deposit (Fig. 9.8). The rate of hydrolysis of sulphamate ions in commercial baths is normally very low provided the temperature and pH are not allowed to vary too much. If the temperature were raised to 80°C, about 8% of a 30% nickel sulphamate solution would be decomposed every hour[66].

A large number of sulphamate bath compositions have been recommended in the literature and a selection of these, given by Hammond[67], is listed in Table 9.4. A temperature of 30-50°C is normally maintained and the solutions are buffered to a pH of 3-4 by means of boric acid.

Although it has been claimed that nickel anodes will actively dissolve in sulphamate solutions in the absence of chloride ions, the majority opinion favours the use of chloride to ensure efficient anode dissolution. Chloride should not, however, be present in excess since this causes an increase in the tensile stress of the deposit. The anode dissolution in sulphamate baths may greatly influence the stress and properties of deposits produced[68]

Nickel sulphamate solutions are very sensitive to the presence of impurities and consequently require careful preparation and purification prior to use. For commercial use nickel sulphamate can be obtained in the form of a pure concentrated solution ready to use. If the unpurified variety is obtained it must be carefully purified according to an elaborate treatment sequence[67] which entails the addition of activated carbon

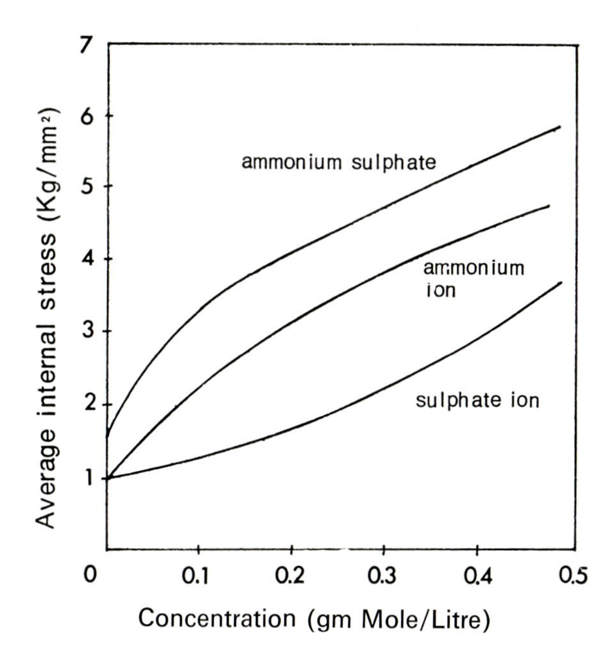

Fig. 9.8. Effect of hydrolysis products of nickel sulphamate on the internal stress of deposits produced from the sulphamate nickel bath[65].

TABLE 9.4

Selected Sulphamate Nickel Plating Solutions[67]

Description	Nickel sulphamate g/l	Boric acid g/l	Nickel chloride g/l	Other additions
Barrett types	450	30	—	
Barrett types	450	30	—	Anti-pit agent 0.4 g/l; SNSR stress reducer
Ollard and Smith	320	27	—	
Diggin – Bath 1	300	30	30	—
Bath 2	300	30	30	Sodium naphthalene 1,3,6-trisulphonic acid.
Albright & Wilson				
Solution 1	350	35	5	
Solution 2	310	31	31	PbNi-6.2 g/l proprietary hardening and stress-modifying agent.
Fanner and Hammond	340	30	3.3	Sodium lauryl sulphate 0.1 g/l as anti-pit, optional.
Kendrick	600	40	5	

filtration and low current density pre-electrolysis. Nickel sulphamate solutions can be contained in rubber- or plastic-lined tanks. Lead linings must not be used since serious contamination of the bath would ensue. The solution must be agitated and this is done by either mechanical means or by using compressed air.

9.5.1. ADDITION AGENTS

In the absence of addition agents, conventional nickel sulphamate solutions give ductile, smooth, low-stress deposits which are fairly soft. Organic hardening agents may be added to the bath to induce slight compressive stress in the deposit and harden it. Naphthalene trisulphonic acid[69] is typically used for this purpose. Cobalt added in small quantities to the bath will also harden the deposit, but gives rise to increased tensile stresses and some reduction of ductility (Fig. 9.9).

Since sulphamate nickel deposits are prone to pitting caused by the adherence of hydrogen bubbles during plating (particularly in unagitated solutions) it may be necessary to add an anti-pitting agent to the solution to reduce its surface tension. For this purpose Newall[70] recommends the use of sodium lauryl sulphate at a concentration of 0.15 g/l.

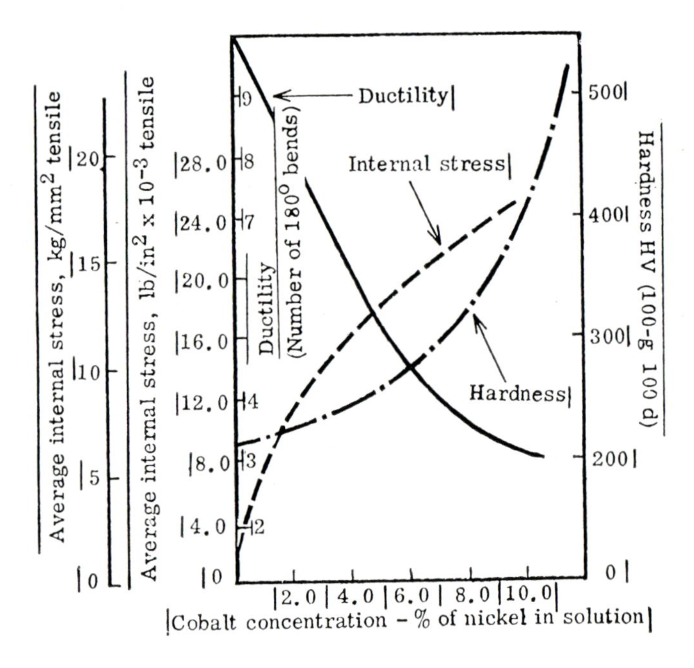

Fig. 9.9. Effect of cobalt additions on the physical properties of deposits from a sulphamate nickel bath.[65]

9.5.2. HIGH CONCENTRATION SULPHAMATE NICKEL BATH

Research work carried out since 1963 in the laboratories of International Nickel Ltd. has resulted in the development of a high-speed plating solution[71–73] (Ni-Speed process).

A high rate of plating is obtained by increasing the rate of diffusion of nickel ions to the cathode. The highest temperature possible without causing hydrolysis is used (60-70°C), as well as vigorous agitation of the solution. The optimum composition was found to be:

600 g/l Nickel sulphamate
40 g/l Boric acid
5 g/l Nickel chloride

Current densities of up to 40 A/dm^2 may be employed, corresponding to a maximum deposition rate of 1.5 mm of nickel per hour, which makes the bath very valuable for rapidly building-up thick deposits during electroforming. The deposits produced at high current densities are hard and dull, whilst those at lower current densities are semi-bright and even lustrous at current densities less than 6.5 A/dm^2. The fine-grained structure of a low current density nickel sulphamate deposit is shown in Fig. 9.10.

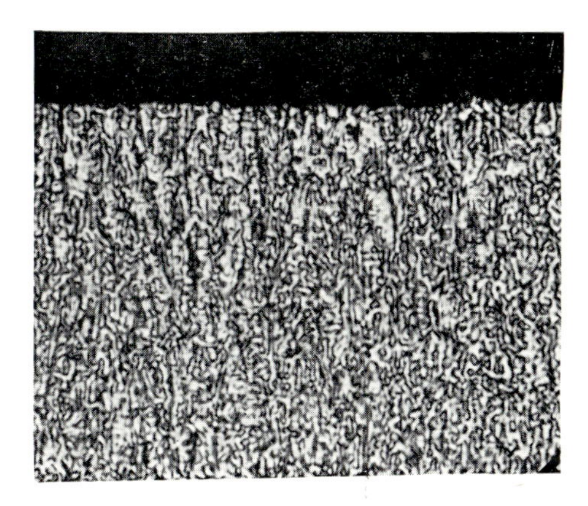

Fig. 9.10. Cross section of a fine-grained nickel from a concentrated sulphamate nickel bath (600 g/l) at 60°C and at a current density of $4A/dm^2$. × 1000.

9.6. Chloride solution

The use of a nickel chloride solution in place of nickel sulphate was advocated by Wesley and Carey[74] in 1939, the bath consisting of:

Nickel chloride $NiCl_2.6\,H_2O$	180 g/l
Boric acid	20 g/l
pH	1–5
Temperature	50–65°C

The pH is adjusted by the addition of hydrochloric acid. It is claimed that the deposit is smooth and easily polished and that the solution can be operated at very high current densities without modulation or treeing.

The chloride nickel solution has the disadvantage however, of being highly corrosive, and is now used only to a limited extent. High-chloride baths have, however, been employed commercially as a basis for bright nickel deposits.

9.7. Fluoborate nickel plating

A nickel fluoborate bath is being used to some extent for electroplating, electrotyping and in plating stereos. It is claimed to produce relatively stress-free deposits, whilst plating rates of up to 0.002 cm in 15 minutes are feasible. A recommended bath contains:

Nickel fluoborate	$Ni(BF_4)_2$	350 g/l
Free fluoboric acid	$H\,BF_4$	4.38 g/l
Boric acid	H_3BO_3	30 g/l
pH		2–3.5
Temperature		45–80°C

To improve anode dissolution, small quantities of nickel chloride may also be added, but normally this is not necessary; nickel anodes will dissolve much more readily in a fluoborate bath than in a chloride-free sulphate solution. Titanium anode baskets cannot be used in fluoborate solutions because of the susceptibility of the metal to rapid attack in the presence of the fluoride ions.

The fluoborate bath is expensive to prepare and must be carefully purified before use.

9.8. Electroless deposition of nickel

A method of auto-catalytically depositing nickel without the use of current was developed by Brenner in 1946. A hypophosphite solution is used in conjunction with nickel chloride, the solution being either acidic(75) or alkaline(76). The function of the hypophosphite is to reduce nickel ions:

$$Ni^{2+}+H_2PO_2+H_2O \rightarrow Ni+2H^{+}+H_2PO_3$$

Gutzeit(77, 78) has suggested that the reduction reaction is more complicated and proceeds via catalytic dehydrogenation of the hypophosphite ion, producing nascent hydrogen gas and metaphosphite ions. Nickel ions are reduced to nickel metal by nascent hydrogen at the catalytic surface (Ni or palladium). This scheme is represented by:

$$(1)\ H_2PO_2^{-} \xrightarrow[\text{Pd or Ni}]{\text{catalyst}} PO_2^{-}+2H$$

$$(2)\ Ni^{2+}+2H \longrightarrow Ni+2H^{+}$$

$$(3)\ PO_2^{-}+H_2O \longrightarrow HPO_3^{2-}+H^{+}$$

The deposited nickel surface can then initiate reaction (1), so that the cycle is continuous. Phosphorous is also produced and up to 15% may be incorporated in the deposit, but the amount is usually about 8%.

The bath is operated at high temperatures (85°–95°C) and the deposit produced is smooth, non-porous and uniform. According to several workers[79–80] the deposit is an amorphous solid, but later work by Graham[81] suggests that electroless nickel deposits have a definite crystalline structure.

As no current is employed, coverage can be obtained in deep recesses and tubes without difficulty. Deposition is at the rate of up to 50 μm per hour, but developments of the process have enabled deposition rates of 10-20 μm per hour to be obtained from baths having greater stability. Many different buffering and complexing agents have been recommended for use in electroless nickel plating solutions, including acetates, citrates, lactates, pyrophosphates, succinates, and amino compounds.

Nickel chloride must be added to the bath from time to time, and also alkali to restore the pH, since hydrochloric acid is liberated by the reaction. Agitation is necessary to dislodge gas bubbles, whilst a high degree of filtration is recommended as particles can act as nuclei for the deposition of metallic nickel, resulting in wastage of metal.

Tank materials for the process present a difficult problem; even glass will scratch and provide a nucleus for the deposition of nickel. Heat exchangers and tanks made of phenolic resin-coated stainless steel have given reasonably good service. Racking presents no problems as shielding does not affect deposition in any way.

The borohydrides or alkyl amine boranes are also used as reducing agents for the production of electroless nickel deposits[82]. The deposition rate is 10-30 μm per hour. The deposits have a hardness of 500-750 Hv as plated, and 1000-1250 HV after heating to 400°C. They contain 7-8% of boron.

9.9. Plating defects

Nickel deposits under some conditions are subject to the formation of pits or slight depressions in the plated surface which reduce the protective value of the coating and are also unsightly in appearance, particularly when the nickel has been polished. These pits do not always penetrate to the basis metal, and the defect is to be distinguished from porosity in the nickel, which will be discussed in greater detail below.

9.9.1. PITTING AND ITS CAUSES

The causes of pitting have been the subject of considerable investigation and some aspects of this defect have already been dealt with. Although, in some instances, apparent pitting may result from a porous condition of the basis metal, true pitting is the result of the attachment of a bubble of hydrogen to the cathode during deposition, which prevents further plating locally. One major cause of pitting is the presence of finely divided basic material in the vicinity of the cathode[83]. These basic compounds of nickel (and particularly of iron, where this metal is present in the solution to any extent), favour the retention of hydrogen bubbles at the nickel surface. This has been emphasised by Hothersall and Hammond[84], who point out that local inclusions in the metal and surface imperfections may cause bubbles to form and adhere, and that these irregularities are, therefore, to be regarded as the primary causes of pitting rather than the hydrogen bubbles. Pitting, when it develops in Watts nickel solutions, may be countered by the addition of small amounts of hydrogen peroxide.

Excessive quantities of oxidising agent must be avoided, however, especially in low-pH solutions containing any appreciable amount of iron since it lowers the cathode efficiency and the throwing power of the solution, and may produce brittle deposits. An excess is also liable to cause burning of the nickel. A typical addition is of the order of 1 litre of peroxide per 10,000 litres of solution twice daily.

9.9.1.1. Use of wetting agents

Wetting agents have also been recommended, and some of these materials are very effective. Very small concentrations, of the order of 0.02 to 0.04 per cent, lower the surface tension of the solution to about half the normal value, thereby reducing the tendency for hydrogen bubbles to adhere to the cathode. Straight chain alkyl naphthalene sulphonates have been found particularly useful, but the amounts employed must be carefully controlled or their presence may result in a substantial reduction in cathode current efficiency and permissible current density[85]. Certain wetting agents must be avoided entirely, as they may lead to the formation of brittle deposits.

9.9.2. LOW ADHESION

Lack of adhesion of the deposit is often due to poor cleaning, but other conditions can also cause the trouble to occur. If a nickel deposit on a zinc alloy die casting, when removed, brings a layer of zinc with it, it is probable that the surface of the casting itself is laminated due to injection into excessively cold dies. If the weak layer is not too deep a longer acid dip prior to copper plating may serve to remove it, although it is best to correct the fault at the casting stage.

The presence of chromium compounds in the cleaners used will cause poor adhesion and blistering. The chromium enters usually by being carried on plating racks used for direct nickel and chromium plating. The addition of sodium hydrosulphite to the cleaner is recommended to eliminate this by reducing the chromic acid, whilst small amounts of sodium bisulphite added to the bright nickel bath have also been recommended[86]. It is noteworthy that chromium is far more deleterious in cleaners which are used cathodically than when they are anodic.

9.9.3. POROSITY IN NICKEL ELECTRODEPOSITS

The protective value of commercial nickel coatings is dependent to some degree on the absence of porosity, particularly as in many instances the nickel coating is cathodic relative to the basis metal. A detailed investigation into the causes of porosity in nickel deposits by Hothersall and Hammond[84] led them to the conclusion that the chief factors are: (a) non-conducting or poorly conducting areas in the surface of the basis metal (e.g. inclusions of scale, polishing composition, etc.); (b) the shielding action of gas bubbles or of particles of solid matter derived from the solution or loosened in etching from the basis metal; and (c) the failure of the deposit to cover the surface of crevices in the basis metal because of poor throwing power or because of exclusion of electrolyte by trapped gas.

Pinner has found that imperfections in steel[87] such as pits and slag inclusions have a most pronounced effect on the corrosion resistance of the plate. These defects exerted the above influence in spite of mechanical polishing or of any normal cleaning procedure. Fragmented metal particles such as are produced in abrasive polishing are very deleterious, and pre-finishing procedures should be designed to remove these as far as possible. A final polishing operation is partially effective in bringing this about, but an anodic etching treatment of about 30 second duration in a sulphuric acid bath of 55^0 Bé at a c.d. of 150 A/dm^2 is more effective. To avoid smut formation as a result of this treatment, Pinner recommends that the etch be followed by a rinse and then in turn by an anodic alkaline electrolytic cleaning operation. When the smut has been removed, the metal is water-rinsed and given a dip in dilute hydrochloric acid. It is then ready for nickel plating.

It is to be expected from the nature of the growth of an electrodeposit that gaps and pores should occur in its structure by virtue of the fact that many crystal forms are simultaneously taking shape in close proximity to one another, so that interference between them is bound to occur. Therefore all deposits are to some degree porous, although this porosity can be greatly influenced by the factors already described.

The work of Wesley and Knapp[88] suggests that the effect of porosity on the weathering of nickel plate has probably been exaggerated and that much of the porosity thought to exist in deposits may have been created by the corrosiveness of the media used for their detection. For example, simple immersion of a piece of nickel foil 0.0004 cm thick for one hour in a solution of 0.5 g/l of potassium ferricyanide and 60 g/l of sodium chloride, as used for porosity testing, either creates or enlarges a multitudinous number of perforations. The authors took the view that a good way to begin a study of the mechanism of corrosion of nickel/chromium coatings would be to observe the weathering behaviour of nickel foils. This was done in an industrial atmosphere and to a lesser extent at a marine site; the factors represented in the tests included microstructure and chemical composition of the nickel, heat treatment, and the presence or absence of the outer film of chromium. The observed behaviour of wrought and electrodeposited nickel foils and of nickel-plated steel were sufficiently alike to raise serious doubt of the significance of as-plated porosity in the weathering of the latter. A chromium film on such foils was helpful in the early stages of weathering in the industrial atmosphere, but increased the number of perforations during the later stages. (cf. p. 361).

9.9.4. ROUGH DEPOSITS

Roughness is a serious defect in nickel deposits and is often difficult to eliminate. Roughness troubles increase with increasing thickness of deposit, and apart from the appearance factor such deposits have a markedly inferior corrosion resistance. Particles of metal and sludge entering from the anodes are a common cause of roughness of a type which is characterised by the fact that it appears mainly on the upper surfaces of articles where the particles have settled by gravity. Metallic contamination of this kind may also enter from the work or from the plant itself, and must be dealt with by filtration.

Polishing can also result in rough work, especially on steel. If the final polishing operation is carried out with a dry wheel, small particles of steel can adhere to the work by magnetic attraction and act as centres for the growth of nodules.

Hart and Tottle [89] have recently discussed the use of various techniques for diagnosing the cause of roughness in decorative nickel-chromium deposits.

9.10. Tests and specifications

In the testing of nickel deposits the thickness of the metal is most important, and can be determined by means of the method of sectioning and polishing a plated sample article. A much more rapid method which is sufficiently accurate for control purposes is the B.N.F. Jet Test, in which a solution of ferric chloride and copper sulphate is directed on to the deposit. The time of penetration of the coating at a particular temperature is proportional to the thickness according to a definite scale.

The thickness of a nickel deposit may also be determined by measuring the quantity of electricity necessary to anodically dissolve a specific area completely[90]. The amount of electricity passed is measured on an integrating meter and the thickness read off a calibration chart. A number of instruments are also available for non-destructive thickness testing. The more important of these are based on β-particle back-scatter, eddy currents, and magnetic attractive or magnetic inductive force.

Methods of thickness testing are summarised in British Standard B.S.S.5411: (Part 2): 1978.

The adhesion can be determined by rubbing hard with the back of a penknife or similar instrument. If adhesion is poor, blistering of the deposit will manifest itself. Ollard[91] has described a method whereby an accurate measure of the adhesion in terms of the pull required to separate the coating from the basis metal can be made. A thick deposit of about 2 mm must be applied on the end of a cylinder of the required basis metal. The cylinder is then turned down leaving the coating to protrude in the form of an annular flange which is then engaged by the ring-shaped steel die in a tensile test machine.

The British Standard specification for nickel plating (B.S.S.1224: 1970) lays down the requirements as regards thickness of nickel with and without copper undercoats on various basis metals.

9.11. Durability of nickel-chromium electrodeposits

The durability of a nickel and chromium deposit depends to a large extent on the thickness of the nickel layer. In the case of dull nickel deposits, it has been found as a result of large-scale tests carried out under the aegis of the American Society for Testing Materials, that when the nickel is deposited in two layers the first being polished, improved corrosion resistance has been found in marine and rural atmospheres but not in industrial atmospheres. This, it is suggested, is due to the fact that in the latter case corrosion results from localised attack on the nickel deposit by corrosive atmospheres, and in the former, primarily from porosity in the nickel which is minimised by the use of a double deposit[92].

It has been assumed that bright nickel solutions after prolonged service might tend to give deposits of inferior durability as a result of the accumulation of impurities in the nickel bath. Decomposition products of brighteners were thought to have similar effects. More recent work has shown, however, that the corrosion resistance of the composite deposits of nickel and chromium do not vary in this way, although large variations in stress and ductility may occur which could affect durability in conditions of service involving flexing or vibration.

In the case of bright nickel coatings, the results are more variable, but the conclusions drawn by inspection of samples from these tests is that the protection afforded by the thickest bright nickel coatings (0.007 cm) is less than would be anticipated from comparable dull nickel deposits.

A series of atmospheric exposure tests on various types of bright nickel and chromium deposits on steel pressings has been carried out by Edwards[93]. Specimens were exposed for one year in London and at a coastal site on the South Coast of England. The results indicated that the resistance to corrosion is virtually independent of additions made to sulphate-chloride solutions to promote levelling, low stress, etc. Unsuitable conditions of nickel deposition or the presence of impurities in the nickel plating bath could give rise to defects leading to premature failure. Failure of all the deposits examined appear to have taken place by penetration of the chromium film, followed by local corrosion of the nickel deposits. Penetration of the nickel deposit was initially relatively rapid whilst there were only a few pits, but became much slower when the number of pits increased. There is no evidence that pre-existing porosity of the nickel layer played any important part in the breakdown mechanism.

The conditions of deposition of chromium, however, play an important part in the protective value of the composite deposit. Thus, plating at a slightly higher temperature (about 55^{0}C) and higher current densities permit the deposition of thicker deposits relatively free from cracks. It has been suggested that a hard, bright nickel coating, perhaps containing unduly high internal stresses might be susceptible to spontaneous hydrogen cracking when chromium plated. It would be reasonable to expect such cracks would be bridged over by the time a thickness of the order of 0.00015 cm was reached.

The subject has been intensively studied by H. Brown[94], who concluded that bright chromium deposited at about 55^{0}C shows minimum porosity and stress cracking at thicknesses of 0.00005 to 0.0002 cm, the preferred range being 0.00012 to 0.0002 cm.

Furthermore the tests indicated that deposits from solutions containing high ratios of chromic acid to sulphate, e.g. 125-150:1, which had been known to exhibit minimum porosity for a given thickness of coating afforded the best protection in the corrosive conditions used in the investigation. A further feature of the investigations was the confirmation that sulphur-free deposits give better protection than those containing sulphur.

9.12. Economic processes – nickel-iron alloy deposits

The growing emphasis on low cost, low energy consumption processes producing minimum pollution has led to the introduction of nickel-iron alloy plating baths, and low temperature baths with reduced metal contents.

A bright, levelling nickel-iron bath capable of giving ductile deposits was introduced in 1973 in the USA, and within three years some 1.5 million litres were in commercial use in both manual and automatic plants.[95] The iron content of the deposit can range from 10 to 30 per cent, with metal cost savings of up to 25 per cent as compared with pure nickel. Nickel-iron alloy coatings are being used mainly as an undercoat for chromium on steel articles such as tubular furniture, and for bathroom fittings for moderate conditions of service.

The solution contains about 45 g/l of nickel and 3 g/l of iron (ferrous) together with a hydroxy-carboxylic acid stabilizer, brighteners and levelling agents. It is operated at 55^0-65^0C at a pH of 2.8-3.6 and current densities of 3-4 A/dm^2. Air agitation is needed to permit higher current densities to be used, and to enable low concentrations of iron to be present in the solution whilst high iron contents are produced in the alloy. For the best corrosion resistance, a final alloy layer with a low iron content (applied in the same bath), a copper undercoat and a thin nickel overplate of the type needed to produce microdiscontinuous chromium coatings are recommended.

Commercial bright and semi-bright nickel plating solutions operating at low temperatures and metal contents are also employed. They contain 30-40 g/l of nickel metal (half that of conventional Watts baths) and are used at 45^0-50^0C as compared with 60^0-65^0C for the latter. The cost savings claimed are comparable with those of nickel-iron alloy deposits; drag-out and heat losses, and initial make up costs are about 60 per cent of those of conventional bright and semi-bright nickel plating baths.[96]

9.13. Stripping nickel electrodeposits

Nickel deposits, when defective, are usually stripped anodically in sulphuric acid; a solution of 10 per cent concentration and a current density of 2 A/dm^2 or thereabouts are used. This method can be used for deposits on steel or non-ferrous alloys and will also remove chromium. Nickel deposits can also be electrolytically stripped from steel by using sodium nitrate or ammonium nitrate solutions. The current density employed should be greater than 5 A/dm^2 to avoid dissolution of the basis metal. Chemical methods for stripping deposits are able to deal more satisfactorily with complex shaped components. Chemical stripping solutions commonly used include cyanide solutions containing aromatic nitro compounds, nitric acid solution (with inhibitors) and alkaline ammonia solutions.

REFERENCES

1. J. Hinde, *Ind. Finishing*, 1972, **24,** No. 285, 52-3.
2. I. Rajagopulan, S. R. Rajagopalan and M. A. Parameswara, *Plating*, 1973, **60,** No. 3, 261-2.
3. O. P. Watts, *Trans. Amer. Electrochem. Soc.* 1916, **29**, 395.
4. R. Weston, U.S. Pat. 211,071.
5. W. A. Wesley and J. W. Carey, *Trans. Electrochem. Soc.* 1939, **75**, 209.
6. E. Raub, *Metalloberfläche* 1955, **9**, 88a.
7. E. Raub and A. Disam, *ibid* 1955, **9**, A54.
8. V. Gluck, *Metal Finishing*, 1974, **72,** No. 5, 96-8.
9. W. M. Philips, *J. Electrodep. Tech. Soc.* 1935, **10**, 134.
10. *Products Finishing*, 1974, **38,** No. 8, 74-5.
11. R. Parkinson, *Electroplating and Metal Finishing*, 1974, **27,** No. 2, 8-13.
12. W. A. Wesley, *Trans. Inst. Met. Fin.* 1956, **36**, 452.
13. I. L. Rogelberg and E. S. Schpichinetskii, *Tsvetnye Metally* 1956, **29**, 68.
14. Brit. Pat. 531, 671.
15. S. A. Watson, *Surfaces*, 1972, **11,** No. 4, 53-58, 59-62.
16. Proceedings of 1st Symposium 'Use of Titanium Anode Baskets for Nickel Plating', Witton, I.M.I., 1962.
17. W. Meyer and R. Kreisel, *Galvano Organo*, 1976, **45,** No. 471, 1101-8.
18. S. A. Watson, *Trans. Inst. Met. Finishing* 1960, **37**, 28.
19. K. S. Indira, S. R. Rajagopalan, M. I. A. Siddigi and K. S. G. Doss, *Electrochimica Acta* 1964, **9**, 1301.
20. A. G. Samartsev and N. V. Andreiyeva, *Zhur. Fiz. Khim* 1961, **35**, 892.
21. U.S. Pat. 2,844,530.
22. E. Raub and M. Wittum, *Z. Electrochem* 1940, **45**, 71.
23. E. Raub, N. Baba and M. Stalzer, *Metalloberfläche* 1964, **18**, 323.
24. H. W. Dettner, *Electroplating and Metal Finishing* 1971, **24**, 9, 13.
25. D. S. Hartshorn, *J. Met. Finishing* 1941, **39**, 561.
26. M. Schlotter, U.S. Pat. 1,972,693.
27. V. Zenter, A. Brenner and C. W. Jennings, *Plating* 1952, **39**, 865.
28. S. Wernick, 'Electrolytic Polishing and Bright Plating of Metals', A. Redman, London, 1948.
29. E. B. Saubestre, *Plating* 1958, **45**, 1219.
30. U.K. Pat. 622761, 1949.
31. Sh. L. Wu, W. Billow and A. R. Garner, *Plating*, 1972, **59,** No. 11, 1033-7.
32. A. D. Du Rose, *Proc. Am. Electroplaters Soc.* 1960, **47**, 83.
33. W. H. Safranek and C. L. Faust, *Plating* 1958, **45**, 1027.
34. M. M. Bedruth, *Plating* 1960, **47**, 402.
35. A. H. Du Rose and W. T. Pierie, *Met. Finishing* 1959, **57**, 40.
36. T. W. Tomaszewski, R. J. Clauss and H. Brown, 'Satin Nickel by Co-deposition of Finely Dispersed Solids', 50th Annual Technical Proceedings, Amer. Electroplaters Soc., 1963.
37. R. A. Baker and N. Christie, *Trans. Inst. Met. Finishing* 1969, **47**, 80.
38. J. A. Hendricks, *Metal Ind.(N.Y.)* 1942, **61**.
39. J. A. Hendricks, *Metal Finishing* 1943, **41**, 134.
40. J. Edwards, *Trans. Inst. Metal Finishing* 1964, **41**, 169.

41. T. P. Hoar, *Trans. Inst. Met. Finishing* 1953, **29**, 302.
42. W. Smith, H. H. Keeler and H. J. Read, *Plating* 1949, **36**, 355.
43. H. J. Read, *Plating* 1962, **49**, 602.
44. D. J. Evans, *Trans. Farad. Soc.* 1959, **54**, 1086.
45. E. Raub, *Plating* 1958, **45**, 486.
46. S. A. Watson and J. Edwards, *Trans. Inst. Met. Finishing* 1957, **34**, 167.
47. E. Raub and K. Müller, *Metalloberfläche* 1963, **17**, 97.
48. G. T. Rogers, M. J. Ware and R. V. Fellows, *J. Electrochem. Soc.* 1960, **107**, 677.
49. J. Edwards and M. L. Levett, *Trans. Inst. Met. Finishing* 1969, **47**, 7.
50. J. Edwards and M. J. Levett, *Trans. Inst. Met. Finishing* 1967, **15**, 12.
51. S. A. Watson, *Trans. Inst. Met. Finishing* 1963, **40**, 41.
52. A. D. du Rose, *Trans. Inst. Met. Finishing* 1961, **38**, 27 and 44.
53. R. J. Kendrick, *Plating* 1961, **48**, 1099.
54. J. K. Dennis and J. J. Fuggle, *Trans. Inst. Met. Finishing* 1970, **48**, 75.
55. *Plating* 1948, **35**, 1122.
56. D. J. Macnaughton and R. A. F. Hammond, *Trans. Farad. Soc.* 1930, **26**, 481.
57. A. W. Hothersall and R. A. F. Hammond, *Trans. Farad. Soc.* 1936, **31**, 1574.
58. L. Weisberg, *Monthly Rev. Electroplaters Soc.* 1939, **26**, 122.
59. M. B. Diggin, *Monthly Rev. Electroplaters Soc.* 1946, **35**, 513, 524.
60. B. F. Lewis, *Proc. Amer. Electroplaters Soc.* 1940, **28**, 64.
61. L. Cambi and R. Piontelli, *Rend. Inst. Lombardi Sci* 1938, **72**, 128.
62. R. C. Barrett, *Electrotypers and Stereotypers Bull.* 1958, **36**, 55.
63. R. L. Moxey, *Compressed Air Magazine* 1962, **67**, 8.
64. F. Siegrist, *Metal Progress* 1964, **85**, 101.
65. J. L. Marti, *Plating* 1966, **53**, 61-71.
66. M. B. Diggin, *Trans. Inst. Met. Finishing* 1954, **31**, 243.
67. R. A. F. Hammond, *Metal Finishing J.* 1970, **16**, 169.
68. B. B. Knapp, *Plating*, 1971, **58**, No. 12, 1187-93.
69. R. Brugger, *Galvanotechnik u. Oberflachenschutz* 1965, **7**, 165.
70. L. L. Newall, Proc. 43rd Ann. Mtg. Amer. Electroplaters Soc. p. 101, 1956.
71. Brit. Pat. 999,117.
72. R. J. Kendrick, 'High-Speed Nickel Plating From Sulphamate Solutions', Proc. 6th International Metal Finishing Conf., 42,235, Disc. 241, 1964.
73. R. J. Kendrick, 'Plating with Nickel Sulphamate' Millan Symposium on Sulphamic Acid and its Electrometallurgical Applications, May 1966.
74. W. A. Wesley and J. W. Carey, *Trans. Electrochem. Soc.* 1939, **75**, 209.
75. A. Brenner and G. E. Riddell, *J. Res. Natn. Bur. Stand* 1947, **39**, 385.
76. A. Brenner and G. E. Riddell, *ibid* 1946, **37**, 31.
77. G. Gutzeit, *Plating* 1959, **46**, Nos. 10-12.
78. G. Gutzeit, *Plating* 1960, **47**, No. 1.
79. A. W. Goldstein, W. Rostoker, F. Schossberger and G. Gutzeit, *J. Electrochem. Soc.* 1957, **104**, 104.
80. W. G. Lee, *Plating* 1960, **47**, 288.
81. A. H. Graham, R. W. Lindsay and H. J. Read, *J. Electrochem. Soc.* 1965, **112**, 401.
82. W. J. F. Andrew and J. T. Heron, *Trans. Inst. Met. Finishing*, 1971, **49**, 105.
83. D. J. MacNaughton, G. E. Gardam and R. A. F. Hammond, *Trans. Farad. Soc.* 1933, **29**, 729.
84. H. H. Hothersall and R. A. F. Hammond, *Trans. Electrochem. Soc.* 1938, **73**, 449.
85. H. Silman, *J. Electrodep. Tech. Soc.* 1944, **19**, 131.
86. R. B. Saltonstall, *Plating* 1949, **36**, No. 12, 1216.
87. W. L. Pinner, *Plating* 1953, **40**,1115.
88. W. A. Wesley and B. B. Knap, *Trans. Inst. Met. Finishing* 1954, **31**, 267.
89. A. C. Hart and L. G. Tottle, *Trans. Inst. Met. Finishing*, 1976, **64**, Pt. 2, 91-6.
90. R. A. White, *Met. Ind.* 1961, **98**, 455.

91. E. A. Ollard, *Trans. Farad. Soc.* 1926, **21**, 81.
92. C. H. Sample, *Soc. Automotive Engineers* 1958, Report 220.
93. J. Edwards, *Product Finishing* 1958, **11**, 58.
94. H. Brown, M. Weinberg and R. J. Claus, *Plating* 1958, **45**, 144.
95. R. J. Clauss and R. A. Tremmel, *Plating* (1973) **60**, 803; Proc. Interfinish '77.
96. E. A. Baker, S. Helmsley and J. R. House. Proc. Annual Conference IMF, (1977).

BOOKS FOR GENERAL READING

1. 'Nickel Plating', R. Brugger, Robert Draper Ltd., Teddington, 1970.
2. 'Fundamentals of Metal Deposition', E. Raub and K. Müller, Elsevier Pub. Co. Amsterdam, London and N.Y., 1967.
3. 'Nickel Plating From Sulphamate Solutions', R. A. F. Hammond, Parts 1-5, Metal Finishing J., June-Oct, 1970.
4. 'Nickel and Chromium Plating', J. K. Dennis and T. E. Such, Newnes-Butterworths, London, 1972.
5. "Guide to Nickel Plating", International Nickel Ltd., London, 1973, 76 pp.

CHAPTER 10

Chromium plating

CHROMIUM plating first came into general use around 1930 following the work carried out in the 1920's by Fink, Sargent and others. Electrodeposited chromium is a very hard metal, having a characteristic bluish cast with very high resistance to tarnish. For decorative purposes it is usually applied as an exceedingly thin deposit (about 0.25-1.0 μm thick) to give non-tarnishing properties to an underlying electrodeposit (usually nickel) which serves as the main protection for the basis metal to which it is applied.

Very much heavier deposits are applied in the engineering industry for imparting wear resistance and low frictional properties to rubbing surfaces such as bearings, moulding tools, inspection gauges and cylinder bores. Heavy deposits of chromium of this type are usually referred to as 'hard' chromium plate although the hardness of the deposit does not differ substantially from that of decorative coatings. For a full account of modern practice in hard chromium plating the reader is referred to the work by Greenwood[1].

Most commercial chromium plating is carried out from solutions of chromic acid containing a small but well defined concentration of a catalyst which is usually sulphate added as sulphuric acid or occasionally as sodium sulphate. Other acid radicals which function as the catalyst have been introduced, particularly hydrofluoric or hydrofluosilicic acids with or without sulphate, in order to obtain specific advantages which are referred to later. Recently, however, there has been a growing interest in trivalent chromium solutions.

10.1. Trivalent Chromium Baths

Chromium can be deposited from chromium sulphate baths without additives, but the deposits are grey and uneven and their use has therefore been confined to electrowinning. These solutions have the advantage of higher efficiencies and the health and effluent disposal problems are much less.

The early work was directed towards depositing chromium from trivalent chromium solutions containing non-aqueous solvents[2, 3]. A solution based on aqueous chromium chloride containing 40% v/v of dimethyl formamide has been developed for decorative chromium plating. The problem, experienced in wholly aqueous solutions of chromic chloride, of maintaining the pH is overcome in the mixed solvent system by adding dimethylamide buffers to stabilise the pH to an acceptable value for plating. The solution thus stabilised is said to possess excellent plating characteristics. It is operated at 25-40°C at current densities of around 10 A/dm^2.

Insoluble anodes cannot be used directly in the bath as they result in oxidation of Cr^{3+} ions, but lead anodes in porous pots containing dilute sulphuric acid are said to be satisfactory. The bath is replenished by small bags containing chromium sulphate suspended near the surface of the solution.

Recently, however, wholly aqueous commercial baths have become available, and are in operation on a fair scale[4]. They contain hypophosphites their outstanding features being:

1. Complete absence of burning at high current density.
2. Elimination of "whitewashing" as a result of current interruption which is an advantage in barrel plating.
3. Very high covering and throwing power.
4. Reduction in effluent disposal problems.
5. Micro-cracked deposits, leading to good corrosion resistance.
6. Carbon anodes, which are more stable and durable than lead, are used.
7. Operation at room temperature saves energy.
8. There is no toxic spray to remove.

As against these advantages, the solution lacks stability and is hence difficult to maintain, several additives being required, including complexants, wetting agents and conductivity salts. The solution is also very sensitive to impurities, especially zinc. The operating voltage is comparatively high (about 9 volts), but the operating temperature is low (20°-25°C). The recommended current density is 8-10 A/dm^2.

Carefully controlled air agitation is required not to prevent "burning", but to maintain uniform metal distribution on the work. The deposit is somewhat darker in colour than conventional chromium.

10.2. The Chromic acid bath

10.2.1. GENERAL

Chromic acid is a strong acid and is supplied for plating purposes in the form of dark red flakes, CrO_3. It must be substantially free from excess sulphuric acid or sulphate because of the critical effect of the concentration of these catalysts on the operation of the solution. A wide range of concentrations of chromic acid can be used, although it is unusual to exceed about 500 g/l.

More dilute solutions containing less than half this concentration are also employed, but almost any concentration of chromic acid can be successfully employed above about 50 g/l of chromic acid provided the correct chromic acid : catalyst ratio is carefully controlled. The normal range of temperatures within which the solutions are operated is from 40 to 50°C.

When CrO_3 is dissolved in water, it will form a mixture of chromic acids, the two principal ones being H_2CrO_4 and $H_2Cr_2O_7$. These are in equilibrium with each other, as follows:

$$2H_2CrO_4 \rightleftharpoons H_2Cr_2O_7 + H_2O$$

and the electrolyte will contain three species of anions, namely $HCrO_4^-$, CrO_4^{2-} and $Cr_2O_7.^{2-}$. The $Cr_2O_7^{2-}$ ion will predominate in more concentrated electrolytes such as are used for electrodeposition purposes, whereas the CrO_4^{2-} ion will form preferentially in dilute solutions. In addition, some Cr^{3+} ions are invariably present.

The deposition mechanism of chromium is much more complex than that of other metals and is still incompletely understood. Muller [6, 7] has postulated that chromium is deposited from trivalent ions contained in a permeable film of basic chromium chromate which is formed during electrolysis on the cathode surface. Trivalent ions are formed from hexavalent chromium species according to:–

$$—HCrO_4^- + 3H_2O + 3e \rightarrow Cr^{3+} + 7OH^-$$

which is followed by the formation of a solid film of $Cr(OH),CrO_4$ on the cathode surface:

$$Cr^{3+} + CrO_4^{2-} + OH^- \rightarrow Cr(OH).CrO_4.$$

Radiotracer studies [8] have confirmed that deposited chromium comes from Cr^{3+} ions originating from $HCrO_4^-$ species and not from hexavalent species present in the bulk of the solution. The function of the catalyst in the deposition process is not understood but it is probable that it modifies the cathode film in such a way that deposition can occur.

Although, as mentioned above, trivalent ions in the bulk of the solution are not reduced to the metal, they nevertheless must be present in order to ensure satisfactory chromium deposits. Their concentration should not be too great, as otherwise the current density range within which bright deposits are obtained will be greatly restricted. For this reason it is essential that the excess Cr^{3+} ions should be re-oxidised to the Cr^{6+} state. This is partially effected at the anode if lead-base alloys are used as the anode material since the oxygen over-voltage on lead is high (see below).

The choice as to whether a concentrated or a dilute chromic acid solution is to be employed depends on working conditions. Low-density baths are naturally much more sensitive to changes in the sulphate ratio such as may be brought about by carry-over from acid dips or by depletion of the chromic acid by rapid working, since the solution is always operated with inert anodes. Dilute solutions have a lower conductivity, so that higher operating voltages are required to maintain an adequate working current density. On the other hand, they have a higher cathode efficiency and a somewhat wider bright plating range.

The high-density solution containing 360 to 500 g/l of chromic acid is more suited to really heavy production because of its higher conductivity and the fact that changes in solution composition are of less significance than is the case with the more dilute solutions. Drag-out losses are necessarily higher with heavier solutions, but this is usually not a serious matter because it is the practice in chromium plating, in view of the relatively expensive nature of the solution and effluent disposal problems, to return the rinsing water to the plating tank for 'make-up' purposes.

The absolute concentrations of chromic acid and sulphate in the bath are of secondary importance to the main factor which is the ratio of chromic acid to sulphate. This ratio is best maintained at about 100:1, the limiting ranges for deposition being within the approximate limits of 200:1 and 50:1. Outside these limits little or no chromium can be deposited from the solution.

Should it be desired to reduce the sulphate content of the bath, this is best carried out by the addition of barium hydroxide, $Ba(OH)_2, 8H_2O$. Rather more than the theoretical quantity of this compound may be required in order to precipitate the requisite amount of sulphuric acid. The theoretical proportion is 3.1 parts of barium hydroxide for each part by weight of sulphuric acid, but not all of the barium compound goes to the precipitation of barium sulphate, as some barium chromate is also formed:

$$Ba(OH)_2 + H_2SO_4 + 2H_2O$$
$$Ba(OH)_2 + CrO_3 = BaCrO_4 + H_2O.$$

The process is best carried out in a separate tank at a high temperature (about 75°C) and the treated solution decanted off. A preliminary test of an aliquot portion of the solution will enable the correct amount of hydroxide required to be calculated.

A new bath does not usually plate fully satisfactorily until it has been aged by plating for some time. The ageing process is apparently connected with the formation of a certain amount of trivalent chromium in the bath, which as stated earlier is essential although excessive concentrations are deleterious. Accelerated methods of ageing, such as boiling the solution with small proportions of citric or tartaric acid, have been proposed but are not widely used.

Details of proprietary decorative chromium plating processes have been given elsewhere[9].

10.2.2. TEMPERATURE OF OPERATION

Owing to the extreme hardness of chromium deposits and the consequent difficulty of polishing them, it is desirable to produce bright coatings directly from the solution. This is of no great consequence where heavy deposits are being applied for industrial purposes, but it is obviously more important in the case of thin coatings applied for decorative purposes. Generally speaking, with the usual types of solution already described the bright plating range is a function of temperature and current density, higher temperatures being required for the higher current densities. If the temperature is too low for the current density being employed, the deposits will be milky or frosted in appearance and can only be polished with difficulty, if at all. If the current density is higher still, burnt deposits will be obtained which are rough or nodular and may be dark in colour. Conversely, if the temperature is raised whilst the cathode current density remains unchanged, the efficiency falls rapidly until deposition is inhibited altogether.

The decrease in cathode efficiency with rise of temperature means that in order to obtain bright deposits, some reduction in cathode efficiency must normally be expected. Bright deposits cannot readily be produced at temperatures below about 25°C, or at current densities of less than about 4 A/dm².

The ranges within which various types of deposit are obtained from a solution of 250 g/l of chromic acid and 2.5 g/l of sulphuric acid are shown in Fig. 10.1.

Fig. 10.2 shows the densities of solutions containing varying concentrations of chromic acid at 25°C. This provides a simple means of testing the strength of a bath by means of a hydrometer, but in the case of old solutions the curve is subject to modification, due to the presence of trivalent chromium, iron, etc.

10.2.3. EFFICIENCY AND CURRENT DENSITY

The efficiency of the chromium plating solution, even under the best conditions, is very low. It is highest when the bath is operated cold, and efficiencies of 35 to 40 per cent can then be obtained. The deposits under these conditions are, however, hard and frosted in appearance. Within the bright plating range the efficiency is much lower and does not usually exceed 10 to 15 per cent. In commercial plating it is usually necessary to plate irregularly shaped articles, so that it is desirable to work with a solution having a wide current density range in order that a bright deposit may be obtained on all significant surfaces; hence the ratio of the minimum to the maximum current densities

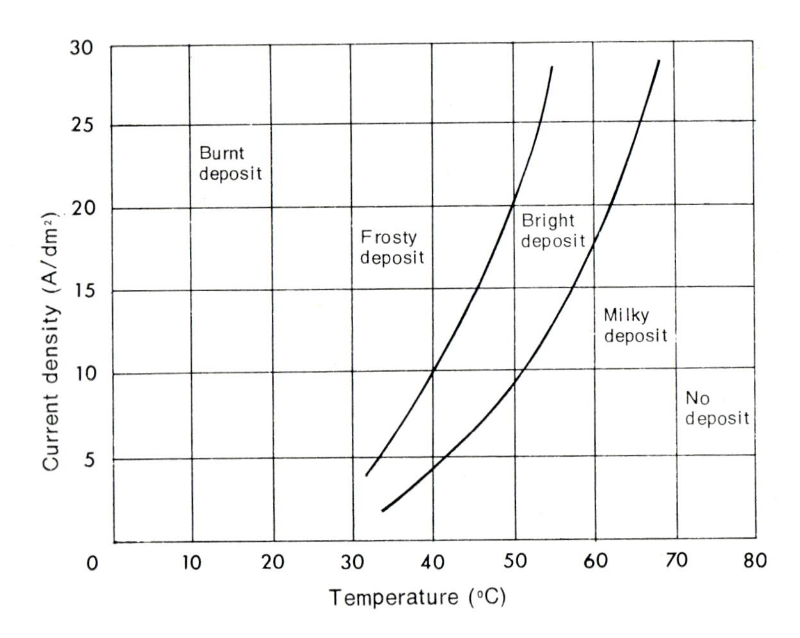

Fig. 10.1. Plating range of low-density chromium solution (chromic acid 250 g/l, sulphuric acid 2.5 g/l).

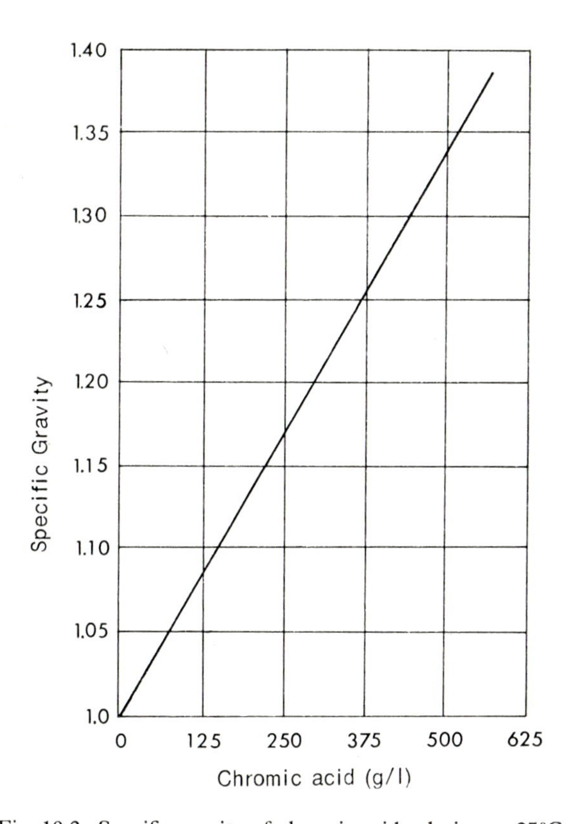

Fig. 10.2. Specific gravity of chromic acid solution at 25°C.

within which bright deposits are obtainable should be as high as possible.

Current densities of 10 to 20 A/dm^2 are most commonly employed in practice, the voltages usually needed being 5 to 8 volts with low-density solutions or 4 to 6 volts with the better-conducting high-density solutions. The high currents used mean that large-capacity 3-phase rectifiers are necessary, whilst plating racks and wires must be of ample dimensions to carry large currents. A good guiding rule is to allow 200 A/cm^2 cross-section of copper conductor.

10.2.4. THROWING POWER

The throwing power of chromium plating solutions with sulphate catalysts is very poor, and in some instances auxiliary anodes in the form of lead strips or wires are necessary to enable recessed areas to be covered. Again, owing to the low efficiency of the solution, large volumes of hydrogen gas are evolved at the cathode and if the articles being plated contain holes, the stream of gas bubbles liberated will effectively prevent deposition over an area around them. Such holes are therefore best plugged during plating.

The throwing power is, again, unfortunately reduced by the elevated temperatures required for bright plating, the cold solutions giving matt deposits having the best throwing power. On the other hand, for a given temperature increased throwing power is obtained by increasing the current density, and it is therefore desirable to work at as high a current density as possible within the bright-plating range if the best throw is to be obtained. It has also been stated that the low-density solutions have better throwing power than have those of high density, while high sulphate ratios also favour increased throwing power.

10.2.5. EFFECTS OF IMPURITIES

The chromium solution, being highly corrosive, tends to accumulate foreign metal ions very readily due to attack on work passing through, on racks, anodes, connecting bars, tanks, etc. The commonest of these extraneous constituents are iron and copper and also trivalent chromium compounds which may build up to an excessive extent in the solution.

In general, the presence of such cations has an undesirable influence on the solution, decreasing its conductivity, although it has been claimed that zinc improves the efficiency. The presence of excessive amounts of iron is known to reduce the bright plating current density range of chromium solutions.

10.2.6. REMOVAL OF TRIVALENT CHROMIUM

In the course of plating, some of the hexavalent chromium content of the solution is reduced at the cathode to the trivalent form; although a degree of re-oxidation takes place at the anode, this is by no means complete so that in due course trivalent chromium compounds (e.g. chromium sulphate $Cr_2(SO_4)_3$ and chromium dichromate $Cr_2(Cr_2O_7)_3$), gradually accumulate in the solution. Some is lost by drag-out, but the concentration should not be permitted to become excessive as trivalent chromium compounds have the effect of narrowing the bright plating range of the solution and decreasing its conductivity. On the other hand, a moderate concentration of trivalent chromium appears to increase the throwing power somewhat.

If the trivalent chromium concentration should become excessively high (more than about 50 to 60 g/l), it is possible to promote its re-oxidation to chromic acid by passing current through the solution, using a lead cathode inside a porous pot. In this way oxidation of the solution at the anode proceeds whilst the reducing action at the cathode is inhibited. The process is, however, a very slow one if large volumes of solution are to be dealt with.

Under the appropriate conditions such re-oxidation can be achieved without the use of a porous diaphragm around the cathode. Electrolysis should be carried out at a high temperature (about 80°C), a high cathode current density, such as 55 A/dm^2, and a low anode current density of about 2 A/dm^2. The cathode surface is a small round steel rod, the anode area being about thirty times as great as the cathode area. Lead or antimonial lead anodes are used. It is not necessary to remove the sulphate from the solution. Under these conditions, with a solution containing 250 g/l of chromic acid and 2.5 g/l of sulphate ion, 22 hours are required to reduce the trivalent chromium content from 22 to 5 g/l.

10.3. Alternative catalysts

In more recent years, there has been a greatly increased interest in chromium solutions containing catalysts other than sulphuric acid, e.g. fluorides, fluosilicic acid and borofluorides with or without the presence of sulphuric acid[10]. The catalyst most commonly employed in this type of bath is hydrofluoric acid or hydrofluosilicic acid, the ratio of chromic acid to catalyst varying from 150:1 to 300:1.

The fluoride-containing electrolytes have the advantage of superior throwing power and greater current efficiency than the conventional sulphate solution. The bright plating range is much wider and hard, bright deposits can, in fact, be obtained in cold solutions of this type. They are also capable of withstanding breaks in current during the plating process, which, in the case of the sulphate-catalysed electrolyte, usually result in dull or poorly adherent deposits. Thus, they are especially useful in barrel plating practice where intermittent contacts constantly prevail, and in fact, fluoride and silicofluoride electrolytes are used almost exclusively for this purpose. They are also extending into fields such as hard chromium plating where the sulphate solution was employed almost exclusively hitherto.

As against their advantages the fluoride-type baths are rather more difficult to control, and analysis is more complicated. They are also more aggressive towards tanks and tank linings, plating jigs, anodes and unplated surfaces of articles being processed. In the absence of sulphate, silicofluorides are unstable in chromic acid solutions and decompose to form silica which adversely affects the deposits. For this reason, silicofluorides are always used together with sulphates as a mixed catalyst system.

A fluosilicic-sulphuric acid catalysed electrolyte has been studied by Wahl and Gebauer[11], who recommend a solution of 250 g/l chromic acid containing 0.6 per cent of sulphuric acid and 1 per cent of fluosilicic acid. The deposits from this bath are harder and more heavily stressed than those obtained from the sulphuric acid solution. At a temperature of 55°C and current densities of around 40 A/dm^2 the cathode efficiency was 26 per cent as compared with 13 per cent for the conventional solution under similar conditions of operation.

10.4. Self-regulating baths

Of great interest, self regulating solutions have been developed which enable the concentration of the catalyst to be maintained constantly at the correct level. Control of the catalysing acid is achieved by the inclusion of a salt mixture which has the solubility needed to maintain automatically the correct concentration of catalyst in the bath.

An excess of the salt is added and this remains in the bath in the undissolved form.

If the concentration of catalyst falls, the requisite amount again goes into solution and cannot be exceeded. Analytical control of the chromic acid content only is therefore all that is necessary.

In practice there is no salt readily available which will function on its own as a self-regulating catalyst. The nearest practicable substance appears to be strontium sulphate[12], which has a solubility of 0.11 g/l at 40°C. This concentration is rather low for plating purposes, so that the difference is made up by the incorporation of sulphates or fluorine-containing salts to give the required concentration of catalyst; this may be around 2.5 g/l for chromium plating solutions containing 250 g/l of chromic acid such as is used for hard chromium plating, down to 2.0 g/l for a bright chromium plating bath containing 400 g/l of chromic acid and having a sulphate ratio of 150:1 or 200:1.

There are numerous patent claims for self-regulating baths. One type of solution (S.R.H.S.) which is used commercially[13] contains 2 per cent of strontium sulphate, 5.5 per cent of potassium silicofluoride and 92.5 per cent of chromic acid. The chromic acid concentration ranges from 200 to 500 g/l depending on the application for which it is required. For decorative plating a temperature of between 40°C and 50°C is employed. Although the covering power of the bath is better at lower temperatures, the higher temperatures favour higher current densities and hence shorter plating times. For 'hard' plating, where considerable thicknesses have to be applied, the bath is operated at a concentration of 250–350 g/l at 50–65°C. The optimum throwing power of the bath is obtained at 400–500 g/l and deposits are bright up to 15 μm thick. This composition can also be used for barrel plating. The highest current efficiencies are obtained at about 325 g/l chromic acid. The S.R.H.S. solution has also considerably better initial covering power than the sulphate solution, and is also claimed to plate on passive nickel or stainless steel more readily.

Owing to the higher cathode current efficiency resulting from the presence of the silicofluoride catalyst[14] and the wide bright plating range of the bath, plating speeds can be up to 50 per cent greater than with the ordinary bath, particularly in the regions of higher current density and temperature. The deposits obtained from the S.R.H.S. bath are generally smoother, finer-grained and more free from imperfections and cracks than those from the conventional baths. This feature is naturally of greater importance in the case of heavy deposits than when decorative chromium coatings are being applied. The relative absence of cracks is probably also the basis of the claim that the fatigue properties of steel are reduced to a lesser extent by chromium plating from the S.R.H.S. bath than from the conventional solution. Fig. 10.3 shows the relative efficiencies of S.R.H.S. and conventional types of solution at 55°C.

It has already been pointed out that the fluoride and fluosilicate baths are more corrosive than the sulphuric acid solutions, and precautions have to be taken accordingly.

More recently[15] succinic acid has been suggested for use, together with strontium sulphate, in self-regulating solutions. The bath recommended contains:

CrO_3	375 g/l
$SrSO_4$	8 g/l
Succinic anhydride	40 g/l

and is operated at 35°C.

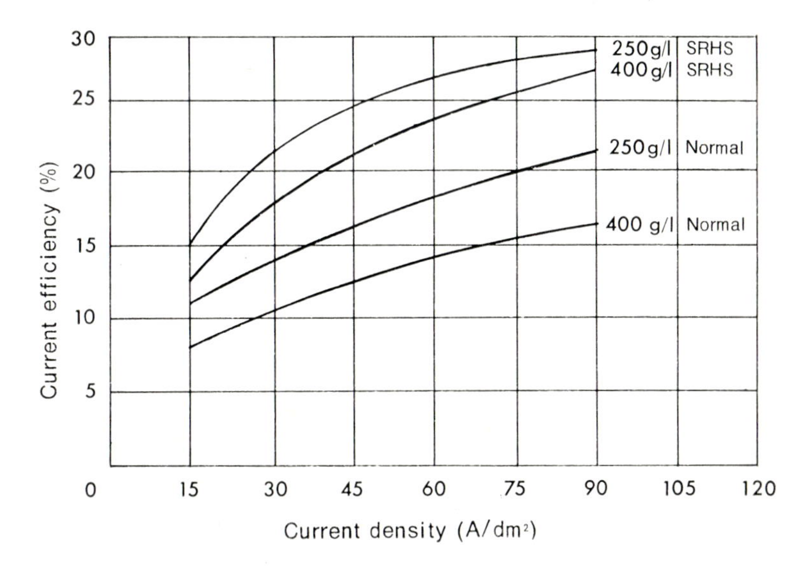

Fig. 10.3. Current efficiency of S.R.H.S. as compared with conventional chromium plating baths (55°C).

10.5. Crack-free process

A modification of the S.R.H.S. bath produces a deposit which is substantially free from the cracks which are a normal characteristic of most chromium electrodeposits. The solution may contain 200–500 g/l of chromic acid and is operated at 65°C and a current density of about 40 A/dm^2. Under these conditions the rate of deposition is between 15 and 40 μm per hour, and the deposits are matt or semi-bright. They can be buffed to a fully bright finish, however, as they are relatively soft, having a Diamond Pyramid hardness of 435-700. The corrosion resistance on testing is superior to that of normal chromium as shown by exposure and salt-spray. The solution is also claimed to have considerable levelling action, unlike sulphate-type chromium plating baths, whilst measurements show that the deposit has only about one-third of the usual internal stress[16].

10.6. Sodium tetrachromate bath

The tetrachromate bath[17, 18] developed by Bornhauser is worthy of mention as it has received application for depositing chromium directly on to inexpensive zinc articles for use indoors. The solution contains caustic soda in addition to chromic acid, so that sodium tetrachromate is formed. A typical solution contains 400 g/l CrO_3, 58 g/l NaOH and 0.75 g/l H_2SO_4. High current densities can be used (20–80 A/dm^2), the current efficiency being 30–35%. Anodes are of antimonial or chemical lead, and a high anode:cathode ratio is recommended. If the bath temperature is maintained at 15–25°C by cooling, current densities of 50 A/dm^2 can be used and deposits of 76 μm obtained in one hour (corresponding to a current efficiency of around 30 per cent). At temperatures in excess of 25°C the sodium tetrachromate begins to decompose, being completely dissociated at 40°C. The deposits from this bath are soft and matt, but can be readily polished, the resulting appearance approaching that of normal bright chromium plate. A wide plating range can be used, since this is not limited by the somewhat narrowly restricted conditions under which the sulphuric acid-catalysed bath will produce a bright deposit.

Deposit thicknesses of the order of 7.5 μm are essential for out-door exposure, but for interior use lower thicknesses can be used. In view of the high rate of deposition the greater thickness is not a real disadvantage.

It has recently been claimed[19] that mirror-bright deposits are obtained from a 12:1 CrO_3:H_2SO_4 bath to which has been added sodium hydroxide. Bright deposits are produced at low current densities in the temperature range 25–50°C, whereas at higher current densities non-bright deposits are obtained which exhibit high corrosion resistance to dilute acids. In addition, it is reported that high-strength steel is markedly less embrittled when plated from such baths at temperatures of between 25 and 45°C. The tetrachromate plating bath has been the subject of a recent review[20].

10.7. Black chromium plating

Black chromium can be employed as an attractive alternative to the usual bright plate and is especially useful for coating optical and other instruments. The deposits as plated contain only about 56% by weight of chromium metal, the remainder being essentially oxygen[21].

Black chromium can be deposited by introducing into the chromic acid bath small amounts of acetates, tartrates or propionates, or alternatively, fluoboric acid. A solution of 300 g/l of chromic acid with about 0.7 g/l of acetic or propionic acid can be used at room temperature to produce a black finish. With a similar concentration of fluoboric acid and a current density of 2 to 2.5 A/dm^2 at a bath temperature of 40 to 50°C a golden chestnut colour can be produced in 10 minutes or so, and a black finish after 20 to 30 minutes.

Black chromium deposits can also be obtained from a bath containing chromic acid and 0.25-0.5 g/l of fluosilicic acid at temperature of 25 to 30°C[22, 23]. A current density of 15 to 30 A/dm^2 is employed.

Baths for black chromium plating must not contain any sulphate since this reduces the current density range over which black deposits can be obtained. Black chromium can be deposited on to copper, stainless steel and nickel, but the best application involves putting the black chromium over a regular chromium deposit[24].

Shenoi and Gown[25] have recently reviewed the production of black chromium deposits.

10.8. Satin chromium finishes

A satin finish may be obtained by lightly abrading or scratch brushing the surface of the work prior to bright chromium plating. However, this method is relatively expensive. It has been claimed that with careful control of the operating parameters, satin chromium deposits can be produced from chromic acid solutions containing 1 or 2% of sulphuric acid. In the absence of any pretreatments, such as anodic etching, satin finishes are only produced at 50°C. As well as high temperatures, high current densities also favour the production of satin deposits.

10.9. Porous chromium

Methods are available for producing porous chromium deposits and such coatings appear to have a greater wear-resistance than the normal type of dense chromium deposits owing to their ability to retain oil[(26)].

Porous chromium finds its chief application in the treatment of engine cylinder bores. The cylinder or liner is first ground oversize and honed to a smooth finish, sharp edges being rounded. Parts not to be plated are treated with a stop-off material after which an internal anode is fitted and the assembly plated with chromium from solutions which are of a similar type to those employed in normal hard chromium plating practice. Etching is then carried out by reverse current either in the same bath or in a separate tank, sometimes using an alkaline solution. The type of pore and its depth can be controlled and may be of the pocket or channel type. In the pocket type the pores are not connected, and this coating is more suitable for diesel and petrol engines and for pumps and compressors.

10.10. Anodes for chromium plating baths

The chromium content of the plating solution must be maintained by regular additions of chromic acid, since insoluble anodes of antimonial lead are usually employed. These may be flat sheets of corrugated construction or round with cast-in copper or brass suspended hooks for maximum conductivity. The shape of the anode has a significant effect on the performance in the plating bath. Friedberg[(27)] found that less voltage was required to maintain a given amperage through chromium plating baths employing round anodes. Sheet anodes have the disadvantage of an inactive back surface.

5–8% antimonial lead is preferable to pure lead because it is mechanically stronger and also resists the action of the solution better. Pure lead is more resistant to the electrolyte than antimonial lead during the passage of current, but is markedly inferior when electrolysis is not proceeding. The resistance of antimonial lead to chemical dissolution by chromic acid increases with increasing antimony content. Tellurium-lead anodes containing about 0.2 per cent of tellurium have also shown certain advantages, whilst the use of 4–7% tin-lead alloys has been found suitable for modern chromium plating baths which contain corrosive fluoride ion catalysts.

In operation the lead anodes become coated on the surface with a chocolate brown layer of lead peroxide, which has the effect of re-oxidising to the hexavalent state any lower oxides of chromium that are produced in the solution. This reduction to trivalent chromium takes place continually and, as has already been stated, excessive concentrations of this compound lead to reduced efficiency of the bath. If the anode area is insufficient, the rate of formation of trivalent chromium will tend to be increased. The anodes may, moreover, become overheated with the high current densities used, and will then disintegrate more readily. In the absence of the peroxide film and during idle periods[28] the anodes become coated with lead chromate as a result of attack by the chromic acid. This film (unlike the lead peroxide) is non-conducting and must be scraped off from time to time. The lead peroxide film can be re-established, if necessary, by working the anodes for a time in a dilute solution of sulphuric acid. By correct operating conditions, the rate of formation of lead chromate can be markedly reduced so that only infrequent cleaning of the anodes will be needed. This anode reaction is an essential part of the chromium-plating process. For this reason chromium anodes could not be used successfully, even if they were available at an economic price.

A further objection to the use of chromium anodes is the fact that the anode efficiency is about ten times as great as the cathode efficiency, so that the metal content of the bath would build up at an inordinate rate.

The anode:cathode ratio is of importance in reducing the tendency for trivalent chromium salts to build up in the solution. The anode area should in no case be less than the cathode area, but the exact ratio depends a good deal on the conditions of operation. Anode:cathode ratios of 2 or 3:1 are advisable.

10.11. Pre-treatment before chromium-plating

The chromic acid solution, being highly oxidising, is capable of plating on to metal surfaces which are slightly greasy and less meticulously free from foreign matter as necessary for the deposition of most other metals. Also, the considerable amount of gas which is liberated during plating serves to break up any slight films which may remain. Nickel deposits after polishing can usually be directly chromium plated without any cleaning provided too long an interval does not elapse between finishing and plating. In handling articles between these operations for wiring or racking it is desirable for the operator to wear cotton gloves since finger marks may in some cases show through the chromium plate. After nickel finishing, a wipe with a dry buff and a little lime suffices to produce a surface clean enough for direct plating, whilst larger surfaces are sometimes cleaned more thoroughly by means of a rouge paste or an organic solvent.

Electro-alkaline cleaning followed by a dilute sulphuric acid dip prior to chromium plating is not often used, but a good case can be made for its introduction in large-scale production. Such cleaning can do much to minimise the incidence of chromium plating defects caused by surface contamination with grease or polishing composition, or resulting from slight surface oxidation. If this procedure is employed, very efficient rinsing arrangements are necessary owing to the extreme sensitivity of the plating solution to the sulphate radical; where the articles to be plated are so constructed that thorough removal of traces of acid cannot be effected, wet cleaning methods of this description are best avoided. Indeed, the problem of acid removal is often an extremely difficult one to solve even under good conditions of rinsing.

Bright nickel deposits are sometimes inclined to be 'passive' especially if any appreciable period elapses between nickel and chromium plating. Unplated or 'white-washed' areas may then manifest themselves on the articles when they are withdrawn from the chromium plating tank. In such cases activation by a dilute acid dip or by electrolytic treatment in a cyanide solution may be helpful.

10.12. Rinsing

After chromium plating, it is usual to have two cold rinses followed by an alkaline neutralising tank containing a dilute sodium carbonate solution, and then a further cold rinse and a hot-water tank for drying off. In the case of large articles drying may be difficult and water stains may then be left on the work, so that sawdust or oven drying has to be resorted to. Infra-red ovens have proved useful for this purpose in some instances[29], but stain-free solvent drying is particularly effective (see p. 91).

Drag-out losses are normally reduced by the use of a drag-out tank for the first rinse, which should be agitated by air; when the chromic acid concentration in this tank has reached about 10 to 15 per cent of that of the plating bath it should be withdrawn and used for making up the main bath either directly or after concentration by evaporation. According to Soderberg[30] drag-out losses can be kept low by: (1) withdrawing the parts slowly from the solution; (2) increasing the draining time above the tank; (3) keeping the withdrawal time as long as possible and the drainage time as short as possible for a given total drag-out time period; (4) racking so that the solution can flow off at a tip or corner and that there are no horizontal surfaces or solution pockets; (5) preventing drip from one part flowing on to another.

10.13. Defects in chromium plating

Amongst the commoner defects in chromium plating practice are:

(a) Streaked deposits. This shows itself in the form of white or milky streaks on the plated surface and is usually the result of plating at too low a current density.

(b) Matt deposits. The bath temperature should be increased to eliminate this defect.

(c) Thin or discoloured deposits. These may be due to excessive bath temperatures which lower the rate of deposition. The sulphate ratio can also cause discoloration, a low ratio giving a yellowish deposit.

(d) Burnt deposits are the result of too high a current density for a given bath temperature and can be corrected either by reducing this or by the use of shields at edges or corners.

(e) Unplated areas. These are usually the result of gas liberation preventing deposition on certain zones. Careful attention to racking position in relation to holes and recesses in the article being plated will correct this condition.

The rate of deposition at various current densities is shown in Fig. 10.4.

10.14. Properties of electrodeposited chromium

Chromium as deposited is a bluish-white metal which has highly non-tarnishing qualities. As is to be expected, its reflectivity in the blue region of the spectrum is good, but over the entire visible range the reflectivity appears to be about 65 per cent. Bright chromium deposits have a hardness range of 700 to 1,000 Brinell, depending on the conditions of deposition. It has been stated that the hardness varies with current density, temperature, chromic acid concentration and the presence of different anions and cations. The crystal structure influences the hardness and the general properties of the resulting deposits. For a concentration of 300 g/l of chromic acid and a sulphate ratio of 50:1 an increase in current density increases the hardness of chromium-plate at temperatures below about 55°C, and reduces it at temperatures above this figure. For all current densities the maximum hardness appears to be obtained at bath temperatures in the neighbourhood of 50°C. No distinction was made in this work between dull and bright deposits. The hardness of the bright range of deposits as obtained from warm solutions appears to be in the region of 800 to 900 Brinell[31]. The hardness of all types of electrodeposited chromium is much greater than that of the cast metal. On annealing the electrodeposited metal, the hardness decreases until values as low as 70–90 Brinell are obtainable.

According to X-ray data, the grain size of bright chromium is of the order of 10^{-6}cm.

The atomic arrangement of electrodeposited chromium can be body-centred cubic (BCC) or hexagonal close-packed (HCP). The deposits usually contain a considerable amount of hydrogen. In HCP chromium the hydrogen is present in the form of an interstitial solid solution, whereas in BCC chromium it is accommodated essentially along the grain boundaries.

According to Snaveley[32], cracks are caused by the formation of unstable hydrides during plating which spontaneously decompose to BCC chromium and hydrogen. This is accompanied by a contraction in volume and formation of internal stress which gives rise to crack formation after a certain thickness is reached, the value of the latter depending on deposition conditions.

Another explanation of the origin of tensile stress in chromium deposits has been given by Gabe and West[33]. According to these authors, the high polarisation during chromium deposition gives rise to numerous vacant sites in the lattice which results in the formation of microscopic stresses if the vacant sites have a common orientation.

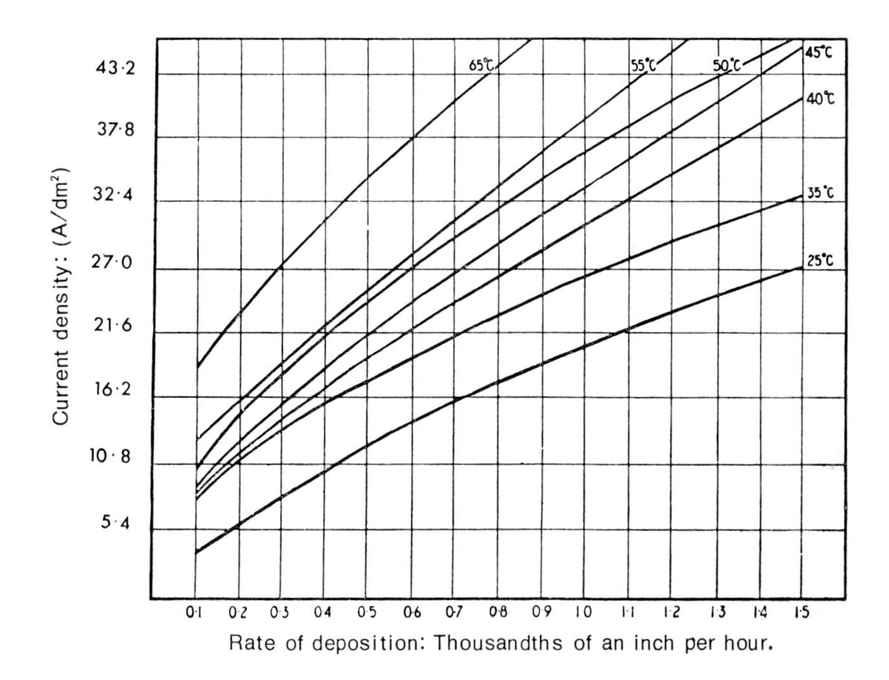

Fig. 10.4. Rate of deposition of chromium from high-density solution (CrO_3 400 g/l, H_2SO_4, 4 g/l).

According to Dennis[34] the surface texture of the substrate influences the type of crack patterns formed in the chromium overlay.

Cleghorn and West[35] ascribe the rapid development of tensile stress in thin chromium electrodeposits on a number of substrates to diffusion of hydrogen in the substrate and coalescence of chromium crystallites formed in the first layer of the deposit.

Bright deposits are characterised by a network of cracks and pores, whereas milky deposits are essentially crack-free. It is assumed that chromium, in the latter case, is deposited directly in the BCC form.

The actual thickness of a bright chromium coating at which cracking will occur depends on the following three factors: (1) bath temperature; (2) CrO_3:H_2SO_4 ratio, and (3) concentration of CrO_3 in the bath. By raising any one of these, but preferably all of them, the maximum thickness to which crack-free bright chromium coatings can be deposited will be increased. By selecting suitable conditions, films of up to 2.5 μm in thickness can be built up in the crack-free condition.

Chromium has a low coefficient of friction. The latter is a number which indicates how easily one surface slides on another, and it is found by dividing the force of friction by the normal pressure. Materials slide over chromium-plated steel surfaces much more easily than over the bare steel surface. Associated with the coefficient of friction are other properties, usually referred to as non-sticking or non-galling properties.

These prevent sticking of force-fitted chromium-plated steel surfaces, or of chromium-plated drills when drilling soft materials (e.g. hard rubber, plastics).

Davies[36] has stressed the importance of the supply wave form in determining the properties of chromium deposits.

10.15. Nickel-chromium plate

One of the most important electrodeposited systems in present day use is nickel plus chromium. Nickel/chromium deposits are almost invariably applied over brass, steel or zinc-base alloys. Plated zinc-base alloy die-castings are extensively used in the motor industry or car door handles, decorative motifs, lamp bodies and many other parts; the possibility of using them at all under severe conditions of exposure is dependent on the fact that they can be plated with a protective deposit of nickel and chromium which must be applied over a copper undercoat since nickel solutions attack zinc. Bumpers, radiators, wheel discs and other components are of plated steel or, less commonly nowadays, of brass. The nickel deposit is much heavier than the chromium, which is normally only about 0.25 μm thick, the nickel, on the other hand, being about 15-40 μm in thickness. Some 40 per cent of all the nickel and chromium plating in the United Kingdom is applied on automobile parts.

When steel is used as the basis metal, it may or may not be copper plated before nickel and chromium. The copper undercoat may be polished before the nickel is applied, since it is easier to produce a good polished finish on copper than it is on steel.

Polishing of the basis metal is a necessary preliminary to almost all decorative nickel-chromium plating and is nowadays increasingly carried out on fully-automatic machines, which present different facets of the articles to a series of rotating fabric polishing buffs. Suitable abrasive compositions are used in conjunction with the buffs, and these can be sprayed on to them in the form of a liquid suspension or applied as a solid bar compound.

After polishing, the articles must be meticulously cleaned; imperfect cleaning can lead to poor adhesion and other defects. The cleaning process has to remove both inorganic and organic contaminants. The inorganic substances are largely composed of corrosion products, metallic oxides, swarf particles and abrasive residues, whilst the organic materials are oils and greases. Cleaning and degreasing are carried out in solvents or alkaline solutions, whilst oxides are removed by pickling in suitable acids.

Nickel-chromium plate is highly durable under most indoor conditions of service, but when it is exposed to the weather a satisfactory degree of weather resistance can only be obtained if rigid standards are maintained. Of all the plated coatings in current use, the behaviour of nickel-chromium deposits has been the most subject to criticism, particularly on automobile parts. As a result of new knowledge as to how chromium plating corrodes in service, a great many preconceived ideas have been changed. The results of these findings are making their impact on industry and methods of obtaining long-lasting chromium plate are now available.

The view is now generally accepted that the inherent corrodibility of the nickel coating itself, accentuated by electrochemical effects resulting from the presence of the chromium layer, is the dominant factor in the failure of nickel-chromium deposits[37, 38]. Because of modern methods of plating, porosity in nickel deposits contributes little to corrosion behaviour, so that the improvement which can be effected by increasing the total plating thickness to reduce porosity is necessarily limited. Hence, the possibilities of improving the protective value of nickel-chromium coatings are dependent on the development of methods of reducing the rate of corrosion penetration of the coatings themselves on exposure to aggressive atmospheres.

The earlier attempts to achieve this were largely based on the use of intermediate deposits of other metals at various stages of the plating sequence. Amongst those experimented with, in addition to copper, were tin, zinc, cadmium, bronzes and brasses. The objective was to attempt to change the porosity characteristics of the composite coating by introducing a discontinuity. The degree of success achieved, however, was very limited, whilst such techniques often introduced practical difficulties such as lack of adhesion between the different layers; the process sequences were also often very long, particularly where alkaline plating solutions were alternated with acid ones. The present use of multiple deposits of different types of nickel, which may be employed in conjunction with similar chromium coatings, has, however, been shown to lead to substantial improvements in the corrosion resistance of the coatings, and these methods are regarded as providing the best means of improvement. Their practical application is also relatively simple. The plating sequences are only marginally longer than those hitherto employed, whilst the problems of adhesion between the separate layers of nickel have been overcome, and do not constitute any real hindrance to the large-scale introduction of processes of this type.

10.16. Chromium-dual nickel deposits

One of the first effective systems for improving the corrosion-resistance of the deposit was the dual nickel system in which a coating of nickel is first applied from a semi-bright nickel sulphate bath, which may contain levelling agents but is free from sulphur-containing brightening chemicals. Since it is the sulphur in the deposit which makes it susceptible to attack, the deposit obtained, although smooth and corrosion-resistant, is not fully-bright. The work is then transferred directly to another fully-bright bath from which the final brilliant nickel is plated. It has been found in this way that the advantages of bright nickel plating can be obtained with a much improved degree of corrosion resistance due to the presence of the underlying semi-bright layer.

The use of a double layer nickel with the usual 0.25 μm thickness of chromium was a major step forward in improving the outdoor corrosion resistance of nickel-chromium plated motor car parts, especially where salt is used to de-ice roads in winter. The general adoption of this dual- or double-layer nickel plating system, especially for bumpers, was greatly hastened in the U.S.A. by the development and extensive use of new accelerated corrosion tests.

With dual nickel coatings a ratio of about 90 per cent semi-bright sulphur-free nickel to 10 per cent bright nickel overlayer gives the best corrosion protection results when about 0.15 to 0.25 μm of chromium is applied as the final layer. As it is difficult to maintain this ratio in practice without resorting to buffing of the final nickel (which is unacceptable) compromises have had to be made; on steel the tendency has been to use an approximately 75:25 ratio, and for zinc die-castings, which are usually of a more intricate shape than steel articles, a 60:40 ratio. It is, however, not easy to obtain adequate brightening in recessed areas unless at least 30 per cent of the nickel plate is bright nickel. The 50:50 ratio gives markedly inferior results.

As outdoor exposure experience with double-layer nickel plate accumulated, it became clear that whilst the results were excellent in marine exposure as well as on the exteriors of automobiles in cities where salt was used in winter, static exposure tests in industrial atmospheres, where acidic moisture condenses on the plate in autumn, winter and early spring, were not entirely satisfactory except when thick plate of at least 40 μm was used. This relatively poor performance of dual nickel coatings in an industrial atmosphere was shown to be due to the fact that most sulphur dioxide attacks pure nickel almost as readily as it does bright nickel containing about 0.03 to 0.08% sulphur as a nickel sulphide. It was also shown that the density of pores in chromium 0.25 μm thick on top of dull nickel (buffed or unbuffed) or on top of semi-bright sulphur-free nickel (buffed or unbuffed) was very much greater than with the same thin chromium deposit on top of bright nickel; this was an important factor in the rather mediocre results obtained with dual nickel in industrial atmospheres as compared with buffed semi-bright sulphur-free nickel or buffed dull nickel, with a subsequent thin chromium layer.

10.17. Chromium-triple nickel deposits

To overcome the difficulty, another method of improving the corrosion resistance of nickel deposits was introduced. This consists in applying a third, very thin deposit of high-sulphide-containing nickel between the layers. As this deposit is readily corrodible, once the top bright nickel layer has been penetrated as a result of corrosive atmospheric conditions, such as those caused by the presence of sulphur dioxide in industrial atmospheres or by chlorides from road salt, attack proceeds laterally between the layers, slowing down further deep pitting corrosion going through to the basis metal which would lead to catastrophic failure[39]. The effectiveness of the method is seen in Fig. 10.5 which shows two plated panels exposed to the corrosive environment of Kure Beach, N. Carolina, for 8 months. The left hand panel was plated with 12.5 μm semi-bright nickel, 1.25 μm of a high-sulphur nickel, 12.5 μm of bright nickel and 0.25 μm of chromium. The right hand panel was plated identically, but without the intermediate high-sulphur nickel layer.

In general, the thin intermediate nickel layer should contain from 0.1 to 0.2% sulphur as nickel sulphide. This allows optimum corrosion protection to be obtained with 50:50 and 60:40 ratios of semi-bright sulphur-free nickel to bright nickel. The high-

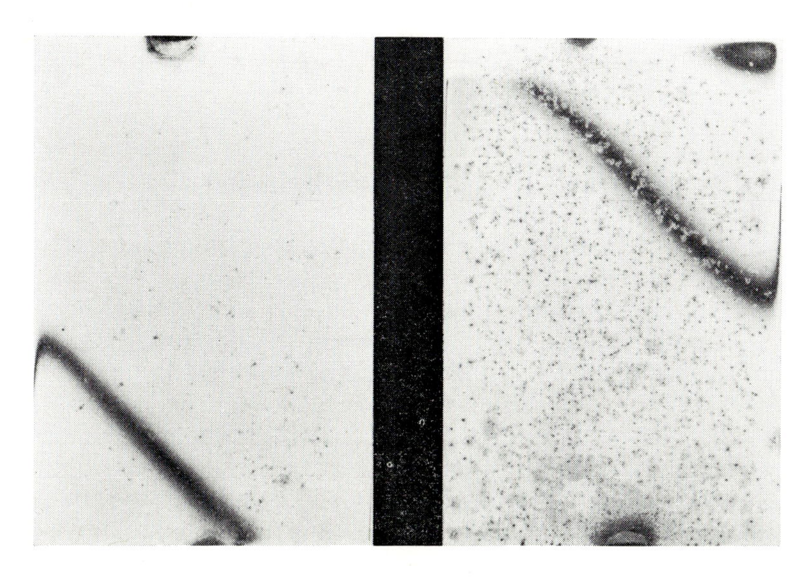

Fig. 10.5. Chromium triple nickel plated panels after exposure to corrosive environment of Kure Beach, N. Carolina. Left hand panel was plated with 12.5 μm semi-bright nickel, 1.25 μm of a high sulphur nickel, 1.25 μm of bright nickel and 0.25 μm chromium. The right hand panel was plated identically but without intermediate high sulphur nickel layer.

sulphide-containing thin nickel layer between the semi-bright sulphur-free nickel and the final bright nickel causes a sharper contrast in anodic activities to occur between the respective nickel layers, thus tending to confine the anodic corrosion attack of the nickel preferentially to the sulphur-containing nickel layers. The semi-bright sulphur-free nickel hence becomes less vulnerable to corrosion attack, with the corrosion pits developing laterally in the sulphur-containing nickel zone once the attack has penetrated to the semi-bright sulphur-free layer. The results obtained by this technique are extremely good and the triple nickel system has found fairly extensive applications, both in the United Kingdom and more especially in the U.S.A. (see also 9.4.4–9.4.8)

10.18. Influence of pores and cracks in chromium plating

The crystal structure of electrodeposited chromium exhibits a cracked appearance which does not appear to be related to any defined crystalline form. Chromium is highly resistant to the atmosphere and to most chemicals, but once corrosion commences, the attack takes place at cracks and fissures in the deposit. When the coating corrodes, the surface of the metal becomes cathodic and the underlying nickel, which is exposed in the pores or stress cracks in the chromium film, becomes anodic; hence very rapid localised attack of the nickel takes place, with penetration to the basis metal and exfoliation and failure of the entire system.

The location and nature of the cracks or pores in the chromium are, incidentally, very important in determining the protective value of a chromium deposit, a large number of fine pores being less harmful than a smaller number of more widely dispersed large ones. This is because the relative sizes of the anodic and cathodic areas are more nearly equal in the former case, so that the rate of corrosion pitting is correspondingly reduced.

It was at one stage thought that thicker bright chromium deposits which were free from stress-cracks might possess very significantly decreased porosity and that this would lead to an important improvement in the corrosion resistance of the nickel-chromium plate. The thicker crack-free chromium deposits gave improved outdoor exposure results both in marine and in industrial atmosphere exposure tests. However, mobile exposure tests in Detroit during the winter season, when about 50 million kg of salt are used to de-ice the roads, showed that insufficient improvement was obtained with the use of thicker bright crack-free chromium as compared to the usual thickness of 0.25 μm of chromium. It was often found during mobile exposure tests in Detroit that whilst eventually less corrosion blisters formed with the thicker crack-free chromium, they often appeared earlier than with the 0.25 μm final chromium plate thickness; in general it was considered that much greater and more consistent improvement in corrosion protection was necessary.

Another disadvantage of the thick crack-free chromium is the residual stress that remains in the plate. Especially with articles which require stove enamelling after plating, it is sometimes found that residual stress with 1.25–2.0 μm crack-free chromium coatings is large enough to crack the underlying bright nickel plate if the latter is not sufficiently ductile.

10.19. Microcracked Double-layer chromium

As a result of work carried out on the corrosion behaviour of various chromium plating systems, a double layer chromium plate which is deposited on copper and bright nickel or on dual coatings, was introduced. The first layer of chromium is deposited from the usual concentrated hexavalent chromium plating solution at 50°–70°C giving a crack-free deposit, and this is followed by a second layer of about equal thickness which is obtained from a solution yielding an extremely fine, lace-like cracked chromium plate (Fig. 10.6). With this system the area of cathodic chromium in the corrosion cells is not very much greater than the anodic area, i.e. the nickel exposed in the microscopic cracks of the chromium plate.

In a recent scanning electron microscope study, Dennis and Fuggle[40] have established that although not all of the micro discontinuities function as effective corrosion sites, there is a sufficient number of them for the anodic current density at any point to be low. To obtain the requisite microscopic cracking, the minimum total thickness of the chromium must be at least 0.6 to 0.75 μm. In general, the thickness of the dual chromium plate ranges from 0.6 to 2.5 μm or even 5 μm. A total plating time of at least 12 to 14 minutes, i.e. 6 to 7 minutes in each chromium plating bath, is necessary.

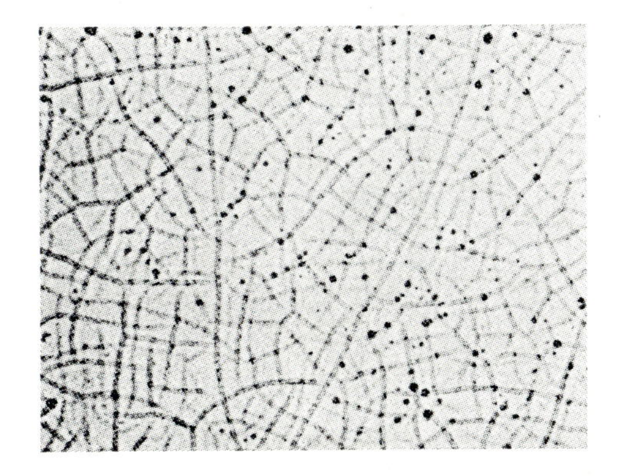

Fig. 10.6. Microcracked chromium after exposure to a corrosive atmosphere. (× 250)

The first chromium layer, which is crack-free and is deposited from a concentrated chromic bath, has good coverage and its residual stress also facilitates the formation of the required fine microscopic cracking in the second chromium layer as deposited from a dilute bath. The stress-cracking of the latter also causes the underlying chromium plate to stress-crack and to provide a network of cracks which penetrates down to the nickel deposit; these are invisible to the naked eye. The nickel plate beneath must not crack and therefore must be of a ductile type.

The use of a copper deposit underneath the nickel definitely has been found to help the overall corrosion resistance of this system considerably. The best corrosion resistance is apparently obtained only when a fine-mesh microscopic chromium crack pattern is produced with a total minimum chromium thickness of at least 0.8 μm and a crack density of 24 per mm. The specified microcrack pattern and minimum thickness are obtained by increased chromium plating current densities or by addition of an agent such as selenic acid[41]. Conditions for applying dual chromium deposits have been described by Lovell, Shotwell and Boyd[42] and also by Serota[43]. The nature of the cracks is also important; coarse patterns are particularly undesirable as they cause crazing to appear after exposure to weathering. Large cracks should also be avoided since they may give rise to a notch effect[44].

Factors aiding the formation of microcracks are: (1) high current densities, (2) greater thicknesses, (3) longer plating times, (4) lower bath temperature, (5) reduced chromic acid concentration, (6) if present, the proper amount of fluoride, (7) the proper amount of sulphate and (8) the presence of an addition agent to promote cracking (i.e. 0.007–0.002 g/l H_2SeO_4) under certain operating conditions.

10.19.1. Microcracked single-layer chromium

More recently single bath chromium processes have been developed for the production of microcracked deposits. In the Harshaw post-nickel strike process[45] a

1 micron thick layer of highly-stressed bright nickel is applied between the chromium deposit and the underlying ordinary bright nickel. The interposed layer, which is obtained from a special nickel bath containing nickel chloride and hydrochloric acid, induces instantaneous microcracking of the chromium layer.

Very significant improvement in corrosion protection has been obtained with micro-cracked chromium in accelerated and in actual exposure tests. It is used widely in the automobile industry, especially on zinc die-castings. It does, however, have one disadvantage in that it is necessary to achieve the minimum of about 0.75 μm of chromium in recessed areas in order to obtain the degree of fine microscopic cracking of the chromium which is essential for the development of the improved corrosion resistance. On some recessed articles this can only be obtained by using conforming anodes in the chromium plating operations. Otherwise the plating time, or the current densities needed would have to be so high that dulling of the chromium deposit might occur on high current density areas, where it could exceed 5 μm in thickness, owing to the relatively poor throwing power of the chromium plating solutions. The use of a multiple nickel deposit underneath the micro-cracked chromium is an important help in giving protection to recessed areas, which are obviously the most vulnerable.

The micro-cracked chromium on top of copper plus bright nickel or on dual nickel gives excellent results on exposure in industrial atmospheres as well as in marine locations. There is also practically no residual stress left in the micro-cracked chromium, whilst any hydrogen embrittlement which may develop as a result of the chromium plating operation is lost quite rapidly by the escape of the gas through the cracks. With parts of complex shape, there is still room for improvement in recessed areas, and the use of an interposed thin nickel plate of relatively high sulphide content in the dual nickel system, as has already been described, is valuable.

10.20. Micro-porous chromium

One of the most widely used methods of obtaining a uniformly porous rather than a micro-cracked chromium deposit, involves the use of an additional, very thin deposit of nickel applied from a special solution immediately before a conventional chromium coating. This constitutes the micro-porous system (known commercially as 'Nickel-Seal'), now embodied in the British Standard BS 1224:1970. It overcomes many of the limitations of micro-cracked chromium. The nickel solution contains a large number of minute, suspended inert particles which do not affect the appearance of the deposit, but being non-conducting, result in the formation of a highly micro-porous chromium deposit (Fig. 10.7). Severe electrochemical attack of the underlying nickel at large cracks or pores in the chromium is thus prevented, and substantial improvements in the corrosion resistance of the combined deposit is obtained. Moreover, the method has the advantage that the distribution of the pores is not greatly affected by the shape of the article being plated, as compared with the methods previously described which depend on producing a micro-cracked chromium coating. A 7.5 μm thick copper deposit underneath the nickel-chromium layer has also been found to add very con-

Fig. 10.7. Surface of microporous chromium deposit. × 150.

siderably to the protective value of the micro-porous nickel. Thereafter, little improvement occurs, although it has been postulated that on rough basis metals, thicker levelling deposits may be needed to give the optimum effect.

When such nickel deposits are followed by about 0.25 μm of chromium, a high degree of corrosion protection is obtained because of a fine porosity pattern which is established in the chromium plate by the presence of the myriads of minute particles on the nickel surface or partially embedded within it. The large number of microscopic anodic nickel sites results in very weak corrosion currents (as in the case with micro-cracked chromium) with extremely slow corrosion penetration. The nickel deposits containing the micro-inclusions of non-metallic particles should be deposited over normal thicknesses of ordinary semi-bright or bright nickel, preferably with an undercoat of copper.

This development came about as a result of work on non-reflecting satin nickel deposits (see Section 9.7) which were required by the American automobile industry to improve road safety. It was found that an acceptable 'satin' nickel could be produced by the use of a dispersion of very fine, bath-insoluble non-metallic powders in semi-bright or bright nickel deposits. It was subsequently discovered that these satin nickels gave quite exceptional corrosion resistance to the combined deposit when subsequently chromium plated. This surprising result was in direct contradiction to previously accepted ideas on techniques for obtaining corrosion-resistant decorative nickel coatings, since occlusions might have been expected to be undesirable.

The next move was to endeavour to produce the improved corrosion resistance without lowering the brightness of the nickel or obtaining a satin deposit. By adding certain powders yielding submicroscopic inclusions which can induce the necessary fine porosity pattern in the usual 0.25 μm thickness of chromium, this was eventually achieved. The chromium does not cover the fine particles embedded in the bright nickel surface completely, and hence the desired porosity pattern is developed in the chromium deposit in an indirect manner. The number of pores can be varied from about 3,000 to several million per square cm. The results in recessed areas are very good with this method, there is no need for auxiliary anodes in the chromium plating operation and only 1 to 3 minutes of chromium plating time are sufficient. The fully-bright nickel deposited with the sub-microscopic inclusion stays very bright for a plating time of up to 5 minutes at 4 to 5 A/dm^2. The highest corrosion protection is obtained when bright nickel of this type is applied on a conventional multiple-layered nickel plate, and is followed by a final chromium coating of about 0.25 μm thickness. The actual plating time in the special nickel solution containing the suspended particles is from 20 seconds to 5 minutes, the deposit being applied over a normal bright nickel. The ratio of the two deposits is determined by the operating conditions needed to produce the optimum results in any particular plant.

The rate of dulling in outdoor exposure of these bright deposits with a thin chromium plate of about 0.25 μm with micro-porosity obtained by depositing it over bright nickel plate containing fine discrete particles in its surface, has been found to be similar to that of micro-cracked chromium of 0.75 to 2.5 μm thickness on normal bright nickel. Thus, by applying a thick micro-cracked chromium (0.25 to 0.75 μm) on multiple-layered nickel, or by the use of the normal multiple-layer nickel in conjunction with a final nickel plate which induces a fine porosity pattern of 150,000–50,000 pores per/cm^2 in the usual thickness of conventional chromium (0.25 μm), the rate of corrosion pitting is slowed down. Moreover, when the sulphur-free semi-bright nickel layer is reached, the corrosion which develops is also diverted laterally, thus delaying very greatly the downward penetration into the vulnerable basis metal.

10.21. Copper undercoats

Since the theory of the mechanism of the microporous chromium system depends on the fact that the occlusions in the underlying nickel provide a large number of sites where nickel can corrode at discontinuities in the chromium deposit, it was originally believed that increasing the chromium thickness too much would be a disadvantage, as it would seal some of the active nickel sites. Recent work by Carter[(46)] indicates that this is not the case provided the porosity in the chromium is not reduced below about 15,000/cm^2. A series of tests was carried out with micro-porous chromium deposits ranging from 0.75 to 2.5 μm thickness both with and without copper undercoats in static and mobile exposure tests. Variations in the chromium thickness was found to be generally without significance, except in the case of steel car hub discs without a copper undercoat exposed at an industrial site. Here the period of full protection was longer where the chromium was thickest.

When copper was present, micro-porous chromium provided full protection of steel in practically all cases for 2 years in all environments. Copper undercoats do not have this effect under conventional chromium deposits. This difference is ascribed by Clauss and Klein[(47)] to the fact that the copper remains cathodic to the nickel because of the large area of nickel involved in the corrosion reaction when the chromium layer is discontinuous. With conventional deposits, however, the smaller area of corroding nickel allows higher current densities in the pits and the copper becomes anodic and is readily penetrated. The effect is reduced when the nickel layer adjacent to the copper is of a less corrodible type (i.e. semi-bright or dull nickel), and hence the advantages of the copper undercoat are less in the systems employing duplex nickel deposits.

There is some deterioration or dulling in the appearance of articles plated with microporous chromium, but this is only of significance in static exposure in the severest environments. Dulling was progressively reduced by increasing the thickness of the chromium deposit within the range studied without adverse effect on the protection of the basis metal against corrosion.

10.22. Plant

Chromic acid solutions are highly oxidising in character, and readily oxidisable materials must not be employed in contact with them. Tanks are generally made of steel, and are usually internally lined with antimonial lead or tin-lead. Lead-lined plant lasts very well, and as steel is only slowly attacked by the chromium-plating solution, local failures in the lead lining do not immediately cause serious trouble. It is necessary to suspend an inner lining, usually in the form of wired or armoured glass sheets, inside the tank to minimise the risk of short-circuits with the anodes of the work during operation. Plastic sheets, such as P.V.C., can also be used for this purpose, but they are liable to distort at the bath temperature used. Sheet containing embedded steel wire mesh is available and is better in this respect, as are also the high temperature polythene plastics.

Rubber-lined tanks cannot be used with chromic acid, but vinyl polymers withstand it well. The plastic lining is applied in sheet form and bonded to the steel tank; such linings have given very satisfactory service.

Chromic acid solutions attack copper only very slowly, so that copper bus-bars are employed for connection and for the wiring-up of small parts. Copper wire mesh trays or baskets can also be used for dealing with very small articles.

The fluoride and fluosilicate baths are more corrosive than the sulphuric acid solutions, and precautions have to be taken accordingly. Although antimonial lead-lined tanks, such as are normally employed in sulphate-catalysed chromium plating baths, can be used, 4–7 per cent tin-lead alloy linings have been recommended as being superior in resisting solutions of this type. There is also a tendency for salts to crystallise out of such solutions at room temperature, so that if the bath is left cold for any length of time it must be heated to about 20°C above the working temperature and the anodes cleaned before use.

10.22.1. HEATING

The established method of heating chromium plating tanks has been by means of iron coils in an outer water-jacket surrounding the tank containing the chromic acid solution. The high density of the chromic acid bath makes it easier to prevent temperature variations within the solution by using a water-jacket, close temperature control being essential if bright deposits are to be consistently obtained. Owing to the high currents used, considerable heat is developed in the solution, so that heating is usually only required when the plant is being started up. In most plants, therefore, cooling coils are installed in the water-jacket in order to keep the temperature of the bath from rising excessively when the plant is being operated continuously for long periods at a time.

In the case of plastic-lined tanks, internal steam heating coils should be used as water-jackets are impracticable. Coils may be made of antimonial- or tin-lead, or of chemically pure lead. One of the principal causes of failure of internal coils is the result of bipolar effects which can develop due to the high currents. Therefore, where such coils are used, their location is of prime importance. A separate heating compartment can be used in which case a pump is employed to circulate the solution, but this is not usual. Titanium heating coils, although relatively expensive, are very successful, being virtually immune to attack by chromic acid when they are anodic, cathodic or neutral. They are, however, attacked to some extent in the fluoride-containing solutions, and are not therefore recommended for this type of bath. Tantalum coils are satisfactory in most types of chromium solutions, but are still more expensive. It is possible to apply a thin film of polytetrafluoroethylene to heating coils; this is unaffected by either the solution or the temperature, and its thickness is too low to interfere appreciably with heat transfer. Heating coils made wholly of bundles of thin tubes of this plastic are also used.

10.22.2 FILTRATION

With the advent of heavy chromium-plating for engineering applications, the advantages of filtering chromium-plating solutions became increasingly apparent[48]. The contaminants in a chromium solution consist to a considerable extent of magnetic particles, when steel is being plated chiefly, calcium and magnesium compounds from water additions and drag-in and lead, tin, or tellurium derivatives from anodes. There is little organic contamination, due to the oxidising character of the bath. Filtration is very helpful in reducing nodular deposits by removing particles upon which local growths can occur. It is also a useful method for removing trivalent chromium; 3.04 parts by weight of potassium permanganate are added to the bath for each part of trivalent chromium (as metal), followed by filtration.

Filters must obviously be resistant to the solutions. Amongst suitable materials which can be employed for pumps, tanks, filter plants and bags, etc., are polyvinyl chloride, chlorosulphonated polythene and polytetrafluoroethylene, either in massive form or as linings on steel. Rubber, both natural and synthetic, is unsuitable as it is attacked by the acid.

10.22.3. AUTOMATIC PLANT

Although much chromium plating carried out in this country is done in manually operated plants, various types of automatic units have been designed, and these present many advantages. Chromium plating in such cases generally constitutes part of combined bright nickel and chromium plating plants. Special-purpose automatic plants for chromium plating alone have been built, but these are used for producing heavier deposits for wear resistance.

10.22.4 BARREL CHROMIUM PLATING

The problems of barrel chromium plating are considerable since it is essential that:

(a) the bright plating range should be maintained,
(b) spray and fume extraction should be adequate, and
(c) there should be high current density and current continuity.

Many attempts to build satisfactory barrels for chromium plating small parts have been made. One of the first successful machines was built in Germany about 1938 and consisted of a partially-immersed horizontal barrel containing a helix, so that the articles travelled horizontally through it, being discharged at the far end. Several types of modern barrels are now available mostly based on a similar principle, and are capable of producing fully-bright deposits. A certain amount of 'burning' or 'missing' is, however, inevitable, and the articles should be individually inspected after plating. The proportion of defective work need not be excessive if care is taken in controlling the plating conditions and selecting the type of articles to be plated in the barrel.

In practice, the articles are fed into the barrel from a hopper by means of a belt conveyor, and after plating for 3 to 5 minutes they are passed into a scoop which lifts them out for draining and discharge into a perforated basket, whence they are removed. Contact is provided by the barrel itself which is connected to the negative electrode of the current supply by means of brushes. A lead anode is located horizontally in the barrel, which has to be removed periodically for de-plating. Heating is effected by means of a water-jacket around the solution tank.

A radically new approach to the problem of chromium plating small articles in bulk is the Fuji machine, shown in Fig. 10.8. It consists essentially of a frame carrying two drums which rotate at 150–200 r.p.m., whilst the drums themselves are rotated in the opposite direction at 3–4 r.p.m. The result is that the articles are held centrifugally against the mesh wall, which forms the cathode, whilst at the same time they are turned over by the slower rotation of the drums.

The electrolyte used is of the silico-fluoride type, which has a wide bright plating range, and is also tolerant to breaks in the continuity of the plating process. A typical solution contains:

Chromic acid, CrO_3	200–250 g/l
Sodium silico-fluoride	12–16 g/l
Trivalent chromium	1.2–5.6 g/l
Temperature	25°–35°C.

A very recent barrel plating machine has been described elsewhere by Silman[49]

Fig. 10.8. 'Rotary Chromer' barrel chromium plating machine.
[Courtesy Fuji Platerite Ltd.]

10.22.5. TAMPON PLATING

An interesting improvement has been made in the technique of applying local deposits of chromium, especially for engineering applications. The method consists in applying the solution by a rubbing action to the article being plated by means of an inert anodic electrode attached to a pad impregnated with the plating solution, and making the article being plated cathodic. Previously this method could not be used in the case of chromium as the available solutions were difficult to control in the small amounts that could be held in the pad, and the deposits were poor or non-existent, but the method has become commercially practicable as a result of the development of improved methods of applying the current and more suitable plating solutions. These are based on an organic amino-oxalate of chromium, such as Gregory's salt $(Cr(C_2O_4)_3).(NH_4)_3$, dissolved in an alcohol such as methyl alcohol with an alkyl amine wetting agent[50].

Very high densities are employed so that the solution boils away in use and is replenished from time to time; hence fresh electrolyte of the correct composition is always being employed for plating. The electrolyte is slightly alkaline and is non-corrosive to the skin or to the metal being plated. Current densities of up to 400 A/dm^2 can be used and the metal is deposited at the rate of 0.25 μm per minute. The deposits are bright and adherent.

Current is supplied by a transformer and rectifier unit containing an arc load release switch which comes into operation when the current density becomes so high that arcing occurs. A press-button re-set switch is provided. The current density is controllable and an integrating watt-hour meter enables the operator to assess the progress of the plating operation.

A graphite rod is used as the electrode to which the pad is attached and this can be water-cooled if need be.

10.22.6. FUME SUPPRESSION

In the course of chromium plating large volumes of hydrogen are liberated because of the low cathode efficiency of the solution. This carries into the air a considerable amount of chromic acid spray which is highly irritant to the mucous membranes of the nose and throat and can cause ulceration and even perforation of the nasal septum unless precautions are taken to prevent its entry into the atmosphere.

In the early days of chromium plating, the tank was often fitted with lids, or a layer of paraffin oil was kept on the surface of the solution in an attempt to prevent liberation of spray. Nowadays cross-ventilation of the tank is employed, the spray being extracted by exhaust fans. Fans and duct-work can be satisfactorily made from steel sheet treated with a suitable paint or, better still, plastisol-dipped to minimise attack by the solution; or else they can be constructed entirely of a suitable plastic material such as polyvinylchloride or polythene. Drain-off valves must be provided at the lowest points of the exhaust installations to enable the condensed spray to be drawn off from time to time. If an adequate rate of air-flow is employed (about 600 m per minute) little spray enters the atmosphere. The above-mentioned rate of extraction, if applied by means of a slot along one side of the tank, is sufficient for a bath about 60 cm wide; wider tanks are best dealt with by slots along both sides. Alternatively, it is preferable to extract even narrower tanks by means of exhaust slots along two sides. The level of the solution should be kept about 25 cm below the top to keep the spray from rising unduly. If possible, the best arrangement is to lead the exhaust ducting down below ground level as in the case of the automatic plant shown in Fig. 10.9.

A layer of small plastic tubes may be floated on the surface of the bath; these tubes prevent the spray from rising mechanically and reduce chromic acid losses considerably. They are of less value in automatic plants as they tend to be carried forward in the direction of movement of the work.

Surface-active wetting agents have been used to reduce losses by lowering the surface tension of the solution, but the vast majority of them are more-or-less quickly oxidised by the chromic acid and must be constantly replenished. Another disadvantage with decomposable wetting agents is that they tend to form thick foam blankets on the bath which trap the liberated oxygen and hydrogen, and can result in unpleasant explosions when these gases are ignited by a spark. The continued reduction of the hexavalent chromium by the additive also tends to lead to an increase in the trivalent chromium content of the bath.

However, it has been found that certain sulphonated perfluorocarbon derivatives have surface-active properties which completely eliminate chromic acid spray when they are added to chromium-plating baths, and are permanently stable in the solution[51].

Fig. 10.9. Automatic plant showing arrangement of exhaust ducting.

The loss is, therefore, limited to 'drag-out' removal, which is important as these compounds are relatively expensive. Also, there is little foam resulting from their use, which eliminates the explosion hazard which can be present if appreciable amounts of foam are present on a chromium bath. The use of one such product ("Zeromist") has had official approval in some countries, including the United Kingdom, as an alternative to mechanical exhaust equipment in chromium plating. Unfortunately, these materials are not recommended for use in 'hard' chromium-plating, as they sometimes cause pitting where the plate exceeds about 25 μm in thickness.

Chromic acid is also irritant to the skin, and good washing facilities must be provided for operators. Dermatitis and ulcers which are difficult to heal are especially prone to occur if the acid is allowed to enter abrasions in the skin.

Robarts[52] has reviewed the chromium plating regulations 1931-73 and also the techniques available for determining the CrO_3 atmospheric concentration.

10.22.7. RACK INSULATION

Rack insulation is very important in chromium plating as the metal is somewhat costly and it is desirable to conserve power as far as possible owing to the poor cathode efficiency of the solution.

The problems of rack insulation are similar to those encountered in nickel plating and to all intents and purposes, the rack coatings currently used are P.V.C. plastisols or fused polythene or similar plastics. Where bright nickel and chromium plating are successively carried out on the same racks in automatic plating plants, the racks after chromium plating are treated in a reducing solution such as sodium sulphite to avoid carrying chromic acid into the nickel bath.

REFERENCES

1. J. D. Greenwood, 'Hard Chromium Plating', 1971, Robert Draper Ltd., Teddington.
2. N. R. Bharncho and J. J. B. Ward, *Product Finishing* 1969, **33**, No. 4, 64.
3. J. J. B. Ward and I. R. A. Christie, *Trans. Inst. Metal Finishing* 1971, **49**, 148.
4. C. Barnes, J. J. B. Ward and J. R. House, *Trans. Inst. Metal Finishing*, 1977, **55**, 73.
5. British Pat. 1,378,883; German Patent DOS 2,558,913.
5. J. C. Crowther and S. Renton, *Electroplating and Metal Finishing*, 1975, **28,** No. 5, 6, 8-9, 12-14.
6. E. Müller, *Z. Elektrochem* 1926, **32**, 399.
7. E. Müller and O. Essen, *Z. Elektrochem* 1930, **36**, 2.
8. F. Orgburn and A. Brenner, *J. Electrochem. Soc.* 1949, **96**, 347.
9. *Metal Finishing Plant and Processes*, 1973, **9,** No. 4, 120-126.
10. J. A. Hood, *Metal Finishing* 1952, **54**, 103.
11. H. Wahl and K. Gebauer, *Metalloberfläche* 1948, **2**, 25.
12. J. E. Starek, F. Passal and H. Mahlstedt, *Proc. Am. Electroplaters' Soc.* 1950, **37**, 31.
13. Brit. Pat. 697,786 (1953).
14. J. L. Griffin, *Plating* 1966, **53**, 196.
15. Netherlands Pat. 6,513,035 (1966).
16. R. Dow and J. E. Starek, *Plating* 1953, **40**, 987.
17. French Pat. 754,299.
18. F. Taylor, *Electroplating and Metal Finishing* 1962, **5**, No. 4, 109.
19. J. C. Saiddington and G. R. Hoey, *Plating* 1970, **57**, No. 11, 1112.
20. *Pinturas y Acabados Ind.*, 1972, **14**, No. 62, 127-8.
21. J. P. Branciarol and P. G. Stutzman, *Plating* 1969, **56**, 37.
22. A. K. Graham, *Proc. Am. Electroplaters Soc.* 1959, **46**, 61.
23. L. Sivaswamy, S. Gown and B. A. Shenoi, *Metal Finishing;* 1974, **72,** No. 3, 48-51.
24. J. E. Longland, *Metal Fin. J.* 1968, July, p. 224.
25. B. A. Shenoi and S. Gown, *Metal Finishing J.*, 1973, **19**, No. 220, 139-42.
26. T. G. Coyle, *Proc. Amer. Electroplaters Soc.*, 1944, **32**, 20.
27. H. R. Friedberg, *Plating* 1959, **46**, 834.
28. D. W. Hardesty, *Plating* 1969, **56**, No. 6, 705.
29. J. H. Nelson and H. Silman, 'The Application of Radiant Heat to Metal Finishing' (2nd Ed.) Chapman and Hall, London, p. 79.
30. G. Soderberg, *Proc. Am. Electroplaters Soc.* 1936, **24**, 233.
31. D. J. Macnaughton and A. W. Hothersall, *J. Electrodep. Tech. Soc.* 1930, **5**, 63.
32. C. A. Snavely, *Trans. Electrochem. Soc.* 1947, **92**, 537.
33. D. R. Gabe and J. M. West, *Trans. Inst. Metal Finishing*, 1963, **40**, 6.
34. J. K. Dennis, *ibid* 1965, **42**, 84.
35. W. H. Cleghorn and J. M. West, *ibid* 1967, **44**, 43.
36. G. R. Davies, *Trans. Inst. Metal Finishing*, 1973, **51,** Pt. 2, 47-55.
37. H. Brown and E. V. Hoover, *Proc. Am. Electroplaters Soc.*, **40**, 32, (1953); *Plating* 1958, **45**, 144.
38. W. A. Wesley and B. B. Knapp, *Trans. Inst. Metal Finishing* 1954, **11**, 267.
39. H. Brown, *Metalloberfläche* 1963, **11**, 133.
40. J. K. Dennis and J. J. Fuggle, *Trans. Inst. Metal Finishing*, 1971, 49.

41. W. H. Safranek and R. W. Mardy, *Plating* 1960, **47**, 1027.
42. W. E. Lovell, E. H. Shotwell and J. Boyd, *Proc. Am. Electroplaters Soc.* 1960, **47**, 215.
43. L. Serota, *Metal Finishing* 1963, **61**, No. 4, 69.
44. D. R. Gabe and J. M. West, *Trans. Inst. Metal Finishing* 1963, **49**, 197.
45. J. E. Longland, *Electroplating and Metal Finishing* 1969, **22**, No. 12, 35.
46. V. E. Carter, *Trans. Inst. Metal Finishing* 1970, **48**, 19.
47. R. J. Clauss and R. W. Klein, *Proc. 7th International Conference on Metal Finishing* 1968, 124.
48. R. F. Ledford and L. O. Gilbert, *Plating* 1955, **42**, 1151.
49. H. Silman, *Metal Finishing Plant and Process*, 1975, **11,** No. 6, 174, 176.
50. Electroplating and Metal Finishing, 1953, **6**, 131.
51. Brit. Pat. 758,025.
52. C. Robarts, *Ind. Finishing*, 1973, **25,** No. 297, 29-30.

BOOKS FOR GENERAL READING

1. 'Chromium Plating', R. Weiner and A. Walmsley, Finishing Publications Ltd., Teddington, 1978.

CHAPTER 11

Tin and tin alloy plating

TIN plating is a process which presents many advantages for a variety of applications, especially in the food industry, because of the non-toxicity of the metal and its excellent resistance to alkalis and detergents. Tin coatings are also useful on parts which have to be soldered. The largest application of tin coatings is in the manufacture of tinned steel sheet on continuous plants for the canning industry and other purposes; such sheet is now largely produced by electroplating, since this method gives uniform coatings of readily controlled thickness, which are relatively free from defects[1]. Refrigerator parts, dairy utensils, kitchen fittings, and numerous domestic accessories are regularly electrotinned, and tin plate is finding increasing use in the electronics industry[2, 3]. The protective value of tin coatings is also useful on non-ferrous materials, e.g. brass. On steel, tin is less protective than zinc or cadmium because of its cathodic characteristics relative to the basis metal.

Tin is deposited from three main types of solution: (a) the alkaline stannate bath, (b) the acid tin sulphate bath, and (c) the fluoborate bath.

11.1. Alkaline stannate plating

11.1.1. SODIUM STANNATE BATH

11.1.1.1. Bath composition

The stannate bath consists of a solution of sodium stannate with a certain amount of free caustic soda. A satisfactory bath can be made up from:

Sodium stannate	$Na_2SnO_3.3H_2O$, 80 g/l
Caustic soda	NaOH (free), 12 g/l

Most of the tin is present as a complex stannate (tetravalent tin) anion. The mechanism by which tin is deposited from this complex is still somewhat uncertain, but it is generally considered that two stages of ionic dissociation are involved:

$$Na_2SnO_3 \rightleftharpoons Na^+ SnO_3^{2-} \quad (1)$$
$$SnO_3^{2-} + 3H_2O \rightleftharpoons Sn^{4+} + 6OH^- \quad (2)$$

From these equilibria it is clear that hydroxide ions play an important part in the deposition process. If the sodium hydroxide concentration of the bath is allowed to fall too low the equilibrium of equation (2) will shift towards the right and the precipitation of insoluble tin compounds will result. Too low a concentration of free caustic soda will also lead to low anode efficiency, although the cathodic efficiency will be increased.

In the early stages of its development the stannate bath was uncertain in its operation due to the tendency towards the formation of dark, spongy deposits. Precipitation of tin salts at the bottom of the bath in the form of insoluble sludges was also liable to occur. This condition was eventually found by Wernlund[(4)] and Oplinger[(5)] to be due to the development of stannite (divalent tin) in the bath. Stannite formation can be prevented by commencing deposition with a relatively high anode current density until a characteristic yellow-green film (probably a suboxide of tin) forms on the anodes, which shows that the tin is all going into the solution in the form of tetravalent stannate; a dull grey appearance indicates that the tin is dissolving as the undesirable stannite.

The operation of a commercial stannate solution containing bismuth has been recently described[(6)]. It is claimed that the presence of bismuth eliminates whiskering of the deposit.

11.1.1.2. Anodes

The anode film is maintained as long as the anode current density is high enough to promote the dissolution of the tin in the tetravalent form, and the free caustic soda content is not excessive. The film forms at current densities as low as 0.5 A/dm^2 with a low caustic soda content in the solution, but with high caustic soda concentrations, greater initial current densities are required. There appears to be a critical anode current density for each solution below which the tin dissolves in the form of the undesirable stannite. This current density varies with the caustic soda content of the solution and with its temperature. It also tends to increase with increasing concentration of stannate. If the anode current density becomes too high, on the other hand, the anodes develop passivity and may become coated with an insoluble and tenacious black film. The polarisation curve shown in Fig. 11.1 illustrates how film formation is related to anode potential and current density[(7)].

Due to the importance of the anodes working in the correct range of current densities, good electrical connections are essential. Faulty contact to the anode may result in the development of sufficient stannite in the solution to cause dark, spongy deposits. Should this occur, the addition of a small amount of either hydrogen peroxide (about 0.5 g/l) or sodium perborate (0.4 g/l) to oxidise the stannite is recommended. The anode area should be slightly greater than the mean cathode area passing through the bath.

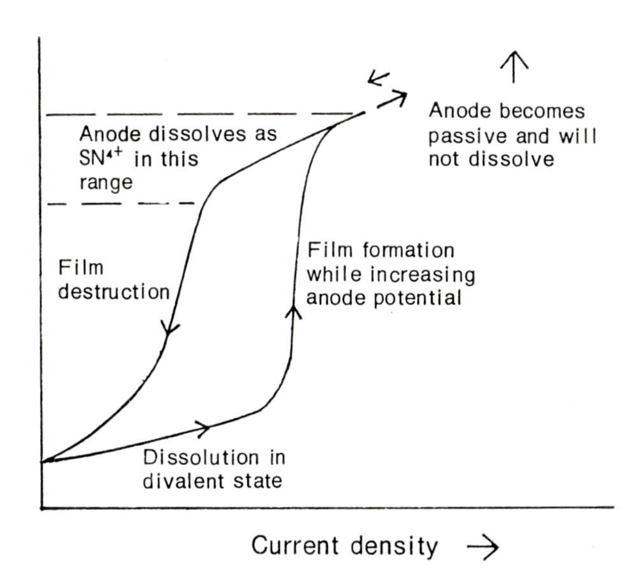

Fig. 11.1. Polarisation curve for a tin anode in sodium stannate-caustic soda solution.

The effect of solution composition in relation to operating conditions is shown in Table 11.1.

11.1.1.3. Operating conditions

The stannate bath has an increasingly low cathode current efficiency as the current density increases above 2 A/dm^2, which limits its application as far as high speed plating is concerned. High operating temperatures are desirable and the solution is commonly maintained at 60-80^{0}C. The throwing power of the solution is very high[9] especially

TABLE 11.1

Operating Conditions of Sodium Stannate Plating Solutions[8]

Solution composition (g/l) Tin	Na_2SnO_3 $3H_2O$	Free NaOH	Cathode current density (A/dm^2	Cathode efficiency	Temperature	Time to plate 25 μm (min)
60	135	15	2	82½	73°C	56
50	112	15	1.6	85%	72°C	73
40	90	12.5	1.6	85%	77°C	73
30	67	10	1	87%	78°C	103
20	45	10	1	87%	78°C	103

if a low metal ion concentration is maintained; increased free sodium hydroxide also favours high throwing power.

In order to minimise the evolution of unpleasant caustic alkali spray into the atmosphere during the working of the bath, a small proportion of sodium oleate is sometimes added. This addition results in the formation of a protective blanket of foam on the surface of the bath. The presence of metallic impurities such as lead, arsenic and antimony in the solution results in dark deposits. Impurities may be introduced by the use of poor quality anodes or salts, but these will tend to plate out eventually.

11.1.2. POTASSIUM STANNATE BATH

Introduced in 1942 by Sternfels and Lowenheim[10], the potassium stannate bath has now replaced the sodium stannate bath for many applications. The advantages of potassium stannate are that the use of equivalent stoichiometric quantities of the latter salt increases the bath conductivity very considerably and raises the cathodic efficiency, whilst the tendency for sludge formation is reduced. The high solubility of the salt is also an advantage since it allows the use of higher current densities (up to 20 A/dm^2) and hence high deposition rates[11, 12]. The only real disadvantage associated with the use of the potassium bath is the rather higher cost of the potassium salt itself.

A solution containing 190-200 g/l of tin and 100 g/l of free caustic potash operates with a cathode current efficiency of 82% at a current density of 16 A/dm^2 and a temperature of 90°C. The corresponding anode current efficiency is 84%. With a tin concentration of 140 g/l and a free caustic potash of 150 g/l the anode and cathode current efficiencies are both 70% at the same temperature and current density. Lowering the free alkali content of the bath reduces the conductivity of the solution, but generally tends to increase the cathode current efficiency so that KOH concentrations of the order of 20 g/l are usually used in commercial practice[12]. Fig. 11.2 shows the comparative efficiencies of equivalent potash and soda baths containing 83 g/l of tin and 28 g/l of NaOH or 38 g/l of KOH[13]. The anode current efficiency in this type of bath at high current densities is said[14] to be improved by the use of an anode containing 1% of aluminium.

11.1.3. ADVANTAGES OF THE ALKALINE STANNATE BATHS

The following advantages can be claimed for the alkaline stannate baths:

1. Cleaning before deposition need not be as thorough as for other tin plating baths since the electrolyte itself acts as a cleaning agent.
2. The throwing power of both sodium and potassium baths is extremely good[9].
3. The electrolyte is essentially non-corrosive, and expensive corrosion-resisting tanks and pumps are not required.
4. The cost of the alkaline solutions is less than that of other tin plating solutions.
5. Maintenance and control of the solution is not critical and little anode sludge is formed.

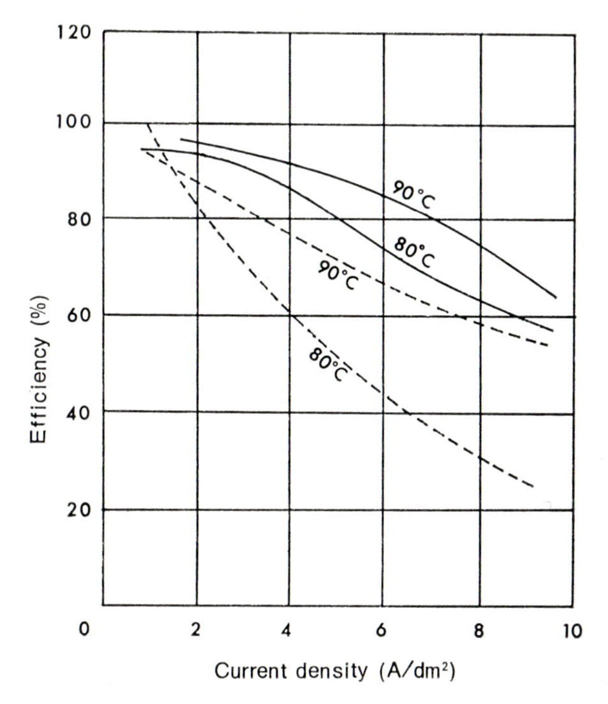

Fig. 11.2. Comparative efficiencies of equivalent potash and soda tin-plating baths.

Potassium stannate bath ————————

Sodium stannate bath - - - - - - - - -

11.2. Acid sulphate bath

11.2.1. BATH COMPOSITION

The acid tin bath consists of a solution of stannous sulphate and sulphuric acid together with various addition agents. The latter are essential since without them the deposits produced are crystalline and non-coherent.

A bath which is widely used in this country[12] contains:

Stannous sulphate	60 g/l
Sulphuric acid (conc)	60 g/l
Phenolsulphonic acid	100 g/l
Gelatin	2 g/l
Beta-naphthol	1 g/l

The solution can be made up either by dissolving stannous sulphate in the mixed sulphuric and sulphonic acids, or by effecting the anodic dissolution of metallic tin in the acid mixture. The gelatin is separately dissolved in hot water and the beta-naphthol, dissolved in alcohol, is added to it slowly, with vigorous stirring. After standing and decanting off any precipitate, the mixture is ready for incorporating in the solution. Instead of phenolsulphonic acid, resorcinol or cresolsulphonic acid[15] may be employed.

11.2.1.1. Operating conditions

The acid sulphate solution must be operated in a lead-lined or rubber-lined tank; alternatively, stoneware or lead-lined tanks with an inner lining of reinforced glass may be used. Room temperatures are generally used, the optimum working range being 20^0 to 25^0C; excessively high temperatures are to be avoided as they tend to cause oxidation of the solution. The cathode current density should not normally exceed about 1 A/dm^2 unless cathode rod agitation is used; air agitation is not recommended as this would lead to rapid oxidation of the tin.

The conductivity of the acid sulphate solution is high, so that only low operating voltages (about 1 to $1\frac{1}{2}$ volts) are required; the throwing power, on the other hand, is fairly low. The covering power of the solution depends largely upon the addition agents and if the concentration of these is allowed to fall too low, poor crystalline deposits are formed with bare areas, particularly at low current density regions. Cathode current efficiency is very high, being almost 100%. The anode current density is not critical, but should not exceed about 2 A/dm^2.

The solution does not require a high degree of filtration, but excessive amounts of sludge should not be allowed to accumulate as they cause porosity in the deposits. The acid tin solution is liable to develop sludge owing to oxidation of tin to insoluble basic compounds, and to gradual precipitation of addition agents. Periodic filtration through kieselguhr is helpful in eliminating sludges.

11.2.2. BRIGHT ACID TIN PLATING BATH

Deposits from the acid bath are, like those from the alkaline solution, matt in appearance. As early as 1936 Schlotter[16] claimed that tin could be deposited bright from acid electrolytes containing additives based upon wood tar or its extracts, with a lyophilic agent such as octyl sulphuric acid. Since this time, many other investigations[17–22] have been carried out to develop bright tin plating baths, and nowadays bright tin coatings have replaced those of nickel for some applications. Virtually all the bright plating baths in this country are proprietary solutions[12]. Although these commercial solutions each have their own particular advantages, the basic compositions are similar and there are many common points in their operation. Essentially, bright tin solutions contain:

stannous sulphate ($SnSO_4$) 50-60 g/l
sulphuric acid (conc.) 70-75 ml/l

and polymerised resin addition agents based on wood tar condensation products, aldehydes, ketones and similar substances, which are matched with appropriate dispersants and surface active materials (e.g. sulphonated aliphatic alcohols). The effect on the appearance of the tin deposits when wood tar (with a dispersion agent) is added to a stannous sulphate bath, is shown in Fig. 11.3.

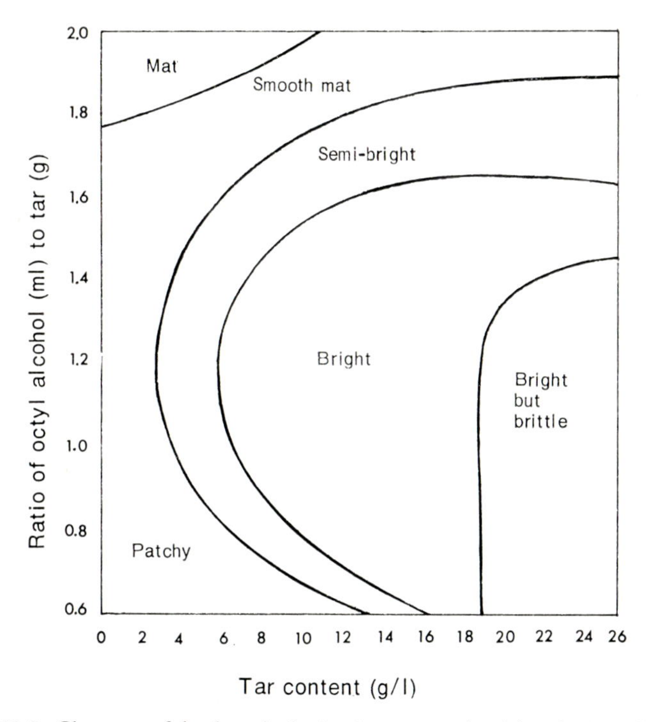

Fig. 11.3. Character of the deposit obtained at current densities of 0.8–20 A/dm^2 as contents of tar and octylsulphuric acid in the electrolyte are varied.

In this case the brightening agent contained in the tar is dispersed by octyl sulphuric acid in the solution. If the amount of added tar is decreased, the quantity of dispersant must be increased to ensure bright deposits. However, below 6 g/l of tar it is not possible to obtain fully bright deposits, whatever quantity of dispersant is added. Increasing the concentration of dispersant enables more of the added acid tar to 'mix' with the water, and hence a greater brightening action is obtained. Layers of organic substances from the tar are adsorbed on to the tin surface and alter the mode of crystal growth. It is most likely that diffusion of tin ions occurs through these layers, and their discharge takes place below the organic layer, giving rise to bright deposition.

11.2.2.1. Operating conditions

The bath is normally operated at room temperature, but can generally be worked within the range 15^0 to 35^{0}C. Moving cathode bar agitation is usually employed, and this enables current densities of up to 4 A/dm^2 to be used.

Bright tin solutions are not particularly susceptible to contamination problems but the deposits are adversely affected by the presence of chromium, nickel, copper and iron and, to a greater degree, chloride ions. However, provided that these impurities are not introduced, the solutions may last for periods of up to approximately three years.

11.2.3. ADVANTAGES OF THE ACID SULPHATE BATHS

The acid sulphate process can be considered to have the following advantages:

1. To deposit a given amount of tin from the acid (bivalent tin) solution requires only half the quantity of electricity that would be necessary for deposition from the tetravalent alkaline bath.
2. Cathodic current efficiencies are generally higher.
3. Much higher anodic current densities are possible and therefore a much smaller area of tin electrode is necessary.
4. Plating is carried out at room temperature, so that facilities for heating the tanks are not required.

11.3. Stannous fluoborate bath

11.3.1. BATH COMPOSITION

This bath is not used to the same extent as the stannous sulphate and alkaline plating baths, but it is preferred for some purposes such as the plating wire and narrow gauge strip[23]. Basically the fluoborate bath contains[24] stannous fluoborate, fluoboric acid and an addition agent to prevent the formation of coarse crystalline non-adherent deposits. Typically, a fluoborate bath contains 80 g/l tin and 50 g/l free fluoboric acid. A higher fluoboric acid content increases the stability of the solution but reduces the maximum current density which can be employed during plating. In addition to stannous fluoborate and fluoboric acid, boric acid will be present in the solution due to hydrolysis of fluoborate. This does not appear to be deleterious with regard to the quality of the deposit, but its presence makes determination of fluoboric acid concentration difficult. The concentration of the addition agents (glue, β-naphthol or gelatin) is best controlled by means of a small-scale plating test. If the deposit obtained is coarsely granular, this indicates that insufficient addition agent is present.

The stannous fluoborate bath is operated at temperatures of between 20^0 and 40^0C, and at the high current densities employed (5-13 A/dm^2) the anode and cathode current efficiencies are close to 100%. A full account of tin deposition from fluoborate solutions is given elsewhere[25].

11.3.2. ADVANTAGES OF FLUOBORATE PLATING SOLUTIONS

1. They are easy to operate and work well at room temperature.
2. Fine-grained deposits can be obtained at very high current densities.
3. Deposition involves discharge of divalent tin ions, so that maximum deposition rates are obtained for a given current density.

The main difficulty associated with the fluoborate bath is the prevention of sludging of the electrolyte due to disintegration of the tin anodes.

11.4. Barrel plating

Tin can easily be deposited by the barrel plating process, and for small articles in bulk this method has found substantial use. Alkaline, acid or fluoborate solutions can be used, employing open-ended barrels, although these are not particularly suited to the alkaline solutions since difficulties arise in maintaining the solution at the high working temperatures. For alkaline solutions it is best to use perforated immersed barrels with external anodes. These barrels are immersed in the plating solution and problems of solution temperature control do not arise.

The solutions for barrel plating are essentially the same as for still vat plating but metal salt and brightener (if added) concentrations may be somewhat changed[12].

11.5. Continuous electrotinning of steel strip

The process of electrotinning of steel strip was originally introduced as a measure of war-time economy to conserve tin, since it is possible to produce tin coatings by this method which are only one third as thick as the thinnest coatings obtainable by hot dipping. It was found, however, that the possibility of high speed plating (up to 1,500 feet per minute or more), the ready control over the thickness of coating and its uniformity, coupled with the fact that the steel can be handled continuously in coil form, presented so many advantages that the process has become the established method of making tin plate, and nowadays many plants are operated all over the world. In view of the large number of innovations and modifications to automatic electrotinning plants, introduced in recent years, only the essential principles can be described here. An account of the operation of a modern tin mill is given by Long[26].

Both the alkaline stannate and the acid sulphate solutions are used for electrotinning steel, but use of the former is predominant in the U.K. The actual electrotinning process can be divided into five stages[27]:

1. Entry
2. Pre-plating treatment
3. Electrodeposition of tin
4. Post plating treatment
5. Coiling and shearing

Entry stage. Separate steel sheets are welded together to enable continuous operation of the bath. In practice it is necessary to have a reserve of strip at the entry stage and to accomplish the welding operation as rapidly as possible.

On entering the *pre-treatment* zone the strip is electrolytically degreased and pickled, and is then passed through a rinse tank before being fed into the *electroplating* zone.

Electroplating is carried out by looping the strip through a series of vertical rubber-lined steel tanks. Each tank is equipped with its own anodes which are narrower than the strip, and tilted so that they are close to the strip at the bottom of the pass rather than the top. This ensures that current distribution is uniform over the entire looped area of the strip. For the same thickness of coating the actual current density is proportional to line speed, so that a compromise between optimum current density and line speed must be sought.

The *post plating zone* involves three separate treatments. First the matt tin coating is brightened by flow melting. This involves heating the tin coating, generally by electrical resistance methods, to a temperature where it just melts and reflows uniformly over the surface of the strip. The next step is to passivate the bright tin coating and seal up any pores and defects. This can be accomplished by immersion of the coated strip in chromate or phosphate solutions[28, 29]. The third surface treatment is the application of a thin film of oil (usually palm oil) for corrosion protection. Finally the tin-plated strip is either wound into coils or sheared into sheets in the *coiling and shearing* zone.

Brown[30] has recently studied the influence of cleaners and rinses on tin plate quality. He found that several common tin plate defects were caused by ineffective cleaning and rinsing.

11.6. Plating tin alloys

The electrodeposition of a tin alloy was first reported by Curry[31] who deposited bronze in 1906. Since that time, and in particular over the last twenty years or so, tin alloy plating has become of increasing importance, speculum, bronzes, tin-zinc, tin-nickel and tin-lead alloys being plated on a commercial scale. Not only does the alloying metal confer some properties of its own to the coating, but often the grain size of the alloy deposit is smaller and the porosity is less than in the case of coatings of the individual metals.

11.6.1. PLATING OF COPPER-TIN ALLOYS

11.6.1.1. Bath composition

During the 1930s a co-deposit of copper and tin was introduced under the name of speculum alloy for articles of jewellery, tableware, bathroom fittings and the like. The preferred composition of the alloy is 42% tin and 58% copper, this being similar to the composition of ancient speculum mirrors produced by the Romans 2,000 years ago. These mirrors were produced by casting owing to the brittle nature of the metal.

Speculum metal alloy has a pleasing appearance, being whiter than chromium, is resistant to washing, and has good wearing properties. It does not tarnish easily in indoor atmospheres, and is not attacked by most foods including those which contain sulphur, such as eggs. The protective value of the deposit against severe atmospheric corrosion is not high, however, and the speculum plating is not recommended for outdoor service. The deposit is also liable to be porous, but this tendency can be reduced by regular filtration of the plating solution.

The most serviceable range of deposits contains from 39% to 55/ of tin. Within these limits, tin-copper in equilibrium[32] consist of intermetallic compounds, and it is to these that the hardness and tarnish resistance of speculum alloy is due.

With tin contents of less then 39% the deposit lacks resistance to tarnish, whilst if the tin content is higher than 55% the coating is soft[33]. The solution used for speculum plating consists of an alkaline sodium stannate bath containing copper cyanide:

Tin (added as sodium stannate)	40 g/l
Sodium hydroxide (free)	15 g/l
Copper added as cuprous cyanide	8 g/l
Sodium cyanide (free)	16 g/l

The copper is initially present in the bath as sodium cuprocyanide together with a specific quantity of free sodium cyanide. The overall equilibrium is expressed by the following equations:

$$2\ NaCN + CuCN \rightleftharpoons 2\ Na^{+} + Cu(CN)_3^{2-} \tag{3}$$

$$Cu(CN)^{2-} \rightleftharpoons Cu^{+} + 3CN^{-} \tag{4}$$

The concentration of cuprous ions is very small and the free cyanide in excess of that required to form the anion complex serves to stabilise it. The tin in the bath is present as sodium stannate, together with free sodium hydroxide; the dissociation equilibrium can, as for the copper, be represented by two equations:

$$Na_2SnO_3 \rightleftharpoons 2Na^{+} + SnO_3^{2-} \tag{5}$$

$$SnO_3^{2-} + 3H_2O = Sn^{4+} + 6OH^{-} \tag{6}$$

The free cyanide and hydroxide concentrations exert opposite effects on the composition of the deposit[29, 30]. Increasing the free cyanide concentration tends to suppress the deposition of copper, as explained earlier (cf p. 303) and to lower the cathode efficiency. In a similar manner, increasing the free sodium hydroxide concentration tends to shift the equilibirum in equations (5) and (6) to the left, and hence to suppress the plating of tin.

11.6.1.2. Anodes

Because tin is more highly polarisable than copper, separate tin and copper anodes have to be employed using individual circuits for each type of anode. Thus, the current density at the copper and at the tin anodes can be separately controlled.

Pure tin anodes are employed and must be kept filmed, as is the usual practice in tin plating from the stannate solution. The anode current density should not fall below 1 A/dm^2 or the film will be lost and stannite will form in the bath. The result will be dark, rough and brittle deposits. For this reason, the tin anodes should be lifted from the bath when the solution is not in use. If stannite should form it can be oxidised to the stannate by the addition of small quantities of sodium peroxide.

The copper anodes are preferably of the flat electrolytic type; the current density used should be about 1 A/dm^2.

11.6.1.3. Operating conditions

The solution can be contained in rubber-lined steel tanks and heating facilities are required to maintain the solution at an operating temperature of 65°C. Temperatures of above 65°C are undesirable since excessively rapid carbonate formation, partially due to hydrolysis of cyanide, is encouraged. Increasing the current density results in a marked reduction in cathode efficiency; current densities should not exceed about 2.5 A/dm^2. The effect of current density on the composition of the alloy is surprisingly small, but close control of bath composition is necessary to ensure good results.

11.6.2. BRONZE PLATING

Investigations on the deposition of bronzes, containing about 10% tin, from stannate-cyanide solutions, have shown that copper is deposited much more readily than tin. Thus the ratio of tin to copper in the bath needs to be considerably higher than that desired in the deposited alloy. A typical modern bath composition comprises[30]:

Copper	25 g/l
Sn	12 g/l
Free cyanide	17 g/l
Free caustic soda	12 g/l

Addition agents are employed to assist anode dissolution, promote grain refinement and brighten the deposit[34, 35].

Recently Menzies and Ng[36, 37] have studied the electrodeposition of bronze from a similar bath. From polarisation curves and other measurements they found that, at low cathode potentials, only Cu ion discharge takes place at all temperatures. At 40°-50°C discharge of stannate and hydrogen ions occurs simultaneously with copper deposition. At the operating temperatures used in commercial practice (65°C) discharge of stannate occurs at lower cathode potentials and the limiting current density for the deposition of tin is also increased.

Unlike the speculum bath, an alloy anode is used in bronze plating, but the anode current density must be carefully controlled since passivation occurs at high values. The cathode efficiency for a bronze solution falls with rise in current density, and to maintain a high efficiency the cathode current density should not exceed about 1.5 A/dm^2. According to Menzies and Ng[36] this current density is also critical with respect to the structure of the bronze deposits. Deposits obtained at current densities of between 0.5-1.5 A/dm^2 have fine-grained structures, whereas deposits obtained at 1.5-2.0 A/dm^2 have undesirable laminated structures.

11.6.3. TIN-NICKEL PLATING

The deposition of tin-nickel alloys has been the subject of many investigations and the process is now used on a limited commercial scale. Electrodeposited tin-nickel alloy is a single-phase intermetallic compound having a composition close to NiSn and a hexagonal nickel arsenide structure[38].

The commercial alloy coating (65% Sn-35% Ni) has a number of advantages over other electrodeposited metal coatings. By contrast with nickel deposition, a top coat of chromium is not required since it is fairly easy to obtain a bright tarnish-resistant finish on tin-nickel coatings. Furthermore, the hardness and durability of the deposit is much greater than that of nickel.

Application of tin-nickel coatings include building-up of brass analytical weights, providing a tarnish-free finish with low frictional properties on watch and other precision parts, imparting solderability and corrosion resistance to fuse caps, printed circuits[2], electrical contacts and terminals, and as an undercoat for precious metals to give good bonding to the basis metal. Other applications of tin-nickel deposits have been described by Angles[39].

The electrolyte used in commercial plants for the deposition of 65/35 tin-nickel alloys is described in the British Standard Specification 3517. It contains stannous chloride, nickel chloride and ammonium bifluoride. Unless the electrolyte concentrations differ widely from those recommended, the composition of the deposit remains within a few per cent of 65% tin and 35% Ni[40].

Rau and Bailar[41] showed that various complexes are formed in these fluoride-chloride electrolytes. Experiments in which the fluoride concentration of the bath was varied indicated that deposition probably occurs from nickel-tin complexes of the type $(NiSnFx)^{4-x}$, with ^{x}=1-4.

The essential purpose of adding fluoride ions is to complex the tin and to suppress its deposition, because in uncomplexed solutions tin is very much more easily deposited than nickel.

At the current densities normally employed (approx. 2.5 A/dm^2) the cathode efficiency is nearly 100%[42] and little or no hydrogen is liberated at the cathode. The dissolution efficiency of the tin-nickel anodes is also very close to 100% but the anodes are often cast from 72/28% Sn-Ni alloy owing to the need to compensate during deposition for the loss of divalent tin ions by oxidation. For the same reason, air agitation cannot be employed and mechanical agitation is used both to stir the electrolyte and to maintain an even temperature throughout the bath. The normal recommended operating temperature for chloride-fluoride electrolytes is 70°C.

The properties of tin nickel deposits and their use in printed circuit production have recently been reviewed[43].

11.6.4. TIN-ZINC ALLOY PLATING

The present-day commercial process for the deposition of tin-zinc alloys is based on the work of Angles and his co-workers in the period 1940-46[44-47]. The alloys deposited contain about 22% zinc, at which concentration the coatings are anodic with respect to steel. Therefore they protect steel sacrificially and have the added advantage over zinc coatings of corroding more slowly, and producing less voluminous and unsightly dissolution products.

The electrolyte employed for plating the alloy contains sodium stannate and zinc cyanide, the ratio of tin to zinc being maintained at 12:1 approximately. The composition and operating conditions are as follows:

Tin (as sodium stannate $Na_2SnO_3.3H_3O$)	30 g/l
Zinc (as zinc cyanide $Zn(CN)_2$)	2.5 g/l
Sodium cyanide (total)	28 g/l
Sodium hydroxide (free)	4 to 6 g/l
Temperature	70°C
Cathode current density	10-40 A/dm^2
Anode current density	7.5-15 A/dm^2

The composition of the complexes in the bath has not yet been completely established, but it is generally thought that tin is present in the form of the stannate ion SnO_3^{2-} and zinc in the form of the zincate ion ZnO_2^{2-} along with some complex cyanide $Zn(CN)_4^-$. The effect of the free OH and free cyanide concentrations on the composition of the deposit can be determined by a consideration of the following equations which govern the overall deposition process:

$$SnO_3^{2-} + 3H_2O + 4e \rightarrow Sn + 6OH^- \quad (7)$$

$$ZnO_2^{2-} + 2H_2O + 2e \rightarrow Zn + 4OH^- \quad (8)$$

$$Zn(CN)_4^{2-} + 2e \rightarrow Zn + 4CN^- \quad (9)$$

The equilibrium in solution between the zincate and zinc cyanide complex ions can be represented by:

$$Zn(CN_4)^{2-} + 4OH^- = ZnO_2^{2-} + 4CN^- + 2H_2O \quad (10)$$

Clearly then, if the concentration of free OH^- is raised, the amount of tin in the deposit will decrease, whereas increase in the concentration of cyanide ions will make the deposition of zinc more difficult and the concentration of tin in the deposit will thus be greater.

Commercially the stannate-cyanide baths are operated at approximately 65°C, since cathode efficiency decreases rapidly with decreasing temperature[48] as shown in Fig. 11.4.

Other conditions of operation in the bath are not critical; current densities of between 1.5 and 4 A/dm^2 can be employed, and cleaning before plating is similar to that which is usual prior to ordinary zinc plating.

Tin-zinc coatings are harder than tin and have been successfully employed on radio chassis, being outstandingly resistant to tropical conditions; they are also readily solderable. They have been used for plating cycle parts, brake units, car accessories and, due in part to their good anti-frictional properties, the rams of hydraulic jacks.

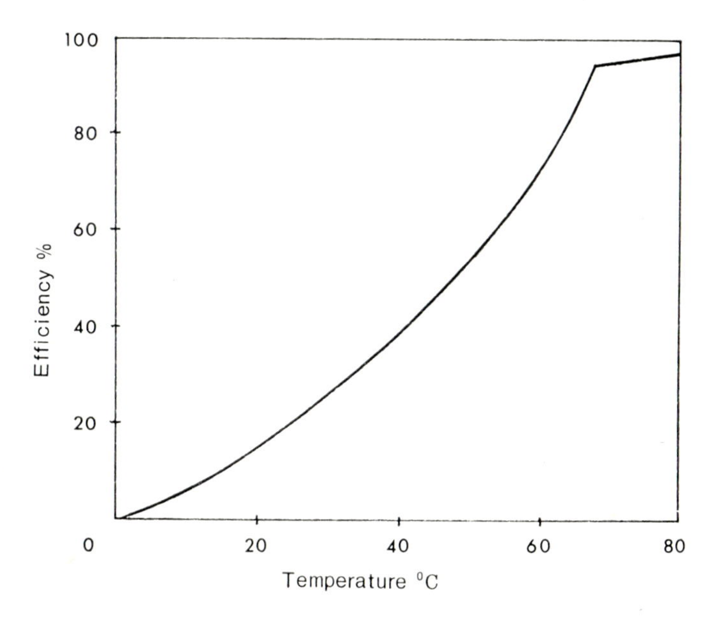

Fig. 11.4. Effect of temperature on the deposition efficiency of tin-zinc alloys from stannate-cyanide baths.

11.6.5. TIN-LEAD ALLOYS PLATING

Like the tin alloys already discussed, tin-lead alloys are deposited from complex ion solutions. Fluoborate solutions are now almost invariably used in commercial practice and the concentrations of tin, lead and fluoboric acid vary considerably according to the desired composition of the plated alloy. Two typical electrolytes which are in current use[11] contain:

1. *Solution for depositing*
7% *Tin-93% lead alloy coatings*

Tin (stannous) as stannous fluoborate	7 g/l
Lead as lead fluoborate	88 g/l
Free fluoboric acid	40 g/l

2. *Solution for depositing*
60% *tin-40% lead alloy coatings*

Tin (stannous) as stannous fluoborate	65 g/l
Lead as lead fluoborate	26 g/l
Free fluoboric acid	40 g/l

Both solutions contain a suitable addition agent (e.g. glue).

Alloy anodes are used having the same composition as that of the desired deposit, and are enclosed in anode bags. The solutions are operated at temperatures of between 25 and 30°C, and cathode current densities of 3 A/dm^2 are employed. The throwing power of the standard commercial baths is rather poor, and recently Rothschild and Sanders[43] have described the development of a high-throwing power solution containing peptone as an addition agent. A major application of tin-lead deposits is for lining bearings of copper-lead; some use of tin-lead deposits is also made in the manufacture of printed circuits[2, 43]. For further study the reader is referred to recent reviews in the literature of tin lead alloy plating[49–52].

REFERENCES

1. R. M. Macintosh, *Ind. Eng. Chem.* 1961, Nov, 909.
2. G. C. Wilson, *Electroplating and Metal Finishing* 1970, No. 12, 15.
3. *Products Finishing* 1968, **33**, 59.
4. C. J. Wernlund and F. F. Oplinger, U.S. Pat. No. 1, 919,000 (1933).
5. F. F. Oplinger, *Metal Industry N.Y.* 1931, **29**, 529.
6. *Products Finishing*, 1974, **27,** No. 10, 10-11, 14.
7. G. Hansel, *Z. Elektrochem.* 1935, **41**, 314.
8. Tin Research Council Publ. No. 92, 1956.
9. F. A. Lowenheim, *Plating* 1954, 1440, Dec.
10. M. M. Sternfels and F. A. Lowenheim, *Trans. Electrochem. Soc.* 1942, **82**, 77.
11. J. W. Cuthbertson, Chapter 5, 'Tin and its Alloys'. Ed. E. S. Hedges. Arnold Ltd., London, 1960.
12. Chapter 24. Canning Handbook on Electroplating, 21st edition. W. Canning & Co., Birmingham, 1970.
13. F. A. Lowenheim, *Trans. Electrochem. Soc.* 1943, **84**, 195.
14. F. A. Lowenheim, *ibid* 1949, **96**, 214.
15. A. W. Hothersall and W. N. Bradshaw, *J. Electrodepositors' Soc.* 1937, **12**, 113.
16. M. Schlotter, Brit. Pat. 843,429 (1936).
17. C. A. Discher and F. C. Mathes, *J. Electrochem. Soc.* 1955, **102**, 387.
18. A. M. Harper, A. Mohan and C. S. Britton, *Trans. Inst. Metal Finishing* 1957, **34**.
19. M. Clarke and S. C. Britton, *ibid* 1962, **39**, 5.
20. D. Jones, *Electroplating and Metal Finishing* 1967, **20**, No. 3, 86.
21. K. Vasantha, K. S. Indira and K. S. G. Doss, *Metal Finishing* 1967, **67**, No. 3, 46.
22. M. Schmitz, *Galvano-Berichte* (*E. German*) 1969, **5**, 36.
23. N. Parkinson, *J. Electrodepositors' Tech. Soc.* 1950, **26**, 169.
24. H. Narcus, *Metal Finishing* 1945, **43**, 188 and 242.
25. 'Electrodeposition of Tin from Fluoborate solutions,' Riedal de Haen A.G., 3016 Seelze. 34p. *Finish Digest* 1974, **3,** No. 4, 94.
26. J. B. Long, *Plating*, 1974, **61,** No. 10, 938-41.
27. G. E. Jones, Iron and Steel Institute Special Report, No. 67, May 1960.
28. M. S. Frant, *Plating* 1958, **45**, No. 3, 34.
29. Brit. Pat. 282,156 (1939); Tin and its Uses, October, 1940.
30. L. J. Brown, *Plating and Surface Finishing*, 1975, **62,** No. 6, 587-92.
31. B. E. Curry, *J. Phys. Chem.* 1906, **10**, 515.
32. G. V. Raynor, The Equilibrium diagram of the System Cu-Sn, Institute of Metals Annotated Equilibrium Diagram Series, No. 2 (1944).
33. R. M. Angles, F. V. Jones, J. W. Price, and J. W. Cuthbertson, *J. Electrodepositors' Tech. Soc.* 1946, **21**, 19.
34. W. T. Lee, *Trans. Inst. Metal Finishing* 1958/59, **36**, 51.
35. H. Schmerling, *Electroplating* 1952, **5**, 115.
36. I. A. Menzies and C. S. Ng, *Trans. Inst. Metal Finishing* 1968, **46**, Pt. 4, 137.

37. C. S. Ng, Ph.D., Thesis, Manchester, 1968.
38. M. Clarke, *Trans. Inst. Metal Finishing* 1965, **38**.
39. R. M. Angles, Tin-Nickel-Alloy Plating, Internat. Nickel Pub. 1965.
40. H. P. Rookesby, *J. Electrodepositors' Tech Soc.* 1951, **27**, 153.
41. R. L. Rau and J. C. Bailar, *J. Electrochem. Soc.* 1960, **107**, 745.
42. N. Parkinson, *J. Electrodepositors' Tech. Soc.* 1950-51, **27**, 129.
43. *Tin and its Uses*, 1974, No. 102, 3-5.
44. V. A. Lowenger, R. M. Angles and S. Baier, Brit. Pat. 548,009 (1942).
45. V. A. Lowenger and S. Baier, Brit. Pat. 533,610 (1941).
46. S. W. Baier and D. J. Macnaughton, Brit. Pat. 525,364 (1940).
47. R. M. Angles, *J. Electrodepositors' Tech. Soc.* 1946, **21**, 45.
48. B. F. Rothschild and D. Sanders, *Plating* 1969, **13**, No. 12, 63.
49. J. B. Mohler, *Metal Finishing*, 1971, **69,** No. 12, 45.
50. C. J. Evans, *Tin and its Uses*, 1972, No. 92, 8-11.
51. B. F. Muller, *Galvanotechnik*, 1974, **65,** No. 3, 180-282.
52. *Surfaces*, 1976, **15,** No. 99, 23-25.

CHAPTER 12

Zinc, cadmium and lead plating

CADMIUM and zinc are widely used for metallic coatings on steel and are similar in the manner in which they protect steel. Both metals sacrificially dissolve, cathodically protecting the steel. This is especially important where the coating may be subjected to mechanical damage in service.

The plating sequences for zinc and cadmium are very similar, so that the main reason for the difference in cost of plating the metals is the intrinsically higher cost of cadmium metal. Clearly, the increased cost of using cadmium can only be justified if the cadmium coating offers specific advantages such as superior corrosion resistance under the designated service conditions. It has generally been considered in the past that, where there is any sulphur in the atmosphere or where the metal is to be used in acidic industrial atmospheres, zinc is preferred to cadmium. Under other circumstances where, for example, precision metal parts are to be plated and the more voluminous zinc corrosion products would be troublesome, cadmium should be used. Cadmium is very much more toxic than zinc and should never be used where it might come into contact with food.

Recently(1) the view has been expressed that, with regard to the cost differential, far more cadmium is plated than can really be justified. It is claimed that cadmium does not protect steel more than zinc under natural corrosion conditions and that it is the less voluminous nature of the cadmium corrosion products which has led to this widespread belief. Cadmium plating was introduced in the 1930's because the baths were easier to operate than cyanide zinc baths, while staining and tarnishing were also less likely to occur on storage. Moreover tests in the neutral salt spray gave considerably better results on cadmium than on zinc. This prejudice in favour of cadmium has remained for a long time, despite the very much higher cost of the metal, although the superior performance of zinc under many conditions in actual service by no means conforms to the results obtained in the now discredited neutral salt spray test. However, cadmium is preferable for some applications particularly those in which parts must be soldered and where the plated parts might come into contact with aluminium components and cause galvanic corrosion.

12.1. Zinc plating

Electrodeposition which, although not the largest consumer of zinc, is nevertheless of special importance because of the characteristics which can be obtained from electro-deposited coatings and which cannot be achieved in any other way at similarly low cost. It is for this reason among others, that the use of electrodeposited zinc for the protection of steel has been growing faster than most other zinc coating methods, and for that matter, faster than most other electrodeposition processes used primarily for protection against corrosion.

Moreover, although zinc is essentially used for protection, rather than decorative purposes, recent developments in bright zinc plating and in the various passivation treatments subsequently applied to them enable finishes of attractive appearance to be obtained at the same time.

It should be mentioned that zinc coatings are not applied exclusively to prevent articles from corroding in service. They play an extremely useful part in preventing corrosion during handling, transport and storage, thus ensuring that the product reaches the user in good condition. This is why, for example, wood screws are being electrogalvanised to an ever increasing extent. Corrosion of such screws is not a serious problem once they have been installed, but even slightly rusted screws may become unsaleable, even though they may be quite satisfactory functionally.

The recently developed bright zinc plating processes are comparatively easy to control and give coatings of better appearance than the earlier zinc plating systems, and often give better protection at much lower cost than cadmium, except possibly in marine or rural environments.

The deposits obtained from different types of plating baths have varying characteristics, some being more suitable for a particular purpose than others, as will be seen later. For example, cast or malleable iron cannot be plated directly in cyanide solutions, but good deposits are obtainable from acid baths. Similarly case hardened, and more particularly carbonitrided articles will frequently fail to accept the zinc deposit. The problem has been studied by Read and Strasser[2] who attempted to correlate the failure with the metallurgical structure or surface composition of the steel. Optical and electron microscope investigations showed definite structural differences between carbonitrided steel screws which zinc plate and those which do not. The non-plateable specimens were harder near the surface and had a higher nitrogen surface.

Hydrogen embrittlement is a serious problem with some types of zinc deposit, the fatigue resistance also being sharply reduced, especially in cyanide baths.

The protection afforded by zinc for steel is based on its position in the electrochemical series, with the result that zinc behaves anodically towards the underlying steel and protects it sacrificially. Zinc has a normal potential of –0.7 V at 18°C as compared with a potential of –0.44 V for iron. The protection is therefore rendered effective even though a small degree of damage may occur to the deposit, or if it is slightly porous. On exposure to humid conditions, zinc forms white corrosion products consisting usually of zinc oxide and basic zinc carbonate, but these have some protective properties. They are however unsightly and may adversely affect the operation of some components. Passivation treatments are used to prevent or reduce their formation as, will be explained later.

Zinc is electrodeposited on a very large number of components and fittings for such products as automobiles, domestic appliances, office furniture, and electrical equipment. In many cases bulk plating in barrels is used, often in automatic plants; this is a very effective and cheap method of corrosion protection. Zinc is also plated on steel sheet and strip in continuous plants. Components can be stamped and formed from such material without the need for the user having his own plating plant. The deposits are also often more uniform than that obtained by plating the finished article, whilst there is sufficient anodic protection afforded at cut edges to prevent corrosion at these locations.

Zinc can be plated from alkaline cyanide, alkaline cyanide-free, acid or neutral baths. Cyanide solutions have excellent throwing power and produce smooth deposits, being superior often to those obtained from acid baths in these respects. Both the current efficiencies and the metal distribution obtained from the different types of baths also vary considerably(3, 4).

In recent years much work has been done on the development of non-cyanide solutions because of the problem of dealing with toxic effluents, and a number of acid baths with much improved characteristics are now available. Nevertheless there are considerable divergencies both in opinions and practice as between the relative merits of the various systems which are by no means resolved. The available methods of zinc plating have been compared by Knaak(5).

12.1.1. CYANIDE BATH

Up till about 1915 all zinc plating was carried out in practice from acid electrolytes(6). The alkaline cyanide solution arrived later with baths containing about 7.5 g per litre of zinc as metal, sodium cyanide, and 30 g per litre of sodium hydroxide(7), which are basically not very different from the newest low cyanide compositions. More concentrated baths enabling higher plating rates to be used were developed by Proctor & Wernlund(8), and by Blum, Liscomb and Carson(9). The Wernlund formulation contained about 15.7 g per litre of zinc, 50 g per litre of total cyanide, and 30 g per litre of sodium hydroxide with a small amount of mercury as an additive. The Blum formulation was still more concentrated containing some 30 g per litre zinc, 100 g per litre of total cyanide and 75 g per litre of sodium hydroxide. This produced excellent results, particularly when correctly purified and used in conjunction with various additives.

A systematic investigation of cyanide zinc electrolytes was carried out by Graham(10) who proposed the electrolyte upon which most cyanide zinc plating processes were based. This also contained about 35 g per litre zinc, 85 to 115 g per litre total NaCN, and 35 to 73 g per litre NaOH. Later the NaOH content was increased to 100 g per litre or more, which gave increased current efficiency. So far as the metal content of the bath is concerned, higher concentrations result in better current efficiency, but reduce the bright plating range and increase costs and pollution problems. The important factor is the relation between the Zn, NaCN, and NaOH contents. For a zinc content of 30 to 40 g per litre a ratio of NaCN:Zn of 3.25 to 2.75, and NaOH:Zn of 2.25 to 2.75 are recommended. The pH of the bath is relatively high (13.0 to 13.5). The solutions are best operated in rubber or plastic lined tanks.

The zinc cyanide solution is prepared by dissolving the caustic soda and sodium cyanide in water and gradually adding the zinc cyanide or oxide with constant stirring. Some pure zinc dust is then added in the proportion of about 1.5 to 2 g per litre, stirring being continued for a further period. After standing for a few hours the bath is filtered and transferred to the plating tank, the bottom portion of the tank being discarded. Before use the bath should be subjected to electrolytic purification at 0.5 A/dm^2 for at least 24 hours, after which it is analysed and the composition adjusted with sodium cyanide and caustic soda. Finally any addition agents required are added.

Zinc oxide may be used in place of the cyanide in which case additional cyanide is required to form cyanide from the oxide:

$$ZnO+2NaCN+H_2O=Zn(CN)_2+2NaOH$$

The oxide dissolves readily in a warm solution of sodium cyanide. Various addition agents are usually added to zinc cyanide baths to improve the throwing power, the appearance and brightness of the deposits.

Chemically, the cyanide solution consists of a mixture of the double cyanide of zinc, $Zn(CN)_2$, 2NaCN, and sodium zincate, Na_2ZnO_2, together with an excess of free caustic soda and/or free cyanide which are essential for deposition to occur. The exact proportions of the various constituents present in the bath are very difficult to determine analytically. An attempt has been made to determine the free cyanide concentration with a silver electrode and the caustic soda with a glass electrode, but the method is subject to considerable error. The indications are that the tendency towards the formation of zinc cyanide complexes as compared with zinc hydroxide complexes is relatively high. Investigation of the chemical changes in composition of a Zn-cyanide plating bath also showed that the equilibrium between cyanide and hydroxy-complexes is shifted in favour of the former. The influence of the NaOH/Zn and NaCN/Zn relationship on the conductivity and viscosity of the solution is similar.

The solution dissociates as follows:

$$Na_2(Zn(CN)_4) \rightarrow 2N_a^{+}+(Zn(CN)_4)^{--}$$
$$(Zn(CN)_4)^{--} \rightarrow Zn^{2+}+4(CN)^{-}$$

The fact that only a small proportion of the zinc is present as anion owing to the low degree of secondary dissociation accounts for the good throwing power of the solution. The deposition of zinc from a cyanide electrolyte in which the complex $(Zn(CN_4))^{--}$, and possibly also $(Zn(OH_4))^{--}$ are present, proceeds through the complex $(Zn(OH)_2$. The following reactions occur in zinc cyanide electrolytes:

$$(Zn(CN))_4^{--}+4OH^{-} \rightarrow (Zn(OH)_4)^{--}+4CN^{-}$$
$$(Zn(OH)_4)^{--} \rightarrow (Zn(OH)_2+2OH^{-}$$

Zinc plating solutions can be conveniently controlled by means of the free caustic soda and the total cyanide contents. An alternative, and in many cases a more satisfactory method of control, is by maintaining the ratio of total cyanide to metal concentration, as the optimum cyanide concentration is dependent on the metal content of the solution, as has already been stated.

While the sodium cyanide-zinc cyanide complex always contains a proportion of zincate, even if no sodium hydroxide be added, the further addition of sodium hydroxide helps to increase the efficiency of the solution and improve the appearance of the deposit. To some extent, the hydroxide is slowly converted to carbonate due to the action of atmospheric carbon dioxide, but further formation of caustic soda can occur in the solution if much hydrogen is liberated on account of reduced cathode efficiency. With high metal and caustic soda contents current densities of 15 A/dm^2 or more can be successfully employed. The high caustic soda types of solution also have good throwing power.

12.1.1.1. Anodes

The zinc anodes tend to dissolve in the solution at an apparent efficiency of more than 100 per cent owing to chemical attack by the solution so that excessive metal concentrations can build up, resulting in a tendency for rough or nodular deposits to occur. Some use has been made of aluminium in the zinc to control the rate of anode dissolution. The aluminium may be alloyed alone with the zinc, or mercury may also be added(11). Aluminium is said to favour uniform anode corrosion with little sludge formation; the aluminium is not co-deposited with the zinc. Magnesium and cadmium have been proposed as alloying constituents of zinc anodes to reduce the rate of dissolution of the zinc(12). To some extent, a proportion of steel anodes is useful in zinc solutions to prevent excessive build-up of metal.

None of these methods, however, reduces the rate of attack during periods when the bath is not in use. This is especially the case when zinc ball anodes are employed in steel cages, which tend to accelerate the rate of attack on the zinc. This problem of zinc build-up in high pH baths when not in use has recently been discussed by Graham and Zurbach(13). The various methods which have been proposed for reducing zinc build-up were summarised as follows:

(1) Removing anodes from the bath.
(2) Lowering solution level below anode level.
(3) Applying a low voltage to the bath (using steel as the anode and the tank anode used as cathode) sufficient to overcome the anode overpotential of zinc ball anodes to steel containers without depositing metal on to the zinc(14).
(4) Plating zinc on both the steel anode baskets and their contained zinc ball anodes during the entire idle period.

According to Graham and Zurbach, method (1) is impractical and method (2) only becomes practical if high speed pumping into an auxiliary holding tank can be made available. Method (3) is difficult to control and method (4) is undesirable because zinc is plated on to the steel during the entire idle period, which consequently necessitates its removal before plating can again commence. The method recommended is to make the anodes cathodic to steel sheet anodes (hung on regular cathode rod) and to plate the coupled zinc and steel anodes with just sufficient zinc to prevent steel being exposed by the much smaller straight chemical solubility during the idle period.

12.1.1.2. Bright zinc plating

Bright zinc deposits having a whiter and more lustrous appearance than the bluish-grey coating normally produced from the usual type of cyanide bath began to make their appearance about 1935. Whilst matt zinc shows finger marks immediately, and provide centres for the initiation of corrosion, bright zinc does not finger mark nearly so badly.

Bright zinc deposits are obtained from a variety of baths by the use of addition agents in the solution. Most zinc plating is nowadays carried out from proprietary solutions containing such additives. The compounds used include, for example, metals such as molybdenum, tungsten and manganese, as well as aromatic aldehydes such as anisaldehyde, vanillin and polyvinyl alcohol. Other organic addition agents worthy of mention which have been patented are coumarin, fluorescein and phenol-aldehyde condensates, often in association with protective colloids such as gelatine. Little use is made nowadays of metallic additives, as they are difficult to control. The subject has been discussed by Weiner[(15)]

These addition agents, besides improving the lustre of the deposit, may also increase the throwing and covering power of the solution, and improve solution stability and operating conditions to varying extents. In general, they add little to the protective value of the zinc deposit, however.

For barrel plating, the bright solutions should be rather more concentrated. The influence of electrolyte composition on the properties of bright zinc cyanide baths consisting of 32 g per litre Zn, 95 g per litre total NaCN and 80 g per litre NaOH have been summarised by Korpium[(16)] (Table 12.1).

Raub[(17)] carried out X-ray diffraction investigations on a number of bright coatings of different metals. The results showed the absence of new lattices which differ from the

TABLE 12.1.

Effect of electrolyte composition on the properties of alkaline cyanide baths

	Current efficiency	Bright range	Throwing power	Recess covering	Anode solubility
Zn content	X	O	O	X	O
NaOH	X	—	O	X	O
Total CN	O	O	X	O	X
Brightener	O	X	X	X	—
Temperature	X	O	O	O	X

X Increase O Decrease

normal lattice structure of the principal metals concerned. Debye-Scherer patterns obtained from matt and bright zinc deposits using commercial electrolytes showed some interference patterns, but no noticeable mutual displacement of the interference; other bright deposits gave similar results. X-ray diffraction revealed that bright deposits generally had a larger number of lattice defects than matt deposits. These defects are a result of the incorporation of impurities, partly in finely dispersed pseudo-isomorphous form, into the crystallites.

When bright deposits are examined under the optical microscope a radial structure with the fibres strictly aligned in the direction of the electric field is often observed. Bright zinc, when deposited from a cyanide electrolyte shows reorientated crystal growth only when hydrogen is simultaneously liberated. Additives which have the effect of suppressing texture also prevent the formation of bright deposits(18). Increasing the amount of impurity which is co-deposited with the metal in the bright deposits greatly inhibits crystal growth in a direction parallel to the substrate surface, growth taking place preferentially in the direction of the current flow.

12.1.1.3. Effects of contamination

It is essential in bright zinc plating to maintain a high standard of purity of the solution and anodes since even slight contamination with heavy metals can cause the deposits to become dark in colour and lose their lustrous appearance. Particularly objectionable are traces of lead, cadmium and to a lesser extent, copper, nickel and iron. Concentration of heavy metals as low as 0.02 per cent are sufficient to affect the appearance of the deposits appreciably. Impure anodes or contaminants dragged in by the work are important causes of contamination.

Chromic acid has a very marked effect if it should find its way into the solution, and as little as 0.01 g per litre of chromium will reduce deposition in low current density areas while higher concentrations may inhibit deposition completely.

Heavy metal contamination can be removed by treatment with zinc powder at the rate of 1 g per litre of solution which precipitates out many foreign metal contaminants, or by the addition of small amounts of polysulphide solution. The latter method is relatively simple and effective. Proprietary sulphide based purifiers are also available which are simpler to use and more controllable than sodium or ammonium sulphide. An excess of the sulphide can be added to take care of newly introduced impurities. If the deposit is then slightly brown in colour it can be restored to the pale bluish tinge of pure zinc by a short dip in 1 per cent nitric acid solution. Sodium hyposulphite is also useful as a purifying agent. In each case the precipitated impurities can be left to settle out, but the most satisfactory procedure is to filter them off.

12.1.1.4. Control of bright zinc solutions

Apart from maintaining the purity of the solution, the most important factor in successful bright zinc plating is accurate control of the ratio of total sodium cyanide to zinc metal content. For a typical bright zinc solution this should lie between 2.5 and 2.7

for operation with high current efficiency at normal current densities and at room temperature. With lower ratios the current density range in which bright deposits are obtained is raised. For barrel plating a lower ratio is desirable than when plating is carried out in still tanks owing to the importance of good conductivity in such solutions; a high caustic soda content is also recommended.

12.1.1.5. Intermediate and low cyanide zinc processes

The first reaction to the sudden realisation of the perils of environmental pollution was a move towards processes which made no use of cyanides at all. For a time, especially in the Scandinavian countries, there was therefore a tendency to replace cyanide zinc by acid zinc processes. However, as will be seen, a major change of this kind can introduce many difficulties unless very careful precautions are taken. Now there is a trend in the United Kingdom towards zinc plating from intermediate cyanide bright zinc baths, since these require little or no change in cleaning procedures as compared with conventional cyanide baths—intermediate cyanide, or 'half-bath' also provides a useful compromise where cyanide treatment facilities are available.

The following are the respective compositions of typical 'full' and 'half' baths. The exact compositions and operating conditions used depend on whether rack or barrel plating is being carried out.

Constituent	*Full bath*	*Half bath*
Zinc cyanide	45 g per litre	25 g per litre
Sodium cyanide	90 g per litre	40 g per litre
Caustic soda	75 g per litre	55 g per litre
NaCN:Zn	2.7:1	2.0:1
NaOH:Zn	2.2:1	2.0:1
Brightener	1.5 to 6.0 ml per litre	

The solutions are operated at 10° to 40°C at cathode current densities of 2 to 6 A/dm^2 and anode current densities of 0.25 to 2.5 A/dm^2.

The low cyanide processes use still lower zinc metal and cyanide concentrations. The composition in this case is approximately:

Zinc oxide	9.5 g per litre
Sodium cyanide	7.5 g per litre
Caustic soda	50 to 75 g per litre
Brightener	4 to 5 ml per litre
NcCN:Zn	1:1
NaOH:Zn	10:1

The solutions are operated at 20° to 35°C at a cathode current density of 0.5 to 5.5 A/dm^2. With these solutions much more thorough cleaning is required to obtain good deposits and minimum rejection rates.

12.1.2. ACID ZINC SOLUTIONS

During the past decade, the growing problems of pollution resulted in special attention being given to the development of improved acid zinc baths, since they would clearly have advantages. The original acid baths which had been known for many years are based on sulphate or fluoborate, but their poor throwing power made them of only limited value for general use. Extensive investigations on chloride and mixed acid electrolytes have led to the introduction of a range of low acid zinc solutions which have made the greatest headway in Germany, where it has been estimated that about 20 to 25 per cent of the total zinc plating solution volume is of this type, the rest mainly cyanide solutions of various types but generally with lower metal and cyanide contents than had previously been employed.

12.1.2.1. Acid (Sulphate) zinc plating

The acid (sulphate) zinc baths are considerably cheaper to operate than the cyanide solutions, but their throwing power is inferior while the deposits tend to be somewhat coarser-grained than those produced from the cyanide baths. On the other hand, deposits from acid baths are rather whiter in colour, while they are also more readily applicable to deposition directly on to cast iron for which purpose cyanide solutions can only be used with difficulty.

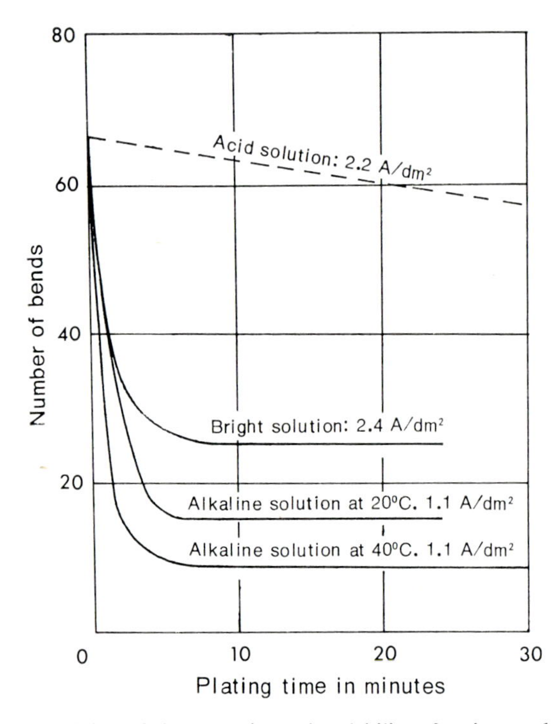

Fig. 12.1. Effect of zinc deposits on bendability of spring steel strip.

An important advantage of acid zinc plating is the fact that deposits obtained from this type of electrolyte are much less prone to produce hydrogen embrittlement. Zinc plating from cyanide solutions is notoriously liable to cause severe embrittlement, particularly of thin section steel of high carbon content (*cf* below).

Fig. 12.1 shows the results of a study in which zinc deposits on spring steel sheets were subject to repeated bending[19]. Bending was carried out around a 5 mm mandrel, the angle of bend being 180 degrees in each case. The number of bends which could be completed before the specimens fractured is seen to be considerably greater in the case of zinc deposits from the acid bath than those from various alkaline solutions.

12.1.2.1.1. Sulphate zinc plating solutions

Acid (sulphate) formulations which have found use include:

$ZnSO_4.7H_2O$	240 g/l
Ammonium chloride, NH_4Cl	15 g/l
Aluminium sulphate, $Al_2(SO_4)_3.18H_2O$	30 g/l

For barrel plating the following composition has been employed:

$ZnSO_4.7H_2O$	480 g/l
Aluminium chloride, $AlCl_3$	18 g/l

and:

$ZnSO_4.7H_2O$	240 g/l
Na_2SO_4	40 g/l
$ZnCl_2$	10 g/l
H_3BO_3	5 g/l

Saubestre, Hajdu and Zehnder[20] have recently recommended the following bath:

$ZnSO_4.7H_2O$	180 g/l
$ZnCl_2$	14 g/l
H_3BO_3	12 g/l

It is claimed that this bath gives very satisfactory deposits with good surface coverage, in the absence of addition agents.

The electrical conductivity of zinc sulphate solutions is low, and hence other salts must be added to increase it. These are normally sulphates or chlorides of aluminium or sodium, and apart from raising the electrical conductivity of the bath, they also somewhat improve its throwing power.

Satisfactory zinc deposits are obtained in the pH range 3.5–4.5 and the pH is adjusted by small additions of H_2SO_4. At lower pH values than 3.5 the anode consumption markedly increases owing to chemical attack by the solution; the zinc deposit may also be affected by the aggressiveness of the electrolyte. At pH values of above 4.5, $Zn(OH)_2$ may precipitate, leading to a rough deposit. Aluminium sulphate and boric acid act as pH stabilisers.

Addition of surface active agents further increases the cathode polarisation, and hence the throwing power, of the bath, refines the structure of the deposit obtained and increases its brightness. However, the maximum throwing power attainable in acid baths is much inferior to that of alkaline electrolytes.

Owing to their relatively poor throwing power, acid baths have found most use for plating products of simple shape, e.g. sheet, strip, wire, etc.

The surface-active agents that are usually added are organic compounds such as gelatin, dextrin, liquorice, sulphonated cresols and alcohols, glycerin, etc. Certain of the addition agents also act as anodic inhibitors, and may help to reduce the tendency for the anodes to dissolve in the acid solution.

12.1.2.1.2. Operating conditions

Acid zinc sulphate solutions are best operated in rubber-lined steel tanks, but lead-lined steel, ceramic or moulded synthetic resin tanks are satisfactory. Heating or cooling coils can be of hard rubber or lead-covered steel. The temperature should not be too high, as at temperatures much in excess of 30°C the deposits tend to become dark and coarse grained when plating is carried out in still solutions. The throwing power is poor, but sometimes adequate covering of deeply recessed articles can be obtained by first briefly plating from a cyanide solution. The range of current densities employed in common practice is of the order of 2 to 4 A/dm^2, but current densities of 10 A/dm^2 and over have been used in the zinc plating of wire in special plant.

The cathode efficiency of the acid solution is very high and may exceed 98 per cent, while the anode efficiency often exceeds 100 per cent.

The zinc anodes should be of high purity since the presence of impurities may result in lowering the cathode efficiency or in the formation of dark deposits of poor corrosion resistance. Also, impure anodes may develop a coating of basic compounds of zinc, iron, lead, etc., which tends to render them passive and necessitates the application of a considerably higher bath voltage. Improvement can be effected either by cleaning the anodes or lower the pH of the solution by the addition of acid.

Improvement can then be effected either by cleaning the anodes or lowering the pH of the solution by the addition of acid.

The solutions are not generally filtered continuously, but periodic filtration is useful to remove suspended matter which can cause rough deposits. Under the usual plating conditions in still vats, however, zinc solutions are not unduly sensitive but in the plating of wire, for example, at high current densities, much more attention has to be given to the purity of the solution if successful results are to be obtained.

12.1.2.1.3. Electro-galvanising of wire, sheet and strip

Electro-galvanising was introduced as a method of zinc coating steel wire in place of the older hot galvanising method. The process has the great advantage over hot galvanising in that no brittle alloy layers are formed between the zinc and the ferrous metal (*cf.* p. 500). These layers make the deposit somewhat prone to peeling when the

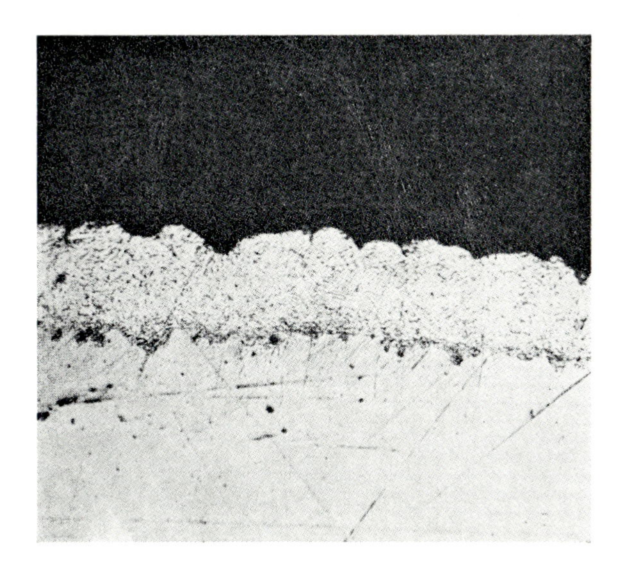

Fig. 12.2. Section of electro-galvanised steel wire, 0.96 oz/ft^2 ($\times$ 300). [Courtesy Rylands Bros. Ltd.]

wire is subjected to excessive deformation. As galvanised wire is usually subjected to considerable bending and twisting during subsequent fabrication operations, the fact that electro-galvanised wire can withstand extensive stressing makes this type of coating particularly suitable for wire. The difference between hot-galvanised and electro-galvanised wire is clearly seen by comparing the hot galvanised structure shown in Fig. 14.1 (p. 501) with that of electro-deposited zinc (Fig. 12.2).

In the Tainton process the zinc sulphate electrolyte is prepared by dissolving roasted zinc concentrates in sulphuric acid, precipitating the iron from the nearly neutral solution by manganese dioxide and then further removing those metallic impurities that are cathodic to zinc by the addition of metallic powder.

During deposition the electrolyte is rather more acid than that employed in still tank plating, since inert anodes are used. It consists of about 70 g/l of zinc and up to 250 g/l of sulphuric acid depending on the stage of electrolysis. The wire is first annealed at 700°C in a lead bath and then carefully cleaned either cathodically in molten caustic soda or in a caustic alkaline solution. It is next anodically etched in a tank of the highly acid spent electrolyte and passes through a strike solution containing about 150 g/l of zinc, before passing to the main plating cells.

The plating cells themselves are about 30 cm long and the electrolyte is circulated continuously through them. They are made of lead-lined steel and are water-cooled to prevent excessive rise in temperature of the bath. The efficiency of an acid electrolyte rises with increasing temperature over a wide range of current densities, as can be seen from the curves in Fig. 12.3, but the deposit is apt to deteriorate in appearance at the higher temperatures.

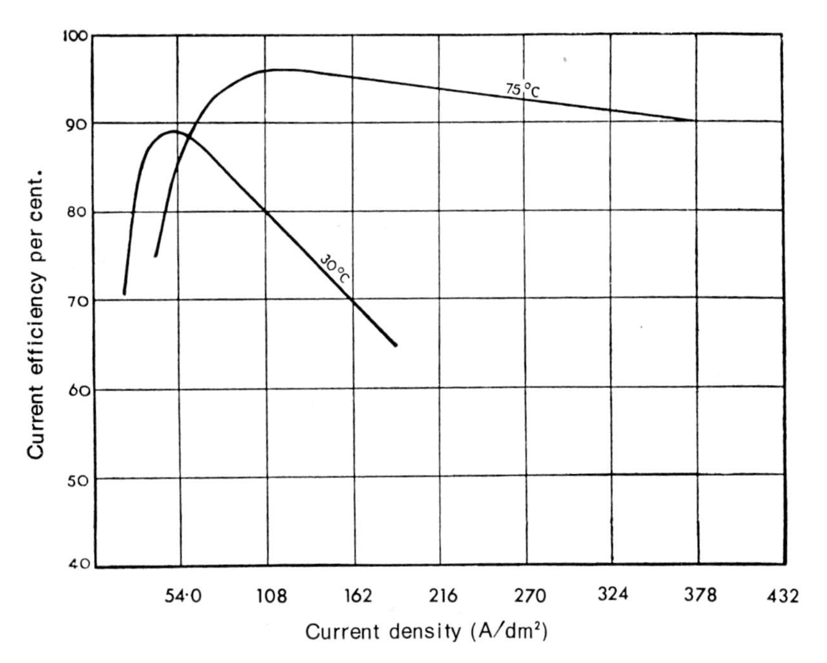

Fig. 12.3. Current efficiency curves at 75°C and 30°C for acid zinc electrolyte consisting of: zinc (metal) 58 g/l; sulphuric acid 220 g/l.

The current densities used are of the order of 80 to 200 A/dm^2, depending on the size of wire being coated. Inert anodes of lead-silver alloy are employed, the silver content being of the order of 1 to 2 per cent. Lead anodes containing 2.5 per cent of silver are up to 25 times as resistant to the electrolyte as pure lead anodes, whilst they are also much more tolerant of such impurities as bismuth, arsenic or antimony, which are very deleterious in lead anodes free from silver. As deposition is carried out the electrolyte naturally increases considerably in acidity, as has been stated, and the metal content is depleted.

Coatings of zinc of up to 5g/dm^2 of surface are regularly deposited and the process has been found particularly suitable where it is desired to draw a thicker gauge coated wire down to a thinner gauge after plating. In this way a compact coating with a high finish can be obtained. The deposit is of high purity (exceeding 99.99 per cent zinc) and is so ductile that the finished wire can be wrapped round its own diameter without fracturing the coating or loosening it from the steel.

The purity of the electrolyte in the electro-galvanising of wire is important. Copper must be kept below 0.009 g/l or the deposit will tend to become spongy, while the current efficiency will be lowered. Iron has a similar effect in concentrations greater than 0.05 g/l. A particularly obnoxious constituent of acid zinc electrolytes appears to be germanium, which in concentrations as low as one part in 10 million will markedly reduce the cathode current efficiency[21]. The process has been fully described by Roebuck and Brierley[22].

An important development has been the increased use of pre-plated metal, both in the form of sheet and strip, for the manufacture of finished products by forming and pressing. Of the materials available, probably the most significant, apart from tin-plate, is zinc-coated steel.

The advantages of using pre-plated metals are the saving of space and capital resulting from the elimination of the need for finishing facilities on the part of the user, and the fact that the coating is generally more uniform than is possible when finished articles are plated. Different coating thicknesses can also be applied on each side.

Against these advantages, however, the use of pre-plated metals results in a certain necessary wastage, whilst only relatively thin deposits can be plated in the case of metals such as zinc if the strip is to form satisfactorily. Other problems such as welding and jointing, and the protection of sheared edges, also arise, but these have been successfully overcome in the case of zinc coated sheet.

Electrogalvanised steel sheet is nowadays an extensively used product and is produced on a large scale in fully automatic plants. Electrogalvanised steel carries a zinc coating generally 2–5 microns thick, which is only about one-tenth of the thickness of a hot dipped coating. However, by virtue of the method of manufacture, no brittle alloy layer exists between the steel and the zinc so that the product can be subjected to a considerable degree of deformation without damage to the coating and can be readily spot-welded.

The plant used for the electrogalvanising of steel sheet and strip is of a special nature, and a number of large installations are in operation throughout the world. Most of these lines are designed to process wide sheets rather than strip, since few users are capable of handling strip up to 150 mm in width.

One method used for transferring the steel sheets from one processing tank to another is to keep them in a horizontal plane throughout the sequence of operations, which consists, in a typical installation, of eight alkaline cleaning tanks, eight sulphuric acid pickling tanks, eleven zinc plating cells and chromate treatment tanks. All the cells are designed with a slot at each end running almost the whole width of the cell, with rubber strips to restrict leakage through the openings. The steel sheets pass horizontally through them between slatted plastic tunnels which act as guides. In the case of the plating cells, provision is made for the zinc anodes which are placed above and below the sheets. The solution which runs out of each cell through the slots is collected into tanks, filtered and pumped back to maintain the correct level of solution above the surface of the steel. Rinsing is carried out between each operation, rubber scrapers being used to remove excess water. Finally the sheets are dried in a tunnel by means of hot air and cooled by cold air to prevent 'sweating' in storage.

In the alkaline cleaning and pickling cells, both of which are electrolytic, the sheets form the anodes; the cathodes in the alkaline cells consists of perforated steel plates with plastic strips acting as insulators and sheet guides, whilst in the pickling cells, lead plates perform the same function. The distance between the sheets and the anodes is only 15 mm so that a high current density can be employed, and uniform deposits obtained.

At each of the transverse points between the cells (approximately 40 in number) pairs of driven rollers serve to move the sheets, the correct pressure being maintained on the rolls pneumatically. In the electrolytic sections of the plant, one of the rollers of each pair adjacent the plating cells is a conducting stainless steel one, the plating current being fed to the sheets through suitable sliprings from the rectifiers; all the other rollers employed in the plant are rubber-covered.

The plants are around 100 to 150 m long, and are designed to operate at up to 50 m per minute, the current being supplied by large rectifiers having a capacity of 60,000 A or more, depending on the size of the installation.

A thin chromate conversion coating after plating is an essential part of the process and ensures that good paint adhesion will be obtained. Phosphating can also be carried out after galvanizing instead of chromate treatment, if preferred.

Other designs of plant, in which each side is plated separately, are in use.

12.1.2.2. THE FLUOBORATE ZINC BATH

The use of fluoborate baths for the deposition of zinc enables high rates of deposition to be obtained both in automatic and manually operated plants. The bath finds extensive use for coating steel strip and wire. Fluoborate solutions are used at room temperature and have high anode and cathode efficiencies. They also give good deposits of excellent colour on cast iron, which is a considerable advantage.

A solution for the high-speed plating of wire, sheet and strip which has been recommended consists of:

Zinc fluoborate	300 g/l
Ammonium chloride	27 g/l
Ammonium fluoborate, $NH_4.BF_4$	36 g/l
Liquorice	1.25 g/l

Boric acid may also be added to zinc fluoborate baths. A pH of 3.5 to 4.0 is used. By increasing the temperature the resistance of the bath is reduced, so that it is more suitable for high-speed plating, although the throwing power is poorer.

Current densities of up to 100 A/dm^2 are being used for this purpose. The deposit is whiter than that obtained from the cyanide solution and is highly ductile. The temperature of the bath is 40° to 55°C at a pH of 3.5 to 4.0. Pure zinc anodes are used, the cathode: anode area ratio being 2:1. The pH is maintained by additions of fluoboric acid, but can be raised if required with zinc carbonate. The liquorice serves to make the deposit smoother and to whiten it, and is controlled by plating tests. The anode and cathode efficiencies are about 100 per cent.

The solution is prepared from zinc fluoborate concentrate to which is added the requisite amount of water followed by ammonium chloride and fluoborate. The carbonate is finally added to raise the pH to the required value. The ammonium fluoborate increases the throwing power and limiting current density, and helps to maintain a fine-grained deposit, whilst the chloride is useful in helping the conductivity of the bath. Excessive concentrations are to be avoided, as they tend to darken the deposit.

Many brighteners have been suggested for use in zinc fluoborate baths including gelatin, destrin, β-naphthol and caffeine. In all cases only semi-bright deposits are obtained. Recently, Grunwald and Varhelyi Cluj[23] have investigated the action of several types of cobalt (111) ammine complexes (hexammine, monoacido-pentammine, diacido-tetrammine and tetra-acido-diammine-salts) and some organic additives on the formation of bright or semi-bright zinc deposits from two acid fluoborate zinc baths.

Bath A	
Zinc fluoborate	200 g/l
Ammonium borate	35 g/l
Boric acid	30 g/l
Operating conditions	temp. 22°C
	pH 5–5.4
	c.d. 4–10 A/dm^2

Bath B	
Zinc fluoborate	230 g/l
Ammonium fluoborate	35 g/l
Ammonium chloride	45 g/l
Operating conditions	Temp. 22°C
	pH 4.0
	c.d. 3–12 A/dm^2

Plating was carried out in a Hull cell with iron cathode panels 100 cm × 70 cm × 0.3 mm and brightener concentrations between 0.1 to 1 g/l. It was found that some complexes, e.g. $[NH_4(CO(DH)_2(NO_2)_2]$ and $[H(CO(DH)_2(NO_2)_2]$ have a negative effect on the deposit. Others, including $[CO(en)_2Cl_2]Cl$, $Co(en)_3Cl_3$, and $(Co(ec)(pyridine)_2Cl$, have a favourable effect resulting in the formation of bright or semi-bright deposits. However, in no case was a mirror-bright electrodeposit obtained.

12.1.2.3. Low Acid Zinc Baths

Low acid zinc solutions are used successfully for both barrel and rack plating. In most modern solutions the zinc is present in the form of an ammonium complex containing 20 to 50 g per litre of metal together with 150 to 200 g per litre of ammonium chloride, which also serves to improve the conductivity of the solution. The baths are formulated to give fully bright deposits, and contain brightening and levelling additives such as aromatic ketones or aldehydes together with solubilising agents for these materials. The bright deposits initially resemble chromium in appearance, but become dull on exposure to the atmosphere with time. The cathode efficiencies are high (95 to 98 per cent), but the throwing power is somewhat inferior to that of the cyanide bath, being comparable with that of the average nickel bath.

A major advantage of these processes is that they produce little or no hydrogen embrittlement on account of their high current efficiencies. Hence no heat treatment is usually necessary after the plating of springs or hardened steel components in these baths. They also have high covering power and can be used for the direct zinc plating of grey and malleable iron castings; this cannot be done from cyanide baths unless an intermediate deposit is first applied.

The main limitations of the acid solutions are their high corrosivity, which means that special materials or protective coatings are needed for plant and equipment, and the fact that the cleaning of the work before plating is also more critical, since acid zinc solutions have little detergent action.

A typical solution for bright plating contains:

Zinc chloride	100 g per litre
Ammonium chloride	200 g per litre
Total chloride	180 g per litre
Cl:Zn ratio	3.6:1
Brighteners	45 ml per litre

The bath is operated at a pH of 5.0 to 5.5. and a temperature of 18^0 to 38^0C for both rack and barrel plating, with cathode current densities of up to 6.5 A/dm^2 in the case of the former and up to 1.5 A/dm^2 for the latter.

The high ammonium content of the bath may interfere with the precipitation of zinc and heavy metals during effluent treatment, but according to Todt[(3)] the precipitation efficiency depends on the degree of dilution. If this is high enough the ammonium complex is dissociated to such an extent that only a very small amount of zinc remains unprecipitated.

If nickel and copper are also present in the effluent, some recomplexing occurs. Copper precipitation may therefore consequently be delayed, but with further dilution this also takes place quantitatively.

Recently a solution available for barrel and rack plating buffered with boric acid and free from ammonium chloride has been introduced which also contains only about 60 g per litre of chloride, so that it is much less liable to corrode plant. The solution contains brighteners and levellers, and is able to produce fully bright solutions comparable with those obtained from the usual ammonium chloride containing solutions. The solution is agitated and filtered, and is used at 18^0 to 25^0C. Cathode current densities are 4 to 5 A/dm^2 for racked work and up to 1.2 A/dm^2 for barrel plating.

Care must be taken to avoid contamination of the bath with iron, copper, cadmium or lead, to which it is susceptible; such contamination can lead to dark deposits and interfere with subsequent chromating, as has already been stated.

The acid zinc solutions cannot be introduced successfully by simply changing the bath. A radical review of the entire plating sequence and plant construction must be made to ensure that much better cleaning methods are used than is normal with cyanide processes, and that proper materials and protection treatments are employed to prevent corrosion of equipment and even of the articles being plated themselves. Much more care must also be taken to avoid contamination. In fact, the entire operation is carried out in a manner more akin to that used for bright nickel plating than to that normally employed in cyanide zinc plating. Unless proper measures are introduced from the start, troubles will certainly occur.

12.1.3. CYANIDE FREE ALKALINE BARREL BRIGHT ZINC

In order to overcome the problems associated with cyanide baths, especially where no cyanide treatment plant is available, and the difficulties involved in introducing acid zinc solutions, attempts have been made over the years to develop successful cyanide free alkaline zinc baths.

The difficulty, however, is that such alkaline solutions can only produce dark and useless deposits. Recently a measure of success has been achieved by the use of solutions containing sophisticated brightening additives systems free from chelating agents. Such solutions have the attraction of being free from the toxicity hazards of cyanide, having little corrosive effects on plant, and considerable volumes of solutions of this type are already in use.

The process can therefore be directly introduced in place of existing cyanide baths without modifications to the plating equipment. Since cyanide free alkaline solutions also have appreciably less detergent action than those containing cyanides, better cleaning is, however, needed than when the latter are used. The pretreatment sequence should incorporate soaking, electrocleaning and acid dipping stages.

The plated work is often somewhat difficult to rinse, since a tenacious film remains after removal from the bath which can interfere with subsequent chromate passivation treatments. Good rinsing and a dilute nitric acid dip are therefore recommended before chromating.

Effluent treatment is comparatively simple; all that is needed is to reduce the pH of the effluent to about 9.0 with hydrochloric acid, when the zinc hydroxide can be precipitated out, leaving a residue of less than 5 mg per litre of metal in solution.

At the present time the solution is only suitable for barrel plating owing to its limited cathode efficiency. Anodes should be removed during long idle periods, since chemical dissolution of the zinc occurs which can result in excessive metal enrichment of the bath. However, this type of bath can be expected to become of greater importance as time goes on as a result of further development.

Like the acid baths the solution is also sensitive to a wide range of contaminants, including in particular, iron, copper, lead, cadmium, nickel, tin and chromium. Should contamination occur it can be dealt with by plating out at low current density or by treatment with zinc dust.

12.1.4. CHROMATE PASSIVATION OF ZINC DEPOSITS

Zinc deposits are prone to develop white corrosion products of basic zinc salts under conditions of humidity. Dipping in an acid solution of sodium dichromate markedly improves the properties of the deposit in this respect. In the Cronak process, which was the first conversion process of this type to be introduced[24], a solution consisting of:

Sodium dichromate	200 g/l
Sulphuric acid	6–9 ml/l

is used for immersion. The time of treatment is 10–20 seconds and the solution is generally used cold. Nowadays many other treatments based on the Cronak process are in use and details of these have been given in a review by Roper[25].

With the Cronak process a film is produced which is highly protective and inhibits the formation of a white corrosion product such as normally appears on zinc in time. In this way it usefully improves the life of zinc or zinc-plated parts, particularly those constituting moving members of mechanisms where corrosion may cause failure. The process can also be applied to cadmium, but its advantages here are much less marked.

The colour of the film depends on the time of immersion. Short dips give a yellowish-green iridescent colour and dips of 10 seconds or longer a golden brown hue; similar colours are obtained on cadmium. The colour is the best guide to the performance of the solution. It is affected by incorrect bath temperature, unsatisfactory cleaning and an excess or deficiency of acid.

Coatings produced by this process may be bleached with some loss in protective value, however, or dyed in such colours as black, blue and green. Parts chromated by this process should be dried in a cool air stream or centrifugal drier and not exposed to hot air or water above 60°C, as heating tends to dehydrate and embrittle the coatings.

A large variety of proprietary treatments which produce a film similar in colour and corrosion resistance to the Cronak process are now available. These are all simple dips, varying in duration from 5 to 60 seconds. Slower operating processes have been developed to enable chromate passivation to be carried out in automatic plants. The solution pH is in the range 1 to 3.5 and there is little or no brightening action on the metal.

The olive drab coatings are considerably thicker and capable of imparting a corrosion resistance to zinc enabling it to withstand up to 500 hours salt-spray before white rust appears. In addition, their greater thickness and absorptivity allow them to be dyed with certain types of organic dye to some extent for decoration. The thinner yellow and bronze films are also absorptive—though to a lesser extent—and have been dyed in pastel shades. There are also a few proprietary processes which yield a black colour, these being specific either for zinc or cadmium. Also noteworthy is a recommended passivating treatment for tin-zinc alloy electroplates[26]. This consists of a 30 seconds treatment in 2 per cent chromic acid at 80°C to give a yellow or brown coating; a 13 seconds immersion at 50°C gives a thin, colourless film which is solderable. The principal object of this treatment is to impart a resistance to finger prints. A solution which can be used cold comprises 1 per cent sodium dichromate and 0.033 per cent sulphuric acid.

The brightening processes provide a much less protective film in most instances, but the degree of protection is nonetheless quite appreciable.

The earlier brightening solutions were simply very weak nitric or chromic acid ones which gave added lustre to a smooth, dull electrodeposit with little loss of plate and no appreciable colouring. More recently developed solutions impart a light yellow film which can be bleached if the loss in corrosion resistance is tolerable. A sodium hydroxide bleach gives a clear, colourless finish, but appreciable iridescence remains when sodium carbonate is used; this can be masked by lacquering.

Other solutions give virtually colourless films without bleaching, but again any unwanted tints can be removed by mild bleaches or lacquer. If the remaining film has a faintly blue cast, and especially if the electrodeposit is a bright one, the finish closely resembles chromium plating in appearance. Protected by a suitably hard and durable lacquer, it has been widely used as a substitute for chromium, although in actuality its performance in no way compares with it.

Another application of the clear, bright finishes has been, particularly in the case of cadmium, the protection against oxidation. Even after long storage cadmium protected in this way can be soldered without difficulty with resin fluxes—a point of great moment to the electronics industry; this is not the case, however, with the heavier types of film.

The effectiveness of the film depends on the liberation of hexavalent chromium to the extent of 0.0001 to 0.0005 g/l in water in contact with it; the film does not appreciably add to the life of zinc exposed to outdoor weathering owing to the rapid leaching out of the chromium compounds which occurs under the latter conditions. The application of the process should, therefore, be limited to those components subject to condensation and humidity but protected from the outside weather.

The term 'passivation' as applied to these films is perhaps a misnomer, as the electrode potential is affected to a negligible extent[27]. The transparent films are highly porous, the yellow or drab ones being substantially non-porous. In sea water (pH 7.5), the pores in the coating become sealed with corrosion products and give good protection.

12.1.4.1. Defects in the film

The colour depends not only on the time of immersion and composition of the solution, temperature, etc., but also on the texture of the zinc surface. Thus, films on bright surfaces tend to be more resistant than those formed on matt surfaces, whilst if the zinc contains copper a strong green colour develops. Within reason such colours are not detrimental to the protective quality of the film. With experience, however, operators treating standard zinc surfaces soon become accustomed to the correct appearance of a good film and can readily reproduce it. If good films are not obtained in a reasonable time, three factors may be responsible:

(a) The temperature of the bath may be incorrect. The films form very slowly at a temperature below 20°C, and something not at all below 15°C but, on the other hand, temperatures which are above 20°C cause chalky, poorly adherent films to be obtained.
(b) If the sulphuric acid content is low, poor films will be obtained; sulphuric acid is consumed during the progress of the treatment, and its concentration must be maintained by suitable additions, as will be described below.
(c) If the work is not properly cleaned, unsatisfactory films will be obtained.

The drying time should not be unduly prolonged since, if the solution dries on the work, an unsatisfactory finish will be obtained, although attack on the zinc ceases when the work is withdrawn from the solution.

The treatment results in a slight removal of metal from the surface, and this loss ranges from 0.000007 to 0.0002 cm, depending on the conditions of immersion. The latter figure is extremely high and is seldom reached.

12.1.4.2. Maintenance of solution

The solution can be maintained by chemical analysis of the constituents. An alternative empirical method is to withdraw an aliquot proportion of the bath and add sulphuric acid until samples dipped into it under the proper conditions become covered with a satisfactory film; the same percentage additions are then made to the main bath; normally, two or three additions of acid can be made before the solution reaches the point where it must be discarded. It is not advisable to add sodium dichromate, since by the time this constituent requires renewal it is no longer practicable to regenerate the bath.

Unsatisfactory films may be stripped by immersion in a boiling solution containing 200 g/l of chromic acid for several minutes. Prior to re-treatment the parts may have to be passed through an alkaline cleaning cycle.

Although the dichromate content of the solution can be varied widely without ill effect, the acid concentration, on the other hand, must be carefully controlled, as too little acid gives films too thin for normal purposes, whilst excess acid concentration results in poor adhesion of the film and unnecessary attack on the zinc.

12.1.4.3. Anodic conversion coating

It is also possible to produce adherent coatings by anodic treatment in chromate electrolytes. In one process the parts are subjected to anodic treatment for 1 minute to produce a yellow coating thought to consist of zinc oxide and chromate. A typical solution consists of:

Sodium dichromate, $Na_2Cr_2O_7.2H_2O$	200 g/l
Na_2SO_4	5 g/l
pH	3–4.5
Temperature	20°C

It is claimed that the coatings obtained in this way are very adherent and are not detached by sharp bending. They will, therefore, withstand cutting and shaping operations on the metal. Furthermore, these coatings are more uniform in colour and harden more rapidly than those produced by the immersion treatment.

12.1.4.4. Plant construction

The tanks used for chromate passivation processes may be made of ordinary black iron. Slow attack may take place, but a reasonable life is to be expected from them. Much better durability is, however, obtained with tanks which are made of stainless steel, earthenware or plastics.

The parts should preferably be racked for treatment, as bulk immersion in baskets may result in the abrasion of the film, which is rather soft while wet; it may also make rinsing more difficult. In some instances, however, such as on parts free from recesses, bulk treatment in baskets is satisfactory.

12.2. Cadmium plating

Cadmium plating is used primarily for the protection of ferrous metals against corrosion. The metal is more attractive in appearance than zinc when electro-deposited, being whiter and more lustrous. It is also more ductile than zinc, which makes it useful for parts which have to be subjected to forming operations after plating. Having a lower coefficient of friction, cadmium deposits on screws make them easier to drive than zinc plated ones. Cadmium also retains its surface conductivity longer than zinc, which makes its use advantageous for electrical applications. It is, however, much more expensive than zinc.

Cadmium has the great advantage of much easier solderability than zinc, while its greater resistance to alkalis makes it useful in domestic equipment. The resistance of cadmium to acids, however, is not much higher than that of zinc and cadmium plating should not be used on parts that are likely to come into contact with foodstuffs (especially those of an acid nature) because of the toxicity of cadmium and its compounds.

Cadmium is sometimes plated on to brass and copper alloy parts where brazed assemblies incorporating steel parts are involved; here the cadmium may serve to reduce the tendency for corrosion to occur at the junction between the two metals.

For outdoor exposure heavier deposits of cadmium are used on steel components, especially on aircraft and marine equipment, because of the good protection such deposits give under these conditions.

Cadmium can be either anodic or cathodic to iron, depending on the environment. It is anodic in solutions containing chlorides, and therefore cadmium deposits are suitable for service under marine conditions. In addition, it is chemically more stable than zinc and resists tropical conditions much better. Where high humidity conditions prevail, the cadmium protects the underlying ferrous metal anodically so that preferential corrosion of the cadmium takes place before the underlying steel is attacked.

Specifications for appearance, thickness adhesion etc. of cadmium deposits on iron and steel have been given in a recent international standards publication(28).

12.2.1. CADMIUM PLATING SOLUTIONS

Cadmium is deposited from the cyanide solution because of the good throwing power of the latter and the fine-grained smooth deposits obtained. Cadmium can be satisfactorily deposited from a fluoborate bath but deposits from acid sulphate electrolytes are rough and nodular (unless some organic addition agent is present) so that the solution has found only limited use in electroplating. However a new, bright cadmium sulphate bath has been described by Laser(29) which, it is claimed, gives fine grained ductile bright deposits with good throwing power. Comprehensive details of cadmium electroplating baths and available commercial solutions are given elsewhere(30, 31).

In the cyanide bath, cadmium is contained in the complex $Na_2[Cd(CN)_4]$, the dissociation of which occurs as follows:

$$Na_2[Cd(CN)_4] \rightleftharpoons 2Na^+ + [Cd(CN)_4]^{2-} \qquad (1)$$

$$[Cd(CN)_4]^{2-} \rightleftharpoons Cd^{2+} + 4CN^- \qquad (2)$$

Owing to the stability of $[Cd(CN)_4]^{2-}$, the deposition potential of cadmium in such a bath is much more negative than in an acid cadmium solution, since the free Cd^{2+} ion concentration is very low. It decreases rapidly with increase in current density (i.e. the cathode polarisation is high). This fact, coupled with the good electrical conductivity of the bath, ensure a good throwing power, a fine grain size and relatively high hardness of the deposit.

The cyanide bath may be made up by dissolving either cadmium oxide or cadmium cyanide in a solution of sodium cyanide. A suitable bath for general work can be made from:

Cadmium cyanide, $Cd(CN)_2$	36 g
Sodium cyanide, NaCN	50 g
Water	1 litre

If cadmium oxide, CdO, be employed, additional cyanide is required to convert the oxide to cyanide in accordance with the equation:

$$CdO + 4NaCN + H_2O = Na_2\,[Cd(CN)_4] + 2NaOH$$

The excess sodium cyanide above that required for the formation of the compound $Na_2\,[Cd(CN)_4]$ may be regarded as the 'free' cyanide content of the solution[32]. Thus, each 1 kg of cadmium oxide requires 1.53 kg of sodium cyanide to form the double cyanide, whilst 1 kg of cadmium cyanide only requires 0.46 kg of sodium cyanide. When cadmium oxide is used, a certain amount of sodium hydroxide is left in the solution, but with cadmium cyanide no caustic soda is formed initially unless it is introduced separately.

The free cyanide content employed can vary widely in cadmium plating solutions. In view of the variations in the metal content of different baths, the solution can be more satisfactorily controlled by maintaining a suitable ratio between the total cyanide and the metal content; a ratio of $3\frac{1}{2}$ or 4:1 is a suitable operating range for most purposes. Sufficient free cyanide must be present to maintain anode corrosion, but excessive amounts reduce the cathode efficiency of the solutions rapidly. Concurrently with this reduction in efficiency, however, high free cyanide gives improved throwing power, but excessive amounts tend to cause the deposit to become spongy.

The pH control of cyanide solutions is helpful and is being used increasingly. The best pH range, particularly from the point of view of adhesion of the deposit, has been found to be 12.0 to 13.0[33].

The metal content adopted depends on several factors. With increased cadmium concentrations (up to 36 g/l of metal) it is possible to plate with higher current densities, but at the same time the throwing power of the solution is reduced and the deposit may become rough and nodular; also the drag-out losses are higher, which is an important consideration in the case of relatively expensive solutions.

Caustic soda is not an important constituent of the bath, although a small amount improves the conductivity of the solution; it is therefore useful in barrel plating. As has already been stated, a certain amount of caustic soda is formed when the bath is made up by dissolving cadmium oxide in the cyanide; it is not usual, however, to add caustic soda, as is the case with zinc solutions, because the advantages are not so marked, although a small concentration increases the cathode efficiency somewhat.

Sodium carbonate gradually accumulates in cadmium solutions owing to decomposition of the cyanide by the action of the carbon dioxide of the atmosphere. Excessive concentrations of carbonate impair the efficiency of the bath; generally, amounts in excess of 35 to 50 g/l should not be allowed to accumulate. The sodium carbonate can be removed by crystallising it out by cooling to 0°C, or more easily by discarding a portion of the solution.

An alternative method for the removal of excess carbonate from plating baths consists in treating the solution with calcium carbide, whereby the carbonate is precipitated as calcium carbonate with the liberation of acetylene:

$$CaC_2+2H_2O=Ca(OH)_2+C_2H_2$$
$$Ca(OH)_2+Na_2CO_3=2NaOH+CaCO_3.$$

This is better than freezing, which only operates successfully when the carbonate concentration is in excess of 60 to 70 g/l. Owing to the liberating of caustic soda, the treatment should be carried out gradually to avoid producing excessive concentrations of free caustic soda.

12.2.2. CADMIUM PLATING PRACTICE

Cadmium plating solutions are usually operated in steel or rubber-lined tanks at room temperature and the current density range under commercial conditions is most generally of the order of 1 to 3 A/dm^2. For the higher current densities, solutions with a greater metal content are necessary (up to 35 g/l of cadmium). The cathode efficiency of a cadmium solution is generally reasonably high, i.e. about 85 to 95%, or even more.

Amongst the troubles that can occur in cadmium plating practice are:

(a) **Burnt deposits.** This is often the result of too low a metal content in the solution, although the bath is being worked at the correct current density.
(b) **Dull deposits.** This condition can result from the solution being out of balance, viz. too high in caustic soda or sodium carbonate content, or too little free cyanide. Excessively high metal content can have a similar effect.
(c) **Pitting of the deposit.** This is sometimes caused by a shortage of caustic soda in the solution and can be corrected by the addition of a small amount of NaOH. More often, however, pitting is due to inadequate cleaning or pickling.

12.2.3. EFFECT OF IMPURITIES

Cadmium deposits are normally of an attractive white colour, but are susceptible to discoloration in the presence of impurities in the solution, particularly lead, arsenic, thallium, antimony, tin and silver[34]. The heavy metals can cause trouble when present in amounts exceeding about 0.05 g/l, leading to dark, coarse and spongy deposits; arsenic is the worst offender and amounts as low as 0.005 g/l can have a marked detrimental influence[35]. Thallium also occasionally causes trouble, being introduced from anode sludges. Should these impurities be introduced from anode sludge or by other means, they can be removed either by prolonged electrolysis at current densities of 0.1 to 0.3 A/dm^2, or by chemical displacement by means of cadmium dust followed by filtration of the solution. Nitrates are also to be avoided in cadmium solutions, as they markedly lower the cathode efficiency of the bath; sulphates have little adverse influence.

Nickel has a brightening influence in cadmium solutions when present to the extent of about 0.2 to 0.3 g/l.

12.2.4. ADDITION AGENTS

Brighteners and addition agents are also added to cadmium plating solutions to some extent, including nickel and cobalt salts of aryl sulphonic acids, starch, destrose, etc., and numbers of them have been patented. Certain addition agents facilitate plating at higher current densities, but their concentration must be kept low or the efficiency of the solution will be reduced.

12.2.5. ANODES

Cadmium anodes are usually cast and should be as pure as possible. The anode current density should not exceed about 1.5 to 2.0 A/dm^2 if uniform anode dissolution is to be obtained. On the other hand, excessively low current densities are also undesirable as they may cause sludging to occur at the anodes.

Cadmium anodes dissolve chemically in the plating solution, resulting in high anode efficiencies which may exceed 100 per cent. The rate of dissolution increases with rising temperature and cyanide content of the electrolyte. An anode/cathode ratio of about 1:3 is therefore best maintained if excessive metal dissolution is to be avoided, additional anode area being provided by the use of steel anodes if necessary.

The presence of even small amounts of impurities in cadmium anodes is harmful since, as pointed out earlier, even small amounts of lead, silver, tin, thallium, arsenic and antimony in solution will cause dark, coarse deposits.

The anodes used may be of the oval type with cast-in hooks, or cadmium balls can be used in steel cages. The latter arrangement has the advantage that additional balls can be added to the top of the cage as dissolution takes place, and ensures that an adequate anode surface is present towards the bottom of the plating tank. It has the disadvantage, however, that poor contact between the balls and the cage can result under certain conditions. Cadmium anode balls in steel cages are not subject to complete polarisation under any current density conditions, however, owing to the low oxygen over-voltage on iron as compared with the over-voltage on cadmium. This means that the cadmium anode does not become inactive while the steel surface is active. It is noteworthy in this connection, that if inert steel anodes are used alone (e.g. for the reduction of the metal content of the solution), they tend to become coated with an oxide deposit and must be cleaned from time to time if a reasonable working voltage is to be maintained.

When the anodes are working correctly their appearance is dull grey, but with excessive anode current densities a black coating may form on their surface. Further increase in current density results in white salt deposits on the anodes accompanied by considerable oxygen evolution. This is due to the fact that above a certain critical anodic current density, anodic corrosion products form at a greater rate than they can diffuse away from the anode into the solution. The anodes thus tend to become passive, and oxygen gas will be evolved when the oxygen evolution potential is reached.

An increase in the free cyanide tends to increase the permissible maximum anode current density while high concentrations of sodium carbonate tend to act in the reverse direction in most types of solution.

Cadmium lends itself to barrel plating for small parts, an excellent finish being generally obtained in a short time. The best type of modern barrel is fully immersed, being made of a suitable plastic such as polypropylene. The internal contacts to the work are carried on flexible, insulated arms, thus facilitating good electrical contact with the work in the barrel and at the same time making it a simple matter to renew the contacts, if necessary, when excessive build-up of metal occurs on them.

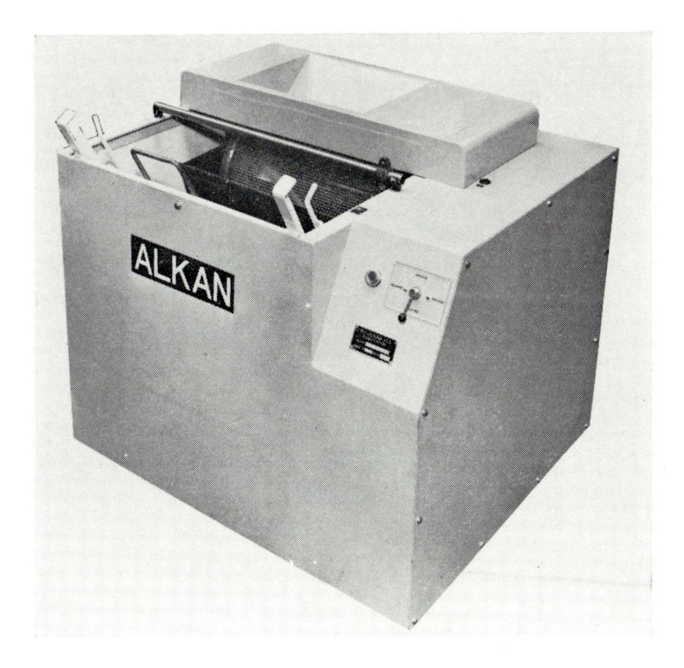

Fig. 12.4. Fully immersed plating barrel suitable for zinc or cadmium plating small parts such as washers. [Courtesy M. L. Alkan Ltd.]

12.2.6. AFTER-TREATMENT OF CADMIUM DEPOSITS

Cadmium, as deposited, presents an attractive appearance, although the actual degree of brightness varies with the brightener used and the nature of the surface of the underlying metal. Sometimes, however, the appearance of the deposit is improved by various bright dips, which consist essentially of solutions of oxidising agents. Cadmium appears to dissolve in these without gas evolution, and a brightening effect is produced by immersion for a few seconds. The commonest bright dips for cadmium are: (a) chromic acid; (b) peroxide dips; and (c) nitric acid. In the first of these, the plated article is immersed for two or three seconds in a solution of chromic acid with about 0.3 to 0.5 g/l concentrated sulphuric acid. The solution attacks the cadmium fairly rapidly, so that immersion must not be excessively prolonged. The peroxide dip consists of a solution of hydrogen peroxide (5 to 10 per cent by volume of 30 per cent (100 vol.) hydrogen peroxide) and about 0.3 to 0.5 per cent of sulphuric acid. The absolute concentrations of the two constituents are of less importance than their relative proportions, a ratio of 4:1 peroxide to sulphuric acid being recommended. Potassium persulphate can be used instead of the peroxide. The nitric acid dip is simply a dilute acqueous solution of nitric acid of 0.5 to 1 per cent strength.

The chromic acid dip reduces the intrinsic corrodibility of the cadmium, while both this and the peroxide dip reduce the tendency for the deposit to stain and finger-mark. The nitric acid dip is less valuable from the latter point of view.

The durability of cadmium plate under many conditions can be improved by chromate passivation, as in the case of zinc (c.f. 12.1.4–12.1.4.2).

12.2.7. HYDROGEN EMBRITTLEMENT

Cadmium plating, like zinc plating, will embrittle steel parts, particularly if they are in a hardened condition. This is due to hydrogen absorption during the plating process; the cadmium itself does not contribute measurably to the embrittlement[36]. Embrittlement due to hydrogen absorption should be relieved by heating to a temperature of up to 200°C for an hour or perhaps more. The exact time and temperature depends on the condition of the metal and its thickness and on the amount of hydrogen absorbed. The effectiveness of the de-embrittling operation can be determined by a bending or flexing test.

As cadmium plating is used on steel components in the aircraft industry, the problem of avoiding or reducing the effects of hydrogen embrittlement is important. Careful handling is necessary in the pre-treatment as well as during the plating operations. Blasting is a safe method of cleaning and this is followed by anodic alkaline cleaning, after which the components may be anodically treated in a sulphuric-phosphoric acid bath to provide a smooth surface.

12.2.8. CADMIUM FLUOBORATE PLATING BATHS[37]

Cadmium can be successfully plated from an acid fluoborate bath and this is used in the United Kingdom, especially for barrel plating components. Apart from the much reduced toxicity as compared to the cyanide bath, the fluoborate bath offers many other advantages including reduced tendency for hydrogen embrittlement of cadmium plated steel components, high cathode current efficiencies approaching 100%, and good solution stability. The throwing power of the bath, although not quite so good as that of cyanide solutions, is adequate for most purposes.

The acid fluoborate solution contains:

Cadmium fluoborate	210 g/l
Sodium fluoborate	25 g/l
Boric acid	25 g/l
Sodium β-naphthalene sulphonate (brightener)	0.8 g/l

The bath is used at 20–30°C and current densities of 2–6 A/dm^2 are employed. In practice, the solution is prepared by adding cadmium carbonate to fluoboric acid until saturation is reached.

$$2HBF_4 + CdCO_3 \rightarrow Cd\,(BF_4)_2 + H_2O + CO_2$$

The solution is then filtered through glass wool and diluted.

12.3. Lead plating

Lead plating has not, in the past, received any great attention as a rust-proofing coating because of the ready availability of zinc and cadmium, which are generally considered to be superior in this respect. It is not normally possible to apply lead to steel satisfactorily by dipping unless some tin or other suitable metal is alloyed with it to promote wetting of the basis metal (*cf.* p. 505). The electrodeposited metal is not, however, subject to this limitation, and is often less porous than hot-dipped coatings of equal thickness. Where resistance to sulphuric acids is concerned, as in the manufacture of lead accumulator fittings and acid-containing fire extinguishers, electrodeposition affords the best means of producing the pure lead covering which is desirable for maximum serviceability, both on ferrous and non-ferrous metals. Lead coatings are not used for decorative purposes, since the coating is not very pleasing in appearance and is easily damaged and worn.

There are three baths commercially utilised for lead deposition, viz. the fluosilicate and the fluoborate baths, and, in the United States, the lead sulphamate solution. Another bath worthy of note is the perchlorate bath[38], but this is little used in commercial practice.

12.3.1. THE FLUOBORATE BATH

The most important bath for lead plating is the fluoborate bath; the following solution is used for the deposition of good lead deposits up to 0.15 cm in thickness, or for barrel-plating[39]:

Basic lead carbonate, $2PbCO_3.Pb(OH_2)$	300 g/l
Hydrofluoric acid (50 per cent HF)	480 g/l
Boric acid, H_3BO_3	212 g/l
Glue	0.2 g/l

A bath of half the above concentration is suitable for thinner deposits at low current densities, but the lead concentration should be kept high if smooth deposits and good throwing power are required.

A high-speed fluoborate bath for lead plating has been developed by Graham and Pinkerton[40]. It is claimed that the deposits produced are fine-grained and cover basis metal defects even at thicknesses of 0.1 mm or less. The recommended solution contains:

$Pb(BF_4)_2$	325 g/l
HBF_4	31 g/l
H_3BO_3	32 g/l
Hydroquinone	6.75 g/l

Fluoborate solutions are best operated in a rubber or plastic-lined steel tank, but lead-lined are also used. Lead is not attacked by the solution but an inner glass or plastic lining is advisable to prevent the lining becoming anodic or cathodic accidentally. Ceramic or synthetic resin tanks are also suitable.

Although the best operating temperature range is between 25°C and 40°C, the bath is commonly used at room temperature. The current density range is between 1 and 3 A/dm^2 depending on the composition of the solution and on the thickness of deposit required. The cathode efficiency of the bath is high, being in the region of 98 to 100 per cent. A low current density range of 0.5 to 0.15 A/dm^2 is preferred where uniformity of coating is required. With light cathode rod agitation the solution can be operated successfully at considerably higher current densities and figures of 8.0 A/dm^2 have been reported. The solution has good conductivity and only relatively low voltages are needed, up to 2 volts being sufficient for plating in still solutions, while 4 to 6 volts may be necessary in barrels. High voltages are to be avoided, as they tend to cause 'treeing' of the deposit.

12.3.1.1. Some characteristics of the solution

It is generally assumed that in the preparation of the solution the hydrofluoric and boric acids interact in accordance with the following equation:

$$4\,HF + H_3BO_3 = HBF_4 + 3H_2O$$

The lead in the fluoborate bath is in the form of lead fluoborate, $Pb(BF_4)_2$, but there is generally an excess of lead dissolved in the solution over and above the requirements of this formula. In operating the bath an excess of fluoboric acid is maintained in the solution: a proportion of boric acid, about 10 to 15 g/l over that required theoretically to form the compound HBF_4, is also usually added. The 'free' fluoboric acid serves to reduce the tendency of the deposit towards 'treeing' and improves the conductivity of the solution. The excess boric acid improves the stability of the bath, reducing the tendency for decomposition to take place with the precipitation of lead fluoride; this can occur if any excess of hydrofluoric acid should develop in the bath.

Control of the pH of the solution is also advisable since it tends to rise during operation. A pH value not exceeding 1.0 is recommended, since at values above 1.5 there is a tendency towards granular deposits and 'treeing'. Glass electrode pH measurements are not reliable in this solution, so that tests should be made using a colorimeter.

The metal content of the bath may be controlled by density. In making up the bath the boric acid is gradually added to the hydrofluoric acid in a suitable tank, after which the basic lead carbonate in the form of a paste is stirred in slowly. After diluting to the correct concentration the solution is carefully filtered, the addition agent is added, and the bath is ready for use. The free fluoboric acid is maintained by an empirical titration with decinormal caustic potash, and additions of boric acid and hydrofluoric acid mixed together in the proportions of 1 to 4 are made from time to time.

12.3.1.2. Addition agents

Various addition agents have been experimented with to reduce 'treeing', to which lead solutions are especially prone, glue being most widely used. The presence of some addition agent is essential if fine-grained deposits are to be obtained and the tendency towards 'treeing' obviated. In some early experiments by Blum[41], it was found that 800 g of lead could be deposited from 1,200 ml of fluoborate solution containing 0.2 g/l of glue before 'treeing' became evident, and even thereafter the tendency towards 'treeing' was much reduced.

Other addition agents which have been proposed are gelatin and phenolic compounds such as pyrogallol and resorcinol, but glue is most commonly used in industrial practice. The grade of glue must be of high quality, and its effect on a small amount of the solution should be tested before it is added to the bulk. A throwing power test is a very satisfactory guide, while a microscopic examination of the deposit under a magnification of about 500 is also of value. The use of an unsuitable grade of glue or an excessive concentration will reveal many dark-coloured inclusions, and in come cases a powdery appearance on the surface of the deposit.

12.3.1.3. Anodes

The anodes used should be as pure as possible, chemical lead being most satisfactory. The anode efficiency is about 100 per cent and an anode:cathode ratio of 2:1 is recommended. Insufficient anode area, and hence high anode current density, causes polarisation of the anodes and a subsequent drop in the metal content of the bath. Impure anodes can cause anode films and sludges which interfere with the formation of smooth and uniform deposits. Copper is particularly to be avoided in the anodes; as little as 0.08 per cent appears to have an adverse effect on the deposit. The anodes may be bagged in Terylene to reduce the entry of sludges into the bath.

12.3.1.4. Cleaning

Cleaning of steel prior to lead plating is similar to that used for other metals, but over-pickling must be avoided. Where a sulphuric acid pickle is employed, thorough rinsing is essential as the introduction of sulphuric acid into the solution results in the immediate precipitation of some of the lead as lead sulphate. Similarly, any water added for topping up the bath should be substantially free from sulphate. Shot blasting is often advisable for badly scaled parts, and on occasion a deposit of copper may be applied to promote adhesion of the lead. Lead plating from this bath has little or no embrittling influence on steel.

12.3.2. THE SULPHAMIC ACID BATH[42]

Excellent deposits, similar to those from the fluoborate bath, are obtained from this solution, which is prepared by dissolving an excess of basic lead carbonate in a 5 per cent solution of sulphamic acid $HSO_3.NH_2$, filtering off the excess and then adding a further 5 per cent of sulphamic acid. The resulting solution has a pH of 1.0 to 1.5. Addition agents are necessary, among those recommended being: (a) 0.07 per cent of beta-naphthol and 0.7 per cent of aloin; (b) 0.05 per cent of casein and 0.1 per cent of aloin; (c) 0.5 per cent of malic acid and 0.1 per cent of aloin; (d) 0.7 per cent of glue and 0.07 per cent of beta-naphthol. The current density which can be employed is of the order of 2 A/dm^2, while the cathode efficiency is over 95 per cent.

12.3.3. RESISTANCE TO CORROSION OF LEAD DEPOSITS

Lead deposits of 0.0010 to 0.0025 cm in thickness give good protection against corrosion if substantially non-porous but, owing to the soft nature of the metal, breaks in the coating occur relatively easily if it be subject to abrasion. Owing to the fact that lead is cathodic to iron, rusting at such breaks occurs more readily than with zinc plating. It is, however, less cathodic than copper to iron, while in marine atmosphere there is some evidence that it may be weakly anodic[43]. With lead deposits heavier than 0.0025 cm the mechanical properties of the coating are improved so that better durability is obtained. In depositing lead on articles of involved shape it should be borne in mind also that the solution has much poorer throwing power than have zinc or cadmium plating solutions.

For the testing of the porosity of lead deposits, the coating is washed in 10 per cent sulphuric acid to remove particles of iron or iron oxide on the metal surface and treated with a solution of 20 g/l of sulphuric acid and 10 g/l of potassium ferricyanide. Bright blue spots appear within a minute to indicate the presence of pores in the coating.

Thin copper coats under the lead do not materially improve the corrosion resistance of thin coatings, but coatings of 0.0015 to 0.0025 cm are valuable as undercoats. A copper undercoat is also useful when soldering has to be carried out on to the lead deposit. Lead does not wet steel, but better wetting results on copper in the presence of the tin in the solder.

In some exposure tests carried out by Clarke[44] at Woolwich, steel sheets carrying lead deposits 0.00025 cm thick were completely rusted in six months, deposits 0.00125 cm thick had small rust patches in one year, while coatings of 0.0025 cm thickness were highly protective, showing no rusting or apparent deterioration at the end of three years.

In contact with zinc or aluminium, in particular, acceleration of the corrosion of the coating tends to occur and such contact is best avoided.

Lead deposits of 0.008 cm or more are relatively easily built up, and have a high degree of durability. For example, a deposit of about 0.01 cm can be produced in one hour at 2 A/dm^2.

12.3.4. PLATING LEAD ALLOYS

Blum and Haring[45] have described the deposition of fine-grained lead-tin alloys from the fluoborate bath. The ordinary lead fluoborate solution is used, the tin being introduced by electrolysis of the lead solution with tin anodes until the requisite amount of tin is present in the solution. It is not practicable to attempt to make tin fluoborate. Owing to the preferential dissolution of tin from the anode it is desirable to work with lead-tin anodes containing rather less tin than that which is required in the deposit.

It is quite practicable to operate this bath within the range of 25 to 75 per cent of tin by weight and maintain a constant alloy composition, but below this tin content the composition of the deposit and the corrosion of the anodes vary so irregularly with slight changes in current density that the operation of the solution becomes impracticable. Glue is used as an addition agent, to prevent 'treeing' and promote smooth deposits. Resorcinol is sometimes added in the proportion of 0.5 to 1.0 g/l as it reduces the tendency for the stannous tin to oxidise.

REFERENCES

1. "Zinc or Cadmium". Editorial, *Electroplating and Metal Finishing*, 1971, **24,** No. 3, 3.
2. H. J. Reed and J. A. Strasser, *Plating*, 1970, **57**, 3, 229.
3. H. G. Todt, *Trans. Inst, Metal Finishing*, 1973, **51**, 3, 91.
4. J. Hajdu and J. A. Zehnder, *Plating* 1971, **58**, No. 5, 458.
5. E. Knaak, *Metal Finishing, J.* 1971, **17,** No. 200, 214.
6. British Pat. 8562 (1907); 4625 (1915).
7. C. H. Proctor, *Metal Ind.* 1918, **16**, 114.
8. C. H. Proctor, C. J. Wernlund. U.S. Pat. 1,435,875 (1923).
9. W. Blum, F. L. Liscomb, C. M. Carson, Nat. Bur. Standards Tech. Paper 195 (1921).
10. A. K. Graham, *Trans. Amer. Electroplaters' Soc.*, 1921, **40**, 257.
11. A. K. Graham, *Trans. Electrochem. Soc.* 1933, **63**, 121.
12. R. O. Hull, *Proc. Am. Electroplaters' Soc.* Conv., 1940.
13. A. K. Graham and E. Zurbach, *Plating* 1971, **58**, No. 4, 349.
14. R. O. Hull, *Met. Fin.* 1944, Sept, 545.
15. R. Weiner, *Galvanotechnik*, 1959, **50**, 10, 535.
16. T. Korpium, *Galvanotechnik*, 1971, **62**, 3, 210.
17. E. Raub, Metalloberflache, 1953, B, **17,** 5.

18. E. Raub and W. Wullhurst, *Mitt. Forschungsges. Blechverbereitung*, 1951, **14**, 279.
19. H. Fischer and H. Barmann, *Zeit. f. Metallkunde*, 1940, **32**, 376.
20. E. B. Saubestre, J. Hajdu and J. Zehnder, *Plating* 1969, **56**, No. 6, 691.
21. U. C. Tainton and E. T. Clayton, *Trans. Electrochem. Soc.* 1930, **57**, 279.
22. H. Roebuck and A. Brierley, *J. Electrod. Tech. Soc.* 1946, **21**, 91.
23. E. Grunwald and Csaba Varhelyi Cluj, *Electroplating and Metal Finishing* 1970, **23**, No. 9, 14.
24. E. A. Anderson, *Proc. Am. Electroplaters' Soc.* Conv. 1943, p. 6.
25. M. E. Roper, *Metal Finishing J.* 1968, **14**, No. 165, 286.
26. J. Whithberm and R. Mangles, *J. Electrochem. Soc.* 1948, **94**, No. 2.
27. S. G. Clarke and J. F. Andrew, Proceedings 1st. Internat. Congress on Metallic Corrosion., 1961.
28. I.S.O. 2082, 1973, 6p.
29. H. Laser, 16th Internat. Electroplating Conference, Gottwaldov, Proc., Pomtechniky SV TS, Bratislava, CSSR, 1973, 7-9.
30. *Metal Finishing Plant & Processes*, 1974, **10**, No. 2, 59-66.
31. R. Walker, *Metal Finishing*, 1974, **72**, No. 1, 59-64.
32. G. Soderberg, *J. Electrodep. Tech. Soc.* 1937, **13**, 1.
33. Z. Irenas, *Monthly Rev. Am. Electroplaters' Soc.* 1943, **30**, 603.
34. L. R. Westbrook, *Trans. Electrochem. Soc.* 1929, **55**, 333.
35. W. B. Meyer, *Met. Fin.* 1949, July, 369.
36. J. A. Gurklis, L. D. McGraw and C. L. Faust, *Plating* 1960, **10**, 1146.
37. L. Serota, *Metal Finishing* 1965, **63**, No. 4, 83.
38. F. C. Mathers, *Trans. Electrochem. Soc.* 1920, **17**, 261.
39. A. G. Grey and W. Blum, Electrochem. Soc., Spec. Vol. 1942, 221.
40. A. K. Graham and H. L. Pinkerton, Proceedings of the 6th International Conference on Electrodeposition and Metal Finishing. 1964, 223.
41. W. Blum, *Trans. Electrochem. Soc.* 1919, **36**, 243.
42. F. C. Mathers and R. B. Forney, *Trans. Electrochem. Soc.* 1939, **76**, 373.
43. *Proc. A. S. T. M.* 1939, **39**, 247.
44. S. G. Clarke, *J. Electrodep. Tech. Soc.* 1939, **15**, 141.
45. W. Blum and H. E. Haring, *Trans. Electrochem. Soc.* 1921, **40**, 287.

CHAPTER 13

The treatment of aluminium and magnesium

THE term 'light alloys' is generally used to refer to aluminium and magnesium and their alloys. Although pure aluminium finds considerable commercial application because of its good fabricating and other properties, magnesium is generally employed as one or other of its alloys either in sheet, rolled or extruded form, or as castings.

13.1. Aluminium

Aluminium is produced commercially in a high state of purity, and is available in many forms. The metal and its alloys are widely used in the production of aircraft, for structural and architectural purposes, and also in the manufacture of kitchen equipment and food-processing plant. Numerous fittings, both decorative and functional, are also made of aluminium alloys.

The pure metal is, however, rather soft, so that various alloys are employed for special purposes. Among these are the casting alloys, which generally contain about 12 per cent of silicon, the forging alloys (containing magnesium, copper and nickel) and the structural alloys of relatively high tensile strength containing copper, manganese and magnesium.

Pure aluminium is reasonably resistant to corrosion, but its alloys are generally less so; special alloys have, however, been developed for resistance to corrosion by sea water, e.g. those containing magnesium (3 to 6 per cent) with 0.25 to 0.75 per cent manganese. The silicon alloys are fairly resistant, but unfortunately the structural alloys are much less satisfactory from this point of view.

Owing to these limitations, finishes are often necessary for aluminium and its alloys both for decorative and protective purposes. Fortunately, aluminium alloys can be finished or protected by a wide variety of highly satisfactory methods. Thus, these alloys are readily etched, scratchbrushed or polished, giving frosted, matt or bright effects, whilst the anodic coatings to be described are hard and highly protective; these latter range from thick opaque coatings to clear and colourless films which can be dyed directly.

13.1.1. MECHANICAL FINISHING AND ETCHING OF ALUMINIUM

13.1.1.1. Polishing

For the treatment of rough surfaces such as castings, etc., rough polishing with a wheel dressed with 60 to 100 mesh abrasive is usual. Emery or alumina is generally used, the peripheral speed of the wheel being about 1500 m per minute. Grease is helpful in preventing overheating. Greased abrasive belts dressed with 80 to 160 mesh emery are also suitable for flat pieces.

For the next operation a felt wheel dressed with 100-180 mesh abrasive is suitable. Lubricants are always used at this stage and may include compositions of tallow, oil and beeswax. To obtain a lustrous finish, loose or stitched muslin wheels are used for final buffing with tripoli compositions, the wheel speeds being of the order of 2000-2400 m per minute. For a really high-grade finish, colour buffing is employed as a final operation, the purpose being to improve the lustre and reflectivity of the metal. Colour buffing is done on open muslin or flannel wheels, at peripheral speeds of 2200 to 2400 m per minute, using Vienna lime or similar composition. Prior to colouring, the work should be clean and dry; sometimes a light etch in caustic soda solution is helpful where the articles are to be anodised, as this removes any embedded surface abrasive.

For fabricated parts the initial rough polishing is not generally required unless the articles are badly marked or scored. Proper lubrication of the abrasive is essential in the polishing of aluminium to prevent loading of the wheels with aluminium particles which can cause scoring of the surface being polished.

Pitting and cloudiness of the finish are usually the result of excessively hard polishing buffs. The softest buffs for colouring are made from materials of low thread count with the minimum of stitching, preferably only around the arbor hole. Three or four spacers may be used between discs or between sections, these being 7.5 cm or 10 cm diameter discs or fabrics. The melting-point of the grease binder used in the buffing compound is important. If the melting-point is too low, there is a danger of pitting of the metal surface and of overheating, which results in a less lustrous finish. The cutting quality of the abrasive is, however, increased so that cloudiness may occur. High peripheral speeds also have the effect of increasing buff hardness and improve the cutting properties. Pressure for polishing should be kept low.

Automatic mechanical polishing machines are widely used for polishing aluminium and its alloys. Machines have been designed to deal with sheet, rod, tube and many wrought and cast shapes. However, careful inspection of the work is necessary before the first polishing stage. Deep scratches which may be present will not be removed by an automatic polishing machine set to apply a uniform overall pressure. For this reason it may be worthwhile to supplement automatic polishing with manual polishing, so that badly scratched work surfaces can be dealt with separately.

13.1.1.2. Etched finishes

Attractive etched finishes can be produced by a short-time immersion in sodium hydroxide solution. In the case of alloys containing copper, however, this results in the formation of a black smut, and a clean surface can only be obtained by subsequent treatment in a bath containing 5 per cent of nitric acid; an addition of 5 per cent of hydrofluoric acid should be made to the nitric acid to remove silicon when this element is present in the alloy.

Where a high rate of etching is required, an accelerator such as sodium fluoride may be added to the sodium hydroxide solution. The temperature of the solution also effects the rate of etching, but if the bath is too hot, staining may occur on drying. For this reason it is sometimes preferable to employ two baths, the second containing a weak solution of about 2 per cent concentration which does not attack the aluminium in the short dipping time used. The first bath may be of 5 to 15 per cent strength and used at a temperature of 50°C to 70°C. Cold water is preferably used for rinsing to prevent staining, and the bath concentration and temperature should not be allowed to fluctuate widely if uniform results are to be obtained.

As well as an accelerator, commercial aluminium etchants also contain a detergent addition to facilitate rinsing, an inhibitor such as gum acacia to prevent the formation of sludge and a sequestering agent which delays precipitation of hard alumina scale on to the sides of the tank. Alumina is formed due to hydrolysis of sodium aluminate (formed by the dissolution of aluminium in the alkaline etching solution):

$$2NaAlO_2 + 4H_2O \rightarrow 2NaOH + Al_2O_3.3H_2O$$

Sequestering agents employed include E.D.T.A. and gluconates.

Castings do not give an attractive finish by alkali dipping, so that scratchbrushing or shotblasting or similar mechanical treatments are to be preferred.

Etching in a hot 10 per cent sodium carbonate solution also produces a pleasing matt surface. Treatment in this solution is rather more easily controlled than in sodium hydroxide solution. Again, silicon- and copper-containing alloys give dark surfaces, so that a subsequent acid treatment in nitric acid or in a nitric-hydrofluoric acid mixture is necessary. In the absence of silicon, a mixture of 5 per cent of nitric and 5 per cent of sulphuric acid proves satisfactory. A mixture of equal parts of saturated sodium fluoride solution and concentrated nitric acid gives a white uniform matt surface on most aluminium alloys, but the solution is unpleasant to use, so that good extraction equipment is essential to remove toxic fumes. Many other etching solutions of diverse composition are currently in use.

13.1.1.3. Scratchbrush finish

Scratchbrush finishes are useful for some purposes, and are produced by treatment with a rotating wire wheel. Typical wheels used are about 25 cm in diameter with stainless steel or nickel wires up to 0.04 cm thick. The speed of rotation is about 2,000 r.p.m. After cleaning the aluminium, the metal is held lightly against the brush until the required degree of finish is obtained. Castings are preferably sandblasted with a fine abrasive before treatment, and the speed of the wheel should be kept at 450 to 600 r.p.m. to prevent tearing of the metal surface and so leading to a non-uniform finish. The direction of rotation of the wheels should be reversed periodically to prevent bending of the wires; they should also be cleaned from time to time by running them against a pumice block or a soft brick.

13.1.1.4. Satin finish

Attractive satin finishes can be produced by the use of brushes with much finer diameter wires (0.005-0.010 cm). For satinising the insides of cooking utensils small nickel-silver wire brushes are employed. The use of nickel-silver ensures that the true white colour of the finish is retained. Castings must first be polished all over with a 120- or 180- mesh abrasive, after which they are buffed with a muslin or felt wheel and an emery composition, and then wire brushed after thorough degreasing. A 15 cm diameter wheel with wires 0.010-0.025 cm in diameter and operating at 450 to 600 r.p.m. will give a fine satin finish. A still finer effect can be obtained by hand finishing with pumice and paraffin.

A satin finish on aluminium and its alloys can also be obtained by blasting the surface. In one process[1] aluminium components are blasted with a mixture of various oxides, including alumina (68%) and manganese oxide (23%). However, it is possible to use many other substances for blast satinising, including materials such as glass beads, barley husks, and coconut shells.

13.1.2. CHEMICAL BRIGHTENING

Aluminium and its alloys are nowadays being brightened chemically and electrolytically on a substantial scale, since the results obtained are not readily produced by other methods. Even when mechanically polished with the greatest care, a reflectivity of only 65-75 per cent in the visible spectrum range can be obtained on aluminium owing to the flow of the metal. Also, abrasive constituents are retained in the surface, so that when subsequent anodising is carried out an opalescent rather than a clear film tends to be produced.

A considerable number of processes have been developed for the chemical brightening of aluminium alloys, and these often give brighter finishes than the electrolytic methods, especially on super-purity aluminium or alloys based on the super-purity metal. The high purity metal generally gives the best results. Also, little or no mechanical polishing is needed prior to treatment as these processes have a levelling action. In many cases the operating costs of these processes are considerably cheaper, not principally because of the low cost of materials employed, but because the short time required for treatment enables a very substantial output to be obtained with a relatively small brightening plant. The most important commercial processes are the concentrated phosphoric-nitric acid solutions and the more dilute nitric acid-ammonium bifluoride types.

13.1.2.1. Phosphoric-nitric acid solution

The phosphoric-nitric acid solutions may also contain sulphuric or acetic acids and are operated at 85°-105°C, an immersion time of 3-5 minutes being necessary. After treatment the metal is anodised in a sulphuric acid bath, but excessively heavy anodic films must be avoided as they result in a reduction of brightness.

A widely used commercial process of this type is 'Phosbrite 159' which is based on a solution containing phosphoric, nitric and sulphuric acids[2]. It gives excellent smoothing and brightening on a wide range of aluminium alloys.

Nitric acid is essential for specular brightening to occur, simple phosphoric acid alone (with or without water) giving only a diffuse appearance. The presence of water in place of sulphuric acid in the ternary mixture, phosphoric acid 75%, nitric acid 5% and sulphuric acid 20%, increases the rate of attack. Phosphoric acid is relatively costly, so that replacing it in part by sulphuric acid and/or water can result in substantial savings.

A solution containing phosphoric acid (Sp. gr. 1.5) 75%, water 20%, nitric acid 5% and copper sulphate 0.25% operated at 100°C is economic and is capable of giving satisfactory results.

Again it is essential that the phosphoric acid content should not be reduced below about 68%. The authors recommend a solution containing technical quality phosphoric acid (sp. gr. 1.5) 75%, water 20%, nitric acid 5% and copper sulphate 0.25% as being a relatively inexpensive bath, operated at 100°C.

13.1.2.2. Nitric acid—bifluoride solution

One of the well known nitric acid-bifluoride solutions is that used in the 'Erftwerk' process[3]. A typical bath consists of:

Nitric acid	13 per cent
Ammonium bifluoride	16 per cent
Lead nitrate	0.02 per cent

Immersion time is only 15-30 seconds, the solution being operated at 65°-75°C. Very high reflectivity can be obtained on super-purity aluminium without mechanical polishing, and this is only slightly reduced by the anodising treatment. The solution may be regenerated by the addition of hydrofluoric acid until the aluminium content becomes too high, when it is discarded. About 0.002-0.005 cm of aluminium is removed by the polishing operation in 15-30 seconds. This rapid rate of removal results in a high degree of smoothing, which is very useful where metal with a poor initial surface is being treated.

The Erftwerk process is more readily applicable to super-purity aluminium and aluminium-magnesium alloys than the phosphoric acid-based brightening treatments; it does, however, result in an appreciably greater removal of metal. The solution is considerably cheaper than the phosphoric-nitric acid baths, there is much less fume, and the disposal of discarded baths presents a less formidable problem. Rolled sheet is more satisfactory for obtaining a bright finish than cast or forged material owing to the compacting of the surface and the small grain size which results from the rolling operation. Amongst the principal impurities to be avoided in aluminium alloys for treatment by the Erftwerk process are iron and titanium, which have particularly adverse effects.

The 'Kynalbrite' process, developed by I.C.I., is applicable to most commercial quality alloys, excluding those having a high silicon content, and has been used for cheap items such as anti-splash tap fittings and pen-tops. The solution contains:

Phosphoric acid (S.G. 175)	78 per cent V/V
Nitric acid (S.G. 1.43)	11 ,, ,, ,,
Sulphuric acid (S.G. 1.84)	11 ,, ,, ,,
Ferrous sulphate	0.86 g/l

The solution is maintained at 95°-100°C in a stainless steel tank; the immersion time is of the order of 15-60 seconds. An immersion time of 15-20 seconds leads to a loss in thickness of super-purity aluminium sheet of between 0.0005 cm and 0.0008 cm. For full brightness lower temperatures are desirable, as at temperatures in excess of 115°C a bright satin type of etched finish tends to be produced. Nitrous fumes are evolved during the treatment, and proper exhaust equipment is essential to remove these together with a suitable scrubbing tower to prevent their discharge to the atmosphere, as they are highly toxic.

The plant needed for chemical brightening is somewhat less expensive than that used for electro-brightening, as the treatment time is short and there is no need for electrical equipment or for special jigging. The short treatment time also means that a small plant can produce a high output. The absence of jigging is not of such importance if anodising has to be carried out after brightening (as is most often the case), for the same jig can be adapted to suit both processes. There are, nevertheless, numerous instances where brightening only, with or without subsequent lacquering, constitutes a satisfactory finish, and it is here that chemical brightening becomes an attractive process to use.

13.1.3. ELECTRO-BRIGHTENING

One of the earliest processes to be used commercially to electro-brighten or electro-polish aluminium was the 'Alzak' process developed by the Aluminium Company of America[4]. This process, which uses a variety of electrolytes based on fluoboric acid, has been widely used in the U.S., but in the U.K. and Europe, alkaline electrolytes ('Brytal' process) have found more favour.

13.1.3.1. Brytal process

The 'Brytal' process[5] makes use of an alkaline solution of 15% wt/vol sodium carbonate and 5% wt/vol trisodium phosphate at a temperature of 75°C-88°C. Stronger solutions may also be used. The parts are immersed in the solution without application of current for ten to thirty seconds, when a rapid evolution of hydrogen takes place; the duration of this operation must be carefully controlled and depends on the alloy being treated. A potential of 10-14 volts is then applied for 5-8 minutes, using stainless steel cathodes, the parts to be treated being made anodic, which results in brightening of the metal. The initial current density is of the order of 2-3.5 A/dm, but this falls to about half the above value as the treatment proceeds. Treatment time is about 5-15 minutes, after which the parts are removed and rinsed in water. The process is reasonably straightforward to operate; one of the main problems encountered is a slight tendency for pitting of the metal to occur particularly at the higher current densities or if the pH falls below 10.0-10.5 (especially if chlorides are present in the anodic bath). The brightening bath must not be stirred during the operation, since otherwise whitish 'gas flow' marks or streaks may occur.

The Brytal treatment leaves a thin oxide film on the metal suıface and this must be removed by immersion in a hot solution of 7% by volume of phosphoric and 4% wt/vol of chromic acids contained in a stainless steel tank before the brightened aluminium is anodised. Treatment time is 5-10 minutes. If the film is not removed it leaves an easily marked layer which impairs the quality of the final product.

Table 13.1 shows some reflectivity figures for aluminium treated by the Brytal process as compared with other reflecting surfaces:

TABLE 3.1

Reflectivity characteristics of some reflecting surfaces

Specimen	Total reflectivity per cent	Specular (pts. in 100)	Diffuse (pts. in 100)
Polished commercial aluminium	73.1	93.1	6.9
Brytal on commercial aluminium	76.8	86.2	13.8
Brytal on A reflector grade aluminium	82.4	96.25	3.75
Brytal on super-purity aluminium	84.1	99.4	0.6
Stainless steel	59.5	97.0	3.0
Chromium-plate	63.0	99.7	0.3
Rhodium on nickel plate	69.1	99.55	0.45
Lacquered silver plate	89.9	96.5	3.5
Silvered glass mirror	90.3	(almost specular)	

13.1.3.2. Alzak process

The 'Alzak' process uses a variety of electrolytes, most of which contain about 2.5% fluoboric acid. Cooling coils may be made of copper and these can also serve as the cathode; they are screened with flannel as far as possible to minimise convection currents and to reduce movement of the solution to a minimum. After removal from the bath, a dip in a hot alkaline solution is necessary to remove the polarised film with which

Fig. 13.1. The office building for the Reynolds Metals Co., Detroit, incorporating much anodised aluminium.

the surface is covered, as in the case of the Brytal process. In both these processes the purity of the aluminium plays an important part in the degree of reflectivity which is obtained, the best results being obtained on super-purity quality metal (99.99%).

After brightening, it is essential that the aluminium should be anodised to provide a hard and durable surface. The anodic film must be of the transparent type.

13.1.4. ANODISING

The inherently good weathering and corrosion properties of aluminium are due to the thin hard oxide film which normally forms on the metal quickly on exposure to the atmosphere, and protects it from further attack. In the anodising process the thickness of the oxide film is increased to something of the order of 10-15 μ to impart superior corrosion and abrasion resistance. The anodic film as formed is hard and stable, and contains pores which can take up dyes to give a range of coloured decorative effects. Anodised finishes are used for an enormous variety of applications ranging from the treatment of very small objects to imparting decoration and corrosion resistance to architectural items. Fig. 13.1 shows the modern office building of the Reynolds Metal Company, Detroit, which was designed by Minoru Yamasaki as a showcase for aluminium. The gold-anodised sun screen covers the second and third office floors on all four sides.

13.1.4.1. Formation and growth of anodic films

When aluminium is anodically polarised in sulphuric, chromic and other acid electrolytes, oxygen gas is produced, some of which remains adsorbed on the aluminium surface and converts the latter to a film of aluminium oxide. If aluminium oxide, Al_2O_3, is insoluble in the electrolyte (e.g. boric acid electrolytes), a dense, compact film is produced which stops growing when a certain thickness is reached. This thickness is related to the anodising voltage, or more specifically to the anodic overpotential. Films of this type are referred to as barrier films or layers. If, on the other hand, the electrolyte exerts some solvent action on the film, as do sulphuric, oxalic, phosphoric and chromic acid solutions, then a micro-cellular structure is produced, each cell having a pore at its centre. An electron photo micrograph of an anodised film showing the porous cellular structure is shown in Fig. 13.2, together with an idealised drawing of the oxide film microstructure.

The diameter of the pores increases with increase in current density, temperature and acid concentration of the anodising solution, but the pore wall thickness is independent of these factors and is controlled solely by the anodising voltage.

The pores in the film do not extend to the aluminium metal surface, so that it is convenient to consider the anodised film as comprising two components[6]:

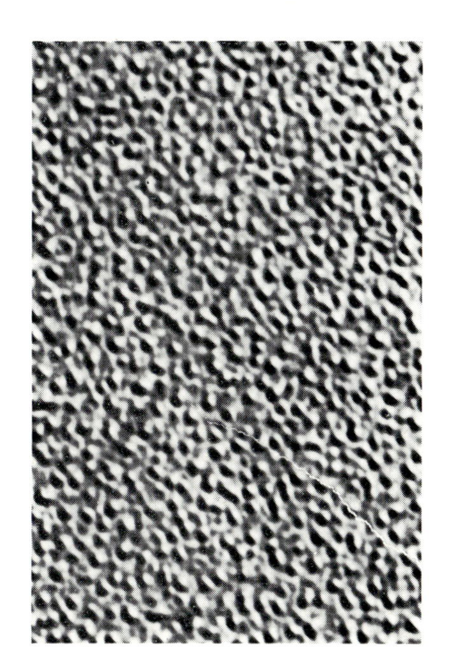

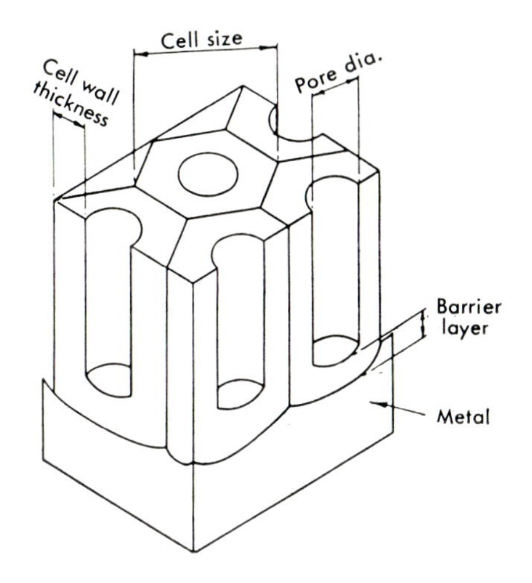

Fig. 13.2. The porous cellular structure of anodic oxide films.

(left) Electron micrograph of a 10 μ aluminium oxide surface anodised at 10 volts sulphuric acid solution. x 200,000.

(right) Microstructure of a porous anodic oxide film[14].

(a) the porous micro-cellular structure, and
(b) an underlying barrier layer.

The chemical and physical properties of the oxide film material are the same in both cases.

The formation of a porous anodic film is indicated during anodising by an initial drop in the current to a minimum, followed by a continual rise to a steady value. The initial current minimum corresponds to the formation of the barrier layer whilst the rise in current to steady state is due to the development of the porous structure.

The pore initiation process does not start until the barrier layer has reached its equilibrium thickness[7, 8]. It has been shown[9] that pore initiation in phosphoric acid electrolytes occurs initially on either side of the sub-grain boundaries. Single rows of cells form at these boundaries where the oxide film is more easily disrupted by the passage of current. The addition of other rows of cells then takes place within the sub-grains until the cell structure is built up. An electron photomicrograph showing the initial cell formation along sub grain boundaries is shown in Fig 13.3.

According to Keller, Hunter and Robinson[9], the porous hexagonal structure of single cells is explained by the solvent action of the electrolyte on the pore base. After formation of the barrier layer, when the current reaches a minimum, the solvent action of the electrolyte thins the oxide at the base of the pores, thus increasing the current and locally raising the temperature. The increase in temperature, which may exceed 15°C[10], results in an increased rate of dissolution of the oxide film so that pores, once started, are perpetuated. In the early stages of growth the cell shape is cylindrical, but as growth continues the shape becomes hexagonal, so that contact is made with the six nearest neighbouring cells.

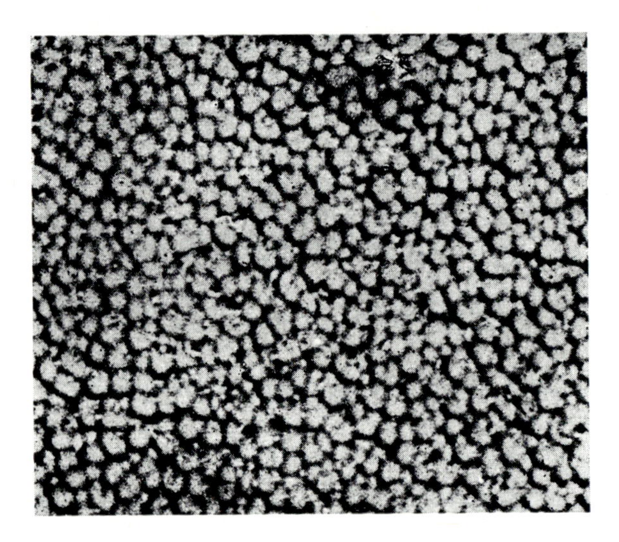

Fig. 13.3. Cell base structure of 30 volt phosphoric acid coating[9]. Electron micrograph. Oxide film replica. × 35,000.

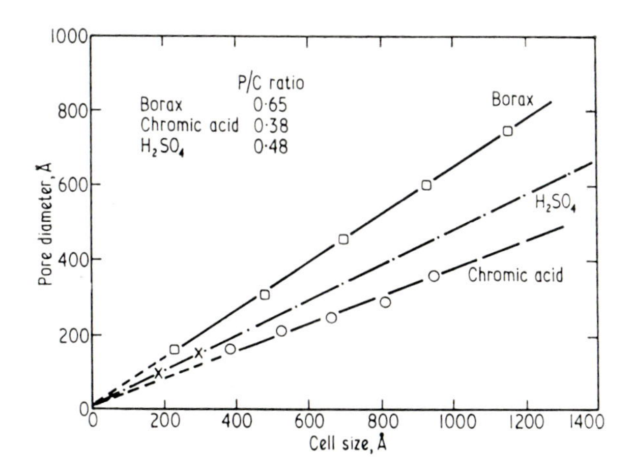

Fig. 13.4. Relationship between pore diameter (P) and cell size (C).[11].

Keller *at al*[9], using electron microscopy, investigated the effect of anodising voltage and concluded that increasing the voltage increased the cell size but did not affect the pore diameter. Great significance has been attached to these results by many workers, since the direct implication is that porosity decreases considerably with increase in anodising voltage. Information of this kind is obviously of importance in estimating the corrosion and dye absorption properties of the film before and after sealing. However, in a recent study, Neufeld and Ali[11] have shown that for any anodising voltage the pore diameter/cell size is a constant value. That is, a plot of pore diameters against the corresponding cell sizes (Fig. 13.4) gave a straight line passing through the origin, showing that the pore diameter and the cell size increase, maintaining the same ratio. Also, since percentage porosity in anodised films can be calculated from the relation:

$$\% \text{ porosity} = 78.5\ (P/C)^2,$$

where P is the pore diameter, C is the cell size, and P/C is constant, it was concluded that porosity is not a function of anodising voltage. In the same investigation it was found that the porosity increases with increasing anodising temperature.

A fuller account of the theories that have been proposed to explain the growth and structure of anodic oxide films is given by Wernick and Pinner[12].

13.1.4.2. Industrial anodising processes

The first industrial application of anodising of aluminium was the process developed by Bengough and Stewart[13] in 1924. It was then applied primarily as a protection for Duralumin alloys against corrosion by sea water. Anodic films produced by the chromic acid process are usually opaque or semi-opaque, and have poorer wear resistance properties than sulphuric acid-anodised films. Although the use of chromic acid has a number of advantages, including its lower corrosivity to aluminium, making it more suitable for components of complex shape (which might entrap anodising electrolyte in crevices, etc.), the sulphuric acid process is generally preferred. The film produced in this process is colourless and transparent on aluminium and on many aluminium alloys not containing manganese and silicon.

The processes involved in industrial anodising practice are shown summarised in the general flow sheet[14] given in Table 13.2. Some of the processes have already been discussed, but others will be dealt with under appropriate section headings.

13.1.4.2.1. Sulphuric acid processes

The sulphuric acid anodic oxidation processes were introduced during the 1930's and 1940's and are now used on a considerably larger scale than other anodising processes. Essentially a solution of sulphuric acid is used as the electrolyte, the concentration employed varying from less than 10 per cent to as high as 75 per cent. Solutions of 6 to 30 per cent concentrations are, however, generally convenient, and maximum solution conductivity is obtained at 30% H_2SO_4, so that electrical power consumption is at a minimum here. Higher concentrations of acid are difficult to control since such solutions absorb water from the atmosphere and become progressively diluted.

Increasing the acid concentration to between 8.5 and 33% results in a reduction in abrasion resistance[15], due in part to the higher porosity of the coating formed. Increasing the temperature of the electrolyte also decreases the abrasion resistance, and also softens[16] the oxide film. Hard deposits are obtained by operating the solution at 0°C, but in normal commercial practice temperatures of between 20-25°C are commonly employed, cooling coils being necessary to prevent the temperature from rising unduly high. British Defence requirements specify the maximum permissible operating temperatures which may be employed with sulphuric acid solutions. The maximum operating temperatures decrease with increase in sulphuric acid concentration, as shown in Fig. 13.5.

Efficient agitation is required to continuously mix hot electrolyte layers near the anodic film with the bulk electrolyte. Local heating of solution near the work results in an increased current density which produces films of inconsistent quality. If there is a choice, it is better to control current density rather than voltage, allowing the voltage to vary to suit a set current value. The film resistance increases during anodising, so that the voltage needs to be increased to maintain a constant current density. The current densities employed are normally in the range 1-2 A/dm^2. The lower current densities produce thin clear films and the higher current densities are used where porous films are required for impregnation with dyes, waxes, etc.

The sulphuric acid process has the disadvantage that it cannot be applied to assembled or riveted parts, as the sulphuric acid electrolyte is corrosive to the aluminium if any traces should be trapped. Very thorough rinsing after treatment is therefore essential.

TABLE 13.2

General Flowsheet for Anodising

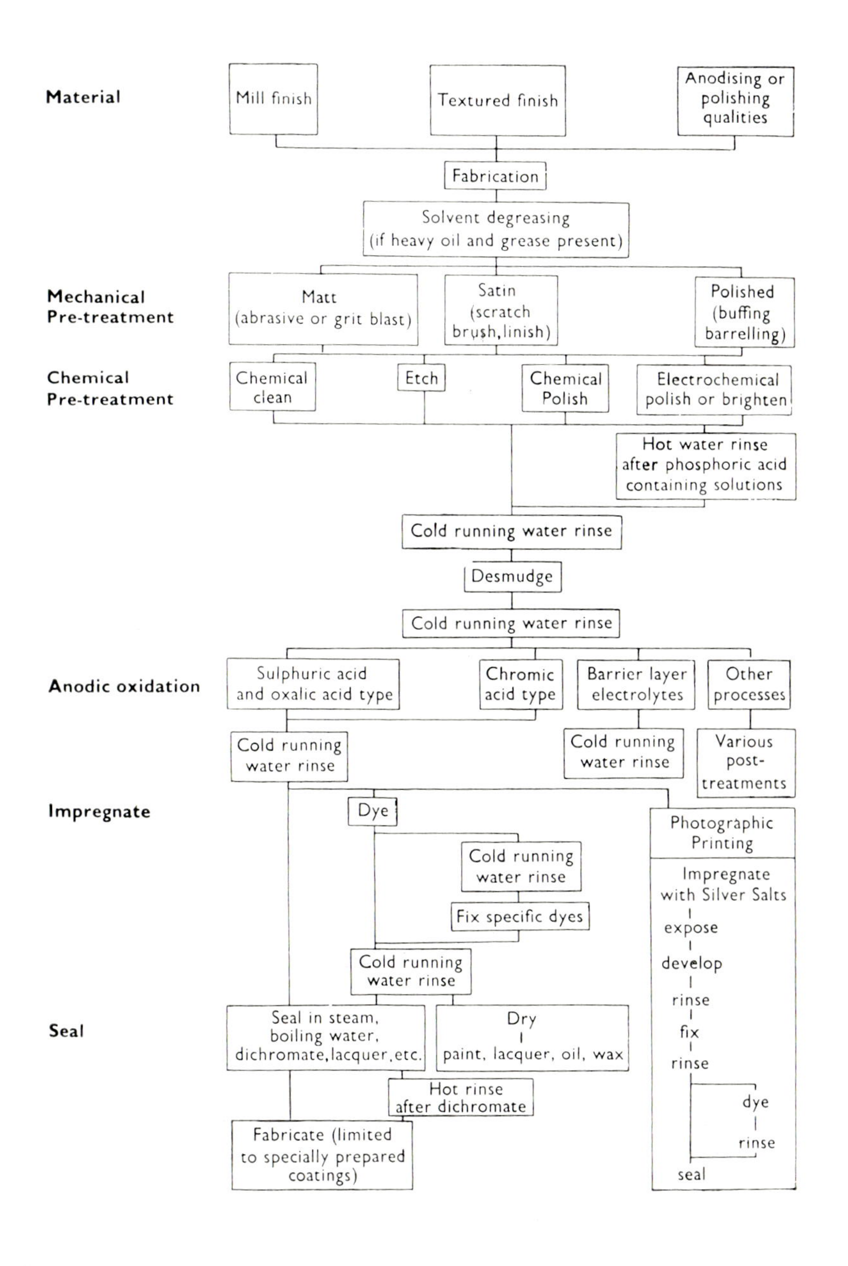

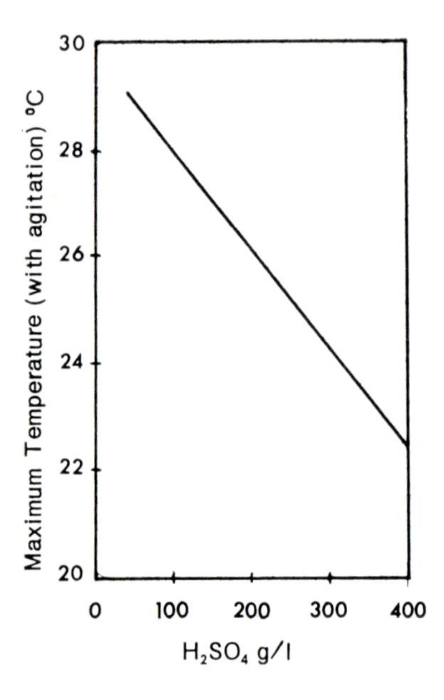

Fig. 13.5. Maximum permissible anodising temperature as a function of H_2SO_4 concentration (D.E.F. 151).

Nevertheless, a proportion of about 15 per cent of sulphate is found in the oxide films even after the most careful washing.

The metal loss is very substantially greater in this process than in the chromic acid method, 0.10 to 0.20 gm/dm^2 being not uncommon in the case of aluminium with perhaps two or three times these amounts when aluminium-magnesium alloys are treated. These losses may be twenty to thirty times greater than those resulting from the chromic acid process[(17)].

13.1.4.2.1.1. Plant

It is normal practice to handle sulphuric acid anodising solutions in either lead- or rubber-lined tanks, but other materials such as PVC and polythene can be used. In lead-lined tanks, the lining itself is sometimes used as the cathode, but care must be taken to ensure that accidental contact between the work and tank sides is avoided. If this occurs, then the work and the lining will be burnt with the possibility of perforating the lining.

Materials that have been used for cathodes in sulphuric acid electrolytes include lead, aluminium and occasionally graphite. Aluminium cathodes give good service providing (a) the tank is in almost continuous use, so that the cathodically polarised aluminium is protected from corrosion and (b) dilute acid solutions are being employed.

Agitation of the electrolyte is provided by air from perforated tubes lying at the bottom of the bath. To ensure that the required degree of agitation is in fact being obtained, a flow meter should be used to measure the rate of air input to the tank.

Some acid spray is liberated during the process, and it is therefore desirable to have extractors fitted to the plant.

13.1.4.2.1.2. Maintenance

During anodising an appreciable amount of aluminium dissolves in the solution and additions of sulphuric acid must be made from time to time or the current density will fall unless the applied voltage is raised. As the aluminium content of the bath increases, further additions of sulphuric acid tend to result in the precipitation of aluminium sulphate, which is only soluble with difficulty. The bath then becomes cloudy and its resistance high. Generally speaking, the bath must be discarded when the aluminium content reaches about 8 g/l. The effect of a falling-off of current when aluminium sheet of 1 dm^2 area is anodised in a solution of 15 per cent by volume of sulphuric acid (sp. gr. 1.83) with a constant voltage of 15 volts and a temperature of 18^0 to 21^0C is shown in Fig. 13.6[18].

13.1.4.2.1.3. Impurities

During the operation of the bath other impurities besides aluminium build up in

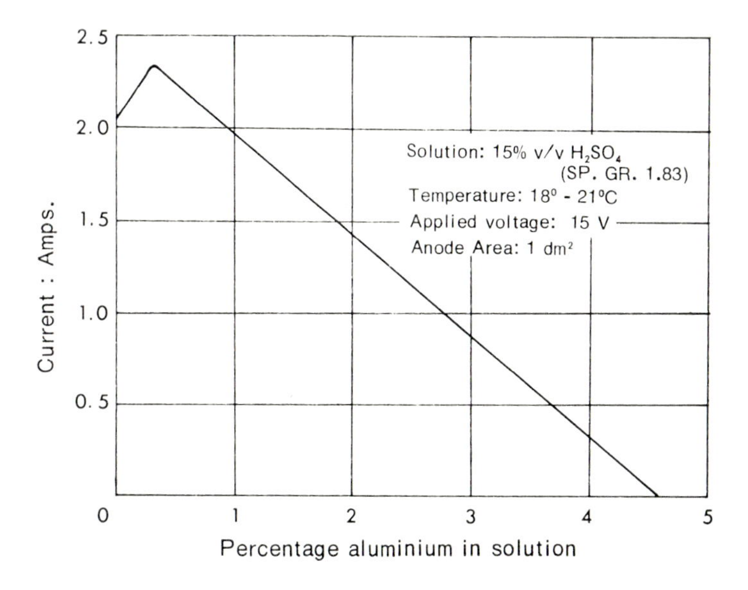

Fig. 13.6. Effect of increasing aluminium content on current passing through sulphuric acid anodising solution.

the solution as various constituents of the alloys being anodised dissolve in it. Silicon tends to accumulate in the form of a light brown suspension and the electrolyte becomes turbid. This may cause slight discoloration of the film, or in worse cases, the silicon may be deposited on the metal surface in the form of a powder which may have to be removed by scouring. Pitted coatings can also result from the presence of suspended matter in the bath.

Copper, when it accumulates in sufficient quantity, is noticeable by the development of a bluish coloration in the solution, while the metal itself is deposited on the cathodes as a spongy deposit which may gradually detach itself and fall to the bottom of the tank or remain suspended in the solution. Although the presence of copper particles is less deleterious to the operation of the process than silicon, it is desirable to remove accumulations of sludge and suspended matter of this type from the bath from time to time either by decantation or filtration. Magnesium or zinc in the solution in small quantity appears to have few adverse effects.

Great care should be taken to avoid contamination of the bath with oil or grease. Particles of oil become strongly adsorbed at the anode surface, resulting in white opaque spots in the films which are not only unsightly but interfere with the subsequent dyeing and sealing of the coating.

In addition to impurities introduced into the bath during anodising, care must be taken to ensure that the sulphuric acid used to make up the solution does not contain high levels of harmful impurities(19). Metals such as Pb, As and Cu should be restricted to concentration levels of the order of 10 p.p.m., although a higher iron concentration can be tolerated (up to 50 p.p.m.).

13.1.4.2.2. The chromic acid process

The chromic acid process has remained essentially unchanged since its introduction in 1924. The parts to be treated are suspended from aluminium wires or on suitable aluminium alloy racks; titanium contact tips are increasingly used as they last longer and are easier to maintain. Copper racks may also be used provided they are insulated. They are made the anode in a tank containing a solution of 3-5 per cent chromic acid in water maintained at a temperature of $40^{0}C \pm 2^{0}C$. The solution should be filtered and gently agitated, and the chromic acid must be substantially free from chlorides or sulphates. A suitable installation for this purpose is shown in Fig. 13.7.

A definite treatment cycle must be followed (Fig. 13.8). An ample source of direct current at about 60V is required, the power requirements being of the order of 0.3-1.0 A/dm^2 of surface being treated, or about 0.02 kilowatt-hours per dm^2. The generator or rectifier should have a means for gradually raising the applied voltage from 0 to 40 over a period of fifteen minutes, after which it is maintained at this value for a further thirty-five minutes. The voltage is then raised once more from 40 to 50 in the following five minutes, and maintained at the latter figure for the last five minutes of the cycle. The total treatment time is therefore one hour. The parts are then removed, rinsed and dried off in hot water.

Fig. 13.7. Installation at Crichley Bros. for the chromic acid anodising of knitting needles. [Courtesy W. Canning & Co. Ltd.]

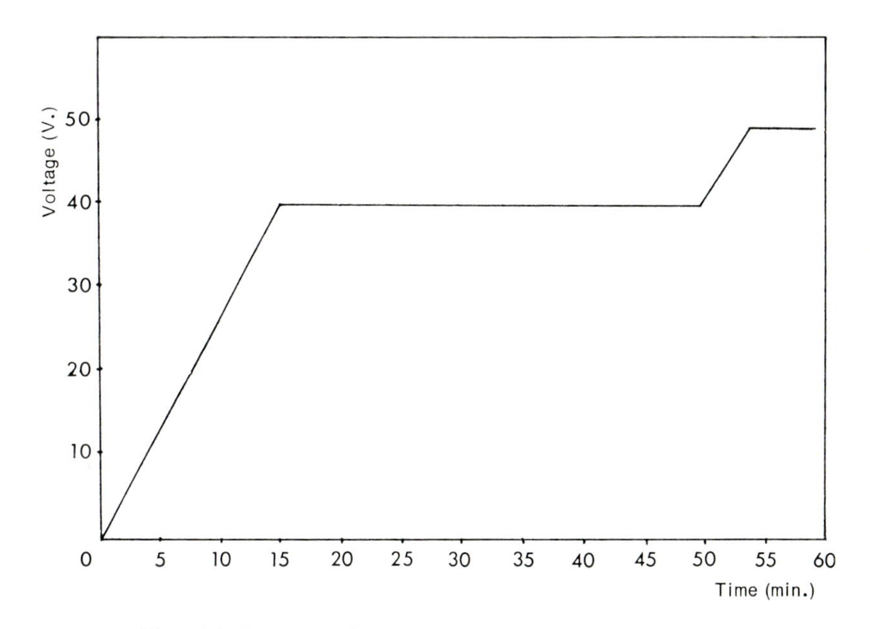

Fig. 13.8. Process cycle for chromic acid anodising treatment.

During anodising the aluminium becomes coated with a dull, semi-opaque layer of aluminium oxide which is relatively hard and has excellent protective properties. The hot-water treatment is an essential part of the process as it serves to seal the pores in the coating, converting some of the oxide to the hydrated form (see p. 474).

The chromic acid content of the electrolyte becomes gradually depleted, but further additions can be made to bring it up to strength. After two such additions, however, the electrolyte should be discarded and a new solution made up, as aluminium chromate accumulates in the bath and inhibits the anodising process. When the voltage is first applied, a relatively high current passes, which gradually falls as the insulating oxide film builds up; as the voltage is stepped up, further passage of current occurs for a short period until finally, after the full voltage has been applied, the entire article is covered with an insulating oxide coat. Coverage takes place over the entire area, even into the interiors of long tubes, since, as the operation proceeds and the oxide coat builds up, the current becomes increasingly concentrated on the uncoated areas. This process has very good throwing power, although in extreme cases it may be advisable to use auxiliary cathodes.

Very good contact must be maintained between the wires or racks and the articles being treated. Imperfect connections can readily form if there is any looseness in the contact owing to the local formation of oxide. The oxide coating must also be stripped from supporting racks after each treatment, caustic soda solution being generally used for this purpose.

13.1.4.2.2.1. Plant

The anodising solution is operated in a steel or suitably lined tank. Cathodes can conveniently consist of stainless steel plates. Steam or some other method of heating the solution is necessary to raise the temperature to the working level in the first instance, but after this, sufficient heat is generated to eliminate the need for further heating. For continuous production, it is essential to have water-cooling coils installed in the tank to enable the operating temperature to be maintained within the specified limits. A small pump and filter unit provide the necessary agitation and filtration of the solution. The articles must be carefully racked to prevent the accumulation of oxygen in recesses, as this gas is liberated during the process; adequate spacing in the tank is also essential. Small parts can be treated in perforated containers which must allow free circulation of the electrolyte combined with firm clamping of the components, sorting is necessary after treatment since a percentage of the parts always remain unanodised due to poor contact.

Owing to the evolution of chromic acid fumes it is essential to have the anodising tank equipped with an exhaust system.

13.1.4.2.2.2. Limitations of the process

The chromic acid process is a versatile method for the anodic oxidation of aluminium and its alloys. The film produced is dense and very thin. Little metal is removed by the treatment, whilst the mechanical properties of most alloys are virtually unaffected. Moreover, owing to the non-corrosive nature of the anodising solution towards aluminium, assemblies and riveted components can be treated with little risk of corrosion due to seepage of electrolyte.

The disadvantages of the process are that it is relatively lengthy (one hour), and the need for controlling the treatment cycle makes it unsuitable for continuous operation. The former disadvantage has been overcome, however, a modified half-hour process having beeen introduced which gives films not very much less protective than those obtained by the original method. In this process the voltage is increased from 0 to 40 volts in the first ten minutes and maintained at the latter value for a further ten minutes; the voltage is finally raised to 50 volts in the next five minutes and kept at this figure for the final five minutes of the cycle.

Casting alloys containing silicon and also the high-copper alloys (above 4 per cent copper) give dark-coloured and somewhat poor films. The high-silicon alloys need very high current densities when anodised in this solution. The appearance of the film produced from the chromic acid bath is not very attractive from the decorative point of view, and cannot be so readily dyed to give a pleasing finish as those produced from the sulphuric acid baths. The chemical composition of the alloy has a decided influence on the appearance of the film, as has already been stated. In addition to this, however, the physical nature of the surface is also important. Thus, abnormal grain size of crystal orientation results in a mottled anodised surface.

Only about 5 per cent of the film is produced in the first fifteen minutes of the treatment cycle, although 12 to 15 per cent of the total energy is consumed during this period. Also, it appears that thicker films are produced from still solutions. The average weight of the film is about 3 g/m^2. This can be determined by removing the film from the aluminium using a stripping solution. Stripping solutions commonly used include phosphoric acid-chromic acid mixtures, sulphuric acid-fluoride solutions, nitric acid-hydrochloric acid mixtures and iodine-methanol solution.

13.1.4.2.2.3. Electrolyte regeneration

As the metal salts build up in the solution the power consumption rises and the quality of the coating deteriorates, so that the bath must be discarded. Regeneration is, however, possible by the use of an ion exchange resin. A portion of the solution is passed through the resin when the heavy metal ions are removed in the form of a strong solution of iron, chromium and aluminium phosphates which are easily disposed of, restoring the chromic acid content of the bath. The quantity of liquor removed for treatment per day is calculated so as to contain the amount of metals introduced during the day's working. The resin is stable in chromic acid up to 12.5% CrO_3 at up to 40°C.

13.1.4.2.2.4. Enamel anodising

Attractive anodic coatings which are opaque and resemble enamel in appearance after dyeing, can be produced by anodising in a 10% chromic acid electrolyte at 54°C and 30 volts[20, 21]. By this means, a film 0.0012 cm thick can be obtained in 20 minutes, and 0.0025 cm thick in one hour.

Films produced in this way have a higher corrosion resistance than those obtained from sulphuric acid electrolytes. The coatings are also ductile and articles can be formed from treated sheet without crazing of the coating. They are much more readily dyed than the normal chromic acid anodic film as obtained by the Bengough process. This is probably due to the fact that the film is about twice as porous as the latter, whilst the thickness, being also about two to three times as great, also serves to increase the density of colour.

The capacity of the film, which consists of pure aluminium oxide, is due primarily to the etching of the underlying metal during processing; the film itself is transparent when detached. Films of similar appearance can be produced by etching aluminium in ammonium bifluoride and then anodising in sulphuric acid, which is known to give a transparent film.

13.1.4.2.3. Hard anodising

In order to obtain very hard anodic oxide films electrolytes are operated at temperatures below 10°C, necessitating the use of refrigeration equipment. Several electrolytes have been used for hard anodising but sulphuric acid solutions give good results. For most purposes, low-concentration sulphuric acid electrolytes can be used for hard anodising, but difficulties arise in the anodising of Duralumin and some other alloys. Duralumin can be successfully anodised if higher concentration solutions are used. Moroz *et al*[22] favours the use of solutions containing 200-250 g/l H_2SO_4. However, the high H_2SO_4 concentration and low temperature results in a decreased solubility of aluminium salts, so that very careful control of the amount of aluminium dissolved in the bath is essential to prevent precipitation of aluminium compounds.

Hard anodised surfaces, when treated with a suitable solid phase lubricant, can give exceptional wear life together with a very low coefficient of friction. A further advantage is that the hard anodic film can be run for some time unlubricated with little wear, so that a safety factor is incorporated should the solid lubricant wear away.*

13.1.4.2.4. Other anodising processes

Other electrolytes have found some application for the anodic oxidation of aluminium, but of these the only one of commercial importance is the oxalic acid process. Anodic oxide films having a natural gold or straw tint can be produced in oxalic acid solutions in sixty minutes, the temperature of the bath being 25° to 35°C. A constant voltage of 50 to 60 volts is used, and either direct or alternating current can be employed. When

*Ministry of Technology, Techlink No. 318 (1969).

alternating current is used the films tend to be yellow or dark, but light straw-coloured coatings are obtained with direct current. The current consumption is somewhat high, being of the order of 3 to 5 A/dm^2.

The solution is employed in lead- or rubber-lined tanks which must be provided with an exhaust system. After treatment the work is rinsed in cold water, followed by a 1 per cent solution of ammonia and further rinsing if there are any joints or crevices.

Gentle agitation of the electrolyte should be carried out, and cooling coils are also required. When using alternating current, process work can be hung on both the anode and cathode rods, but for direct-current treatment separate cathodes, which may be of carbon, lead or stainless steel, are required.

13.1.4.3. Properties of anodic oxide films

The characteristics of anodised films depend to a considerable extent on the methods used for their production. The films obtained by the sulphuric acid processes are colourless and transparent, as has already been stated, but are harder than those from the chromic acid bath. They are also rather thicker and more absorbent prior to sealing, and can therefore take up dyes more readily. Generally, thicker films are less flexible and will withstand less deformation without crazing than thin films. Anodic films are insulating, thick films having naturally higher dielectric properties than thin ones; the water content of the film probably plays a greater part in determining its dielectric properties than the actual method of producing it. Breakdown voltages of up to 1000V are obtainable. A number of patents have been taken out for the production of films of high dielectric strength for use in the manufacture of condensers[23]. Anodic oxide films provide an excellent base for paint, anodised and painted aluminium surfaces having a high degree of weather resistance. A comprehensive review of the chemical, physical and mechanical properties of anodised films is given by Brace[24].

13.1.4.4. Dyeing of anodic oxide films

Anodised articles must be thoroughly rinsed in cold running water after treatment, and in the case of the sulphuric acid process a neutralising rinse containing about 1 per cent of ammonia is often employed. After a further rinse, the articles can be immersed in a suitable solution of dyestuff when, owing to the mordanting action of the freshly-formed aluminium oxide film, permanent coloration can be obtained. The anodic film must not be allowed to dry before transfer to the dye baths. It is desirable to have relatively thick and colourless films in order that clear bright colours should be produced, and for this reason practically all articles to be dyed are anodised by the sulphuric acid process. Also, films produced from the latter bath are more translucent and abrasion-resistant than those produced from the chromic acid solution, and are therefore more suitable for colouring. Where dark colours or yellow, however, are concerned, films from the chromic acid bath can be successfully dyed. Generally speaking, a high degree of freedom from porosity is necessary where castings are to be dyed, since any pores may lead to local oxidation by retained traces of the anodising electrolyte during treatment in the dye solution, with consequent local weakening or even removal of the anodic film. This results in white spots being left on the dyed surface. Casting alloys containing 5 to 13 per cent of silicon are only suitable for dyeing with dark colours, owing to the fact that the films themselves tend to lack whiteness and translucency.

13.1.4.4.1. Dyes

Substances used for colouring anodic films can be divided into two main categories; organic dyestuffs and inorganic pigments.

A list of some commercial dyes having adequate fastness is given in Table 13.3, the light fastness of full shades on sulphuric acid film being on the scale of 1 to 8. Dyestuffs with a light fastness of less than six cannot be considered satisfactory.

TABLE 13.3

The Colour Index numbers and Light Fastness of some commercial dyestuffs

Dyestuff	Bright Colour Index No:	Light Fastness
Alimax Yellow 2G	—	7 to 8
Alimax Orange GR	274	7 to 8
Solochrome Brown RG.200	—	7
Alimax Red A	1,034	6
Alimax Fuchsine R	—	8
Solway Blue SE.200	1,053	8
Alimax Blue BGN	1,053	8
Alimax Green G	1,078	7 to 8
Acid Dyes		
Alimax Yellow N	640	6 to 7
Croseine Scarlet 2 BS	252	6
Alizarin Sapphire Blue CB	1,054	8
Alimax Green BN	6	8
Acid Black G	247	–
Alimax Black 2Y	—	8

Among the most useful dyes are those of the direct acid type; they must have good fastness to light. Only a relatively small number of colours have been found acceptable for this application, black, blues, reds, greens and yellows being the most satisfactory. Among suitable acid dyes are the Solway Blues, Naphthalene Scarlet and Fast Yellow. Alizarin Green and Red and Solochrome Yellow have also been recommended. Nigrosine is widely used for dyeing the film black, the time of treatment being about half an hour.

The dyes are employed in the form of relatively strong solutions (0.1 to 0.5 per cent), the temperature of the dye bath being kept at 60° to 90°C, depending on the nature of the dye. Temperatures in excess of 75°C introduce a weak sealing effect, which slows down the dyeing rate.

In making up the dye solution it is important to ensure that no undissolved particles remain, or spotted and patchy colours will result. Most dyestuffs have an optimum pH range within which the best results are obtained, so that it is desirable to control the pH of the bath within specified limits. Excessive drag-out of acid into the dye bath must be avoided, as it can cause flocculation of the dyestuff. The dye bath should be kept in movement preferably by air agitation, and surface scum removed from time to time. The time of treatment varies, but must be sufficiently prolonged to ensure complete penetration of the anodic film; usually twenty to thirty minutes are required for good impregnation.

Primary colours can be obtained by the use of single dyestuffs but other shades can be produced by using mixed dyestuffs in the same solutions (provided they do not interact with one another) or by treatment in successive baths of different colours, rinsing after each immersion.

Cold dyeing can also be carried out, the colours being more readily controlled. It is, however, necessary to carry out sealing in steam to prevent leaching of the colour.

The affinity of the aluminium for dyestuffs can be increased by immersing the article immediately after anodising in a solution of 5 cc of sulphuric acid per litre of water at 80°C for one minute. The process enables the concentration of dyestuff to be reduced, so that cold dyeing can be carried out more easily. The time of treatment is also somewhat less. The method is also useful if dyeing is to be carried out some time after anodising.

Ferric ammonium oxalate is one of the most widely used inorganic pigments[25]. Anodic films immersed in solutions of this compound can take up a range of colours varying from red to pale brass, depending upon the operating conditions. The palest colours are obtained with small concentrations of the pigment and low solution temperatures. The pH of the solutions should be controlled at about 4.5; if it exceeds 5, there will be a tendency for the precipitation of hydroxides. Anodic films may also be coloured with inorganic pigments by means of a double dipping technique, causing the precipitation of insoluble coloured salts within the pores before sealing[26]. A bright yellow, for example, is obtained by immersion in a 5 per cent potassium dichromate solution, rinsing, and then immersing again in 5 per cent lead acetate bath so that lead dichromate is precipitated. Cobalt acetate and potassium permanganate solutions give a bronze precipitate. Red is obtainable by precipitating lead permanganate or silver chromate in the film, whilst potassium ferrocyanide gives blue colours with iron sulphides. Some of these colours, notably the chromates, have a high degree of resistance to fading, but others are less satisfactory in this respect.

Dyestuffs are susceptible to metals, particularly copper and iron, so that the dye container should be made of a material that will not interfere with the dyeing process. Stainless steel is very satisfactory, but vitreous enamelled iron and Monel tanks have also been successfully employed. Equal attention must be given to the material of which heating coils are constructed. The articles being treated should not be allowed to come into contact with the walls of the vessel during dyeing.

The alloying constituents present in the aluminium play an important part in the colour obtained. Pure aluminium gives the brightest shades of colour; the presence of copper (as in alloys of the Duralumin type) results in duller and somewhat darker shades. Casting alloys containing 12 to 14 per cent of silicon are best dyed black as the anodic film itself is dark grey in colour and rather thin. With lower silicon contents (up to 5 per cent) other colours can be obtained, but they are also rather dull. The effect of magnesium in small amounts is negligible, but when more than about 3 to 4 per cent is present the shades obtained are darker. Manganese-containing alloys tend to give colours which are of low intensity and somewhat dull.

13.1.4.4.2. Intrinsically coloured films

Completely light-fast colours can be obtained by anodising alloys containing specific constituents which give coloured coatings without the use of dyestuffs. Such constituents include 3-5 per cent of silicon and intermetallic compounds (e.g. Al_3Fe and Al_6Mn). The presence of manganese results in a grey to black coating. By the use of electrolytes containing organic acids as well as sulphuric acid, integral coloured anodising can be performed on pure aluminium as well as on alloys. Light and medium brown films are obtained by the use of 8 per cent oxalic acid in the Alcanodox process. The Kalcolor process, using sulphosalicylic acid and sulphuric acid, produces coatings ranging from brown to black, whilst the Duranodic process, based on sulphophthalic acid, produces similar colours. Other processes use dibasic acids such as maleic and tartaric acids in conjunction with oxalic and sulphuric acids (Veroxal and Permalux processes).

13.1.4.4.3. Multi-colour effects

Attractive effects can be obtained by the combination of two or more colours. Thus, by the use of dilute nitric acid and an acid-resisting mask it is possible to remove the colour from a portion of a dyed article, after which it can be re-dyed locally with another colour or left uncoloured. For this process to be successfully carried out, the first dye impregnation should be a relatively short one, or decolorisation may be difficult. Another procedure is to protect predetermined areas from the dye by impregnation or printing with an oil or with a non-absorbed lithographic printing ink. The protective agent is applied from a rubber pattern or by means of an offset press, being later removed. A still more satisfactory local 'stop-off' is black cellulose lacquer, which prevents the dye from being absorbed at the areas protected by it. The lacquer can then be readily removed later by means of organic solvents.

Marbled or angular-patterned effects can be produced by mixing separate dyestuffs with glycerin and applying these solutions haphazardly or by means of a patterned rubber roller on to the surface of the anodised dried article. On lightly spraying with water the colours are locally absorbed, after which the glycerin is removed by heating in an oven and the residual dry dye removed by rinsing rapidly in running water. Sealing is then carried out in the usual way. Multiple colours can also be produced by applying litho-compound to the areas to be protected through a silk screen cut out to the required design. After allowing the ink to dry, the unprotected parts of the plate are bleached by means of dilute nitric acid, and the palest colour applied from a suitable bath. By further application of the stop-off followed by bleaching, four or five different colours can be applied in this way and attractive patterns built up. Blocks can be used instead of silk screens where large quantities of the same pattern are to be produced.

13.1.4.4.4. Theory of dyeing

According to Giles[27] the first step in the dyeing process is the etching of the surface layer of the aluminium oxide to increase the surface area. Rapid absorption of dyestuffs then occurs in the external interface of the coating followed by diffusion along the pore channels. This last diffusion step is thought to be slow and therefore responsible for the longish process times needed for dyeing. Although all dyes probably diffuse into the coating in this manner, it seems fairly clear that different classes of organic dyes are held or bonded to the aluminium oxide surface in different ways.

Speiser[28] considers that the acid and substantive dyestuffs are only physically adsorbed in the film and that this is the reason why their fastness to light is poor. On the other hand, dyestuffs which have good fastness are considered to be chemically incorporated into the coating. Alizarin Bordeaux R forms a lake with the aluminium oxide:

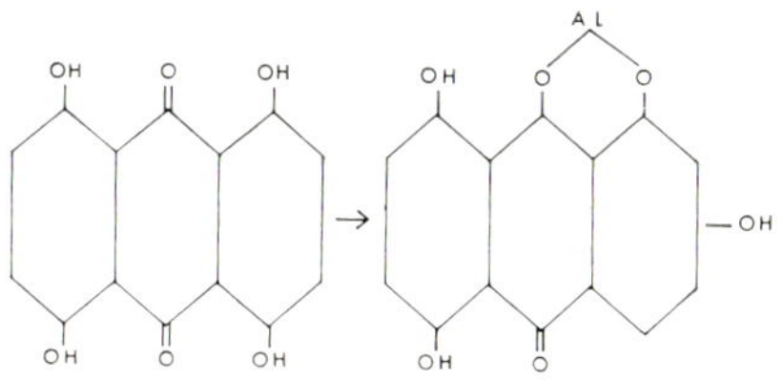

Metal complexing acid dyes such as Chrome Fast Orange R are probably fixed in the oxide film by the formation of an aluminium-bearing complex, but subsequent complex formation with metal cations contained in the sealing solutions might also be involved.

Detailed information concerning the action of a variety of organic dyestuffs is given by Giles and School[27, 29–32], Speiser[28] and Cutroni[33].

13.1.4.4.5. Photographic processes

The anodic film on aluminium can also be impregnated with a photo-sensitive gelatine emulsion[34]. Photographs or diagrams can be printed on to the metal surface by exposing, followed by developing and fixing of the image by methods similar to those employed in normal photographic practice. The image can be toned by means of gold or platinum salts, if necessary, after which the finished article is buffed, producing a hard, glossy panel. In this manner[35], calculating rules, trade-marks as well as instruction and identification plates for motors and appliances can be reproduced accurately and rapidly. The hardness and corrosion resistance of the coating makes such reproductions very durable and wear-resistant after sealing; they will also resist the action of organic solvents, and withstand remarkably high temperatures. Coloured reproductions can also be obtained by dyeing unexposed portions of the surface.

13.1.4.5. Sealing of anodic films

Whichever anodising process is employed, it is essential to seal the film. This applies particularly after any dyeing or photographic process has been carried out on the anodic coating. The sealing treatment is applied after soluble constituents have been removed by rinsing in cold water. Prior to sealing, the parts must not be handled or allowed to becomes contaminated in any way, or the final finish will be patchy and irregular.

The simplest method of sealing consists in immersing the anodised articles in boiling water for twenty to thirty minutes. At this temperature alumina combines with water to form the monohydrate, $Al_2O_3.H_2O$, and sealing occurs by virtue of the expansion that takes place on hydration. At low temperatures the hydrate is not formed and water is held mechanically in the pores of the film.

According to Hoar and Wood[36] pores are sealed by inward movement from their sides whilst they are also plugged at the surface. This process is shown schematically in Fig. 13.9.

The amount of silica present in water used for sealing must be very carefully con-

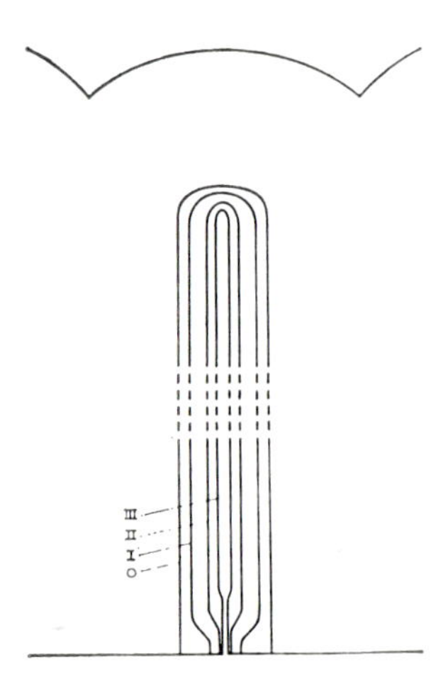

Fig. 13.9. Schematic illustration of the sealing process. O is the wall of the unsealed pore while I, II and III represent intermediate stages of sealing[36].

trolled. Scot[37] investigated the effect of silica on sealing and reported that more than 20 p.p.m. of silica adversely affects the sealing process. However, as reported by Sheasby[38], the amount of silica that can be tolerated is dependant on the pH of the sealing water. The results obtained by Sheasby are given in Table 13.4. Weight losses higher than 0.20 mg/cm^2 are indicative of unsatisfactory sealing.

TABLE 13.4

Effect of pH and silica content of sealing water on weight loss (mg/cm^2) in Kape test (see p.476) on Al-Mg-Si alloy anodised to 25 microns[38]

SiO_2 p.p.m.	pH 5.0	pH 6.0	pH 7.0
0	0.22	0.07	0.05
2.5	0.80	—	0.02
5	1.09	—	0.04
10	1.22	0.13	0.04
20	1.22	0.19	0.09

13.1.4.5.1. Salt sealing

Another method of sealing which is widely used is termed salt sealing and consists of immersing the articles in solutions of weakly dissociated salts, such as nickel or cobalt acetate. The adsorbed salts are hydrolysed and a finely divided hydroxide of the heavy metal is deposited in the film. The amount of hydroxide is so small that no coloration results; hence, the process is applicable to dyed or colourless anodic finishes. A suitable solution for sealing contains:

cobalt acetate	1 g/l
nickel acetate	5.6 g/l
boric acid	8.5 g/l

A pH of between 4.5 and 6.5 at 80°C is best, and a degree of turbidity of the solution should be evident on boiling.

In the dichromate method of sealing, the articles are immersed for ten to fifteen minutes in a boiling solution of 5 per cent sodium dichromate. This method has the disadvantage of staining the film a brownish yellow, and should therefore only be used on yellow or orange dyestuffs, but has the effect of imparting very high corrosion resistance to the coating. In chromate sealing, the chromate is first adsorbed and converted into aluminium oxychromate or oxydichromate, which in turn becomes hydrated. The pH of a dichromate solution is 3.7, but as this is raised the rate of hydration increases, and hence the speed of sealing. For this reason the sealing solution is best operated at a pH of between 6 and 7 when not only is the sealing time substantially reduced, but the effectiveness of the seal is greater[39]. Excellent results can be obtained with a sealing time of as little as four minutes.

Another method of sealing is by immersion in a solution of sodium silicate followed by treatment in weak acetic acid and thorough rinsing. In this way silicic acid is deposited in the pores of the film, the sodium acetate resulting from the interaction of the silicate and the acid being removed by the washing operation. The method is especially suitable where castings having some degree of porosity have to be treated. It has the disadvantage, however, of tending to leave the surface somewhat uneven in appearance. It has been shown by Whitby[40] that the silicate used should possess a molecular ratio of SiO_2 to Na_2O of not less than 2.65:1, the optimum ratio being 3.3:1; excessively alkaline silicates attack the oxide film. The sodium silicate sealing treatment gives a protective value of the same order of efficiency on aluminium-magnesium-type alloys as that resulting from sealing in nickel or cobalt acetate solutions, i.e. considerably better than that obtained by sealing in boiling water or steam but inferior to the protection obtained by dichromate sealing. After sealing, a powdery residue often appears on the surface of the metal. This can be removed by light polishing with chalk powder before being finally rinsed and dried.

A simple method of testing for the efficiency of sealing has been developed by Kape[41]. It consists in immersing the specimen in a solution of 10 g/l of anhydrous sodium sulphate adjusted to pH 3.75 with glacial acetic acid and then to pH 2.5 with 5 N sulphuric acid for about 10 minutes at 85°C. The resulting bloom, if any, gives an indication of the quality of the sealing obtained. Poor or partially sealed pores develop heavy or light bloom during the test, but properly sealed articles are unaffected.

A variation of this method consists in exposing the specimens to an atmosphere containing sulphur dioxide at a temperature slightly above ambient for twenty four hours[42]. The degree of white bloom and the amount of deposit formed during the test period decrease as the effectiveness of the sealing increases.

Marran and Wood[43] advocate an electrical sealing test. The basis of the test is to measure series resistance (R) and series capacitance (C) from which the impedance (Z) of the anodic oxide film can be obtained from the vector expression:

$$Z = R + \frac{1}{JWC}$$

where W is the angular frequency of the current. The test involves the use of a small cell containing an auxiliary electrode and filled with a salt solution. Measurements are thus easily made with a resistance and capacitance bridge. For industrial purposes a 'go-no go' impedance figure is used to identify components which will later show poor corrosion resistance. It is claimed that a test of this nature gives results in agreement with those from a C.A.S.S. (copper accelerated-acetic acid-salt spray) test of 6 hours.

13.1.4.5.2. Sealing by impregnation

Finally, sealing of the film can be carried out by impregnation with oils or waxes.

Prior to this treatment, thorough drying at a low temperature (50°C to 60°) must be carried out to avoid patchy results. After this, the article is immersed in linseed oil or in solutions of wax, lanolin, or petroleum jelly in a suitable solvent such as naphtha or white spirit. Reasonably good results are obtained particularly if drying oils, such as linseed or tung oil, are employed. This method of sealing is comparatively little used nowadays, except in some instances where dyed films are concerned.

In special cases as, for instance, when aluminium engine pistons are anodised for wear resistance, sealing is not carried out, as it tends to soften the film to some degree and because the absorbent qualities are required for oil retention. For this purpose impregnation with colloidal graphite preparations in oil are very useful.

In producing reflectors for electric heaters, it is recommended that the metal be heated to 200°C after treatment to drive off absorbed water; following this, the coating may be impregnated with a suitable wax. After sealing, the parts can be lightly polished with a swansdown mop. In this way a brilliant clear finish results.

13.1.4.6. Automation in anodising

Numerous plants have been constructed for anodising. Parts can be cleaned in alkali, chemically brightened, anodised, dyed and sealed during the automatic process cycle. One of the problems that must be dealt with is cooling, since considerable heat is generated during anodising. To maintain uniformity of the coating it is essential that this heat should be quickly removed, usually be refrigeration. Agitation is also employed to maintain a uniform electrolyte temperature and to prevent the surface temperature of the work from rising.

The thickness of anodic coating formed in any given anodising time is directly proportional to the current density. The current density is influenced by the temperature, the concentration of the electrolyte and the aluminium content. These latter factors are bound to vary with the amount of work processed, and it is desirable to use some system of constant current density control as described on page 479. The acid concentration and aluminium content should be checked twice daily and maintained within suitable limits.

When dyeing of the film is carried out, consistency of colour is facilitated by the closer control possible in an automatic plant. One problem is that there is always a tendency to design plants so that the anodising, dyeing and sealing times are identical. In the case of a 10-minute anodising time this can give rise to difficulties, since most dyes are absorbed rapidly in the first five minutes and thereafter more slowly. While some dyes reach saturation in 10 minutes, many require 15 to 20 minutes (black 30 minutes) to give maximum absorption. If the dyeing time is too short the light fastness of the dye will be adversely affected and it will be more likely to give rise to colour variations due to minor changes in operating conditions.

A large automated anodising plant installed at the Bank Quay, Warrington, plant of the British Aluminium Co. Ltd. is capable of dealing with extrusions or tubes up to 28 feet (8.53 m) in length and occupies an area of 10,300 sq. ft. (957 m^2). Automatic effluent treatment is provided to deal with up to 20,000 gallons (91,000 l) per hour of rinses from the anodising operations.

The tank sequence and details of handling are as follows:[44]

1. *Hot Alkaline clean.* A welded mild steel tank, 9m×1.8m×0.9m, contains an inhibited alkaline solution, thermostatically controlled at 65°C.
2. *Cold rinse*
3. *Hot alkaline etch* maintained at 40°C with fume and steam extraction.
4. *Cold rinse*

5. 6. and 7. *Desmut and cold rinse tanks*

8. and 9. *Anodising.* Two rubber-lined tanks, each 1.2m wide, hold an air-agitated solution of 7½% sulphuric acid. Refrigeration equipment with a capacity of 164 kW keeps the bath at 20°C.

10. and 11. *Cold rinse*

12. *Dyeing.* Two stainless steel agitation coils are installed in a stainless steel tank with bath temperature thermostatically controlled at 60°C.
13. *Boiling seal.* This final process is carried out in boiling deionised water in a 1.5m stainless steel tank.

Effluent overflow is run into an underground divided tank for treatment with automatically dispensed sodium carbonate additions. A proportional plus integral controller

Fig. 13.10. Large chromic acid anodising installation at Hawker Siddeley Aviation for the anodising of aircraft wing panels. [Courtesy W. Canning & Co. Ltd.]

unit is employed for controlling the amount dispensed. pH signals from a pair of electrodes mounted in the neutralisation stage are relayed via an amplifier to the control unit. From there, signals are sent to a motorised proportioning valve which dispenses sodium carbonate into a recirculation loop. If the pH decreases then the valve opens to an extent governed by the difference between the actual and desired pH.

13.1.4.6.1. Automatic current density control in anodising

The anodising current density can be controlled automatically and suitable equipment is available for this purpose.

Reference for the control systems used is obtained from the anodising tank by a small probe, or probes, suspended in the acid solution. These probes are accurately made to a known surface area and are usually machined from tantalum, this having been found to be the most suitable for anodising plants operating with sulphuric acid at concentrations of between 6 and 10 per cent.

In order to make the complete anodising cycle entirely automatic, the auto-control may be supplemented by a smaller unit incorporating a timer and current density meter. Current passing through the probe is registered by this meter in amps/unit area. The current density previously found to be the most suitable for the alloy and quality of anodising required is set on the meter by means of a moveable contact, and the required process time to give a pre-determined anodic coating thickness is governed by a timer. The process is started by a press button switch which activates both the timer and the motor-driven regulator, bringing up the current until the required density is attained.

The delay between operation and attainment of the pre-set current density varies from plant to plant, but generally this has been found to be most suitable at approximately 60 to 90 seconds. When the current density has reached the pre-determined level, it is held constant by the control mechanism until the process time has elapsed. At this point a relay is operated which can sound a warning bell or switch on a warning light, either of these indicating to the operator that the process is complete. The unit then automatically cuts off the current and the timer resets itself to zero.

Control of the current density is normally available between 0.4 and 4 A/dm^2, whilst provision is made for the process to be controlled, if required, at a pre-set constant voltage, as well as at constant current density. A motor-driven regulator or variable transformer is necessary to enable the equipment to be installed on existing plant.

13.1.4.6.2. Continuous anodising of wire and strip

The problem of anodising wire and strip presents special difficulties, and custom-designed plant has been developed for this purpose. In the continuous anodising of wire the current that can be employed is limited by the fusing current of the wire itself. In the case of 18 s.w.g. (1.219 mm dia) wire in air, this is about 80 amps. Also, sparking and burning of the wire are liable to occur at the points of contact. These problems may be overcome by feeding the current to the wire via the electrolyte in a chamber when it passes to the main anodising tank, the contruction being such that the wire remains immersed continuously from its point of entry to its final exit. Means must be provided for circulating and cooling the electrolyte and for coiling and uncoiling. Wire can thus be treated at the rate of 600 to 900 metres per hour in sulphuric acid electrolytes.

Similar considerations apply to the treatment of foil and strip, but in this case the rate of treatment is lower, i.e. of the order of 150 metres per hour.

A detailed account of the technology underlying modern strip and wire anodising techniques has been given by Barkman[45].

13.1.5. CHEMICAL CONVERSION COATINGS

In many instances the anodising process is unnecessarily expensive and lengthy when the surfaces concerned are subsequently to be painted. In such cases, simple immersion treatments can be employed to produce coatings which, although of less protective value than anodised films, are valuable either for improving the appearance of weather resistance of the aluminium surface or for increasing the adhesion and durability of paint films.

The chromate-oxide or chromate-phosphate coatings are primarily used for decorative or anti-glare treatments for aluminium, whilst the chromate-fluoride films are mainly used as bases for paint[46].

The available processes for treating aluminium are summarised in Table 13.5. The coatings can be applied by dipping or by spraying with chemical solutions which serve to increase the thickness of the natural oxide film or to produce a surface oxide film of a complex nature containing chromates or phosphates. The most important coatings may be grouped by colour. *Grey* coatings are produced by the alkaline chromate solutions which give chromate-oxide films. *Green* coatings result from chromate-phosphate type processes, the films consisting chiefly of amorphous phosphates. *Brown* coatings are produced in acid chromate baths, the films obtained from these consisting of fluoride-chromate mixtures. *Light grey* crystalline phosphate films are formed by the zinc phosphate processes.

13.1.5.1. Chromate-oxide coatings

In the MBV (Modified Baur-Vogel) process, an alkaline chromate solution of the following composition is used:

Na_2CrO_4	1-3% w/v
Na_2CO_3	5% w/v

The solution is prepared with deionised water. The bath is operated at a temperature of 90⁰ to 100⁰C and the time of treatment is three to fifteen minutes. A uniform, somewhat soft and velvety film is produced, varying from light grey to almost black in colour, depending on the composition of the alloy. It is desirable to clean the articles prior to

TABLE 13.5

Chemical conversion processes for aluminium

Process	Main constituents	Operating conditions	Coating weight mg/dm^2
Chromate-oxide (MBV, E.W., Pylumin)	Alkali metal chromate/carbonate	3-15 min at 92-100°C immersion (unsuitable for spray application)	5-50
Chromate-fluoride	Acid chromate/fluoride	$\frac{1}{2}$-9 min at 15-55°C. Spray or immersion	3-15
Chromate-phosphate	H_3PO_4/chromate/ fluoride	1-5 min, 25-60°C. Spray or immersion	5-100
Zinc phosphate	Zinc phosphate/ fluoride	4-10 min immersion at 55-65°C. 1-2 min spray at 43-50°C	5-120
Bohmite	Distilled or deionised water.	30-90 min at 100°C	5-15

treatment, and they are finally dried off in boiling water. This last treatment lightens the colour of the film and helps to improve its durability. The standard process is not suitable for copper-containing alloys and aluminium-magnesium type alloys which give powdery coatings, while in the case of heat-treated cast alloys it is desirable to remove the oxide film on the surface before treatment, either by shot-blasting or by means of caustic soda. The film thickness produced by the M.B.V. process is considerable and appears to increase fairly regularly with time of treatment. A coating of approximately 0.0001 cm thickness can be produced in five minutes, whilst thicknesses of about 0.0006 cm are formed in one hour. The coatings consist mainly of a mixture of aluminium and chromium oxide (Cr_2O_3).

The first step in the formation of the mixed oxide film is the anodic dissolution of aluminium together with a Faradaically equivalent production of hydrogen gas:

$$2\ Al \rightarrow 2\ Al^{3+} + 6e$$
$$6H_2O + 6e \rightarrow 6OH^- + 3H_2$$

The trivalent aluminium ions are then considered[47] to react with hydroxyl ions to form hydrated aluminium oxide:

$$2Al + 6OH^- \rightarrow Al_2O_3.nH_2O + (3\text{-}n)\ H_2O$$

Chromium is incorporated into this coating by reduction of chromate, probably according to the following reaction:

$$2CrO_4^{2-} + 3H_2 \rightarrow Cr_2O_3 + 4OH^- + H_2O$$

Many modifications of the M.B.V. process have been introduced, perhaps the most important being the Erftwerk[48] and Pylumin[49] processes. The Erftwerk process is similar to the M.B.V. treatment but the rate of coating formation is reduced by addition of fluoride or silicate. The coatings produced are almost transparent, and their low porosity confers greater corrosion resistance. However, the lower porosity reduces the effectiveness as a paint bond coating.

The Pylumin process has been widely used in the U.K. since its introduction in 1936. The bath contains a heavy metal chromate, such as chromium, which makes the process easier to control and good results can be obtained even on the high-strength aluminium and copper alloys.

The paint-holding value of the Pylumin process is shown in Fig. 13.11 in which two sheets of aluminium, one of which has been Pyluminised and the other untreated, have been sprayed with one coat of lacquer and subjected to an indentation test.

The M.B.V.-type films can be dyed, but organic colours have poor durability even indoors. Inorganic dyeing is sometimes carried out for identification of parts, for example, but the water fastness is poor after dyeing; the film is sealed in boiling water. Yellow films are obtained by a dichromate seal followed by ammonium sulphide or silver nitrate whilst yellowish-brown to black colours can be produced by treatment in 1 per cent potassium permanganate containing 0.3 per cent of nitric acid and 1-2 per cent of copper or cobalt nitrate.

13.1.5.2. Acid-chromate process

Acid-chromate processes are mostly proprietary and give yellow to golden brown films 0.0002-0.0008 cm thick consisting of mixtures of trivalent and hexavalent chromic

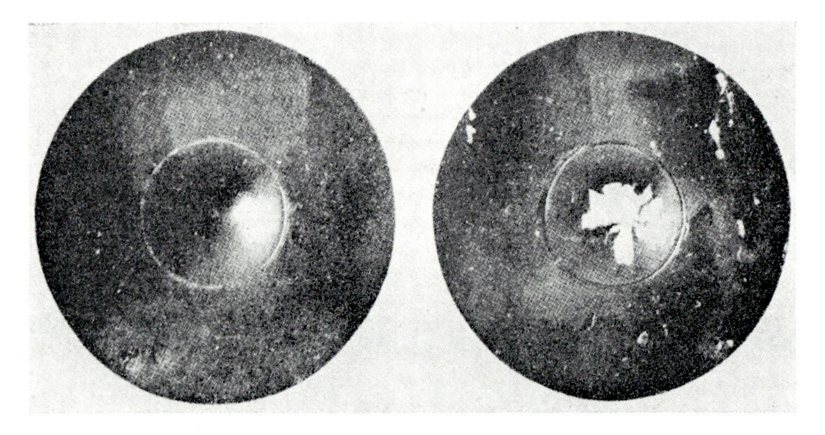

Fig. 13.11. Influence of Pylumin pre-treatment on paint-holding properties.
Left – Indentation test on Pyluminised aluminium finished with one coat of lacquer.
Right – Indentation test on plain aluminium finished with one coat of lacquer.
[Courtesy Pyrene Chemical Services Ltd.]

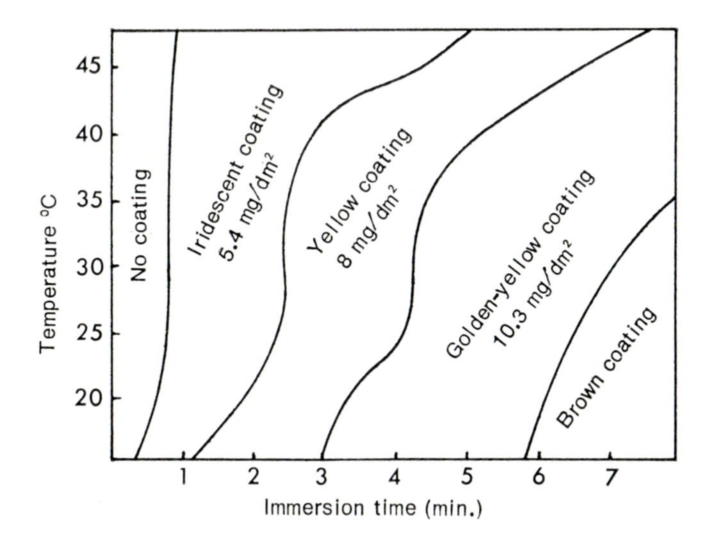

Fig. 13.12. Variation of weight and colour of acid chromate coating with time of immersion and temperature of solutions. (From data published by Freeman and Triggle[80])

(a) irridescent coating. 5.4 mg/dm^2
(b) yellow coating. 8 mg/dm^2
(c) golden-yellow coating. 10.3 mg/dm^2
(d) brown coating.

compounds including basic chromium chromate and hydrated chromic oxide. The solutions are essentially based on acid chromates with various activators and other constituents such as boric acid and ammonium bifluoride. Treatment time can range from $\frac{1}{2}$ minute to 9 minutes, the bath temperature from 15-55°C, and the pH between 1.2 and 2.2, depending on the process. The time of immersion and the temperature of the solution control the weight and colour of the coating[50] as shown in Fig. 13.12.

Thorough degreasing is necessary before acid chromate treatment as the solution has little cleaning action, but is less critical after rinsing than is the case with the alkaline chromate baths. Drying of the rinsed films can be carried out at quite high temperatures (up to 130°C). Defective films can result from too high temperatures, long processing times, too low pH or impurities in the bath, all of which can lead to powdery coatings. On the other hand, light or iridescent coatings are caused by too low a bath temperature or solution concentration, too short an operating time or an excessively thick oxide skin on the metal.

13.1.5.3. Chromate-phosphate processes

These processes have been used in a considerable number of commercial forms; the solutions use phosphoric acid and chromic acid or a dichromate together with fluorides. The films formed are amorphous and are 0.0005-0.005 cm thick[51]. They consist essentially of chromium phosphates with traces of aluminium compounds and fluorides, and range in colour from iridescent green to a uniform opaque green. For the opaque effect, control of the trivalent chromium and aluminium in the bath is essential, and cation exchange equipment is desirable for this purpose[52]. The aluminium content of the bath can also be controlled by addition of small amounts of sodium fluoride to precipitate it as sodium aluminium fluoride, but this gives rise to sludge and is expensive.

The processes are generally operated at up to 50°C, although they can be applied at room temperature. In the latter case, the treatment time is about 5 minutes, but this can be reduced to ½ minute by raising the temperature. Spray processes can also be used, when the treatment is complete in about 30 seconds. Cleaning before treatment is imperative, and an after-treatment, such as a 15 second dip 0.05 per cent chromic acid at 38° to 48°C followed by drying at 38° to 65°C, is recommended. Tanks and jigs are usually made of stainless steel, but aluminium can be used; fume extraction is necessary.

Chromate-phosphate coatings are much less extensively used than previously; although the process is still used to produce coatings which are not intended to be painted, it has been largely supplanted by the acid and alkaline methods.

13.1.5.4. Zinc phosphate process

Crystalline phosphate films can be produced on aluminium from acid zinc phosphate solutions similar to those used in the phosphating of steel. The coatings are up to 0.0007 cm thick and the solution is applied by dip or spray for ½-10 minutes at 40–70°C. The main application of the zinc phosphate process is for coating fabricated articles made of both aluminium and steel. In most other cases chromate coatings are generally preferred.

13.1.5.5. Böhmite process

Films of Böhmite, a crystalline hydrated aluminium oxide having the formula $Al_2O_3.H_2O$, are formed when aluminium reacts with boiling distilled or deionised water or steam at temperatures in excess of 75°C. Various methods are available for forming the films which are about 0.001 cm thick after one or two hours' treatment. Böhmite coatings have a fair resistance to weak acids, behaving similarly in this respect to anodic films[53].

In practice, the films are produced by immersing the aluminium in a boiling 1.5 per cent solution by volume of triethanolamine in de-ionised water for about 15 minutes, subsequently rinsing and air drying. Alkaline cleaning is carried out before the treatment followed by etching in a 5 per cent sodium hydroxide solution for 2 minutes at room temperature. The aluminium is then rinsed, dipped in 50 per cent nitric acid for 30 seconds and finally rinsed in deionised water before the triethanolamine treatment.

13.1.6. PAINTING ON ALUMINIUM

Most normal paint systems can be successfully employed on aluminium, but the metal surface must be suitably roughened, by etching, by mechanical means, by anodising or by treatment by one or other of the immersion processes described above, if maximum durability is to be obtained[54]. After pre-treatment the articles should be kept clean and not handled prior to painting owing to the tendency for the surface to absorb grease and dirt readily. Stoved resin-based primer coatings have better adhesion then air-drying paints. Amongst the most satisfactory primer paints for aluminium and its alloys are the zinc-chromate types, but red oxide and red and white lead primers can also be employed successfully.

The finishing coat may be based on cellulose or synthetic resins, but for some severe applications bitumen or asphaltic black paints are applied directly to the metal without an undercoat.

13.1.7. CHEMICAL AND ELECTROLYTIC DEPOSITION OF METALS ON ALUMINIUM

Plating is carried out on aluminium primarily for decorative purposes, but other advantages include improved scratch resistance and increased resistance to corrosion by alkalis.

Plating on aluminium presents special difficulties due to the very adherent oxide film which rapidly forms on the metal. Further complications are caused by the reactivity of aluminium towards many plating solutions; in this respect each group of the wide variety of aluminium alloys behaves characteristically. Alloying elements may be present as micro particles of intermetallic compounds such as Mg_2Si or AlMg which give rise to sites of differing chemical activity. The difference between the expansion coefficients of the aluminium and the coating metal, and also the fact that aluminium often contains occluded gases, may result in failure in service even when the deposit appears to be satisfactory immediately after plating.

The problem of plating metals on aluminium has been attacked by five general methods[55]. These can be classified as follows:

1. A bonding metal (usually zinc) is plated by chemical deposition on to aluminium, replacing the natural oxide film which is simultaneously removed. Electroplating using conventional techniques can then be carried out.
2. Plating is carried out on to anodic oxide films, the pores in the oxide film providing a key for the deposited metal[55, 56].
3. The aluminium surface is prepared by etching and is plated with zinc, brass, nickel and then chromium. Adhesion is promoted by heat treatment (Vogt and Ore processes).
4. Special plating solutions which have been developed for the electrodeposition of metals such as chromium, nickel, zinc and cadmium directly on to aluminium are used. Control of such solutions is difficult and the resulting adhesion may be poor.

5. Molten salt baths may be employed to apply coatings. The high temperatures involved greatly increase the cost of the process and may also impart undesirable mechanical properties to the deposit.

Of these methods, the first have the greatest commercial importance. Most electroplating on aluminium is carried out over a zinc bonding coat applied by means of chemical immersion process.

13.1.7.1. Immersion deposition of zinc onto aluminium

Immersion processes for producing adherent metal coatings on aluminium must incorporate two essential constituents(57): a salt of a metal such as zinc which is more noble than aluminium, and an aggressive anion capable of removing the air-formed oxide film without dissolving the underlying aluminium basis metal. Heiman(58) has shown that the ability of anions to remove the oxide film decreases in the order OH^-, F^-, SiF_6^{2-}, BF_4^{2-}, PO_4^{3-} and Cl^-. Only hydroxyl and fluoride ions can remove the film at a suitable rate and of these OH^- is by far the more widely employed.

In practice zinc is deposited on to aluminium by immersing the metal in an alkaline solution of sodium zincate. After removal of the oxide film two net reactions take place at the aluminium surface:

(i) the anodic dissolution of aluminium leading to the formation of aluminate:

$$Al + 3OH^- \rightarrow Al(OH)_3 + 3e$$
$$Al(OH)_3 \rightarrow AlO_2^- + H_2O + H^+$$

and (ii) the cathodic deposition of zinc, most probably(59) from the zinc complex $Zn(OH)_4^{2-}$:

$$Zn(OH)_4^{2-} \rightarrow Zn^{2+} + 4OH^-$$
$$Zn^{2+} + 2e \rightarrow Zn$$

The alkaline zincate solution is prepared by dissolving zinc oxide in sodium hydroxide solution, the ratio of the two constituents normally being of the order of 1:10 (e.g. 50 g/l ZnO and 500 g/l NaOH).

Increasing the hydroxide concentration or temperature increases the initial rate of zinc deposition(60) as shown in Figs. 13.13 and 13.14. However, after about 15 seconds the weight gain of the deposit is more or less independent of alkali concentration and little affected by temperature.

The replacement zinc deposit is thick and not always sound, probably owing to its mode of growth. In order to control its structure, the process has been modified so as to complex the zinc in an alkaline solution in which nickel and copper ions are also present(59, 60). A film of controlled thickness is produced, the chemical composition of which changes progressively during its growth. The complexing agents are NaCN and sodium tartrate; the nickel and copper ions are responsible for arresting the film

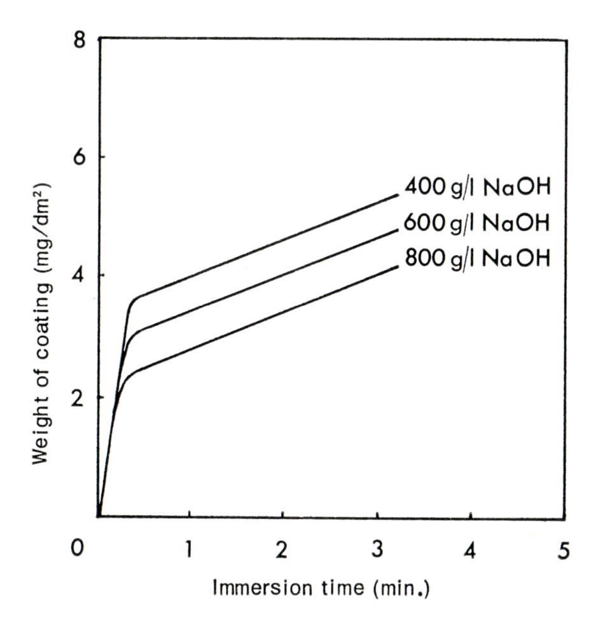

Fig. 13.13. Effect of NaOH concentration on the weight of zinc deposited from zincate solution on commercial aluminium sheet at 21°C. From data published by Keller and Zelley[60].

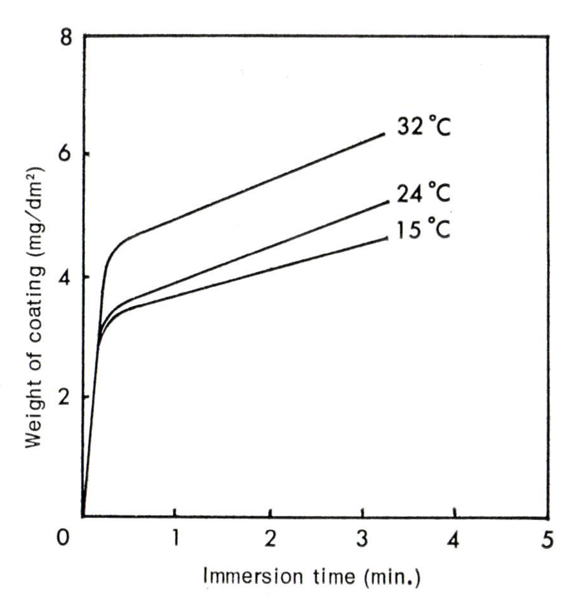

Fig. 13.14. Effect of temperature of the zincate bath on the weight of the deposit produced on commercial aluminium sheet. From data published by Keller and Zelley[60].

growth in its final stages, thus preventing filamental growth which would make the film mechanically weak.

The most important factor in promoting adhesion of subsequent nickel electrodeposits to the zinc coating is the presence of nickel in the zincate solution[61]. The addition of nickel ions to the immersion bath enables bright nickel to be deposited directly and adherently on to aluminium without any plated undercoat of copper being necessary.

The modified solution[61, 62] has been in commercial use[63] in the form of the 'Bondal' bath for several years. The bath is tolerant to changes in composition and has a fairly long life before it starts to deteriorate. A typical film produced in the Bondal bath contains 86% zinc, 8% copper and 6% nickel. The film readily reduces nickel from a sulphate solution and incorporates it into the coating.

For commercial purity aluminium and alloys containing a total of less than 1.5% of alloying elements, a simple process sequence is recommended[59, 61] for preparing the metal for subsequent electrodeposition:

1. Trichloroethylene degrease or soak clean in a hot non-silicated non-etch cleaner
2. Water rinse
3. Cathodically clean in a non-silicated caustic-based cleaner
4. Water rinse
5. Nitric acid dip (50% by volume)
6. Water rinse
7. Immerse in modified zincate solution for 1 to 2 minutes at 16-30°C
8. Water rinse
9. Electroplate with nickel or the desired metal using conventional or bright plating baths

13.1.7.2. Double immersion zincate process

Alloys such as those containing large proportions of copper or silicon are difficult to plate with zinc by the single immersion technique, and can only be satisfactorily plated by a double immersion process. The cleaned aluminium surface is treated in the zincate solution, the zinc deposit produced is stripped in nitric acid or a mixture of sulphuric and nitric acids, and the aluminium treated in the zincate solution a second time. The result is that adhesion of a subsequently applied zinc electrodeposit is much improved and the grain size reduced. This can be explained as being due to the greater uniformity of the oxide film found on the aluminium after stripping.

The double immersion process is also suitable for use in conjunction with the Bondal bath.

Many metals can be electroplated onto the immersion zincate coating. Zinc can be deposited directly either from acid or alkaline electrolytes using conventional methods, but other metals such as tin and nickel are usually plated onto an undercoat of copper deposited from a cyanide electrolyte. If the modified zincate process is used, then adherent nickel coatings can be applied directly to the zinc deposit.

Chromium, especially where its purpose is mainly decorative, should be plated onto a nickel undercoat. If, however, chromium is deposited directly onto the zinc coating, then it is best to maintain the chromic acid plating solution at a temperature of about 18–20°C. At this temperature the deposit is quite adherent and may be readily polished to a bright finish.

13.1.7.3. Bronze and brass strike pre-treatment of aluminium

In this method[64] of pre-treatment, aluminium articles, after being degreased, cleaned and acid dipped, are transferred to a stannate-containing activating solution for 30 seconds at 30°C. They are then given a 3 minute 'strike' in a bronze electrolyte, water rinsed and electroplated as desired.

The bronze strike solution contains Cu^{+} ions, Sn^{4+} ions, free cyanide, free KOH and certain additives. The parts are plated in this electrolyte at a cathode current density of 1.5-12 A/dm^2.

Schwartz and Newkirk[65] have reported a new technique employing the deposition of brass onto aluminium using an ac biased current. They claimed that the process produces superior adhesion than the zincate process.

13.1.7.4. Anodic pre-treatments

Anodic coatings have been employed to promote the adhesion of electrodeposits on to aluminium alloys. The articles are first anodised so that a fairly thick oxide coating is produced, after which they are directly immersed in the electroplating solution. The anodic film serves to protect the metal and also provides the necessary key for the electroplated coating.

The Elytal process[65] makes use of an anodic treatment for ten to fifteen minutes in a phosphoric acid bath, followed by cathodic treatment in an alkaline solution (when partial reduction of the film takes place), after which plating is carried out in a copper or nickel bath.

The operating conditions of the phosphoric acid bath have to be varied with the type of alloy being treated. In the case of work-hardening aluminium alloy sheet (containing 1.25 per cent manganese) the conditions recommended are:

Electrolyte: Phosphoric acid (H_3PO_4)	354 g/l
Temperature	30°C
Current density	1.5 A/dm^2
Time	10 minutes

The work is next either directly nickel- or copper-plated, the control of the anodising conditions being much more critical before nickel- than before copper-plating; the nickel solution must be of the buffered type. Copper cyanide baths, however, must be used with care since the high alkalinity of these solutions may result in the aluminium being attacked before it has become covered by the copper. In all processes of this type the efficiency of the bond obtained is dependent on the uniformity of the anodic film, and for this reason it is desirable to clean the metal prior to anodising and then to subject it to a mild etching treatment. A satisfactory method is to degrease and then clean cathodically for a minute or two in a solution of 40 to 50 g/l of sodium cyanide. This is followed by rinsing, dipping in nitric acid of 30° Bé density, and a further rinse prior to anodising. In the case of silicon-containing alloys, immersion in a hydrofluoric acid solution is needed to remove the black film left by the nitric acid treatment.

As an alternative to the electrolytic cyanide cleaning bath, immersion in a solution containing about 30 g/l of sodium carbonate and an equal quantity of trisodium phosphate can be carried out. This should be at a temperature of 65° to 70°C, immersion time being 3 to 10 seconds.

In plating over the oxide coating, the initial deposit appears to be formed within the structure of the oxide. During the first few moments of deposition the anodic film becomes dark in colour, but this effect soon disappears. There is no doubt that deposition actually occurs over the oxide itself since its presence can be readily demonstrated under the plated layer. Thus, copper plated in this way over the anodic film can be dissolved in concentrated nitric acid and the re-exposed oxide film dyed by immersion in a suitable dye bath.

As the presence of the oxide film reduces the conductivity through the metallic coating, this method of pre-treating aluminium for plating is not recommended where good electrical conductivity is required.

13.1.7.5. Vogt process

One of the first commercially practicable processes for plating on aluminium was that developed by Vogt in 1929 in England, and it is still in use in a variety of modifications. Essentially, the process[66] consists of electrolytic and acid etching followed by zinc, brass and finally nickel plating.

The first electrolytic cleaning operation is carried out in a solution consisting of:

$NaOH$	30 g/l
$NaCN$	30 g/l
Na_2CO_3	14 g/l

The work is made cathodic for 3 to 5 minutes at a voltage of about 7 and a solution temperature of 20°C. The acid etch which follows consists of equal volume of commercial sulphuric and acetic acids used at room temperature, immersion time being 2 to 5 seconds. Another electrocleaning operation follows in a solution containing 14 g/l each of caustic soda and sodium cyanide for 10 to 30 seconds at room temperature. The voltage used is again around 7.

The zinc deposit which follows is applied from a very dilute bath consisting of:

Zinc chloride	0.5 g/l
Sodium cyanide	0.5 g/l
Caustic soda	10 g/l

The bath is operated at room temperature and a current density of 0.5 A/dm^2, the plating time being 20 seconds. A uniform colour should be aimed at when carrying out this operation.

The work is then brass plated for about 8 seconds in a special dilute solution consisting of:

Copper acetate	12.5 g/l
Zinc chloride	12.5 g/l
Sodium cyanide	30 g/l
Sodium carbonate	10 g/l
Sodium bisulphate	12.5 g/l
Current density	0.7–1.0 A/dm^2
Temperature	25^0–32^{0}C

70:30 brass anodes are used and a pale yellow brass deposit should be obtained. Nickel plating is then carried out directly after thorough rinsing and without intermediate acid dips. The air-agitated nickel solution used is of the low chloride type to ensure that a soft, unstressed deposit is obtained. The following composition is used:

Nickel sulphate	250 g/l
Magnesium sulphate	100 g/l
Sodium chloride	45 g/l
Boric acid	25 g/l
Current density	1.0–1.5 A/dm^2
pH	5.6
Temperature	45^{0}C

After nickel plating, the articles are stoved for half an hour at 230^{0}C. This greatly improves adhesion and also serves as a test for the quality of the deposit, which will blister during the operation if adhesion is likely to prove unsatisfactory in subsequent service. Finally the articles are chromium plated.

13.1.7.6. Ore process

A modified version of the above process has been described by Ore[67] for the plating of aluminium hollowware. This process is basically similar to that used by Vogt, but a 10 per cent tartaric acid dip is added between the brass and nickel plating operations, whilst heat treatment is also carried out after chromium plating for 5 minutes at 175–200^{0}C.

13.2.2. ANODIC TREATMENT

Several anodic treatments for magnesium have been proposed, and many of these give good protection.

An anodic process known as 'Manodyzing', which can also be coloured by dyeing, has been described by Cutter[70]. It consists of an anodic treatment in a caustic soda solution containing various silicate and phenolic additives. The film produced is an oxide-silicate of magnesium and varies in depth from 0.001 cm (produced in ten minutes) up to a maximum of 0.0025 cm obtained in thirty minutes at a current density of 1 to 2 A/dm^2. Alternating current can be used also for producing the film, in which case the build-up in thickness is less.

It is claimed that the coating produced is hard and durable and is also sufficiently robust to insulate the magnesium from galvanic corrosion when it is in contact with other metals. The colour of the coating varies from greenish to light or dark grey, depending on the alloy being treated; it can be dyed by a technique similar to that used for aluminium and its alloys.

The coating serves as a good base for paint, especially if it is treated by means of a two-minute dip in hot chromic acid solution at a pH of 2.3 before painting. A zinc chromate primer should be used, followed by an aluminium-pigmented final varnish.

The fluoride anodising process developed in the U.K.[71, 72] is a combined cleaning and protective treatment for unmachined magnesium alloy coatings. The treatment is effective in removing foundry sand and oxides adhering to the metal and replacing them with a coating of magnesium fluoride.

A solution containing 10–20% ammonium bifluoride is used at room temperature. The voltage is gradually raised to about 100V and treatment then carried out for 10–15 minutes. The coating can be sealed with a resin after driving off moisture.

13.2.3. PAINTING

The pre-treatments described above do not in themselves provide adequate protection for the magnesium where severe conditions are concerned; further protection by the use of a paint system is essential. Particularly satisfactory finishes are oil-based undercoats pigmented with zinc chromate or a mixture of zinc, barium and strontium chromates, followed by a finish coat. The priming coat should be free from lead driers (which are particularly deleterious) and from chlorides and soluble salts, other than chromates. Pigments recommended for use in conjunction with zinc chromate include barium sulphate, ferric oxide, lithopone and titanium dioxide. Dyes or lakes used should be free from chlorides.

13.2.4. PLATING

The difficulty with plating on to magnesium arises from the fact that a hydroxide film forms on the magnesium surface as soon as it is immersed in an aqueous bath.

One method whereby magnesium can be plated is to form a deposit of zinc by immersion in an acid zinc solution by a technique similar to that used in the plating of aluminium. The process is described by de Long[73]. The parts are first solvent-degreased and then cleaned in a strongly alkaline bath for 3 to 15 minutes at 80°C to 100°C. They are then pickled, if at all oxidised, in a solution containing 20 per cent by volume of glacial acetic acid and 5 per cent by weight of sodium nitrate. This is followed by immersion for 3 to 5 minutes in a solution containing:

Tetrasodium pyrophosphate	120 g/l
Zinc sulphate	40 g/l
Potassium fluoride	10 g/l
Potassium carbonate	5 g/l
Temperature	75°–82°C
pH	10.2–10.4

The bath must be made up with water substantially free from iron salts and the chemicals used should be free from heavy metal contamination. After treatment, the parts are thoroughly rinsed and plated with copper from a Rochelle salt bath, first at a low current density of 0.5 to 1 A/dm^2 for 2 to 3 minutes, followed by a further 5 minutes or more at 1.5 to 2 A/dm^2. After this, plating is carried out in a conventional bath, the metal deposited being dependent on the application of the article. A list of coatings applied to magnesium and classified with respect to their properties has been given by Hepfer[74].

Corrosion resistance only
1. Copper–electroless nickel
2. Copper–heat-flowed tin
3. Copper–electroless nickel–heat-flowed tin
4. Copper–cadmium

Hard or Wear-resistant surfaces
1. Copper–electroless nickel
2. Copper–chromium
3. Copper–electroless nickel–chromium

Solderability
1. Copper-tin with or without heat flowing
2. Copper–electroless nickel–tin which is usually heat-flowed
3. Copper–electroless nickel–gold
4. Copper–cadmium

Emmissivity
1. Copper–silver–gold
2. Copper–electroless nickel–gold
3. Copper–nickel–black nickel

Surface conductivity

1. Copper–silver
2. Copper–electroless nickel–silver.

Racks for plating should be insulated and should be copper plated on exposed areas. This is because magnesium does not take the zinc coating satisfactorily at areas where it is in contact with other metals as a result of electrolytic action. Copper, however, becomes zinc-coated in the bath immediately it is in electrical contact with the magnesium so that this effect is eliminated. The best method is to give the racks a copper strike after each cycle.

Nickel can be stripped from magnesium anodically in a bath containing 15 to 25 per cent of hydrofluoric acid and 2 per cent of sodium nitrate. As soon as the magnesium is exposed the reaction stops, due to polarisation with the formation of magnesium fluoride. Copper can be removed in an alkaline polysulphide dip, followed by a dip in a cyanide solution. The corrosion resistance with nickel deposits of 0.0012 to 0.0035 cm in thickness appears to be good, as is the adhesion of the coatings.

REFERENCES

1. *Electroplating and Metal Finishing* 1960, **13**, No. 1, 27.
2. W. K. Bates and C. D. Coppard, *Met. Fin. J.* 1958, **4**, No. 37, 5-10.
3. Brit. Pat. 693,776, 693,876 and 738,711.
4. U.S. Pat. 2,101,603.
5. Brit. Pat. 449,162.
6. S. Setch and A. Miyata, *Sci. Pap. Inst. Phys. Chem. Res. Tokyo* 1932, **17**, 189.
7. T. P. Hoar and N. F. Mott, *J. Phys. Chem. Solids* 1959, **9**, 97.
8. J. S. L. Leach and P. Neufeld, *Corrosion Sci.* 1969, **9**, 413.
9. F. Keller, M. S. Hunter and D. L. Robinson, *J. Electrochem. Soc.* 1953, **100**, No. 9, 411.
10. R. B. Mason, *J. Electrochem. Soc.* 1955, **102**, 156.
11. P. Neufeld and H. Ali, *Transactions of the Institute of Metal Finishing* 1970, **48**, 175.
12. S. Wernick and R. Pinner, 'The Surface Treatment and Finishing of Aluminium and its Alloys', Robert Draper, Teddington 1973.
13. G. D. Bengough and J. M. Stuart, Brit. Pat. 223,994.
14. 'Anodic Oxidation of Aluminium and its Alloys' Information Bulletin No. 14, published by the Aluminium Federation, 1966.
15. W. C. Cochran and F. Keller, *Proc. Am. Electroplaters' Soc.* 1961, **48**, 82.
16. C. Th. Speiser, *Aluminium* 1966, **42**, No. 7, 422.
17. N. D. Pullen, *J. Electrodep. Tech. Soc.* 1939, **15**, 69.
18. 'Anodic Oxidation of Aluminium'. A. Jenny, Trans. W. Lewis, 1940, p. 177.
19. J. Else, *Galvanotechnik* 1962, **53**, 374.
20. B. C. Lewsey, *Electroplating and Metal Finishing* 1952, **5**, 250.
21. R. Feek and A. W. Brace, *Trans. Inst. Metal Finishing* 1957, **34**, 232.
22. I. I. Moroz, I. P. Kharlamov and L. S. Sogolava, *Stankii Instrument* 1961, **32**, No. 11, 32.
23. Brit. Pat. 397,538 and 398,825.
24. A. W. Brace 'The Technology of Anodising Aluminium'. Robert Draper, Teddington, 1968.
25. U.S. Pat. 2,290,364.
26. E. Hermann and W. Hubner, *Aluminium Suisse* 1955, **5**, 134.
27. C. H. Giles, Aluminium Development. Ass. Conference on Anodising, Nottingham, 1961, p. 174.

28. C. T. Speiser, *Electroplating and Metal Finishing* 1956, No. 4, 109.
29. C. H. Giles, H. V. Mehta, C. E. Stewart and J. Subrahamian, *J. Chem. Soc.* 1954, 106, 4360.
30. C. H. Giles, H. V. Mehta and C. E. Stewart, *J. Appl. Chem.* 1957, **9**, 457.
31. C. H. Giles and K. V. Datye, *J. Appl. Chem.* 1963, **13**, 473.
32. C. H. Giles, 'Review of Dyeing Processes', *Chem. Ind.* (*London*) 1966, p. 92.
33. A. Cutroni, *Galvano Tecnica* 1961, **12**, No. 7, 148.
34. N. Budeloff and K. Hahn, *Metallwirtschaft* 1941, April 18, 387.
35. V. F. Henley, *Met. Ind.* 1943, **63**, 386.
36. T. P. Hoar and G. C. Wood, Aluminium Development Ass. Conference on Anodising. Nottingham, 1961, p. 186.
37. B. A. Scott, *Electroplating and Metal Finishing* 1965, **18**, 47.
38. P. G. Sheasby, *Electroplating and Metal Finishing* 1966, **19**, 104.
39. N. D. Tomashov and M. N. Tyukina, *Bull. Acad. Sci., U.S.S.R.*, (*Chem. Sci. Sect.*) 1944, 5.
40. L. Whitby, *Met. Ind.* 1948, **72**, 400.
41. J. M. Kape, *Met. Ind.* 1959, **95**, 115.
42. A. W. Brace and P. Pocock, *Trans. Inst. Met. Finishing* 1958, **35**, 277.
43. V. J. J. Marran and G. C. Wood, *Electroplating and Metal Finishing* 1970, **23**, 17.
44. *Metal Finishing Journal* 1968, **14**, 48.
45. E. F. Barkman, Aluminium Federation Symposium on Anodising Aluminium, Birmingham, 1967, 115, discussion 153-163.
46. S. Spring, *Product Finishing* 1965, **29**, No. 8, 34.
47. J. F. Murphy, *Tech. Proc. Am. Electroplaters' Soc.* 1961, **48**, 60.
48. W. Holling and H. Neunzig, *Aluminium* 1938, **20**, 536.
49. E. C. F. King, Brit. Pat. 441,088.
50. D. B. Freeman and A. M. Triggle, *Trans. Inst. Met. Finishing* 1960, **37**, 56.
51. A. Douty and F. P. Spruance, Jr., *Ann. Proc. Am. Electroplaters' Soc.* 1949, **36**, 193.
52. G. Gubnelson, *Met. Finishing Journal* 1959, **5**, (49) 19.
53. D. A. Henpohl, *Metall* 1958, **12**, No. 6, 503.
54. 'Finishing on Aluminium,' p. 93, S. J. Ketcham, Ed. G. H. Kissen, Reinhold Pub. Co., N.Y., 1963.
55. K. S. Indira, R. Subramanian and B. A. Shenoi, *Met. Finish*, 1971, **69**, 4, 53.
56. A. Akiyama, K. Yasurhara, K. Tanaka and T. Saji, Plating, 1971, **58**, 594.
57. Luis, J da, V. Lobo, P. S. Kinnerkar, G. M. Ganu and W. L. Ray, *Electroplating and Metal Finishing* 1969, **22**, No. 9, 21.
58. S. Heiman, *Trans. Electrochem. Soc.* 1949, **95**, 205.
59. T. P. Dirkse, *J. Electrochem. Soc.* 1954, **101**, 328.
60. F. Keller and W. G. Zelley, *J. Electrochem. Soc.* 1950, **96**, 143.
61. T. E. Such, and A. E. Wyszynski, *Plating* 1965, **52**, 1027.
62. A. E. Wyszinski, *Trans. Inst. Met. Finishing* 1967, **45**, 147.
63. 'Canning Handbook on Electroplating', p. 301, W. Canning & Co., Birmingham, 21st Ed., 1970.
64. J. C. Jongkind, *Trans. Inst. Met. Finishing* 1967, **45**, 155.
65. B. C. Schwartz and J. B. Newkirk, *Plating*, 1972, **59**, No. 5, 432-6.
66. A. Wallbank, *J. Electrodepos. Tech. Soc.* 1951-52, **28**, 209.
67. R. Ore, *Bull. Inst. Met. Finishing* 1953, **3**, 163.
68. M. Sutton and F. L. LeBrocq, *J. Inst. Met.* 1935, **57**, No. 2, 199.
69. G. D. Bengough and L. Whitby, *J. Inst. Met.* 1932, **48**, 147.
70. P. R. Cutter, *Proc. Am. Electroplaters' Soc.* 1946, p. 257.
71. W. F. Higgins, *Light Metal Age* 1959, **18**, No. 8, 11.
72. J. K. Wilson and F. D. Waldron-Trowman, *Light Metals* 1958, **21**, 186.
73. M. K. de Long, *Met. Finishing* 1948, 7, 46.
74. I. C. Hepfer, *Products Finishing* 1965, **29**, No. 8, 51.

BOOKS FOR FURTHER READING

1. S. Wernick and R. Pinner, 'The Surface Treatment and Finishing of Aluminium and its Alloys,' Vols. 1 and 2, Robert Draper, Teddington 1973.
2. A. W. Brace, 'The Technology of Anodising Aluminium', R. Draper, Teddington, 1968.

CHAPTER 14

High-temperature and mechanical methods of coating

14.1. Hot-dipped coatings

14.1.1. GENERAL

THE process of coating application by hot dipping is one of the oldest, simplest and cheapest methods of metal finishing. However, it has many limitations. Since the coating metal must have a considerably lower melting point than the metal to be coated, hot dipping is generally confined to low-melting metals, viz. tin, lead, tin/lead alloys, zinc and aluminium. Although steel is by far the most widely used basis metal, cast iron, copper and, for some purposes, titanium, molybdenum and special alloys are also often hot-dip coated.

The coatings are relatively non-uniform in thickness, and it is difficult to coat parts to close dimensional tolerance requirements.

In order to obtain a well-adherent coating, the molten metal must 'wet' the surface of the substrate metal evenly, and mutual alloying must occur, i.e. the two metals must diffuse into each other to a shallow depth at the interface. The coating normally consists of several layers of different compositions and physical properties. The innermost layer is richest in the substrate metal, whereas the outermost is richest in the coating metal. The thickness of the individual layers, and that of the coating as a whole, depend on the particular molten metal used, bath temperature and time of application. Certain additions made to the molten bath may profoundly affect the rate of diffusion and the composition of the alloy layer formed. The thickness of the coating may be controlled by rolling or wiping it prior to solidification. In the continuous zinc-coating of steel strip, the thickness is frequently controlled by a 'gas knife'—a jet of high-pressure air or steam impinging on the molten coating from a narrow slot (see p. 503).

Articles for hot dipping must be pretreated by degreasing, pickling and rinsing. Cast and malleable iron are usually shot blasted. They are then fluxed by immersion in an aqueous fluxing bath and dried. Alternatively, they may be passed through a molten flux floating on the metal bath; sometimes both processes are used. This is followed by dipping. Fabricated articles are dipped in batches, mounted on suitable jigs, whereas very small articles such as bolts, nuts, washers, etc. are held in perforated baskets during dipping. Strip and wire are coated continuously, whilst sheets are dipped singly either mechanically or by hand. In the continuous coating of steel strip with zinc or aluminium, the metal usually enters the molten bath from a furnace with a reducing atmosphere such as cracked ammonia (25% nitrogen plus 75% hydrogen), thereby eliminating the need for a flux.

The corrosion resistance of hot-dipped zinc coating may be improved by phosphating or chromating, and that of hot-dipped aluminium coatings by chromating. Heat treatment is carried out where complete conversion of the coating into an alloy between coating metal and substrate metal is desired. Aluminium coatings on steel acquire a high degree of heat resistance as a result of such treatment.

14.1.2. GALVANISING

Zinc protects steel electrochemically against corrosion, and in fact the term 'galvanising' is derived from the concept of galvanic protection of steel by the more base zinc.

Iron forms three compounds or phases with zinc, viz., θ containing 6.25% Fe, $FeZn_7(\gamma)$ containing 11% Fe and an ε phase containing 22–28% Fe. A normal galvanised coating will consist of layers of these phases with an outermost layer of virtually pure zinc. The structure and composition of a galvanised coating is shown in Fig. 14.1, together with the relevant portion of the Fe-Zn equilibrium diagram. Both of the compounds θ and γ are very brittle, so that when a certain crucial coating thickness is exceeded the outer iron-zinc layer is liable to detach itself and sink to the bottom of the bath. The 'spangled' surface finish on galvanised coatings is due to the presence of small amounts of foreign metals, notably lead[1] and tin. When both of these metals are present in quantity large 'spangles' are produced[2] and the ductility of the coating is reduced.

The molten bath is operated at between 430° and 470°C. At temperatures above 480°C a sudden sharp increase in the rate of alloy formation occurs and for this reason the bath temperature should be kept below about 470°C. Also, the thickness of the zinc layer formed in too hot a bath decreases whilst the life of the container is drastically reduced at elevated temperatures. On the other hand, so-called 'cold' baths (those operated at below 430°C) give rough deposits, with an accompanying increase in zinc consumption.

The iron content of the bath should not be allowed to build up to above 0.05%, since otherwise the mechanical properties of the galvanised coat will be adversely affected. Solid alloy crystals must be periodically removed from the bottom of the

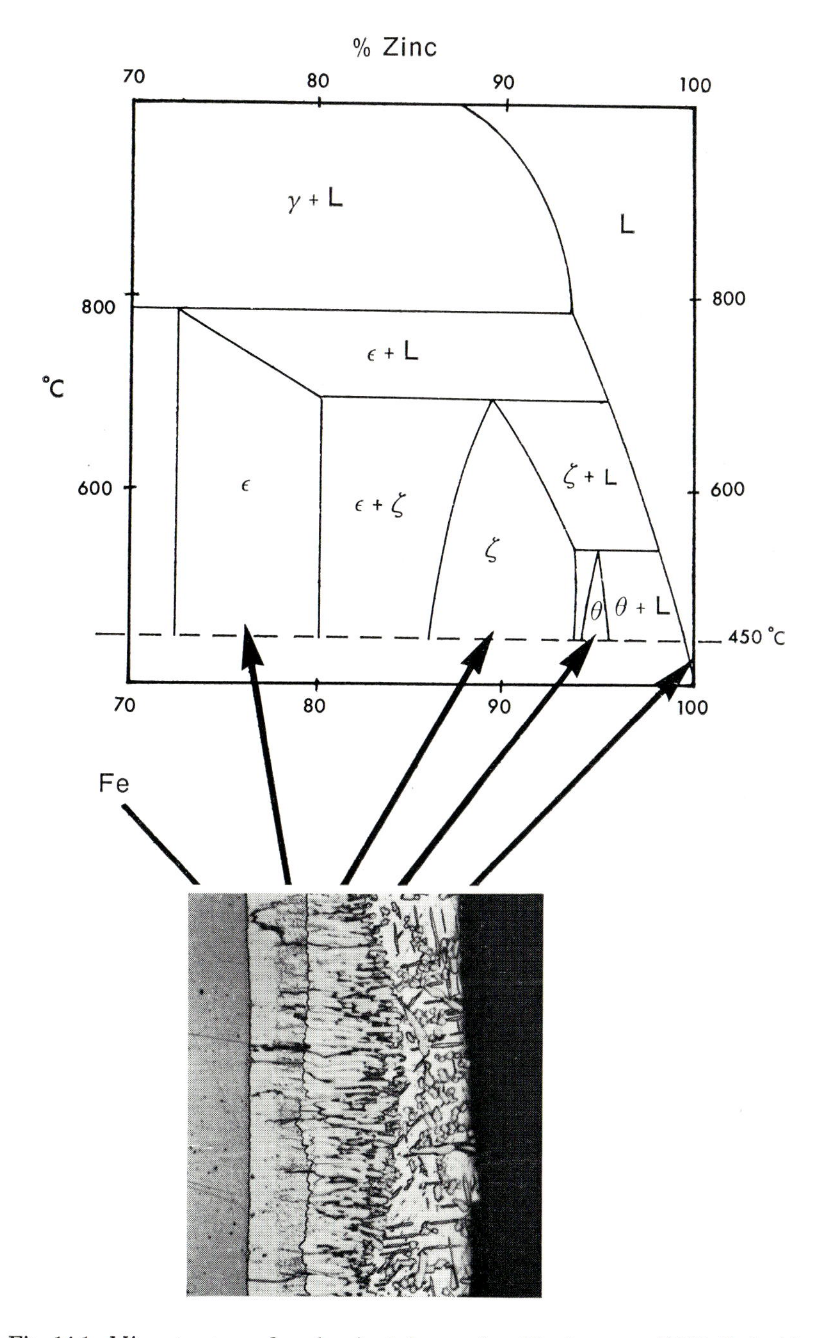

Fig. 14.1. Microstructure of a galvanised zinc coating. 20 minutes at 450°C. Etched in 4% nital. × 200

container. They must not be stirred up or they may settle on the work, producing rough, unacceptable deposits. The bath gradually becomes contaminated with ZnO during operation due to oxidation of the zinc by air at elevated temperatures. This increases the viscosity of the bath and embrittles the coat. Between 0.15 and 0.25% aluminium is sometimes added to the bath in order to reduce the solubility of iron in zinc and to minimise the thickness of the iron-zinc intermetallic compound layers in the coat[3–7]. The addition of aluminium also gives rise to the formation of bright zinc coatings and increases their corrosion resistance. Small amounts of tin reduce the viscosity of the bath and improve the surface appearance and adhesion of the coating.

A considerable tonnage of steel sections is galvanised, largely for the building industry, but the bulk of the useage is in the form of galvanised sheet and strip. Over 400,000 tons of galvanised sheet were produced in the U.K. in 1970[8]. Galvanised sheets are classified into grades according to the weight of zinc coating per unit area. Coating thickness may vary from approximately 0.02 to 0.12 mm, according to specific requirements.

Before arriving at the galvanising shop the sheets are cold-rolled to give them a good surface finish, and then annealed and sheared to size. They are then thoroughly etched for 30 minutes in a 5–10% sulphuric acid solution, water-rinsed for 30 minutes, dipped in a 1–3% hydrochloric acid or acidified zinc chloride solution, washed and usually stored under water until galvanised.

14.1.2.1. Sheet plant

The plant for galvanising individual steel sheets consists of a rectangular kettle (typical dimensions: 5 m × 2 m × 1.5 m deep) made of mild steel and supported over a furnace which is usually gas-fired. The galvanising equipment consists essentially of three pairs of rolls, a flux box and a frame. The first pair of rolls is inclined at an angle and feeds the strip down to the flux box. The latter is a rectangular box, open at top and bottom, extending the width of the kettle at the feed end. A second pair of rolls is placed vertically in the centre at the bottom of the kettle, and a third pair is positioned horizontally at the exit end. Guides are provided, and the operation is continuous. The flux box extends into the molten zinc and is filled with fused ammonium chloride.

The sheet is brought from the wash tank, dropped into the dilute hydrochloric acid pickling solution and fed into the first pair of rolls. These force it down through the flux and under the surface of the molten zinc. The flux removes the film of water and acid out of contact with air. The sheet is then directed by the guides to the second pair of rolls, and after passing through them, moves towards the exit rolls. During its passage through the bath the sheet becomes heated to the bath temperature, at which alloying between the iron and the zinc occurs. The speed of travel of the sheet through the bath is adjusted so that when it is withdrawn by the exit rolls, a good bond has formed. The exit rolls regulate the thickness of the coating by removing surplus molten zinc; they also help to even out irregularities in the coating. From the exit rolls, a guide bends the sheet forward, so that it falls on to a conveyor and thence proceeds to a roller leveller and a cooling rack.

14.1.2.2. Dross formation

One of the main sources of loss of zinc is the formation of dross. This is a zinc-iron alloy containing 3–7% Fe, which is pasty at the bath temperature. It is very brittle when cold and spoils the surface of the sheet if it comes into contact with it. Being heavier than zinc, it sinks to the bottom unless convection currents, due to non-uniform heating, keep it in suspension.

There are various ways in which the amount of dross can be reduced, the principal method being the use of neutral fluxes. When the sheet is removed from the dilute hydrochloric acid solution prior to galvanising, it retains a film of iron salts which combine with zinc. One of the methods uses a slightly acidified zinc chloride ($ZnCl_2$) solution in place of hydrochloric acid and a mixture of zinc chloride and ammonium chloride (NH_4Cl) as a flux.

14.1.2.3. Continuous plant

The sheet dipping plants have been largely replaced by continuous high-speed lines; over 10 million tons of zinc-coated steel are now produced annually, three-quarters of which comes from continuous mills. In continuous plants, the strip is first cleaned either chemically or by heating in an oxidising atmosphere, and then passed through an annealing furnace containing a protective reducing atmosphere (usually 'cracked' ammonia, consisting of 75% hydrogen and 25% nitrogen). From this it passes directly into the zinc bath through an enclosed tunnel which reaches below the surface of the molten metal, thus preventing any contact with the air. The strip passes under a roller immersed in the bath, whence it emerges vertically, and is cooled and coiled. The thickness is controlled by exit rolls at the bath surface.

A recent development(9) is the use of gas knife jets to control the thickness of the zinc coating more accurately. This has resulted in a wider range of coating thicknesses, more uniform coatings and the possibility of producing coatings with different thicknesses on each side of the steel strip. The gas knife consists of a narrow slot extending the full width of the strip, of carefully designed width and shape, through which high pressure air or steam are directed at the molten surface as it emerges from the bath.

In the continuous galvanising process no flux is needed, which makes control of the process easier and also results in a more corrosion-resisting material, since the possibility of occluding flux particles in the deposit is eliminated.

14.1.3. HOT-DIP TINNING

Tin is electrochemically more noble than iron in many media. However, it forms complexes with many organic acids, and hence under anaerobic conditions it may become anodic to iron. This, together with its easy solderability and the fact that tin compounds are non-toxic, makes tin an eminently suitable coating metal for food canning.

For this reason, some 40% of the world's tin output is used for the production of tin plate, which is almost entirely made by applying the tin electrolytically. (See p. 399).

Tin coatings are impermeable to nitrogen, and are hence used to stop-off those selected areas on steel surfaces during nitriding which are not to be surface-hardened. Tin readily alloys with iron to form the compound $FeSn_2$, but the alloy layer formed during hot tinning grows slowly as compared with the more complex alloy layers formed in galvanising and aluminising. The solid solubility of iron in tin is very low (a fraction of 1 per cent). For this reason, a hot-tinned coating on steel will consist of a thin alloy layer in direct contact with iron and an outer layer of tin. The thickness of the latter can be made to vary according to requirements.

In the production of tin plate by hot dipping, mild steel sheets are annealed and etched in a hot 4–5% sulphuric acid solution (black etch); they are then re-annealed and etched in a cold 2% sulphuric acid solution (white etch) in order to remove thin oxide films formed during re-annealing. Etching is complete when the sheet is uniformly grey in appearance. This stage should be reached in 3 to 5 minutes. The sheets are then thoroughly rinsed in running water, dipped in a 0.5% hydrochloric acid solution, and tin coated in tinning machines. These are similar to those used for hot-dip galvanising, but are normally smaller. Also, there is a tower at the exit end filled with palm oil which floats on top of the molten tin.

The sheets are passed through the flux box and into the molten tin bath. They are then picked up by vertical rolls and passed upwards through the layer of oil. The oil slows down solidification of the tin coating, making it bright and preventing oxidation. The tin bath is maintained at a temperature of 300°C. This superheat (tin melts at 231.5°C) compensates for heat losses due to entry of the cold, wet sheet. The bath is kept ebullient by the film of acidified water on the entering sheet.

During its passage through the molten tin bath the temperature of the steel sheet is raised to above that of the melting point of tin. The temperature of the palm oil bath must be rigidly controlled because it is essential that the temperature of the emerging tin plate should be only just above the melting point of tin, since otherwise oxidation of the coating is liable to occur. In practice, the plate emerges from the oil bath at around 240°C. The thickness of the coating is controlled by the last pair of rolls in the oil bath. The oil consumption is approximately 3 kg per ton of tin plate. A fuller description of the hot tinning process is given by Hoare[10].

High-quality tin is essential for hot tinning; the total impurity content should preferably not exceed 0.1% and the maximum permissible lead content is 0.04%. In contrast to galvanising, no other metals are added to tin to improve the quality of the coating. The flux is normally a mixture of zinc chloride and ammonium chloride.

When protection against atmospheric corrosion is required for steel, the tin coating must be free from pores as otherwise severe corrosion will occur at the bottom of a pore, the substrate acting as a small anode and the tin coating as a very large cathode.

Different grades of tin plate carry different amounts of tin. Those with the least tin are known as coke plates, and those with the thickest coating as charcoal plates. Hot-dipped tin coatings are also applied to steel wire, steel strip, fabricated steel articles, cast iron and copper products.

14.1.4. LEAD COATINGS AND TERNE PLATE

Lead coatings of appropriate thickness are impervious and can give good protection to steel components against corrosion. On sheet, they can withstand considerable deformation before rupturing, and hence are used to a limited extent for the manufacture of components made by stamping.

Lead does not wet the surface of most other metals, and therefore a 'binder' metal must be used with it. This can either be added to the lead bath or applied to the surface of the components by a convenient method. In practice tin is used since it alloys readily with both lead and iron. The compound $FeSn_2$ is formed with iron, whilst with lead, tin forms a eutectic alloy at 60% Sn and 183°C. The resulting hot-dipped coatings containing up to 25% Sn are known as 'terne' coatings. When the tin content is very low (e.g. 2%) the coatings are referred to as lead coatings. Although the process is similar to hot tinning, the bath temperature must be raised to suit the freezing range of the particular alloy used.

14.1.5. HOT-DIP ALUMINISING

Hot-dip aluminising resembles hot-dip galvanising, but the process is complicated by the following factors:

(1) the melting point of aluminium is higher than that of zinc (660°C as against 419°C), necessitating a bath temperature of above 700°C. This makes it desirable to use a ceramic-lined container instead of an iron pot, which would be rapidly attacked by the molten aluminium. The rapid reaction between iron and aluminium at the bath temperature leads to dross formation.

(2) Fluxing is more difficult than in the case of zinc or tin.

(3) There is danger of contamination of an article leaving the bath with streaks of Al_2O_3 film or with globules of metal entangled in the oxide film.

Hot-dip aluminising is applied essentially to steel. Cleaned steel articles can be coated by first immersing them in a bath of molten flux maintained at a temperature similar to that of the molten aluminium bath containing sodium and potassium chlorides with varying additions of zinc chloride, calcium chloride, sodium fluoride, or mixtures of these, or by immersion in glycerin or alcohol as a protection against oxidation[11].

Most steel strip is aluminised in continuous plants essentially similar to those used for zinc coatings. Many plants, especially in the U.S.A., are designed to coat steel strip at speeds of up to 30 m per minute or more. The design and installation of a plant for the large-scale aluminising of steel strip has been described by Silman[12, 13]. The steel strip is directly immersed in the molten metal from a reducing atmosphere after first heating in an oxidising furnace to about 500–600°C, which removes lubricants and surface contaminants and leaves a thin oxidised film on the metal. The oxide is then preferably removed by a light pickle in hydrochloric acid. The strip is next subjected to a reducing atmosphere containing hydrogen in a second furnace at 700–800°C, where a certain amount of annealing also takes place. Cracked ammonia, consisting of 75% hydrogen and the rest nitrogen, is convenient for this purpose, but pure hydrogen can also be used.

After cooling, and whilst still in the reducing atmosphere, the strip enters the bath of molten aluminium maintained at 700–750°C via a submerged snout and then emerges vertically after passing under a roller immersed in the molten aluminium. It must be rapidly cooled after leaving the bath, and is finally cut into sheets. The emerging strip may be passed through a pair of driven rolls, the speed and direction of rotation of which can be varied to control the type and thickness of coating in accordance with requirements. The lay-out of a line for aluminising steel strip up to 1.5 m in width and from 2–0.5 mm in thickness has been described[12, 13]. The design can also be adapted to the construction of a dual-purpose plant, the aluminium pot in this case being replaced by a separate zinc pot and a different exit mechanism.

Steel wire can be prefluxed in a hot aqueous solution of fluorides for aluminising, or treated in a manner similar to that used for strip.

Molten aluminium rapidly forms intermetallic compounds, the most important of which is $FeAl_3$ (41% Fe). The higher the carbon content of the steel, the thinner and more uniform will be the layer of alloy formed. Since the compound $FeAl_3$ is very brittle, the alloy should ideally not exceed 10^{-3} cm in thickness. An addition of 1.5–4% silicon to the bath[14] reduces the thickness of the alloy layer by more than 50% and considerably lowers its hardness. It also reduces the total coating thickness and increases its formability. Silicon present in the basis metal also has some effect in reducing the thickness of the alloy layer.

The alloy cannot be eliminated, whatever additions are made to the bath, and therefore, if the coating is to be subsequently subjected to deformation, the time of immersion and bath temperature must be kept to a minimum. Coatings of 0.025 mm thickness can be deformed satisfactorily, but their resistance to atmospheric corrosion will not be very high. The durability of an undeformed coating is proportional to its thickness, and if it is to withstand atmospheric corrosion it must be made thicker, with consequent loss in formability.

Thick aluminium coatings have an excellent resistance to heat, because at high temperatures the coating is converted to a heat-resistant alloy.

Aluminium is normally anodic to iron in the electrochemical series, but owing to the dense and protective nature of the aluminium oxide film forming on its surface and its polarising influence, anodic protection does not usually occur. Rust stains can develop at cut edges or at discontinuities in the coatings on exposure to the atmosphere, but the resulting attack proceeds relatively slowly, being stifled by the oxide formed. In marine or highly industrial atmospheres under moist conditions some cathodic protection can develop, but the rate of attack on the coating is slower than with zinc. Aluminium coatings are thus more durable under such conditions than might be expected, and are, in practice, comparable with galvanised coatings.

The outdoors durability of the product can be improved by a chromate treatment.

This delays the development of white corrosion products and also has an inhibiting effect at any local defects that may occur. Many chromate treatments are available, including proprietary ones; these are applied by simple immersion and colour the coating yellow or greenish-yellow, depending on the solution used (see 12.1.4).

The actual and potential scope for the use of aluminised steel is considerable, chiefly because it combines corrosion resistance with the ability to withstand considerable temperatures over long periods. It is in the latter respect that it is superior to galvanised coatings, to which it is in other ways similar. A further important advantage is the non-toxicity to food products of aluminium coatings as compared with zinc. It is used for automobile silencers, gas fired heat exchangers, baking trays, agricultural machinery and cladding for buildings.

14.2. Diffusion coatings

14.2.1. GENERAL

The chemical composition, and hence the properties, of a metal surface can be modified, or even completely changed, by diffusing another metal or a non-metallic element into it. This will result in the formation of an alloy between the parent metal and diffused metal, or non-metal, at the surface, the dimensional changes accompanying this diffusion being normally slight.

Although copper and its alloys, nickel, cobalt, titanium and its alloys, molybdenum, tungsten, niobium and tantalum have been used as parent metals for the deposition of diffusion coatings, the main application of the latter is to ferrous metals. Zinc, aluminium, chromium, silicon and boron are most frequently used as the diffusing metals. All of them (except boron, unless subsequently treated) protect steels against atmospheric corrosion, but chromium also affords protection against high-temperature oxidation and reduces friction. Aluminium and silicon diffusion coatings raise the thermal oxidation resistance of ferrous metals, whereas boron imparts high hardness to other surfaces.

Diffusion coatings are deposited either by heating the components to be treated in contact with the coating metal powder in an inert atmosphere (solid-state diffusion), or by heating them in an atmosphere of a volatile compound of the coating metal (gas-phase deposition).

Diffusion of metals atoms in the solid state is caused by thermal oscillations; the atoms leave their positions in the crystal lattice, thereby creating vacancies behind them. Thus, movement of both atoms and vacancies (holes) occurs. The equilibrium concentration of vacancies and straying atoms is temperature-dependent—the higher the temperature, the greater their concentration. Lattice distortion occurs during the diffusion process when substitutional or interstitial solid solutions form in the surface layer. These distortions lower the plasticity and increase the hardness of the metal surface.

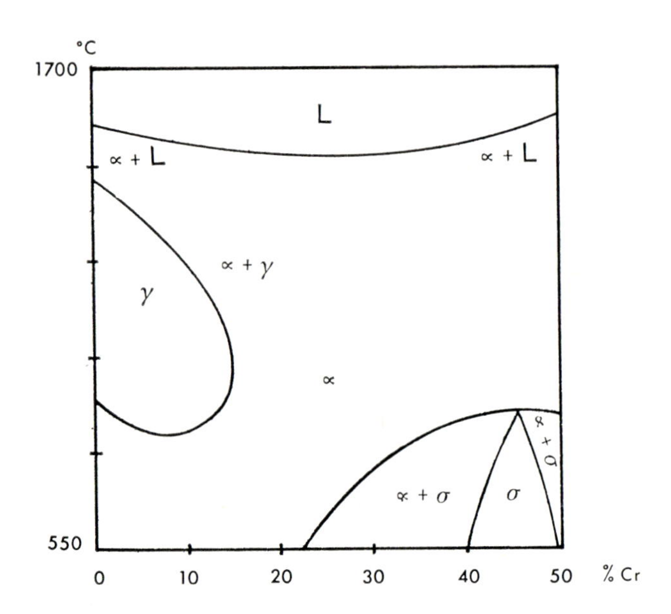

Fig. 14.3. Iron-chromium equilibrium diagram.

The microstructure of chromised coatings is usually columnar. The change in composition with distance from the surface is gradual, and this integration of the coating with the basis metal imparts valuable properties to it.

Coatings produced by chromising have three principal fields of utility:

1. They confer resistance to oxidation to steels at elevated temperatures (e.g. a diffusion depth of 0.04 mm will protect mild steel for long periods at 750°C).
2. If conditions are such that iron-chromium carbides are formed, excellent resistance to wear results, and on certain cast irons a surface hardness of up to about 1,800 VPN has been attained.
3. If an iron-chromium solid solution can be produced with a minimum of chromium carbide, excellent corrosion resistance will result. However, should the coating consist of grains of iron-chromium solid solution delineated by chromium carbides, preferential attack may occur in certain environments on the solid solution. This is due to depletion of the latter constituent in chromium which has been used up in the formation of chromium carbides. This danger can be avoided by using steels containing small proportions of niobium or titanium. The latter, being stronger carbide-formers than chromium, prevent the chromium depletion of the iron-chromium solid solution.

14.2.3.2. Siliconising

Additions of silicon to normal carbon steels (within the limits of solubility of silicon in iron in the solid state) raise the high-temperature oxidation resistance of the metal to about 800°C, and increase its resistance to attack by aggressive acid media (e.g. nitric and hydrochloric acids). Siliconised articles made from carbon steel acquire the surface properties characteristic of iron-silicon alloys. Bulk alloys of this composition are brittle and cannot be worked. Hence, components made from this material can only be made by casting. The advantage of siliconising is that mild steel components of complex shape can be produced by processes such as stamping or forging prior to diffusion of silicon into their surface.

Chlorine gas is passed over silicon-bearing powder contained at one end of a retort and heated to about 980°C. Silicon chloride vapour is formed, which passes over the articles to be processed at the other end of the retort. The following reactions occur:

$$Si + 2\,Cl_2 \rightarrow SiCl_4\ \text{(vapour)} \qquad \text{(i)}$$
$$4Fe + 3\,SiCl_4 \rightarrow 3\,Si + 4\,FeCl_3\ \text{vapour} \qquad \text{(ii)}$$

Equation (ii) represents a displacement reaction. At the steel surface $SiCl_4$ is decomposed and atomic silicon diffuses into the steel. $FeCl_3$ vapour and unreacted Cl_2 gas pass from the retort into an exhaust fan through a water trap.

As in chromising, there is a slight increase in volume.

14.2.3.3. Mixed diffusion coatings

Chromocalorising and chromosiliconising as well as many other combinations are used, the coating being produced by the gaseous method. The advantage of mixed coatings is a somewhat greater high-temperature oxidation resistance.

14.2.4. ELECTROLYTIC DIFFUSION COATINGS (METALLIDING)

N. C. Cook [15] has reported work carried out at the General Electric Research and Development Center (U.S.A.) on the simultaneous electrodeposition and diffusion of metals from molten fluoride salt baths. Altogether, 25 different metals were electrodeposited on, and diffused into, 40 different metal substrates, the total number of combinations so far produced being 400. Examples are the diffusion of (i) boron into the surface of molybdenum, resulting in a surface with a hardness approaching that of diamond; (ii) silicon into molybdenum, producing a material capable of withstanding white heat for hundreds of hours, (iii) beryllium into copper, making it stronger, harder and more oxidation-resistant without impairing its electrical conductivity. Borided steel is as hard as tungsten carbide, titanided copper resists boiling nitric acid and tantalided nickel becomes almost as resistant to corrosive oxidation as pure tantalum.

This simultaneous electrodeposition and diffusion process has been given the name 'metalliding'. Molten fluorides of alkali metals and alkaline earth metals are used as solvents. The fluorides have melting points ranging from about 400° to 1,350°C. These salts are more stable chemically than any of the fluorides dissolved in them for metalliding. Thus, they act not only as solvents but also as media from which metal ions can be deposited onto metallic surfaces. In addition, the fluxing action of the molten fluorides dissolves the oxide film which forms in air on all but the noblest metals from the surface of the cathode metal. Diffusion of the electrodeposited metal into the substrate is rendered possible by the high bath temperature, which provides the necessary activation energy. Only very small quantities of the metal to be deposited need to be dissolved in the solvent, which acts as a very efficient electrolyte. Since there is no hydrogen in the bath, highly electronegative metals such as aluminium, titanium, etc., can be deposited.

A number of methods of producing wear and corrosion-resisting coatings by electrodeposition followed by thermal diffusion have been developed, and have been used to some extent. Tin deposits diffused into iron and steel at 580°–600°C are used in marine engines and on cranes and trucks. Bronze deposits diffused into copper alloys and steel at about 550°C produce coatings 20–40 μm thick which can replace solid bronze, whilst copper-indium alloys applied on aluminium and diffused at about 130°C have been used on motor cylinders, cycle wheel sprockets, and are especially useful for dry running and in high vacuum. The processes have been described by Caubet[(16)].

14.3. Metal spraying

14.3.1. GENERAL

The object of applying sprayed metal coatings to a metallic substrate is to protect the latter against atmospheric corrosion, and in many instances also to improve its appearance. Sprayed metal coatings may also be applied to repair worn surfaces, to correct rejects due to faulty machining or casting faults, or to impart specific properties to a substrate. The advantages of this method of metal film application over others are that (1) large, assembled structures can be coated *in situ;* and (2) non-conducting materials can be spray-coated, e.g. paper, wood, cellulose, plaster of Paris, etc. Its disadvantages are that (1) even relatively thick coatings are porous; (2) the adhesion is poorer than that of coatings applied by other methods; and (3) a considerable proportion of the sprayed metal is wasted.

The degree of adhesion of the sprayed globules of liquid metal depends (a) on their size, (b) on their velocity of flight, and (c) on the degree of deformation they undergo when impinging on to the work. When a droplet impinges on the work surface it flattens out and merges with neighbouring particles by partly diffusing into them. Eventually a hard, laminar coating is formed.

14.3.2. FLAME SPRAYING

Two principal methods of flame spraying are available, using materials in two different physical forms—wire and powder. The spray guns, or pistols, can be either gas-fired or electric.

In the wire process, a wire of the coating metal is melted by means of an oxy-acetylene or oxy-hydrogen flame; in the case of electric pistols it is melted by means of alternating current. Such pistols consist of a wire feeder and a nozzle; the important features of the gun are shown in Fig. 14.4. The metal wire is fed to the atomiser, and the drop of molten metal formed by the heat generated is disintegrated by a stream of compressed air, which is directed against the surface of the work. In gas-fired pistols the heat is produced by combustion of the gas, and in electric pistols it is generated by two current-carrying wires melting at the point of contact due to formation of an electric arc. The wire melts 1 to 2 mm from the nozzle orifice, which is held at a certain distance (typically 15 to 20 mm) from the work surface.

The powder process invented by Schori[17] depends on the melting of fine metallic particles suspended in a gaseous medium being blown through a blow pipe flame. Many materials which can be supplied in the form of wire or powder can be sprayed. This includes metals, plastics and a range of lower-melting refractory materials. High-melting refractories can be sprayed by the use of thermal plasmas[18]. A thermal plasma, the heat source of a plasma torch, is produced by partially ionising a gas in an electric arc and passing it through a small orifice to produce a jet of hot gas moving at high velocity. The energy supplied by the arc to dissociate and ionise the gas is released to the gas stream in the form of heat as ions capture electrons and atoms combine to form molecules. Nitrogen is normally used for the plasma flame, but for those materials that react with it, argon with a small proportion of hydrogen is usually employed.

14.3.3. SURFACE PREPARATION

The degree of adhesion of a sprayed metal coating depends largely on the quality of surface preparation of the basis metal. The latter must be free from rust or scale, and is normally grit-blasted with angular steel grit or aluminous abrasives. There is insufficient heat present in the particles at the time of their arrival at the basis metal surface for incipient welding to take place, and adhesion therefore depends essentially on the mechanical interlocking of the metal particles with a rough surface.

14.3.4. APPLICATIONS OF SPRAYED METAL COATINGS

Zinc and *aluminium* are normally used as coatings for steel structures and components in order to protect them against corrosion. Both are anodic to iron and hence protect it sacrificially. Although aluminium is thermodynamically more reactive than zinc, the impervious thin oxide film formed on its surface greatly reduces its reactivity and makes it more noble than zinc in most environments. Hence, an aluminium coating will give less sacrificial protection to steel than zinc. Nevertheless, the protection offered by a sprayed aluminium coating to a ferrous substrate, as well as to many other substrates, is very high. This is due to the fact that sprayed metal coatings, immediately

Fig.14.4. (above) Modern wire-type spraying pistol.

[Courtesy Metco Ltd.]

(below) Schematic diagram of gas-fired wire gun.

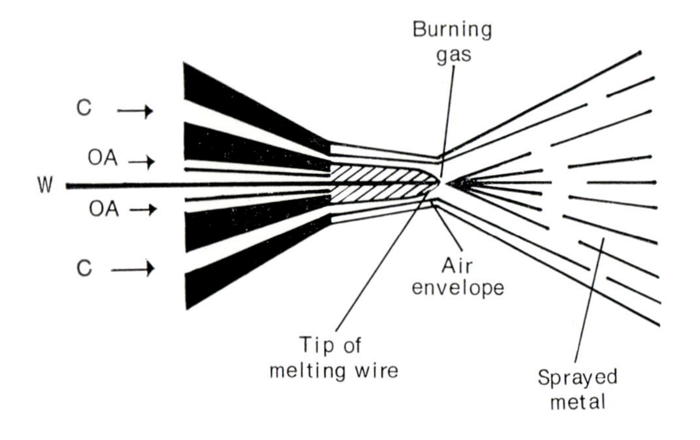

after application, are porous. Initial corrosion of the coating results in the formation of insoluble corrosion products which clog the pores, thereby converting the coating into an impervious layer and stifling further attack. The steelwork of the Severn and Forth road bridges were zinc-sprayed and primed before erection.

Whereas aluminium-sprayed coatings give good protection in sea water and sulphurous atmospheres, zinc has a superior resistance to the combined action of sulphur and chlorine.

The thickness recommended for a sprayed coating depends on the conditions of service. For service under atmospheric conditions, the average thickness of a sprayed aluminium coating should be about 0.1 mm, and that of a sprayed zinc coating 0.075 mm. If the surface is subsequently to be painted, an average thickness of 0.05 mm is satisfactory. For immersed conditions, the coatings should be thicker.

If aluminium-sprayed steel components are sealed with bitumen or water glass and annealed at 850°C for approximately half an hour, diffusion of iron into the aluminium coating will result in the formation of alloy layers of different compositions within the coating below a skin of aluminium oxide. Such deposits will withstand oxidation at up to 900°C for very long periods of time. In the case of Al–0.75% Cd coatings, sealing may be omitted.

Sprayed *lead* coatings offer protection against sulphuric acid-containing atmospheres. Lead sulphate forms due to initial attack, which fills the pores and prevents penetration of the corrosive medium to the substrate. Lead is cathodic to steel, and hence at points

Fig. 14.5. Powder spray metallising gun. [Courtesy Metco Ltd.]

Fig. 14.6. Plasma spray metallising gun. [Courtesy Metco Ltd.]

of mechanical failure corrosion of the steel substrate will be accelerated, often resulting in detachment of the lead coating.

Sprayed *tin* coatings are used for food containers. Since tin is even more cathodic to steel than lead, the pores must be completely sealed. This is done by depositing a thick (0.4 mm) coating and polishing it. Since tin is soft, it will flow easily when polished and the pores will be filled in.

Stainless steels, copper, nickel, Monel metal and nickel/chromium alloys are sprayed for various purposes. Nickel/chromium coatings, when heat-treated at 1,100°C, will give protection against high-temperature oxidation in air. Pump rods are sprayed with thick coatings of stainless steel, nickel or Monel metal. These must subsequently be ground or polished.

Sprayed coatings of *copper* and its alloys are used for ornamentation, but since they are very strongly cathodic to steel, they are suitable for service only under conditions of very low humidity (e.g. interiors of buildings in dry atmospheres).

Chromium/nickel/beryllium alloys are used in a spraying process known as hard facing. The coating is subsequently heated by a torch until fusion and welding to the substrate takes place. For this reason, the process is also known as spray-welding. Such coatings combine an excellent wear resistance with a high corrosion resistance.

14.4. Vacuum evaporation

When a metal is vaporised in a chamber evacuated to pressures of 10^{-4} to 10^{-5} mm Hg, the metal vapour will fill the chamber evenly and condense on any cool surfaces irrespective of whether they are metallic or not. Thus, a deposit will be obtained, not only on the articles intended to be coated by the metal, but also on the walls of the chamber.

The articles for coating are normally pickled, degreased and cleaned, dried and, for decorative purposes, often lacquered. The lacquer prevents outgassing of foreign matter during evacuation, and at the same time provides a smooth basis on which a bright coating can build up. The articles, particularly if they are of irregular shape, should be placed in the coating chamber in such a position as to avoid 'shadowing' and thereby reduce to a minimum the wastage of coating metal. The coatings can be controlled to be of fairly uniform thickness, and there is comparatively little build-up on edges.

Where hydrogen embrittlement is to be avoided, pickling is replaced by shot- or grit-blasting. Adhesion can be further improved by post-evaporation heat treatment.

The rate of deposition depends on the particular coating metal, e.g. cadmium and zinc coatings of 0.025 to 0.05 mm thickness can be deposited in less than ten minutes, whereas considerably longer times are required for aluminium and copper, and much longer times still are needed for higher melting metals.

The process is relatively simple, and as compared with electroplating has the advantage of being a dry one, thus obviating the necessity for subsequent rinsing and drying. The temperature rise of the articles being plated is very small. In addition, the deposit is free from inclusions and normally also free from pores.

Vaporised aluminium coatings are applied over painted steel pressings for motor-car headlamp reflectors, and onto metal and plastic articles for decorative purposes[19]. The aluminium vapour is produced by heating tungsten filaments covered with aluminium[20] in the vacuum chamber. Gold finishes are produced by subsequently applying a yellow lacquer.

Limitations of the process are the comparatively restricted capacity of the vacuum plant and the time required to deposit an appreciable thickness of high-melting metals.

Whereas aluminium is widely used as a deposit metal for decorative purposes, tin, zinc and cadmium are used for their protective properties. Where protection rather than decoration is the main requirement, the intermediate lacquer coating is normally omitted.

Vacuum evaporation is used for depositing protective coatings on very high-strength steels where hydrogen embrittlement is to be avoided. It finds application in aviation and space craft, as well as for lamp reflectors and for the preparation of specimens for inspection under the electron microscope.

14.5. Cathodic sputtering

The plant and procedure used are similar to those used in vacuum evaporation. The atmosphere of the coating chamber may be either air or an inert gas. The pressure inside the chamber is reduced and an arc struck. The deposit metal is vaporised and the vapour deposits on all surrounding cool surfaces (as in the case of vacuum evaporation).

The rate of deposition is controlled by the temperature and pressure of the coating chamber atmosphere, by the arc voltage, cathode current density and the geometry of the cathode and of the articles to be coated. When an inert gas is used, a high vacuum is not essential. The rate of disintegration of the coating metal is affected by the atomic weight of the coating chamber gas.

The rate of deposition is slower than in vacuum evaporation and hence this method has been confined to special purposes, e.g., rendering crystal vibrators electrically conducting, coating parts in some selenium rectifiers, etc. However, recent developments have speeded up the process, so that it is being used on external automobile parts.

14.6. Peen plating

When powders of certain soft metals are tumbled with solid articles, a film of the soft metal gradually builds up on the latter. This is due to the powder particles attaching themselves to the solid metal surface, initially mechanically and subsequently by welding on to the surface and to each other. A film of metal so deposited is referred to as a peen-plated coating.

Since initial keying action is involved, adhesion will be the better the rougher the surface of the article to be coated. For this reason, articles are abraded or pickled prior to being peen-plated. A 'strike' with copper or another soft metal will facilitate initial welding-on.

'Promoters' and wetting agents are used to remove traces of grease and oxide films, which would prevent adhesion.

Since the process works only with soft metals, it is in practice restricted to tin, zinc, cadmium, lead, aluminium and certain soft alloys. The geometry of the articles to be coated must be simple, i.e. their entire surface must be directly accessible to the powder particles and must be capable of making intimate contact with them during tumbling.

The method is empirical in that the quantities and grades of coating powder, as well as the speed of rotation of the barrel, must be ascertained experimentally. The rate of deposition is slow (approximately 0.007 mm per hour), but relatively thick deposits (0.05 mm) can be built up. Subsequent heat treatment will improve adhesion.

Peen-plated coatings are used essentially because of their good corrosion resistance rather than for their decorative properties, since they are not bright. Applications include plating high-tensile steel springs and other parts where hydrogen embrittlement must be avoided, and the coating of ferrous nails, bolts, washers and chain links.

Zinc peen-plated steel articles are often given a chromate passivation treatment to further increase their corrosion resistance.

14.7. Metal cladding

If a 'duplex' ingot is made in such a way that the central portion of one metal is surrounded by a thick layer of another, then, provided that the two metals have similar rolling characteristics, the ingot can be rolled into plate, bar or sheet in which the outer-layer metal will form a continuous coating over the core metal. Coated wire can also be drawn from a duplex billet in this way.

Structural aluminium alloys clad with commercially pure aluminium are widely used in aircraft construction in the form of sheet or plate where a combination of strength, lightness and good resistance to corrosion is required. The latter is due to the layer of aluminium.

Stainless steel-clad copper is used for the manufacture of cooking vessels. The outer layer of stainless steel provides durability, whereas the copper core ensures the required lateral conductance[(21)]. Aluminium-coated steel strip is also made on a commercial scale by hot-rolling the two metals together.

Nickel and Monel are readily bonded to steel; the bonding of stainless steel to steel surfaces is facilitated by nickel plating.

Explosive bonding has found some application, as it enables virtually any metal to be bonded to another without the formation of an intermediate alloy layer. Explosives are used to produce shock waves which cause the metals to weld together. Two types of explosives are used, i.e. the high velocity type detonating at a speed of 15,000 to 20,000 ft. per second, such as TNT, and the cheaper low and medium velocity types, having detonation velocities of 5,000 to 15,000 ft. per second, such as dynamite. The explosive can be used in the form of a flexible plastic sheet, cord or a variety of cast shapes, as well as powder or granules. Clad sheets are made by placing the two panels to be joined, either parallel to one another, or at a small angle. One of the main problems with explosive forming is the difficulty of carrying it out in a factory or generally in inhabited area, because of the noise problem, which not only constitutes a nuisance, but can create a good deal of alarm. Experiments to reduce noise have included carrying out the operation under water or sand or in a vacuum chamber. Explosively cladded plate can be formed for any of the wellknown methods such as welding, rolling, bending, pressing, flanging and spinning. They can also be flame cut, ground and welded by suitable techniques. Vessels 8 metres high and 3 metres in diameter have been constructed from titanium clad steel sheet. Boiler ends and sections of pressure vessels are produced, as well as circular stainless clad steel doors. On a smaller scale, dental plates and surgical implements constitute another fruitful field of application.

REFERENCES

1. W. G. Imhoff, *Wire and Wire Prod*, 1952, **27**, 356, 420, 630, 692.
2. D. L. Phillips, Iron and Steel Inst., Sp. Rept. No. 49, 1954, 1.
3. H. Bablik, *Metal Finishing J.*, 1955, **1**, 5.
4. H. Bablik, F. Gotzl and E. Nell, *ibid*, 1955, 347.
5. M. L. Hughes, *J. Iron and Steel Inst.*, 1950, **166**, 77.
6. B. Ulrich, Internat. Conf. Hot Dip Galvanising, 5th Benelux, 1958, (Sidney Press Ltd.), p. 185.
7. A. T. Baldwin and W. H. McMullen, *Iron Age*, 1957, **179**, No. 13, 115.
8. F. C. Porter, Focus on Zinc I, *Anti-Corrosion*, 1970, **17**, No. 10.
9. K. J. Lewis, Focus on Zinc 5, *Anti-Corrosion*, 1970, **17**, No. 10.
10. W. E. Hoare, 'Hot Tinning', 2nd Ed. Tin Research Inst. (1964).

11. M. L. Hughes and D. P. Moses, *Metallurgia*, **48**, 287, (1953).
12. H. Silman, Trans. Inst. *Metal Finishing*, **40**, Pt. 2, 85 (1963).
13. H. Silman, *Surface Coatings*, 1966, April.
14. D. O. Gittings, D. H. Rowland and J. O. Mack, *Trans. Amer. Soc. Metals*, **43**, 587 (1951).
15. N. C. Cook, *Scientific American*, 1969, August, p. 38.
16. J. J. Caubet and G. Salomon, *Schmierlechnik & Tribologie*, 1974, (5) 106-110.
17. F. Schori, Brit. Pat. 432, 831 (Jan. 8 1934).
18. 'Protective Coatings for Metals', R. M. Burns, W. W. Bradley, p. 90. Reinhold Publ. Corp., N.Y., London, 1967.
19. S. Tolansky, *J. Electrodepositors' Tech. Soc.*, 1951-52, **28**, 155.
20. L. Holland, *J. Electrodepositor's Tech. Soc.*, 1951-52, **28**, 167.
21. J. Kinney Jr., *Product Eng.*, 1952, **23**, No. 4, 129.

GENERAL READING

In view of the very wide subject matter covered in this chapter a Bibliography of further general reading is given below.

I. *Hot-dipped Coatings*
1. Burns, R. N., and Bradley, W. W. 'Protective Coatings for Metals', A.C.S. Monograph No. 163 (3rd Edition). Reinhold Publishing Corporation, 1967.
2. First International Conference on Hot Dip Galvanising, Copenhagen, 1950, Zinc Development Assn., Oxford, 1950.
3. Fourth International Conference on Hot Dip Galvanising, Milan, 1956, Zinc Development Assn., London, 1957.
4. Hoare, W. E. and Hedges, E. S., 'Tinplate', Arnold, London, 1945.
5. Bablick, H., 'Galvanising (Hot Dip)', Spon, London, 1950.
6. Bakhalov, G. T. and Turkovskaya, A. V., 'Corrosion and Protection of Metals', Pergamon Press, 1965 (translated from Russian).
7. 'Corrosion', Ed. by L. L. Shreir, George Newnes Ltd., 1963.

II. *Diffusion Coatings*
1. Samuel, R. L., in 'Corrosion', Vol. 2, ed. by L. L. Shreir.
2. Burns, R. M. and Bradley, W. W., 'Protective Coatings for Metals', A.C.S. Monograph No. 163 (3rd ed.), Reinhold Publishing Corporation, 1967.
3. Bakhalov, G. T., and Turkovskaya, A. V., 'Corrosion and Protection of Metals', Pergamon Press, 1965.

III. *Metal Spraying*
1. Ballard, W. E., 'Metal Spraying and Sprayed Metal', Griffin, London, 1948.
2. Ballard, W. E., 'Metal Spraying and Flame Deposition of Non-Metallic Materials', Griffin, London, 1963.
3. Burns, R. M. and Bradley, W. W., 'Protective Coatings for Metals,' A.C.S. Monograph No. 163 (3rd ed.), Reinhold Publishing Corporation, 1967.
4. Bakhalov, G. T. and Turkovskaya, A. V., 'Corrosion and Protection of Metals', Pergamon Press, 1965.

IV. *Vacuum Evaporation*
1. Burns, R. M. and Bradley, W. M., 'Protective Coatings for Metals', A.C.S. Monograph No. 163 (3rd ed.), Reinhold Publishing Corporation, 1967.
2. 'Corrosion', ed. by L. L. Shreir, George Newnes Ltd., 1963.

V. *Cathode Sputtering*

1. Holland, N., 'Vacuum Deposition of Thin Films', Chapman & Hall, London, 1956.
2. 'Corrosion', ed. by L. L. Shreir, George Newnes Ltd., 1963.

VI. *Metal Cladding*

1. Burns, R. M. and Bradley, W. W., 'Protective Coatings for Metals', A.C.S. Monograph No. 163 (3rd ed.), Reinhold Publishing Corporation, 1967.
2. Bakhalov, G. T. and Turkovskaya, A. V., 'Corrosion and Protection of Metals', Pergamon Press, 1965.

CHAPTER 15

Testing methods

THE testing of coatings must form an integral part of the quality control system. Visual appeal is always an important aspect of quality but it is regrettable that all too often it has tended to be the only criterion of respectability. This has two disadvantages, since it may result in the rejection of finishes of greatly superior durability on account of a slightly reduced lustre, for example, whilst on the other hand finishes of very low quality may be accepted because they look excellent. Whilst visual examination provides useful data as to the quality of a finish, such inspection can be greatly misleading when not coupled with other inspection tests.

The essential characteristics of a finish which are of primary importance are:

1. Appearance
2. Thickness
3. Porosity
4. Corrosion resistance
5. Adhesion
6. Other properties, e.g. hardness, ductility, stress- and abrasion-resistance, composition and structure.

15.1. Thickness determination

To a very considerable extent the protective values of many coatings, and particularly of electrodeposits, are a function of their thickness and all specifications include a thickness requirement.

The measurement of the thickness of an electrodeposit can be carried out in several ways.

15.1.1. MICROSCOPIC METHOD

The most accurate method of determining the thickness of a coating is to prepare a section of the specimen by normal metallographic procedures, after which the thickness is measured under the microscope. The sample is best mounted into a plastic or resin matrix or embedded in a low-melting alloy to prevent flow of the coating during polishing when soft metals such as zinc or cadmium are being dealt with. It is essential, however, to ensure that the section is truly perpendicular to the deposit, otherwise considerable errors may result. This method is, however, slow and expensive and is only employed for special purposes such as checking the accuracy of other methods. Under the best conditions and taking precautions[1], the method is capable of giving an accuracy of $\pm 0.8\ \mu m$.

15.1.2. MESLE CHORD METHOD

In this method, which is little used nowadays, a circular grinding wheel is applied to the coating until it is cut through (Fig. 15.1). The thickness can then be calculated from the formulae:

$$t = c^2/8r$$

where t is the thickness of the deposit, c is the width of the cut, and r the radius of the wheel. The method is limited by the fact that a precision grinder is required for flat surfaces, but it can be applied to curved surfaces by using a fine flat file if the radius of curvature of the surface is known. It cannot readily be used on concave surfaces.

15.1.3. CHEMICAL METHODS

It is possible to dissolve the coating from the basis metal and to determine the weight of metal present in the solution by chemical analysis, and hence to calculate the thickness of the deposit. This is seldom carried out, however, since it is often simpler to select a treatment which will dissolve the deposit without affecting the basis metal.

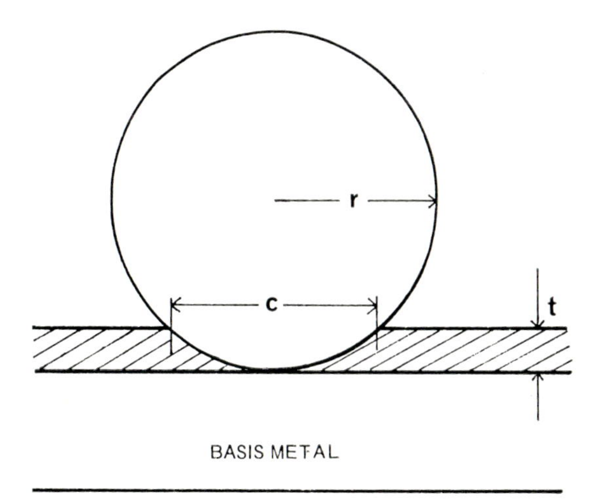

Fig. 15.1. Principle of the Mesle chord method.

By weighing the specimen before and after stripping, the weight of the deposit is obtained which is divided by the surface area of the part and the density of the metal to give the average thickness.

Nickel can be estimated by anodic dissolution in 70 per cent sulphuric acid using a current density of about 2 A/dm^2. This method is applicable to deposits on steel, copper, brass and zinc-base alloys. It is, however, difficult to ensure absolute freedom from attack, especially on steel, so that this method is not very accurate. Another method has been suggested by Mears[2] in which the deposit is made the anode in a 20 per cent solution of sodium cyanide. The steel is not attacked but passivity may develop in the nickel; should this occur the current is momentarily reversed and the process continued.

Chromium is best stripped anodically in caustic soda solution from all basis metals except zinc, when sodium carbonate solution should be used.

Zinc, cadmium, and tin are stripped from steel by means of a strong solution of hydrochloric acid containing a small amount of antimony trioxide which acts as an excellent inhibitor of attack on the basis metal. Cadmium and zinc are also readily removed from steel by a 5 per cent aqueous solution of ammonium persulphate to which is added 10 per cent by volume of ammonia (sp. gr. 0.880). In the case of zinc, a hot sodium polysulphide solution made up from 250 g/l of sodium sulphide ($Na_2S,9H_2O$) and 25 g/l of sulphur is satisfactory.

Copper deposits are removed from steel by means of a chromic acid solution containing about 500 g/l of chromic acid and 5 g/l of sulphuric acid. The solution is best used hot.

Anodic stripping of copper can be satisfactorily carried out in a solution consisting of:

Copper cyanide	40 g/l
Sodium cyanide	80 g/l
Trisodium phosphate	60 g/l
pH	12.2

The solution deteriorates rapidly, however, if large anodic areas of steel are exposed due to the formation of carbonates and the decomposition of cyanides. The former must be removed and the latter replaced if the bath is to be regenerated. The pH should be controlled at between 11.9 and 12.5 by the addition of caustic soda; the presence of *cyclo*hexylamine in combination with sodium sulphite has been found to be effective prolonging the life of the bath[3].

Nickel or copper or both can also be removed from steel by anodic treatment in sodium nitrate solution. The solution is fairly concentrated (500 g/l) and is kept at a pH of 7.6 to 8.2. It is used at a current density of 20 A/dm^2 at 6 volts, and a steel cooling coil is required in the solution to prevent overheating by the large currents used.

Copper, brass, bronze, cadmium, zinc and also gold and silver can be stripped anodically in a solution containing caustic soda 15 g/l and sodium cyanide 100 g/l.

Silver deposits on brass or nickel silver can be stripped by immersion in a bath of hot concentrated sulphuric acid containing 5 per cent by volume of concentrated nitric acid.

Lead can be removed from steel by dissolution in a mixture of acetic acid and hydrogen peroxide used at room temperature. A suitable bath consists of:

Acetic acid	10 per cent
Hydrogen peroxide	10 to 20 per cent

Small additions of hydrogen peroxide should be made from time to time to keep the solution active.

Gold can be dissolved from copper, brass, steel or nickel alloy surfaces by treatment in a sodium cyanide-hydrogen peroxide bath, small amounts of hydrogen peroxide being added as required to a solution containing about 50 g of sodium cyanide per litre of water. The thickness of some gold or other precious metal deposits can sometimes be determined by dissolving away the basis metal and directly weighing the deposit.

15.1.4. B.N.F. JET TEST

The B.N.F. Jet Test is a widely used method for the routine measurement of plating thicknesses[4]. It has the advantage of being readily carried out and requiring very inexpensive apparatus. In this method a jet of a specified reagent is directed on to the coating under carefully standardised conditions and the time for penetration determined. Suitable solutions are available to enable this method to be applied to nickel, copper, zinc, cadmium and silver deposits on most of the commonly employed basis metals. The jet is directed on to the carefully cleaned article under a constant head of 25 cm (10 in). The apparatus consists essentially of a glass capillary tube of 1.5 mm circular bore, not less then 10 cm in length, drawn out into a smooth taper at the nozzle end and ground off flush at the tip of the nozzle. The dimensions of the tapered portion should be such that $\frac{D-d}{L}$ must be between 0.030 and 0.045 where D is the internal diameter of the capillary tube, d the internal diameter of the nozzle tip and L the length of the taper. The jet must deliver 10 ml of water at 20°C in 27 to 29 seconds under a 25 cm (10 in) head. Fig. 15.2 shows a suitable apparatus for the test consisting of a container for the fluid with a stop-cock, a constant head device with thermometer, a shielded jet of the dimensions cited above and a clamp for holding the specimen.

For *nickel*, the test solution consists of:

Ferric chloride ($FeCl_3,6H_2O$)	300 g/l
Copper sulphate ($CuSO_4,5H_2O$)	100 g/l

When penetration has occurred, a copper spot appears in the case of nickel deposits direct on steel. As this is not visible whilst the jet is running, the latter must be stopped

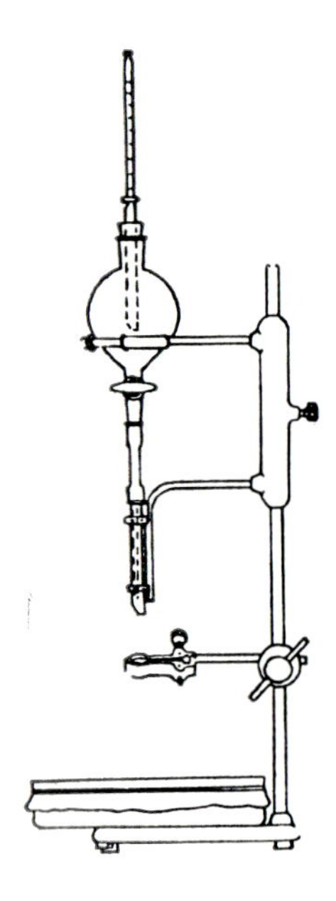

Fig. 15.2. B.N.F. jet test apparatus.

from time to time to ascertain whether penetration has occurred; a seconds clock is started and stopped during the test to obtain a measure of the total time during which the jet is actually running. The time of penetration is dependent on the temperature of the solution as shown by a thermometer immersed in the reservoir. When bright nickel deposits are being tested, allowance has to be made for the influence of the particular brightener on the time of penetration. Different brighteners give rise to different penetration times so that the test needs to be calibrated for each particular type of bright plating solution.

For *cadmium* the solution used consists of:

Ammonium nitrate	17.5 g/l
Hydrochloric acid (normal)	17.5 ml/l

Whilst for a *zinc* deposit, a solution four times as concentrated as the above is used. In this case penetration is complete when steel shows through the coating. This is a little more difficult to see.

The difficulty in observing the end-point in the case of cadmium deposits on steel can be overcome by applying one or two drops of a solution consisting of 10 grams of mercuric chloride and 50 ml of concentrated hydrochloric acid per 100 ml of water to the test spot. The effect produced is a bright mercury spot at the point of exposure of the steel, the surrounding cadmium being blackened. If the solution has been applied before perforation has been reached, the jet test should be recommenced at another spot.

Charts showing rates of penetration are given by Clarke (*loccit*). At 25°C 4.2 seconds are required to penetrate 0.00025 cm of nickel and 2.2 seconds for the same thickness of copper with the ferric chloride solution. In the case of the ammonium nitrate solution, 0.00025 cm of zinc is penetrated in 4.0 seconds and a similar thickness of cadmium in 9.4 seconds at 25°C.

For *silver* deposits[5], the solution used consists of 250 g/l of potassium iodide and 7.44 g/l of iodine. Silver is dissolved at the rate of 0.00025 cm in 5.6 seconds at 25°C, and in 6.6 seconds at 18°C. The end point is shown by the development of a bright crescent-shaped area on the underlying metal.

The thickness of *lead* deposits can be determined by the use of a solution consisting of:

Glacial acetic acid	1 vol.
Hydrogen peroxide (5% H_2O_2)	1 vol.
Distilled water	3 vol.

The strength of the peroxide is standardised by titration against potassium permanganate from time to time, since the solution is somewhat unstable. The method is applicable to lead coatings on steel; the specimen should be free from grease or oxide films before test[6].

The jet test is accurate to about 15 per cent, which is sufficient for most inspection purposes.

15.1.5. DROPPING TEST

A similar method employed in the U.S.A. consists in using an apparatus designed to give a series of drops instead of a jet of solution. The drops fall at the rate of 100 per minute, and the thickness of deposit is again a function of time and temperature.

15.1.6. COULOMETRIC THICKNESS TESTER

The principle of this method is the anodic dissolution of an area of the plated metal, the current-time product (coulombs) being determined and used as a measure of deposit thickness. The apparatus and method of measurement are fully described in a recent International Standard[7]. With one type of circuit the current through the test cell is controlled to a constant value and the end point denoted by a sharp rise in cell voltage which occurs when the coating metal has been removed. A relay is incorporated into the circuit which stops a timing clock and causes an indicator bulb to light up.

A more recent version of this type of instrument employs a sensitive electrical gate to stop a digital counter running at constant speed[8]. Alternatively, an integrating meter may be used to determine the number of coulombs passed, in which case it is not necessary to accurately control the value of the current or measure the time of dissolution.

The electrolyte used must show no reaction with the coating metal in the absence of

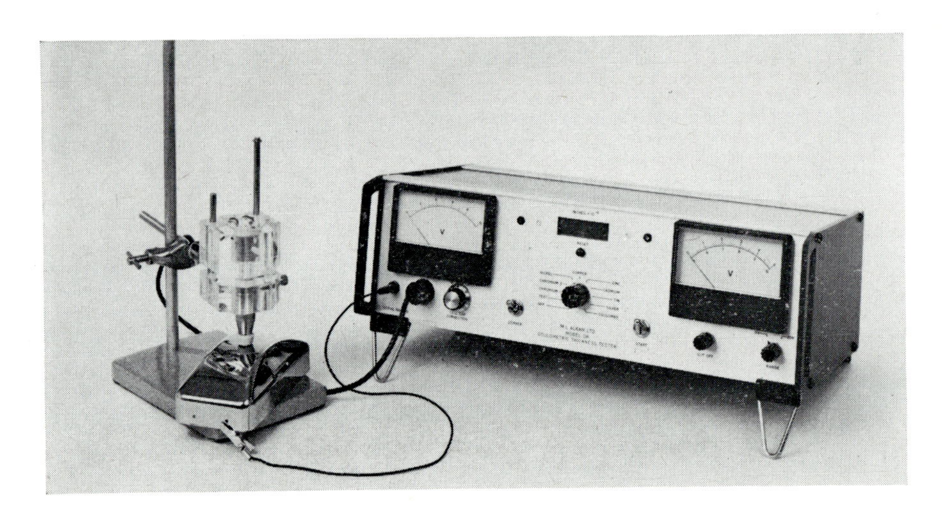

Fig. 15.3. Coulometric plating thickness meter. [Courtesy M. L. Alkan Ltd.]

an applied current and must be suitable for dissolving the deposit with 100% anodic efficiency. A wide variety of solutions, recommended for different purposes are given in the relevant International Standard[7]. The method is normally accurate to within ±10%.

15.1.7. MAGNETIC METHODS

Several types of magnetic gauges have been developed for thickness testing. The method, which is non-destructive and quickly performed, is nowadays very popular. It is applicable in many cases and has recently found use in the thickness determination of galvanised coatings on steel[9]. Coatings of thickness between 50 and 250 microns are said to be readily measurable to an accuracy within ±10%.

The principle of operation is based on either measuring the attraction between a magnet and the basis metal, which is affected by the thickness and nature of the coating, or by determining the reluctance of a magnetic flux passing through the basis metal and coating. Methods based on the first principle enable thickness determinations to be carried out on non-magnetic coatings on magnetic basis metals and *vice versa*, and also on magnetic coatings on magnetic basis metals such as nickel on steel.

One instrument which operates on this principle is the Magne-Gage shown diagramatically in Fig. 15.4. This consists of a spring-controlled torsion balance from which a small magnet is suspended. The test piece is clamped with the magnet first touching the surface and the dial rotated until the magnet is lifted from the surface. The thickness of the deposit is usually obtained directly from the position of the calibrated dial. In using this instrument care must be taken to minimise vibration of the test piece and consideration must be given to the roughness of the coating on the substrate surface which influences the measured results.

Instruments which operate by determining the variation of magnetic reluctance are limited to measuring the thickness of non-magnetic (metallic or non-metallic) coatings on magnetic basis metals. They are not widely used, are expensive and generally not very accurate.

Magnetic test devices are subject to some limitations. Curvatures of the surfaces being tested, both concave and convex, may influence the accuracy of the results seriously if at all pronounced, whilst at small thicknesses the sensitivity of the instrument tends to be low. The thickness determination of non-magnetic coatings on magnetic basis metals is dealt with in a recent International Standard[10].

15.1.8. EDDY CURRENT METHODS

By these methods the thickness of the coating is obtained by determining the difference in electrical conductivity between the coating and basis metal. The method is principally employed for the testing of non-conducting coatings on metals, but at the present time the appropriate International Standard[11] only recommends the method for the determination of thickness of non-conducting coatings on non-magnetic but conducting materials.

The principle of the method is to induce eddy currents in the conducting basis material via a probe through which is generated a high frequency electromagnetic field. The magnitude of the eddy current developed is dependent on the thickness of the non-conducting coating interposed between the probe and basis metal. The measuring

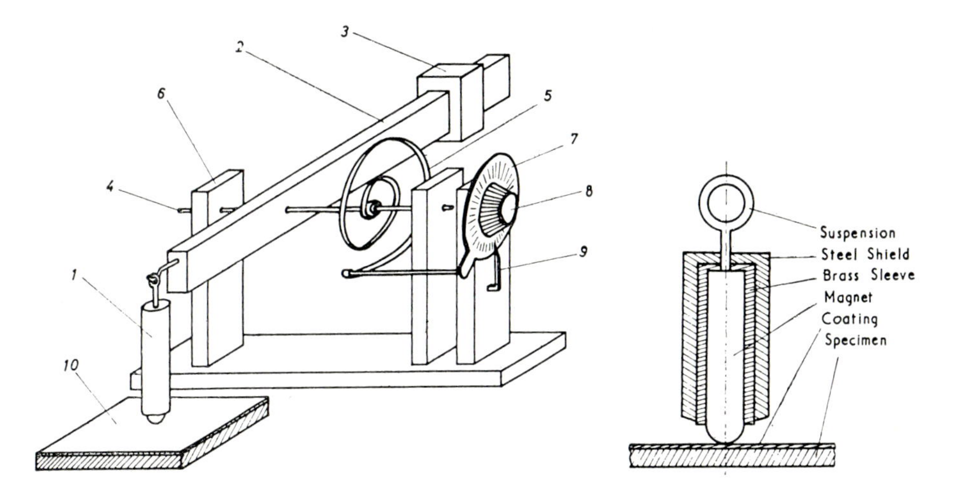

Fig. 15.4. Schematic diagram of the Magne-Gage.

accuracy is normally of the order of ±10% of the coating thickness, but at small coating thicknesses (<2, microns) the accuracy is somewhat decreased.

15.1.9. BETA-BACKSCATTER METHOD

A radio isotope which emits β-rays is employed in this method together with a detector for measuring the intensity of backscattered radiation from the specimen. The principle of the method[12] is shown in Fig. 15.5. β-radiation is backscattered from the coated metal into an ionisation chamber which registers the radiation intensity on an associated indicator. The radio isotope source must have sufficient energy to penetrate the coated metal and must be carefully shielded so that no direct radiation reaches the ionisation chamber. The degree of backscattering is influenced by the thickness of the metal; it increases with increasing thickness up to a limiting value which depends on both the type of material and the source energy. The intensity of backscattering also increases with increasing atomic number of the metal. Thus, more radiation is backscattered from tin than iron, so that in the case of tinplated steel the intensity of backscattering will increase with increasing thickness of tin from the limiting value for steel up to the limiting value for tin.

It is claimed[13] that the technique offers several advantages over other thickness testing methods:

(a) high precision measurements can be ±4% of plating thickness at 95% confidence level;

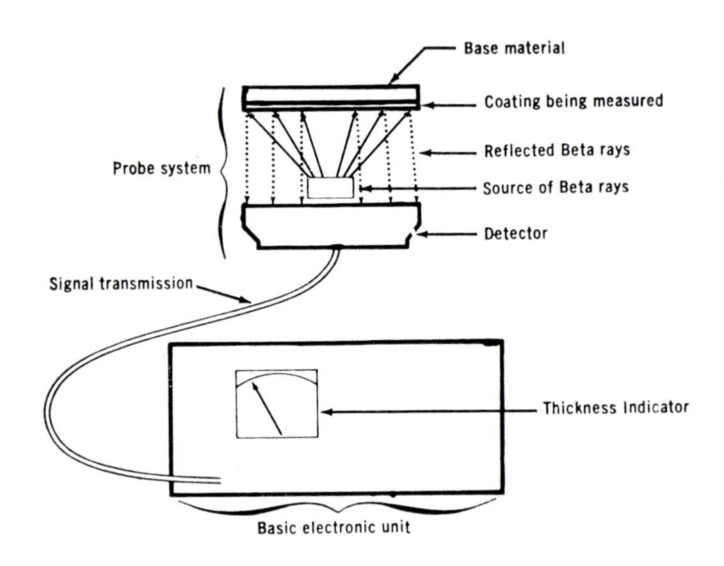

Fig. 15.5. Principle of the ß-backscattering technique.

(b) the test takes only 10–30 seconds, so that several tests can be carried out making it feasible to carry out true statistical quality control;
(c) only a small surface area is required (< 1 mm^2) which is important in view of the construction of electronic components in small sizes;
(d) the technique is very useful for the determination of thin coatings such as precious metal deposits of 0.2–5 μm.

The β-backscattering technique is particularly suited to the continuous measurement of deposit thickness since there is no contact between the instrument and coating metal.

15.1.10. THERMO-ELECTRIC PLATING GAUGE

The B.N.F. thermo-electric thickness meter, shown schematically in Fig. 15.6, consists of a copper probe terminating in a small hemispherical steel tip which is heated to a constant temperature by means of a small electrical winding. When the probe is applied to the plated article, a thermo-electric potential is developed between the hot probe and the cold part of the article, and this potential depends on the thickness of the plated coating beneath the probe. This potential is amplified electronically and shown on a meter with three scales calibrated directly in terms of nickel thickness on steel, brass or zinc.

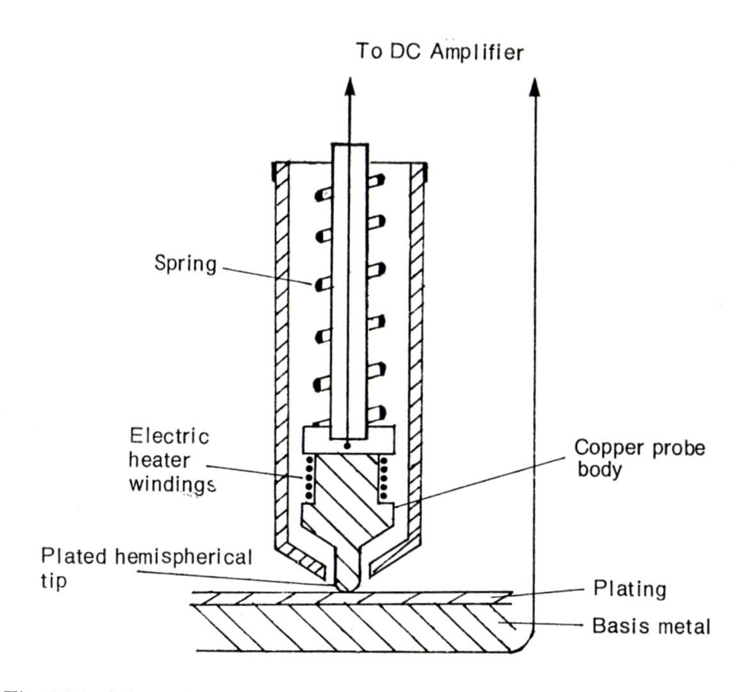

Fig. 15.6. Schematic arrangement of the B.N.F. thermoelectric thickness tester.

Once adjusted for a particular coating and basis metal combination, it requires no skill in operation. Nickel thicknesses can be measured with an accuracy of ± 0.0002 cm. The probe tip is a hard steel hemisphere less than 3 mm in diameter, so that small localised areas can be measured. Curved surfaces present no difficulties, and the positive spring-loaded probe can be used at least 100,000 times before it requires replacement. The test is, of course, non-destructive.

15.1.11. PROFILOMETER METHOD

In this method the thickness of a deposit is determined by traversing a stylus over the edge of the coating, or alternatively by traversing across a groove or step height of the deposit obtained by dissolving a portion of the deposit away. Vertical movement of the stylus is detected by means of a suitable transducer which provides an electrical signal for amplification and recording. This method is only suitable for deposits above a certain thickness and for coatings on relatively smooth substrates.

15.1.12. INTERFEROMETRY METHOD

An ordinary optical microscope is used in this method to examine interference fringes formed between an optical flat and the surface of the coating metal. Discontinuities in the coating displace the interference pattern from which can be determined the depth of the discontinuity. As with the profilometric method, in order to determine deposit thickness a portion of the deposit must be removed, which allows the fringe pattern to develop at the coating substrate step. The method is said to be particularly applicable to the thickness determination of very thin coatings.

A more detailed account of the principles on which this technique depends has recently appeared elsewhere[13].

15.1.13. X-RAY METHODS OF DETERMINING COATING THICKNESS

There are several specialised X-ray techniques for the determination of deposit thickness. The most important of these methods involves the measurement of the intensity of secondary radiation produced when a coated metal is bombarded with X-rays. The coating thickness can be calculated from the intensity of secondary radiation with an accuracy greater than $\pm 1\%$.

15.2. Porosity

The porosity of a coating is less significant than its thickness from the durability point of view since in general the degree of porosity decreases with increasing thickness. Porosity is, however, important in the case of deposits which are more noble than the basis metal, since in such coatings the rate of attack at the pores is likely to be accelerated.

Thus it is more important that tin and nickel deposits on steel should be free from porosity than, say, zinc or cadmium coatings.

Traditional tests for porosity are based on the application of a reagent which will react with the basis metal through pores in the coating and yet will not attack the deposit itself.

15.2.1. HOT WATER TEST

The hot water test for porosity was originally introduced for the examination of tin-plate but has also found some application in the testing of nickel and other deposits[14]. The specimens are first thoroughly cleaned, preferably by cathodic treatment in a cold 1 per cent solution of sodium carbonate; trichloroethylene degreasing alone is inadequate. They are then immersed in distilled water at a pH of 3.5 to 5.0 for a period of six hours. It is necessary to add very dilute hydrochloric acid to reduce the water to this pH value. The water is kept at 95°C in a tinned copper vessel, as glass vessels tend to introduce alkali into the water. At the end of the test period, which lasts from three to six hours, the samples are removed, rinsed and dried, and the number of rust-spots counted. The test is applicable to the testing of large areas for gross porosity. It is not readily applicable to electro-tinning, since the rust-spots are more diffuse and are not readily counted; in this case it has been recommended that the tin surface be treated in an oxidising solution (e.g. a 10 per cent chromic acid solution at 90°C) for five minutes to form an inhibiting film prior to the hot water test[15].

Another version of the hot water test, developed particularly for phosphate coatings, consists of boiling the sample in aerated water for 15 to 20 minutes.

15.2.2. FERROXYL TEST

For determining the porosity of coatings which are more noble than steel, such as nickel, chromium, tin, silver or copper on steel, the ferroxyl test has been used. The ferroxyl solution may be made up from:

Potassium ferricyanide	10 g/l
Sodium chloride	5 g/l

A piece of soft smooth rag paper (e.g. filter paper) is applied to the coating and dampened with the above solution by means of a brush in such a way that the paper is in close contact with the surface to be tested. It is allowed to remain in this position for 10 minutes (further solution being applied if there is any tendency for the paper to dry) and is then removed, washed and dried. Any pores will show themselves as blue spots on the test paper, which can also be used as a permanent record of the degree of porosity. The test has also been used for the testing of the porosity of silver and tin coatings.

The ferroxyl solution may have a little agar jelly usefully incorporated in it to prevent spreading of the spots.

The ferroxyl solution itself attacks nickel slightly, so that thin deposits may show more porosity than is really the case. It has therefore been suggested that for thin coatings a paper dipped in a sodium chloride solution containing a little gelatin should be used in contact with the test piece. After this has been in contact with the latter for ten minutes it is immersed in the solution of potassium ferricyanide, when blue spots will appear wherever pores are present. The latter method is also applicable to copper coatings on steel, as porosity in such deposits is not readily revealed by the ferroxyl paper test. The results obtained by the ferroxyl test can only reveal a small proportion of the total pores, since pure ferricyanide solution has no action on polished iron. It is only by the addition of chloride in increasing quantity that blue spots are produced. This indicates that the surface is reactive only at discrete points, so that the reagent will only detect a pore if that pore happens to coincide with an active spot on the basis metal.

The ferroxyl tests are considered to be somewhat suspect owing to the likelihood that the solution can itself attack the deposited metal and lead to the development of pores not originally present. For this reason most porosity tests are nowadays carried out by subjecting the coated metal to an accelerated weathering or corrosion test.

15.2.3. ABSOLUTE MEASUREMENT

However, attempts have been made in the past to find an absolute means of measuring porosity. One procedure has been described by Thon and Keleman [(16)]. In this method the intrinsic porosity is obtained by measuring the rate of flow at extremely low pressure and under a very low overpressure of a non-corrosive gas through a plated coating which has been detached from the basis metal either by peeling it off a passive basis-metal surface such as stainless steel, or by dissolving the basis metal in a reagent which will not attack the coating.

Experiments with this test indicate that the finer the grain structure of the deposit, the less is its intrinsic porosity. Atmospheric corrosion rapidly increases the porosity of thin deposits, probably by selective attack at local areas which open up numerous channels and increase the permeability of the coating. This may occur without any apparent massive damage to the deposit or measurable loss of weight by corrosion.

In determining the porosity of a foil by this method, the apparatus is first evacuated on both sides to a pressure of about 0.001 mm of mercury, after which one side of the apparatus is opened to atmospheric pressure. If the foil is porous, the low pressure on the vacuum side will gradually rise with time at a rate dp/dt (where p is the pressure and t the time) which is directly proportional to the surface area of the foil F, directly proportional to the overpressure p, inversely proportional to the volume v on the low-pressure side, and directly proportional to a constant k, characteristic of the given foil and expressing its particular degree of permeability; the higher the porosity the greater the value of k.

$$dp/dt = k \times F \times p/V.$$

The overpressure p in this procedure does not change in the course of the experiment, whether the high-pressure side is left open to the outer atmosphere or whether it is closed after admission of the atmospheric pressure, since in the latter case the decrease of the high pressure is negligible. Consequently p is constant and remains equal to the initial overpressure $p_0=1$ atmosphere. As all factors on the right-hand side of the above equation are constant, on integrating we find that at any given time, t, the pressure p on the low-pressure side is given by the equation:

$$p=k\times F\times p_0\times t/V$$
$$\text{or} \quad k=V\times p\ (F\times p_0\times t).$$

If V is expressed in litres, F in sq. in., and t in minutes, the characteristic permeability constant k is expressed in litres per cm^2 per minute.

Further work described by Thon[17] indicates that under various corrosive conditions the rate of attack is very slow as measured by the increasing permeability of the foil – much slower, in fact, than would be expected. Also, there are very great differences between the rate of corrosive attack on electrolytic nickel deposited from different solutions, e.g. nickel from a warm chloride bath is attacked very much more slowly than from a sulphate solution.

15.2.4. AUTORADIOGRAPHIC POROSITY DETERMINATION

An interesting method of assessing the porosity of a nickel deposit consists in plating the nickel over an electrodeposit containing radioactive iron and then evaluating the discontinuities by using a photographic film suitably exposed to radiation emanating through the nickel plate from the iron. The method is non-destructive, and does not induce porosity. It also has the advantage of reproductibility, whilst a permanent record is made at the same time. The radiation hazard is negligible in view of the low emanation type of isotopes used[18].

15.2.5. ELECTROGRAPHY

Electrographic methods have been proposed as a means of showing porosity[19]. In the original procedure, paper impregnated with the test electrolyte is applied to the coating and a counter electrode of a suitable metal, such as aluminium or lead, is placed on the other side of the paper. With ferroxyl solution the article to be tested is made the anode and with an applied e.m.f. of 5 to 6 volts; any porosity is indicated in fifteen seconds. In the case of nickel on brass, porosity can be tested for by applying a filter paper moistened with a 5 per cent solution of sodium nitrate, a sheet of platinum, lead or nickel being placed on top. The test specimen is made the anode and the metal plate the cathode whilst a current of .03 milliamps per sq. cm. is passed for three minutes. The filter paper is then developed for one minute in a solution of 50 g/l of potassium ferrocyanide and 30 g/l of glacial acetic acid. Porosity in shown by brown spots on the paper.

A later form of this test is particularly useful for showing up the pore distribution or crack pattern in chromium deposits[20]. The method used is to dampen a piece of photographic paper, without emulsion, on to the chromium plated surface and to hold it under a pressure of about 10 lb/sq. in. (6.89476×10^4 N/m^2; 0.70307 kg f/cm^2) by means of a metal plate. A potential of 7.5 volts is applied between the plate and the article, which is made the anode. After $2\frac{1}{4}$ minutes the paper is removed, when cracks and pores in the chromium layer are revealed as red lines and spots. There appears to be good correlation between the discontinuities as shown by this test and those found after Corrodkote testing (see below). In the case of one set of tests where blister formation on zinc die-castings was investigated, it was found that 78 per cent of the blisters could be associated with porosity or discontinuities in the chromium layer as shown by the electrographic printing tests.

15.3. Corrosion testing

Corrosion resistance and porosity are closely inter-connected; some indication of the expected life of a protective coating is obviously desirable and a great deal of attention has been given to the development of accelerated tests which will give some indication of durability. Actually, exposure tests, either accelerated or natural, should only be used for the comparison of finishes under identical conditions whilst the interpretation of the results demands the greatest care or they may lead to entirely erroneous conclusions. Exposure tests of plated coatings have been carried out on an extensive scale, the specimens being suspended usually on an inclined frame in the type of atmosphere (e.g. urban, marine, industrial, etc.) in which it is desired to carry out the test. Edges are generally painted with a suitable protective lacquer to prevent edge effects. The methods of attaching specimens to the supporting frame and the angle at which they are suspended also demand careful attention. Recording results largely depends on personal interpretation of the appearance of the samples at the end of the test period. This period also must be carefully chosen, as clearly if the time of exposure is either too long or too short there will be little divergence in the appearance of the different specimens.

The chief difficulty with exposure tests is the variability of the conditions of exposure, since no two environments or successive periods of tests are identical. There is also a considerable difference of opinion as to whether samples should be cleaned during the exposure period.

Several large-scale exposure testing stations have been set up for the assessment of the durability of materials and finishes. One of the best known is the large test station at Kure Beach, South Carolina.

Industrial locations such as Euston, in London, have also been extensively used for exposure testing.

A more satisfactory means of assessing results is by the use of statistical methods which ensures that the maximum information is obtained without undue multiplicity of specimens. In itself the mere duplication of tests on a large scale is no guarantee of reliable information being obtained. The use of change in weight of specimens as a criterion of the rate of corrosion has been used, but is not always satisfactory. The use of loss of tensile strength as a measure of the rate of corrosion, particularly on wire, has been employed, but again the method can give inconsistent results.

15.3.1. RATING SYSTEMS

The rating of specimens after exposure is commonly done visually, but attempts have been made by various authorities to introduce mathematical methods of assessment. They are, however, difficult to apply and there are objections to most of them. A method of examining test specimens has been specified by the British Standards Institution (BS 3745). The surface of the test specimen is covered with a graticule and the number of sequences are counted which have corrosion spots in them.

Another rating system is that introduced by the American Society for Testing Materials. In this method a numerical rating is arrived at from the formula

$$R = 3\,(2 - \log W_A) \text{ where } W_A = \sum {}^{w}{}_{a} \text{ and } {}^{w}{}_{a} = a.f.$$

In this formula a is the area on the test panel, appropriate to a particular defect and f is the weighting factor. W_A is hence the sum of all the weighted areas corresponding to each defect. A list of defects has been drawn up, with a weighting factor relating to each, as follows:

Pinhole rusting	1.0
Light iridescent stain	0.01
Crater rusting	4.0
Blisters	2.0
Peeling and flaking	4.0
Light surface pits	0.01
Crows foot cracking and moderate to heavy surface pitting	0.25
Dark stain	0.10

It will be appreciated that this weighting system must inevitably be largely a matter of opinion, and agreement on the respective values is by no means universal. Also the method is tedious and lengthy to apply, so that in many quarters simpler procedures are preferred, such as the use of photographic standards as bases for comparison. A much fuller discussion of rating systems has appeared elsewhere[21].

15.3.2. SALT SPRAY TEST

The use as accelerated corrosion tests of spray tests based on salt solutions to which other corrosion accelerators are added is very widespread. Various types of apparatus have been described, but the actual design of the equipment is not of great importance provided it meets the following basic requirements:

1. The chamber should be constructed of non-corrodible material.
2. Spray jets should be arranged to produce a salt-spray fog in the chamber without direct impingement on to the specimen being tested.

3. The salt solution should not be re-sprayed.
4. Wide variations in temperature should be avoided.

Arrangements are provided in the apparatus for the air supply to the nozzles to be saturated with moisture by passing it through a container filled with water at a higher temperature than the chamber in order to counteract the drop in temperature which tends to occur by expansion at the nozzle. By maintaining the air entering the chamber at not less then 35°C it is possible to ensure that the air is at 84 to 90 per cent relative humidity. The specimens are supported at about 15 degrees from the vertical. The spray is run continuously and the amount of fog entering the chamber should be such that 0.5 to 3.0 ml of solution per hour are deposited for each 30 cm^2 of horizontal area. Baffles must be provided to prevent direct impingement of the spray on to the specimens. The air used for producing the spray should be saturated by bubbling through water prior to entering the chamber. The jet tubes should be of glass, ebonite or Monel metal, with an internal orifice of 0.038 cm. The temperature of the chamber should be within the range of 18-25°C. In another version of the test the operation of the spray is intermittent. The chamber is left open during a predetermined part of the cycle so that the spray dries on the surface.

The original salt spray test in which a solution of up to 20 per cent NaCl was used has been the subject of considerable criticism as being unreliable and misleading(22). An investigation carried out by Section B on Porosity Tests of Sub-Committee III of Committee B8 of the A.S.T.M.(23) revealed that there was little agreement between the four laboratories involved in the tests. It has been shown that the 20 per cent salt spray test can be misleading owing to nozzle clogging and dry spray caused by poorly humidified air and inadequate liquid supply at the nozzle.

For testing plated parts the addition of acetic acid to the salt solution to maintain it at a pH of around 3.2 is very useful in reproducing atmospheric service failures which are considered to be caused by industrial atmospheres containing sulphur dioxide rather than by the use of salt on the roads, for instance. The addition of acetic acid to the salt solution was suggested by C. F. Nixon in 1945 as a modification specially intended to reproduce the characteristic blistering in service of plated zinc alloy diecastings, and after further development, the test was included in the British Standard 1224 for nickel and chromium coatings, in 1959. In that form the test has been retained in BS 1224: 1970. A salt solution containing $5 \pm 0.5\%$ NaCl and glacial acetic acid to give a pH of 3.2 to 3.5, maintained at a temperature of 35^{+} 3°C, is used for the test, which is applied for 72 hours.

An International Standard specifying acetic acid-salt spray tests is at the time of writing in course of preparation.

15.3.3. CASS TEST

One of the most widely used modified salt spray tests in use in the U.S.A. and U.K. which appears to give good correlation with outdoor exposure is known as the CASS test (Copper chloride modified acetic acid-salt spray).

In this test[24] a 5 per cent salt solution is used to which is added 0.25 g/l of cuprous chloride, $CuCl_2.2H_2O$, and enough glacial acetic acid to maintain the solution at a pH of 3.0 to 3.2. The test is carried out at 50°C. This test is more rapid than the acetic acid-salt spray, whilst the attack is also more severe. Before the test, the specimens should be cleaned by rubbing with a paste of MgO and water, so that they are free from water breaks after rinsing. This treatment accelerates corrosion, but is considered an essential part of the test. Articles are exposed to the salt fog for 16 to 18 hours, two cycles (i.e. 32 to 36 hours) being commonly employed for a full test. On completion of the test, the articles are gently washed in warm running water to remove salt deposits before being dried and inspected.

The CASS test reproduces fairly accurately the corrosion pattern produced in outdoor exposure although there is sometimes excessive spreading of corrosion product. Rusting of a ferrous basis material is shown up clearly and the test also produces blisters on plated zinc diecastings.

The test, which was developed by the General Motor Corporation (U.S.A.), is incorporated in BS 1224: 1970 for nickel and chromium coatings.

15.3.4. CORRODKOTE TEST

The Corrodkote test[25], which also originated in the U.S.A., is used widely on automobile parts. In this test a slurry consisting of 30 g of kaolin and water to which has been added 7 ml of 5 g/l cupric nitrate solution, 33 ml of 5 g/l ferric chloride solution and 10 ml of 100 g/l ammonium chloride solution, is used. About 250 ml of water are added to prepare the slurry, which is applied evenly to the parts to be tested either by spray or brush. After drying for one hour the articles are placed in a humidity cabinet under non-condensing conditions for one hour. Cleaning is necessary before examination. The composition was selected after analysis of road mud, which was considered to be one of the agents mainly responsible for the corrosion of plated automobile components. It must be made up fresh each day.

The test produces a pattern of corrosion similar to that developed out-of-doors. It is capable of developing blisters on zinc diecastings as well as producing a clear indication of penetration to the basis metal in the case of steel articles.

15.3.5. SULPHUR DIOXIDE TEST

In view of the importance of sulphur dioxide in corrosive atmospheres, tests are carried out in which articles plated with gold or other noble metals (except silver) are exposed to high-humidity atmospheres containing carbon dioxide with small concentrations of sulphur dioxide. A large gas-tight test cabinet of approximately 200 litres capacity is used in which are contained the test specimens mounted on a rack moving at a rate of 40 metres/hour. The air mixture entry and exit system should be such as to permit several complete changes of atmosphere per hour. Claims are made for this test that it closely simulates service failures, since the pattern of corrosion which develops after exposure is very similar in appearance to that which develops when the plated articles are exposed in service. The test is used for a variety of purposes, including testing the corrosion resistance of plated electrical contacts and connections, but not generally for nickel-chromium coatings.

15.3.6. ELECTROCHEMICAL METHODS

Potentiometric measurements have been employed as a method of assessing the protective values of different coatings in various media, the changes in potential between the basis and the coating metals being measured.

Pierce and Pinner[26] have developed a test cell which is designed to provide an accelerated corrosion test which would give results which could be related to those obtained in service. They employed a cell attached to the plated article which formed the anode. A cradle of copper wire formed the cathode, a 3 per cent aqueous sodium chloride solution being the electrolyte. A potential of 0.3V was applied for a time when perforation occurrred through preferential attack. The number and type of perforations indicated the corrosion resistance of the plating.

Experiments on automobile parts showed a close degree of correlation between the accelerated and the practical tests. Plate thickness, the preparation of the metal before plating, and plating bath operation were found to be important factors in determining the corrosion resistance of electrode parts. Sandblasting was inferior to polishing and cold rolling superior to either. The test is claimed to be useful for quality control and results may be extrapolated. Tests of this type must, however, be used with great discretion.

15.4. Adhesion tests

The adhesion of an electrodeposited coating is an important measure of its serviceability, but satisfactory tests are difficult to devise. Since only a brief discussion of the numerous tests which have been proposed can be given here, readers requiring a fuller account are referred to the excellent review by Davies and Whittaker[27]. These authors have tabulated the different types of qualitative adhesion tests which are largely self-explanatory, and this is reproduced here in Table 15.1.

15.4.1. QUANTITIVE TESTS

One of the simplest ways of quantitatively determining the adhesion of a coating is to attach a grip to its surface via which known forces can be applied to separate the deposit from the substrate. This can be accomplished using solder or epoxy resins, or electroforming a nodule on to the surface[28]. Whichever method is used the adhesion of the grip to the deposit must clearly be greater than the adhesion of the deposit to the substrate, and the grip material and fixing substance must themselves be strong enough to resist the applied force.

TABLE 15.1

Qualitative adhesion tests[27]

Tests	Comments
MECHANICAL	
1. Bend, twist, and wrapping	A range of coating thicknesses can be tested. Ductile coatings may reduce the stress by plastic flow, so that poorly adherent coatings may not be detached. Brittle deposits may crack. The tests are in common use since they are easily performed and the procedure can be standardised, particularly for strip or wire.
2. Squeezing in compression	Destructive tests, essentially similar to (1), which are limited by the geometry of the part.
3. Burnishing, buffing, and abrasion	The tests are based on severe local deformation, probably assisted by heating. They can be applied locally to most production parts, and are considered non-destructive where the adhesion proves acceptable. No attempt has been made to standardise the various factors such as pressure, surface area of contact, nature of burnishing tool, and speed of the operation.
4. Heating and quenching	Differential expansion between coating and basis metal, or the accumulation of gas at the interface, may cause detachment of the coating. The heating may modify the coating or substrate structures, or the nature of the bond. Interdiffusion can result either in reduced or improved bonding. The test is largely independent of the geometry of the part and may be non-destructive.
5. Scribing, chiselling, filing, and grinding	These tests essentially involve cutting through the coating in some way and they are, therefore, always destructive, unlike those in (3). The manner of carrying out the tests differs widely, but scribing is confined to thinner coatings and chiselling to thicker coatings. While all are to some extent operator-dependent, the grinding test is particularly so and is also confined to non-ductile coatings.
6. Impact and hammer	These tests cover the repeated hammering or impacting with a single tool or the use of grit or shot propelled at the surface. Attempts to standardise the test were made with the B.N.F. impact tester.
7. Cupping and indentation	These tests have largely evolved from, on the one hand, the Erichsen cup test, and on the other metallurgical hardness testing. As they are essentially bend tests they are subject to the same limitations with respect to ductile and brittle coatings as those in (1).
NON-MECHANICAL	
1. Fluorescent	This test involves preparation of a section through a coated part and the testing of the section by a conventional dye-penetrant method used in crack detection.
2. X-ray	Non-destructive test similar to conventional technique for detecting defects in castings or welds.
3. Ultrasonic	Again, similar to the non-destructive techniques for detecting defects.
4. Cathodic treatment	The coated part is made the cathode in an electrochemical cell, and the hydrogen evolved diffuses through the coating to collect at the interface and cause blistering. It is obviously of limited application.

15.4.2. OLLARD ADHESION TEST[29]

This is a specialised test requiring considerable time and equipment to enable it to be carried out. It is, moreover, not applicable where soft metal deposits such as zinc and cadmium are concerned. A measure of the adhesion is obtained in terms of the pull required to separate the coating from the basis metal. A thick deposit of about 2 mm must be applied, after which the specimen is machined as shown in Fig. 15.7. The hatched area shows the deposit, which is built up on the end of a cylinder of the required basis metal. Pressure is then applied to the hollowed part of the specimen by means of a tensile test machine while it is held in a suitable die. The force required to separate the ring of deposit can then be expressed in terms of standard stress units. The test has been criticised on a number of counts and several attempts have been made to modify and improve it[30–32].

15.4.3. ULTRA-CENTRIFUGE TECHNIQUE

This is an interesting direct method of measuring adhesion which consists in plating the metal on to a small rotor and spinning this in an ultra-centrifuge in vacuum until the plate is thrown off by the centrifugal force developed[33]. An absolute measure of the adhesion strength can then be calculated. The high speeds are obtained by suspending the small rotor, which is made of steel or nickel, freely in a glass vacuum chamber by the axial magnetic field of a selenoid, when speeds in excess of 10^6 revolutions per minute can be obtained and centrifugal forces of the order of 10^9 times that of gravity obtained.

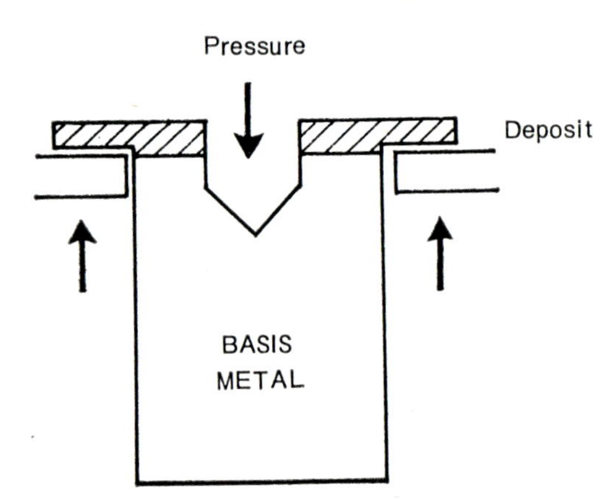

Fig. 15.7. Specimen for testing adhesion of deposits by the Ollard method.

15.5. Hardness and ductility tests

The hardness of a coating can be determined using a micro-hardness tester. A diamond pyramid is pressed into the deposit under a load of between 1 and 40 g and the indentation diagonal measured after the load is removed. The diamond pyramid must have a square base and the angle at the vertex between two opposite faces must be 136 ± 0.5^{0}. The load applied is such that the diamond pyramid does not penetrate the deposit to a depth greater than 10 per cent of its thickness. This can readily be ascertained since the penetration depth is equal to 1/7 of the indentation diagonal. An International Standard defining the technique and the procedure to be adopted is, at the time of writing, in course of preparation.

In many cases the ductility of a deposit is an important characteristic. This is particularly so if the plated article is to be subsequently stressed or formed, in which case poor ductility may cause cracks to appear in the coating. The ductility of a deposited metal can be assessed by depositing it on to a standard test piece which is then pulled in a tensometer machine. The elongation is noted at which cracks first appear in the stretched coating.

REFERENCES

1. I.S.O., R.1463.
2. R. B. Mears, *J. Electrodep. Tech. Soc.*, 1934, **9**, 55.
3. *Plating*, 1948, **35**, 569.
4. S. G. Clarke, *J. Electrodep. Tech. Soc.*, 1937, **12**, 157.
5. R. A. F. Hammond, *J. Electrodep. Tech. Soc.*, 1940, **15**, 69.
6. R. A. F. Hammond, *ibid*, 1948, **23**, 113.
7. I.S.O., 2177 (1972).
8. *Product Finishing*, 1970, **23**, No. 6, 45.
9. *Metal Finishing J.*, 1968, **14**, No. 163, 228.
10. I.S.O., 2178 (1972).
11. I.S.O., 2360.
12. H. Langel, *Metal Finishing J.*, 1969, **15**, No. 169, 9.
13. *Metal Finishing J.*, 1971, **17**, No. 195, 84.
14. A. W. Hothersall and R. A. F. Hammond, *J. Electrodep. Soc.*, 1938, **13**, 449.
15. R. Kerr, *J. Soc. Chem. Ind.*, 1942, **61**, 181.
16. N. Thon and D. Keleman, *Plating*, 1948, **35**, 917.
17. N. Thon, *Plating*, 1949, **36**, No. 9, 928.
18. R. H. Wolffe, M. A. Henderson and S. L. Eister, *Plating*, 1955, **42**, No. 5, 537.
19. H. D. Hughes, *J. Electrodep. Soc.*, 1945, **20**, 17.
20. W. H. Safranek, H. R. Muller and C. L. Faust, *Proc. Am. Electroplaters Soc.*, 1960, **47**, 573.
21. 'Electroplating Engineering Handbook', Ed. A. K. Graham, p. 314, 1962, Reinhold Publishing Corporation.
22. F. L. Labue, *Materials and Methods*, 1952, **35**, 77.
23. Report of Section B on Porosity Tests of Sub-committee III A.S.T.M., *Proc. Am. Soc. Testing Materials*, 1954, **54**, 299.
24. W. L. Pinner, *Plating*, 1957, **44**, 763.
25. W. E. Cooke and R. C. Spooner, *Plating*, 1961, **48**, 42.
26. W. J. Pierce and W. L. Pinner, *Plating*, 1954, **41**, 1034.
27. D. Davies and J. A. Whittaker, *Metals and Materials*, 1967, **1**, No. 2, 71.
28. A. Brenner and V. D. Morgan, *Proc. Am. Electroplaters Soc.*, 1950, **37**, 51.
29. E. O. Ollard, *Trans. Farad. Soc.*, 1926, **21**, 81.
30. A. W. Hothersall and C. J. Leadbeater, *J. Electrodepositors Tech. Soc.*, 1938, **13**, 207.
31. E. J. Roehl, *Iron Age*, 1940, **146**, No. 13, 17 and No. 14, 30.

32. W. Bullough and G. E. Gardam, *J. Electrodepositors Tech. Soc.*, 1947, **22**, 169.
33. J. W. Beams, *Proc. Am. Electroplaters Soc.*, 1956, **43**, 211.

BOOKS FOR GENERAL READING

1. J. A. von Fraunhofer, '*Instrumentation in Metal Finishing*', Elek Science, London 1975.
2. A. Kutzelnigg, '*Testing Metallic Coatings*', Robert Draper, Teddington 1963.
3. G. Isserlis. Ed. *Quality Control in Metal Finishing*, Columbine Press, London 1967.
4. T. Biestek and S. Sekowski, '*Methods of Testing Metallic Coatings*', Finishing Publications Ltd., Teddington, 1978.

CHAPTER 16

Effluent treatment

To meet the requirements of water authorities it is necessary to employ properly designed chemical procedures to deal with cyanides and toxic heavy metals, especially chromium and chromates. A variety of plants and techniques are available for the purpose.

Essentially, the contaminated water is collected and treated either in suitable tanks or in continuous plants to destroy the toxic materials before the water is discharged. In the absence of chromates and cyanides, heavy metals can be precipitated as hydroxides in many cases, and removed simply by settling or filtration. Nowadays, the emphasis is on plants which will enable the water to be re-used as far as possible, as this results in substantial cost savings.

16.1. Package plants

One of the difficulties associated with effluent treatment plants for the electroplating industry is the fact that they have, for the most part, to be designed specifically for each user. This makes for long deliveries and high cost. One way of overcoming these problems is to instal a package unit which can be supplied virtually off the shelf; this can considerably simplify the problem in suitable cases.

In one system designed to deal with cyanides and hexavalent chromium, the cyanide unit is constructed of two 2,700 litre tanks each topped with a reagent tank. It is designed to convert up to 25 kg of sodium cyanide per hour to carbon dioxide and nitrogen at a load of 140 litres per minute by means of chlorine and caustic soda. The hexavalent chromium unit is similar, and will convert some 5 kg per hour of hexavalent chromium to the trivalent form by means of sulphur dioxide, followed by addition of caustic soda to precipitate chromium hydroxide. The solids are removed by settling in each case, and the liquid residue is then suitable for discharge. A two-stage unit for the treatment of chromium effluents is shown in Fig. 16.1. Safeguards are built into the plant to avoid accidental discharge of untreated material. Thus the units provide for twice the retention times needed for complete destruction, and the chlorinator and

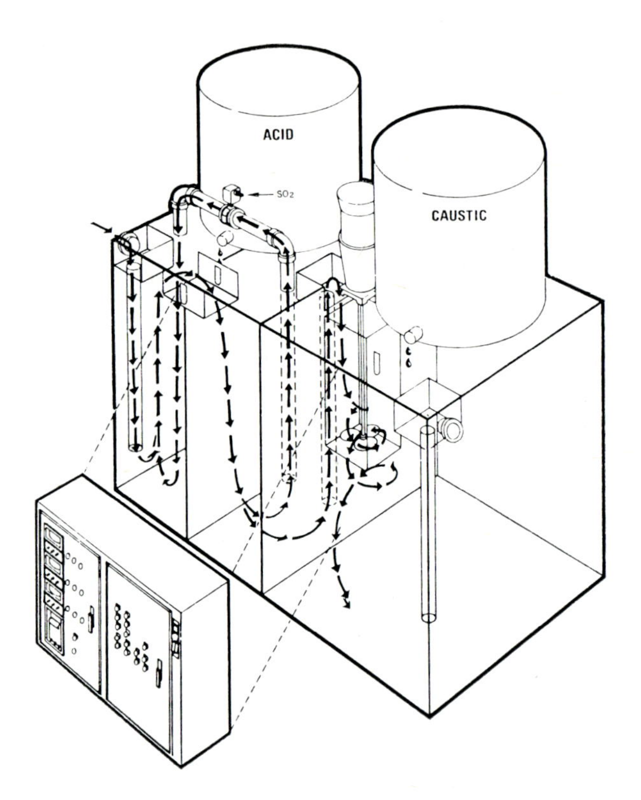

Fig. 16.1. General arrangement of a two-stage unit for the treatment of chromium effluents. [Courtesy Oxy Metal Industries Inc.]

sulphonator are designed to accommodate surges. Complete instrumentation and recording equipment is incorporated, and every attempt has been made to reduce maintenance to simple proportions.

Units of this kind do not, of course, provide for water conservation or recirculation.

16.1.1. CHEMICAL REACTIONS

The basic equations governing the reactions are as follows:

Cyanide

(1) $NaCN+Cl_2 \rightarrow CNCl+NaCl$

(2) $CNCl+2NaOH \rightarrow NaCNO+NaCl+H_2O$
$NaCN+Cl_2+2NaOH \rightarrow NaCNO+2NaCl+H_2O$
$2NaCNO+3Cl_2+4NaOH \rightarrow 2\ CO_2+N_2+6NaCl+2H_2O$

Hexavalent chromium

(1) $SO_2+H_2O \rightarrow H_2SO_3$

(2) $3H_2SO_3+2H_2CrO_4 \rightarrow Cr_2(SO_4)_3+5H_2O$
$Cr_2(SO_4)_3+6NaOH \rightarrow 2Cr(OH)_3+3NaSO_4$

Packaged units of this type are open to the objection that the standardisation of size may make them too small or too large for a specific requirement. However, additional units are readily added to increase capacity, if necessary. A packaged treatment plant is not a complete installation in itself, and it is necessary to provide effluent collection and precipitate settling tanks, as well as pumps for delivery of the solution to the unit.

16.2. Controlled recirculation

Another method, which is increasingly popular for the removal of toxic constituents from rinse waters enabling them to be re-used, is the Controlled Recirculation Method[1]. In this, the excess of treatment chemicals is present in the first rinse station by confining the treatment operation to automatically-controlled reagent addition. Streams of different chemical types are separated from all others, and each stream is subjected to an automatically-controlled treatment to replace any toxic ions present by acceptable types through chemical or ion exchange methods to the calculated degree. Care is taken to avoid reagent overdosing, so that no excess reagent is returned to the rinse system. A proportion of the treated stream is returned continuously to the rinse system from which it comes, and the remainder bled off to discharge. The total water requirements are maintained by continuous make-up with fresh water.

It is claimed that the Controlled Recirculation Method can save 50 to 70% of the water used with single systems and 60 to 80% with respect to the normal water requirements. This is similar to the savings obtainable with the Lancy system. (See 16.5)

As compared with ion-exchange methods (see below) it is to be noted that capital costs are not affected by effluent loads at a given flow rate, so that it can be used to complement deionisation.

16.3. Purification by ion exchange

A method of effluent treatment which also has the advantage of facilitating the re-use of water, is the ion exchange system where the water in the plant is re-circulated continuously, both the anions and cations accumulating in it being removed by suitable ion-exchange materials[2]. These are complex organic resins, e.g. styrene-divinyl benzene or cross-linked polystyrene resins, in bead form, which are able to exchange hydrogen or hydroxyl ions for the positive metallic ions or the negative ions, such as chromate, sulphate or cyanide, in the solution which is to be treated. These resins are specific towards either anions or cations, but mixed beds can be employed where complete deionisation of the water is required. Regeneration of the exhausted resins can be carried out a considerable number of times by treatment with either alkali or acid, according to the type of resin.

The reactions involved can be expressed as follows:

$$R^-.H^+ + C^+ = R^-.C^+ + H^+$$
$$R^+.OH^- + A^- = R^+.A^- + OH^-$$

where $R^-.H$ and $R^+.OH^-$ are the respective cationic and anionic exchange resins, C^+ and A^- being the foreign cations and anions to be removed. These resins thus replace contaminating ions by hydrogen or hydroxyl ions so that the purified effluent can be discharged with safety or the water re-used.

It is also possible to replace toxic ions such as copper or chromate by harmless ones (e.g. sodium or chloride) by the use of ion exchange resins, this procedure being somewhat cheaper than complete de-ionisation. The only fresh water that needs to be added, in theory at any rate, to a closed system such as this is that required to replace evaporation losses and to treat the ion-exchange materials during the regeneration process. With a properly designed and operating plant, the recovered water can be used for most rinsing processes and for the maintenance of solution levels.

A continuous form of ion-exchange treatment making use of a moving bed has also been introduced. In this, water for treatment enters at the bottom of an exchange column and passes upwards against a counter-current of fluidised ion-exchange resin moving gently downwards from the top of the column. The exhausted resin arriving at the bottom goes to the top of the second column for regeneration. Here it moves downwards in counterflow again, initially against an upward flow of rinse water. It finally passes out from the bottom of the column in a fully-regenerated state to return to the top of the exchange column. The action is maintained by a pump circulating the rinse water supply, metering pumps injecting regenerant and rinse water, a proportional control system and a single valve for control of the relative countercurrent rates.

16.3.1. Large scale installations

The possibilities and limitations of a large recirculation system have been described by Pearson and Parker[4]. The plant which is in operation at a large motor works was designed with several objects in view; among the most important of the relevant factors were: (1) the need to avoid surcharging the city sewers at peak periods. (2) the very high hardness of the local water, and (3) the considerable cost of water and effluent disposal. The part of the plant for dealing with rinse waters consists essentially of two mixed-bed ion exchange units fed through one of a pair of sand filters from a sump holding 25,000 litres. The anion exchange resin is regenerated by means of caustic soda solution and the cation exchange resin by sulphuric acid, while town water is used for washing out the filters and resin beds. When the ionic load in the rinse water exceeds a predetermined limit, it is automatically diverted to a strong acid sump where it is dealt with separately, together with oxidised cyanide residues and regeneration liquors.

Strong cyanide-containing solutions are segregated and passed to a 4,000 litre sump for chlorine-caustic soda treatment, when it is necessary to dispose of them, after which they are transferred to the high ionic load treatment unit described above. Strong chromic acid-containing solutions again are treated separately by means of sulphur dioxide reduction and lime precipitation before passing to the settlement tank where the sludge is removed and the supernatant liquor passed to the effluent drain after filtration. The precipitation process also applies to the regeneration effluent from the ion exchange beds themselves. The anion resin effluent is treated like the chromic acid, while the cation resin effluent is neutralised with lime and passes into the precipitation and settling system. Considerable amounts of sludge are produced and their disposal presents quite a difficult problem in itself.

The plant, which caters for the greater part of the consumption of fresh water, which would otherwise be required by the plating installation which it serves, is based on a flow of 30,000 litres/hr. in a 90 hr. week.

Ion exchange plants are somewhat sophisticated pieces of equipment and require a degree of skilled supervision. They are also fairly expensive to purchase and to maintain. The resins are not cheap, and can become permanently contaminated by certain materials, such as nickel—cyanide complexes. Furthermore, the regeneration of the resin produces a concentrated effluent itself, which will require treatment before it can be discharged. There is also the possibility that the water temperature can build up. This is a factor common to most closed systems and can cause difficulties due to the drying out of the water film on plated articles during rinsing.

16.4. Electrolytic methods of effluent treatment

A number of electrolytic processes have come into use in effluent treatment, although on a limited scale.

Cyanide can be oxidised anodically in concentrated solutions, including spent stripping and derusting solutions, cyanide hardening salts, etc. Attempts to use this technique for destroying cyanide in very dilute solutions, e.g. rinse water following plating, have usually failed since the anode efficiency is too low. An interesting new approach to the problem of the electrolytic destruction of toxic cyanides and chromic acid constituents in rinse water has, however, been made recently which makes use of a semi-conducting bed of carbonaceous particles between the electrodes. The minute particles present to the extent of about a quarter of a million per cubic metre occupy about 50% of the cell volume. Each particle functions as a bipolar electrode. It is claimed that 600 ppm cyanide solution can be treated to reduce its cyanide content in 30 minutes. Besides treating cyanides and chromates, the system also reduces metals such as copper, zinc and silver; they react with the surplus hydroxyl ions present to form metallic hydroxides, or plate out on the cathode in a powder form. However, the method has not yet been used on any substantial scale.

Recovery by electrodeposition of metals lost by drag-out is also possible. Tin, silver and gold are examples of metals which are deposited from dilute solutions at sufficiently high cathode efficiencies to make the process worthwhile. Laminar cells are used in integrated systems, i.e. closed loop recirculation with the first rinse following plating.

Units for this purpose are available commercially. The recovery process can be incorporated in an integrated system[3]. For other metals, such as copper and nickel the problem is more difficult although recent work with more sophisticated cell designs appears to be promising.

16.5. Integrated effluent treatment

In the Integrated Treatment (Lancy System)[5] which is used extensively for any operation involving rinsing, effluent treatment is combined with the rinsing operation itself. The rinses are passed to a treatment and solids removal stage where an excess of chemicals is added to effect the treatments. For this purpose, instead of a pure water rinse the workpieces are immersed in a chemical rinse solution which destroys or precipitates the toxic cations or anions in the drag-out solution on the surface. In many cases the chemical rinse is then followed by a normal water rinse, the function of which is then relegated to washing off the dilute, harmless treatment chemicals before the work enters the next stage.

Such chemical rinse systems are in use not only after process solutions containing cyanide and hexavalent chromium but also for a variety of other metal finishing processes, including the great majority of plating and pickling processes, aluminium anodising, etching, electro-chemical or chemical polishing, etc.

The Integrated Treatment System is a self-contained closed-loop system in which any number of chemical rinse stations are linked to a reservoir tank. The reservoir is generally between 2,000 and 15,000 litres in size and acts as a settling tank for metallic sludges as well as a buffer tank to reduce concentration changes in the solution. Such a system is generally emptied and renewed every 3 to 4 months by pumping the solution and sludge into a gravity sludge filter bed. This is possible because in the prolonged dwell time of several weeks or months the sludge is de-watered to occupy between 1/10th and 1/5th of the volume of fresh sludge which is produced in conventional systems.

When Integrated Treatment is employed it is usual to add a batch treatment plant to the system to treat the concentrated spent process solutions which are discarded from the department and any accidental floor spillage. Such batch treatment plant is quite small, and is usually manually operated once every week or two weeks.

The method has a number of important advantages:

(i) It is possible to meet effluent standards far higher than by other methods, e.g., such standards as 0.05 ppm cyanide and chromium and 0.1 ppm of any other toxic metal can easily be achieved by integrated treatment. The reason for this is that the system has a large inbuilt dilution factor since the concentration of metal or anion which is in the effluent from the plant is no longer the residual concentration of the ion in the actual treatment system (as in continuous automatic treatment) but the residual concentration multiplied by a factor corresponding to the drag-out volume divided by the total volume of water leaving the plant in unit time (in practice between 0.001–0.005) of the residual concentration.

(ii) Separation of the different metal is built into the system. As a result recovery of metals is simple and such methods are now widely employed for recovering nickel, copper, cadmium, tin, silver, gold, etc., often by fully automatic recovery units.

(iii) Since the chemical rinses operate on a closed loop system rinse water is used mainly for washing off dilute chemical rinse solution, and water consumption is significantly reduced. It is therefore now common practice to include rinse water recirculation systems with such plant which recovers on the average of between 80 and 90% of the rinse water. At the same time since water flow through rinse tanks can be greatly increased independent of water consumption, it is possible to gain very significant improvements in rinsing efficiency. In many cases also the chemical rinse itself is considerably more efficient than water since, in removing impurities from the surface of the work, this decreases the risk of staining, e.g. after chromium plating.

(iv) Sludge handling is minimised (a) because its volume is greatly reduced (see above), (b) because metals are often recovered instead of being transformed into useless sludge, and (c) the sludges which are uneconomic to recover, e.g. iron and zinc, are much reduced in volume (see above).

(v) Since toxic material is not permitted to enter the rinse water system it is unnecessary to have large reaction and settling tanks. Consequently the plant is often considerably smaller and less expensive than conventional treatment plant. In general the larger the plating, pickling or anodising plant the more significant the saving which can be obtained by using Integrated treatment and recovery plant.

There are an increasing number of cases in which this type of plant is installed purely because of the financial benefits it brings, and the fact that it also enables the factory to meet effluent standards is of secondary importance.

The Integrated System is subject to the limitation that it is not suitable for small plants doing a large variety of work and using less than approximately 4,000 litres of water per hour. There is also the problem of accommodating the necessary additional tanks in the case of existing plants.

REFERENCES

1. G. Mattock, *Chem. and Ind.*, 1970, 46.
2. P. R. Price, *Metal Finisihng J.*, 1972, **18**, No. 212, 279.
3. B. Surfleet and V. Crowle, *Trans. Inst. Metal Finishing*, 1972, **50**, Pt. 5, 227.
4. A. A. L. Pearson and G. G. Parker, *Trans. Inst. Metal. Finishing*, 1961, **38**, Pt. 5, 159.
5. R. Pinner, *Electroplating and Metal Finishing*, 1967, **20**, No. 7, 208, 248, 280.

BOOKS FOR GENERAL READING

1. 'Effluents in Electroplating and the Metal Industry', Portcullis Press, Redhill, 1978.

APPENDIX

TABLE I

Equivalent Thermometer Scales

°F.	°C.	°F.	°C.	°F.	°C.	°F.	°C.
0	—17.7	51	10.5	102	38.8	250	121.1
2	—16.6	52	11.1	104	40.0	300	148.9
4	—15.5	53	11.6	106	41.1	350	176.7
6	—14.4	54	12.2	108	42.2	400	204.4
8	—13.3	55	12.7	110	43.3	450	232.2
10	—12.2	56	13.3	112	44.4	500	260.0
12	—11.1	57	13.8	114	45.5	550	287.8
14	—10.0	58	14.4	116	46.6	600	315.6
16	— 8.8	59	15.0	118	47.7	650	343.3
18	— 7.7	60	15.5	120	48.8	700	371.1
20	— 6.6	61	16.1	122	50.0	750	398.9
22	— 5.5	62	16.6	124	51.1	800	426.7
24	— 4.4	63	17.2	126	52.2	850	454.4
26	— 3.3	64	17.7	128	53.3	900	482.2
28	— 2.2	65	18.3	130	54.4	950	510.0
30	— 1.1	66	18.8	132	55.5	1000	537.8
32	0.0	67	19.4	134	56.6	1050	565.5
33	0.5	68	20.0	136	57.7	1100	593.3
34	1.1	69	20.5	138	58.8	1150	621.1
35	1.6	70	21.1	140	60.0	1200	648.9
36	2.2	72	22.2	142	61.1	1250	676.7
37	2.7	74	23.3	144	62.2	1300	704.4
38	3.3	76	24.4	146	63.3	1350	732.2
39	3.8	78	25.5	148	64.4	1400	760.0
40	4.4	80	26.6	150	65.5	1450	787.8
41	5.0	82	27.7	152	66.6	1500	815.5
42	5.5	84	28.8	154	67.7	1550	843.3
43	6.1	86	30.0	156	68.8	1600	871.1
44	6.6	88	31.1	158	70.0	1650	898.9
45	7.2	90	32.2	160	71.1	1700	926.7
46	7.7	92	33.3	170	76.6	1750	954.4
47	8.3	94	34.4	180	82.2	1800	982.2
48	8.8	96	35.5	190	87.7	1850	1010.0
49	9.4	98	36.6	200	93.3	1900	1037.8
50	10.0	100	37.7	212	100.0	2000	1093.3

TABLE 2

Density Conversion Tables

Sp. Gr.	°Bé,	°Tw.	lb. per cu. ft.
1.005	0.72	1	62.682
1.025	3.54	5	63.929
1.050	6.91	10	65.488
1.075	10.12	15	67.048
1.100	13.18	20	68.607
1.125	16.11	25	70.166
1.150	18.91	30	71.725
1.175	21.60	35	73.285
1.200	24.17	40	74.844
1.225	26.63	45	74.603
1.250	29.00	50	77.962

TABLE 3

Conversion Factors

	Multiplication factor
Inches to centimetres	2.54
Centimetres to inches	0.394
Gallons to litres	4.5
Litres to gallons	0.22
Pounds to kilogrammes	0.45
Kilogrammes to pounds	2.2
Square feet to square metres	0.093
Square metres to square feet	10.76
Cubic feet to cubic metres	0.028
Cubic metres to cubic feet	35.3
Ounces per gallon to grammes per litre	6.25
Grammes per litre to ounces per gallon	0.161
Grammes per litre to Troy ounces per gallon	0.146
Troy ounces per gallon to grammes per litre	9.8
Pennyweights per gallon to grammes per litre	0.49
Grammes per litre to pennyweights per gallon	2.73
Ampères per square foot to ampères per square decimetre	0.108
Ampères per square decimetre to ampères per square foot	9.29
Grammes per square decimetre to ounces per square foot	0.328
Ounces per square foot to grammes per square decimetre	3.04

Note: The Imperial gallon is equal to 1.20 U.S. gallons.

TABLE 4

Conversion Table (Avoir-Troy-Metric)

	Oz.			Dwt.	Equivalent thickness	
	Avoir.	Troy	Gm.	Troy	in. per sq. ft.	μm per sq. m.
1 oz. (avoir.) .	1.0	0.911	28.35	18.23	0.000633	0.0000136
1 oz. (troy) .	1.097	1.0	31.10	20.0	0.000683	0.000146
1 dwt. (troy) .	0.0549	0.05	1.555	1.0	0.0000341	0.0000734
1 gm. . . .	0.0353	0.0322	1.0	0.643	0.0000219	0.0000472

TABLE 5

Chemical and Electro-Chemical Constants of Metals
(based on 100% cathode efficiency)

Metal	Symbol	Atomic weight	Valence	Specific gravity	g/amp hr.	Lb/1,000 Ah	Oz/sq.ft. for 0.001″	Kg/m² per 25μ	A/h to deposit 0.001″/sq.ft.	A/h to deposit 25μ/m²
Cadmium	Cd	112.4	2	8.64	2.097	4.6226	0.72	0.22	9.73	105
Chromium	Cr	52.01	6	7.1	0.323	0.7129	0.591	0.18	51.8	557
Cobalt	Co	58.94	2	8.9	1.099	2.4236	0.74	0.226	19.0	204
Copper II	Cu	63.54	2	8.92	1.186	2.6142	0.74	0.226	17.7	190
Copper I	Cu	63.54	1	8.92	2.372	5.2283	0.74	0.226	8.84	95
Gold	Au	197.2	3	19.3	2.4522	5.406	1.61	0.491	18.6	200
			1	19.3	7.3567	16.2187	1.61	0.491	6.2	67
Hydrogen	H	1.008	1	—	0.0376	0.0829	—	—	—	
Indium	In	114.76	3	7.31	1.4271	3.1461	0.608	0.186	12.1	130
Iridium	Ir	193.1	4	22.42	1.8001	3.9704	1.869	0.57	29.4	316
			3	22.42	2.4012	5.2938	1.869	0.57	22.1	238
Iron	Fe	55.84	2	7.9	1.042	2.2963	0.66	0.201	17.9	193
Lead	Pb	207.2	2	11.3	3.865	8.5210	0.94	0.287	6.91	74
Nickel	Ni	58.69	2	8.9	1.095	2.4135	0.742	0.226	19.2	207
Oxygen	O	16.00	2	—	0.2985	0.6580	—		—	
Palladium	Pd	106.7	2	12.0	1.9903	4.3878	0.998	0.305	14.2	153
Platinum	Pt	195.23	4	21.4	3.6416	8.0283	1.78	0.543	13.85	149
Rhodium	Rh	102.9	3	12.5	1.2797	2.8213	1.04	0.317	23.1	249
Silver	Ag	107.88	1	10.5	4.0245	8.8726	0.875	0.267	6.16	66.5
Tin IV	Sn	118.7	4	7.3	1.1070	2.4406	0.61	0.186	15.63	168
Tin II	Sn	118.7	2	7.3	2.2141	4.8812	0.61	0.186	7.82	84
Zinc	Zn	65.38	2	7.1	1.2195	2.6886	0.59	0.18	13.7	147

Grams per ampere hour $= \dfrac{\text{Eq wt} \times 3{,}600}{96{,}500}$
$= \text{Eq wt} \times 0.0373$
$= \text{Mg per coulomb} \times 3.6$

Ampere hour per gram $= \dfrac{96{,}500}{\text{Eq wt} \times 3{,}600}$
$= \dfrac{26.806}{\text{Eq wt}}$

Lb. per 1,000 ampere hours $= \dfrac{\text{Eq wt} \times 3{,}600 \times 1{,}000}{96{,}500 \times 453.6}$
$= \text{Eq wt} \times 0.082$

Ampere hours per lb. $= \dfrac{96{,}500 \times 453.6}{\text{Eq wt} \times 3{,}600}$
$= \dfrac{12{,}158.7}{\text{Eq wt}}$
$= \text{Coulombs per mg} \times 126$

TABLE 6
Metal content of Common Plating Salts

Salt	Chemical formula	Per cent metal
Antimony trichloride	$SbCl_3$	53.4
Cadmium cyanide	$Cd(CN)_2$	68.3
Cadmium oxide	CdO	87.5
Chromic acid	CrO_3	52.0
Cobalt sulphate (anhydrous)	$CoSO_4$	38.0
Cobalt sulphate, crystals	$CoSO_4.7H_2O$	21.0
Copper carbonate (basic)	$CuCO_3.Cu(OH)_2$	57.5
Copper chloride (II)	$CuCl_2$	47.3
Copper cyanide (I)	$Cu_2(CN)_2$	71.0
Copper fluoborate	$Cu(BF_4)_2$	26.8
Copper sulphate (II), crystals	$CuSO_4.5H_2O$	25.5
Ferrous chloride, crystals	$FeCl_2.4H_2O$	28.1
Ferrous ammonium sulphate	$FeSO_4.(NH_4)_2SO_4.6H_2O$	14.2
Gold chloride (III)	$AuCl_3$	64.9
Gold chloride (III) crystals	$AuCl_3.2H_2O$	58.1
Gold chloride (I)	$AuCl$	84.7
Gold cyanide (I)	$AuCN$	88.3
Gold potassium cyanide	$KAu(CN)_2$	68.3*
Gold potassium cyanide, crystals	$KAu(CN)_2.2H_2O$	60.8
Gold sodium cyanide	$NaAu(CN)_2$	72.5*
Indium chloride	$InCl_3$	51.8
Indium cyanide	$In(CN)_3$	59.4
Indium sulphate	$In_2(SO_4)_3$	44.3
Lead carbonate (basic)	$Pb(OH)_2.2PbCO_3$	80.1
Lead fluoborate	$Pb(BF_4)_2$	54.4
Nickel ammonium sulphate (double nickel salts)	$NiSO_4.(NH_4)_2SO_4.6H_2O$	14.9
Nickel carbonate (basic)	$2NiCO_33Ni(OH)_2.4H_2O$	50.0
Nickel chloride, crystals	$NiCl_2.6H_2O$	24.7
Nickel sulphate (single nickel salts)	$NiSO_4.6H_2O$	22.3
Platinum chloride, crystals	$H_2PtCl_6.6H_2O$	37.7
Potassium stannate, crystals	$K_2SnO_3.3H_2O$	39.6
Rhodium phosphate, crystals	$RhPO_4.3H_2O$	29.9
Rhodium sulphate, crystals	$(Rh_2(SO_4)_3.12H_2O$	29.0
Silver chloride	$AgCl$	75.3
Silver cyanide	$AgCN$	80.5
Silver potassium cyanide	$KAg(CN)_2$	54.2
Silver sodium cyanide	$NaAg(CN)_2$	59.0
Silver nitrate	$AgNO_3$	63.5
Sodium stannate, crystals	$Na_2SnO_3.3H_2O$	44.5
Tin chloride (II) crystals	$SnCl_2.2H_2O$	52.6
Tin fluoborate	$Sn(BF_2)_2$	40.6
Tin sulphate (IV)	$SnSO_4$	55.3
Tungstic acid	H_2WO_4	73.6
Tungstic oxide	WO_3	79.3
Zinc chloride	$ZnCl_2$	48.0
Zinc cyanide	$Zn(CN)_2$	55.7
Zinc fluoborate	$Zn(BF_4)_2$	27.3
Zinc oxide	ZnO	80.3
Zinc sulphate, crystals	$ZnSO_4.7H_2O$	22.7

TABLE 7

Conversion Factors for Cyanide Plating Baths

Multiply weight of	by	to obtain weight of
Cadmium	1.14	CdO
	1.46	$Cd(CN)_3$
	0.875	NaCN equivalent to (CN) in $Cd(CN)_2$
	1.75	NaCN equivalent to (CN) in $Na_2Cd(CN)_4$
CdO	0.875	Cd
	1.62	$Cd(CN)_2$ equivalent
	1.53	NaCN to convert to $Na_2Cd(CN)_4$
	0.625	NaOH formed by adding NaCN to form $Na_2Cd(CN)_4$
$Cd(CN)_4$	0.683	Cd
	0.693	CdO equivalent
	0.596	NaCN equivalent to (CN) in $Cd(CN)_4$
	0.479	NaCN to convert to $Na_2Cd(CN)_4$
	0.940	NaCN equivalent to (CN) in $Na_2Cd(CN)_4$
Copper	1.43	CuCN equivalent
	0.77	NaCN equivalent to (CN) in CuCN
	2.31	total NaCN equivalent to (CN) in $Na_2Cu(CN)_3$
	1.02	KCN equivalent to (CN) in CuCN
	3.07	total KCN equivalent to (CN) in $K_3Cu(CN)_3$
CuCN	0.70	Cu
	0.546	NaCN equivalent to (CN) in CuCN
	0.725	KCN equivalent to (CN) in CuCN
	1.09	NaCN to convert to $Na_2Cu(CN)_3$
	1.45	KCN to convert to $K_2Cu(CN)_3$
Zinc	1.82	$Zn(CN)_2$ equivalent
	1.24	ZnO equivalent
	1.50	NaCN equivalent to (CN) in $Zn(CN)_2$
	3.0	total NaCN equivalent to (CN) in $Na_2Zn(CN)_4$
$Zn(CN)_2$	0.55	Zn
	0.83	NaCN to convert to $Na_2Zn(CN)_4$
	0.687	ZnO equivalent
ZnO	0.80	Zn
	1.44	$Zn(CN)_2$ equivalent
	2.41	NaCN to convert to $Na_2Zn(CN)_4$
	0.985	NaOH formed by adding NaCN to form $Na_2Zn(CN)_4$

TABLE 8

Approximate Cathode Efficiencies of Some Plating Solutions

Solution	Efficiency (per cent)
Cadmium	90–95
Chromium	10–15
Copper (sulphate)	96–100
Copper (cyanide)	40–65
Gold	70–90
Lead (fluoborate)	95–100
Nickel	95–98
Silver	98–100
Tin (acid)	85–95
Tin (Stannate)	70–85
Rhodium	10–20
Zinc (sulphate)	90–95
Zinc (cyanide)	85–90

TABLE 9

Metal Contents of Some Metallic Salts

Salt	Formula	Metal Content per cent
Cadmium cyanide	$Cd(CN)_2$	68.3
Cadmium oxide	CdO	87.6
Chromic acid	CrO_3	52.0
Chevreul's salt	$Cu_2SO_3, CuSO_3, 2H_2O$	49.3
Cobalt sulphate	$CoSO_4, 7H_2O$	21.0
Copper carbonate (basic)	$CuCO_3, Cu(OH)_2$	57.5
Copper cyanide	$CuCN$	71.0
Copper sulphate	$CuSO_4, 5H_2O$	25.5
Gold chloride	$AuCl_3$	64.9
Lead carbonate (basic)	$2PbCO_3, Pb(OH)_2$	80.1
Nickel sulphate	$NiSO_4, 7H_2O$	20.91
Nickel ammonium sulphate	$NiSO_4, (NH_4)_2SO_4, 6H_2O$	14.9
Nickel chloride	$NiCl_2, 6H_2O$	24.7
Silver cyanide	$AgCN$	80.6
Sodium Stannate	$Na_2SnO_3, 3H_2O$	44.5
Stannous sulphate	$SnSO_4$	55.1
Stannous chloride	$SnCl_2, 2H_2O$	52.7
Zinc cyanide	$Zn(CN)_2$	55.7
Zinc sulphate	$ZnSO_4, 7H_2O$	22.8

TABLE 10

Steam Table

Temperature °F.	°C	Absolute* Steam Pressure lb./in.^{2}	kN/m^2
110	43.3	1.274	8.782
120	48.9	1.692	11.665
130	54.4	2.221	15.311
140	60.0	2.887	19.092
150	65.6	3.716	25.618
160	71.1	4.739	32.671
170	76.7	5.990	41.295
180	82.2	7.510	51.774
190	87.8	9.336	64.362
200	93.3	11.525	79.453
210	98.9	14.123	97.363
220	104	17.188	118.49
230	110	20.78	143.25
240	116	24.97	172.14
250	121	29.82	205.58
260	127	35.43	244.25
270	132	41.85	288.51
280	138	49.20	339.18
290	143	57.55	396.75
300	149	67.01	461.97
310	154	77.68	535.52
320	160	89.65	618.05
330	166	103.03	710.29
340	171	117.99	813.42
350	177	134.62	928.07
360	182	153.01	1054.8
370	188	173.33	1194.9
380	193	195.70	1349.2
390	199	220.29	1518.7
400	204	247.25	1704.5

*Atmospheric (gauge) pressures are approximately 14.7 lbs./in.^{2} (101.3 KN/m^2) less than the absolute pressures.

TABLE 11

Rate of Deposition of Silver

(*Cathode Efficiency* 100%)

Thickness of Deposit		Weight		Time (mins.) required for deposition at various current densities (A/ft.^{2})*			
μm	in.	oz./ft.^{2}	gm/m^2	2	3	4	5
0.0025	0.0001	0.079	0.228	18.6	12.4	9.3	7.4
0.0050	0.0002	0.158	0.456	37.2	24.8	18.6	14.9
0.0075	0.0003	0.237	0.684	55.8	37.2	27.9	22.3
0.0100	0.0004	0.316	0.913	74.4	49.6	37.2	29.8
0.0125	0.0005	0.395	1.141	93.0	62.0	46.5	37.2
0.0150	0.0006	0.474	1.369	111.6	74.4	55.8	44.6
0.0175	0.0007	0.553	1.598	130.2	86.8	65.1	52.1
0.0200	0.0008	0.632	1.826	148.8	99.2	74.4	59.5
0.0225	0.0009	0.711	2.054	167.4	111.6	83.7	67.0
0.0250	0.0010	0.790	2.282	186.0	124.0	93.0	74.4

*To convert to A/dm^2 multiply by 0.108.

TABLE 12

Rate of Deposition of Gold

(*Cathode efficiency* 100%)

Thickness of deposit		Weight		Time (mins) required for deposition at various current densities (A/ft^2)*			
μm	in	oz/ft^2)	g/m^2	1	2	3	5
0.00025	0.00001	0.015	0.043	3.7	2.5	1.2	0.7
0.00050	0.00002	0.029	0.084	7.4	4.9	2.5	1.5
0.00075	0.00003	0.044	0.127	11.1	7.4	3.7	2.2
0.00100	0.00004	0.058	0.168	14.8	9.8	4.9	3.0
0.00125	0.00005	0.073	0.211	18.5	12.3	6.2	3.7
0.00150	0.00010	0.146	0.422	36.9	24.6	12.3	7.4
0.00175	0.00015	0.219	0.633	55.4	36.9	18.5	11.1
0.00200	0.00020	0.293	0.846	73.8	49.2	24.6	14.8
0.00225	0.00025	0.366	1.157	92.3	61.5	30.8	18.5
0.00275	0.00050	0.732	2.115	184.5	123.0	61.5	36.9

*To convert to A/dm^2 multiply by 0.108.

TABLE 13

Rate of Deposition of Cadmium

(Cathode Efficiency 95%)

Thickness of Deposit		Weight		Time (mins.) required for deposition at various current densities (A/ft.²)*			
μm	in	oz./ft.²	g/m²	10	15	25	40
0.0025	0.0001	0.071	0.205	6.1	4.1	2.5	1.5
0.0050	0.0002	0.142	0.410	12.3	8.2	4.9	3.1
0.0075	0.0003	0.213	0.615	18.4	12.3	7.4	4.6
0.0100	0.0004	0.284	0.820	24.5	16.4	9.8	6.1
0.0125	0.0005	0.355	1.026	30.7	20.5	12.3	7.7
0.0150	0.0006	0.426	1.231	36.8	24.5	14.7	9.2
0.0175	0.0007	0.497	1.436	42.9	28.6	17.2	10.7
0.0200	0.0008	0.568	1.641	49.0	32.7	19.6	12.2
0.0225	0.0009	0.639	1.846	55.2	36.8	22.1	13.8
0.0275	0.0010	0.710	2.051	61.3	40.9	24.5	15.3

*To convert to A/dm² multiply by 0.108.

TABLE 14

Rate of Deposition of Chromium

(Cathode Efficiency 15%)

Thickness of Deposit		Weight		Time (mins.) required for deposition at various current densities (A/ft²)*			
μm	in.	oz./ft.²	g/m²	80	90	100	150
0.00025	0.00001	0.0059	0.015	2.6	2.3	2.1	1.4
0.00050	0.00002	0.0118	0.031	5.2	4.6	4.1	2.8
0.00075	0.00003	0.0177	0.046	7.8	6.9	6.2	4.1
0.00100	0.00004	0.0236	0.062	10.4	9.2	8.3	5.5
0.00125	0.00005	0.0295	0.078	13.0	11.5	10.4	6.9
0.00150	0.00006	0.0354	0.093	15.5	13.8	12.4	8.3
0.00175	0.00007	0.0413	0.109	18.1	16.1	14.5	9.7
0.00200	0.00008	0.0472	0.124	20.7	18.4	16.6	11.0
0.00225	0.00009	0.0531	0.134	23.3	20.7	18.6	12.4
0.00275	0.00010	0.0590	0.155	25.9	23.0	20.7	13.8

*To convert to A/dm² multiply by 0.108.

TABLE 15

Rate of Deposition of Copper from Copper Cyanide Solution

(*Cathode Efficiency* 95%)

Thickness of Deposit		Weight		Time (mins.) required for deposition at various current densities (A/ft.^{2})*			
μm	in.	oz./ft.^{2}	g/m^2	10	15	25	40
0.0025	0.0001	0.074	0.194	5.6	3.7	2.2	1.4
0.0050	0.0002	0.148	0.390	11.1	7.4	4.4	2.8
0.0075	0.0003	0.222	0.585	16.7	11.1	6.7	4.2
0.0100	0.0004	0.296	0.780	22.2	14.8	8.9	5.6
0.0125	0.0005	0.370	0.974	27.8	18.6	11.1	7.0
0.0150	0.0006	0.444	1.169	33.4	22.3	13.3	8.3
0.0175	0.0007	0.518	1.364	38.9	26.0	15.5	9.7
0.0200	0.0008	0.592	1.560	44.5	29.7	17.8	11.1
0.0225	0.0009	0.666	1.754	50.0	33.4	20.0	12.5
0.0275	0.0010	0.740	1.950	55.6	37.1	22.2	13.9

*To convert to A/dm^2 multiply by 0.108.

TABLE 16

Rate of Deposition of Copper from Acid Copper Solution

(*Cathode Efficiency* 100%)

Thickness of deposit		Weight		Time (mins.) required for deposition at various current densities (A/ft.^{2})*			
μm	in.	oz./ft.^{2}	g/m^2	10	15	25	40
0.0025	0.0001	0.074	0.213	10.6	7.1	4.3	2.7
0.0050	0.0002	0.148	0.428	21.2	14.2	8.5	5.3
0.0075	0.0003	0.222	0.641	31.9	21.2	12.8	8.0
0.0100	0.0004	0.296	0.855	42.5	28.3	17.0	10.6
0.0125	0.0005	0.370	1.069	53.1	35.4	21.3	13.3
0.0150	0.0006	0.444	1.283	63.7	42.5	25.5	16.0
0.0175	0.0007	0.518	1.496	74.3	49.6	39.8	18.62
0.0200	0.0008	0.592	1.710	85.0	56.6	34.0	21.3
0.0225	0.0009	0.666	1.924	95.6	63.7	38.3	23.9
0.0275	0.0010	0.740	2.138	106.2	70.8	42.5	26.6

*To convert to A/dm^2 multiply by 0.108.

TABLE 17

Rate of Deposition of Nickel

(*Cathode Efficiency* 95%)

Thickness of Deposit		Weight		Time (mins.) required for deposition at various current densities (A/ft.²)*			
μm	in.	oz./ft.²	g/m²	10	15	25	40
0.0025	0.0001	0.074	0.194	11.81	7.9	4.7	3.0
0.0050	0.0002	0.148	0.390	23.6	15.7	9.5	5.9
0.0075	0.0003	0.222	0.585	35.4	23.6	14.2	8.9
0.0100	0.0004	0.296	0.780	47.2	31.5	18.9	11.8
0.0125	0.0005	0.370	0.974	59.1	39.4	23.7	14.8
0.0150	0.0006	0.444	1.169	70.9	47.2	28.4	17.7
0.0175	0.0007	0.518	1.364	82.7	55.1	33.1	20.7
0.0200	0.0008	0.592	1.560	94.5	63.0	37.8	23.6
0.0225	0.0009	0.666	1.754	106.3	70.8	42.6	26.6
0.0275	0.0010	0.740	1.950	118.1	78.7	47.3	29.5

*To convert to A/dm² multiply by 0.108.

TABLE 18

Rate of Deposition of Palladium

(*Cathode Efficiency* 100%)

Thickness of deposit		Weight		Time (mins). required for deposition at various current densities (A/ft.²)*			
μm	in.	oz./ft.	g/m²	2 Amp	4 Amp	6 Amp	8 Amp
0.00025	0.00001	0.009	0.026	4.1	2.0	1.7	1.0
0.00050	0.00002	0.019	0.555	8.1	4.1	2.7	2.0
0.00075	0.00003	0.027	0.078	12.3	6.1	4.1	3.0
0.00100	0.00004	0.036	0.104	16.2	8.1	5.4	4.1
0.00125	0.00005	0.046	0.132	20.3	10.1	6.8	5.1
0.00150	0.00010	0.091	0.263	40.5	20.3	13.5	10.1
0.00175	0.00015	0.137	0.396	60.8	30.4	20.2	15.2
0.00200	0.00020	0.182	0.526	81.0	40.5	27.0	20.3
0.00225	0.00025	0.228	0.659	101.3	50.6	33.8	25.3
0.00275	0.00050	0.455	1.304	202.5	101.3	67.5	50.6

*To convert to A/dm² multiply by 0.108.

TABLE 19

Rate of Deposition of Zinc from Acid and Cyanide Zinc Solutions

(*Cathode Efficiency* 95%)

Thickness of Deposit		Weight		Time (mins.) required for deposition at various current densities (A/ft.^{2})*			
μm	in.	oz./ft.^{2}	g/m^2	10	15	25	40
0.0025	0.0001	0.059	0.155	8.7	5.8	3.5	2.2
0.0050	0.0002	0.118	0.311	17.3	11.5	6.9	4.3
0.0075	0.0003	0.177	0.466	26.0	17.3	10.4	6.5
0.0100	0.0004	0.236	0.621	34.6	23.1	13.8	8.6
0.0125	0.0005	0.295	0.777	43.3	28.9	17.3	10.8
0.0150	0.0006	0.354	0.932	51.9	34.6	20.8	13.0
0.0175	0.0007	0.413	1.088	60.6	40.4	24.2	15.1
0.0200	0.0008	0.472	1.243	69.2	56.2	27.7	17.3
0.0225	0.0009	0.531	1.399	77.9	51.9	31.1	19.4
0.0275	0.0010	0.590	1.554	86.5	57.7	34.6	21.6

*To convert to A/dm^2 multiply by 0.108.

Index to authors

Subject Index

Advertisers Announcements

Phosphating of Metals

By Guy Lorin
(Société continentale Parker, France)

230 pp 65 figs Published 1975 ISBN 0 904477 00 2

Price: £14.50 ($34.00) post free

This book consists essentially of two parts, the first half is devoted to theoretical considerations and the second discusses the practical applications of phosphate coatings for corrosion protection, paint adhesion, cold extrusion, tube and wire drawing and bearing surface lubrication.

It should be readily understood by anyone possessing a moderate knowledge of inorganic and physical chemistry and, although written primarily for practical workers in the field, it will also be of interest to researchers, laboratory personnel, students, etc.

The book, by an undoubted authority in the field with long practical experience, is the first major work on the subject in 25 years. It was first published in the French language and has been freely translated into English by Mr. F. H. Reid and edited by Mr. D. B. Freeman, Technical Manager of Pyrene Chemical Services Ltd.

CONTENTS

FINISHING PUBLICATIONS LTD.

28 High Street, Teddington, Middlesex, England TW11 8EW

Telephone: 01-943 2610

Forthcoming Books—

Methods of Testing Metallic Coatings

By T. Biestek and S. Sekowski

480 pages approx. To be published 1979.

Completely updated and revised English language translation of the second edition of the book 'Metody Badan Powlok Metalowych' published in Poland in 1973 and translated into German also in 1973. After an introductory chapter a further twelve chapters are concerned with properties and applications; visual evaluation; classification of coatings and processes for their removal; destructive and non-destructive methods of thickness testing; investigation of levelling, adhesion, porosity, physical properties, methods of corrosion testing. There is also a chapter on testing of conversion coatings and a section of tables.

Dr. Biestek is Head of the Department of Evaluation and Testing at the Institute of Precision Mechanics in Warsaw where research in electroplating is centred. He is also secretary of the International Standards Organisation Technical Committee 107 on corrosion testing.

Chromium Plating

By R. Weiner and A. Walmsley.

230 pages approx. To be published 1979.

Originally written by R. Weiner and published in German, this book has been translated into English and updated and revised by A. Walmsley in the U.K. After a section on historical development there are chapters on theory of chromium plating; structure and properties of chromium deposits; constituents of chromium baths and their action; influence of the operating conditions on the properties of chromium deposits, method of operation of the chromium solution; technical chromium plating; corrosion protection by chromium electrodeposits; plant considerations; control and testing methods.